Renewable Energy and Power Supply Challenges for Rural Regions

Valeriy Kharchenko
Federal Scientific Agroengineering Center VIM, Russia

Pandian Vasant
Universiti Teknologi PETRONAS, Malaysia

A volume in the Advances in Computer and Electrical Engineering (ACEE) Book Series

IGI Global
DISSEMINATOR OF KNOWLEDGE

Published in the United States of America by
IGI Global
Engineering Science Reference (an imprint of IGI Global)
701 E. Chocolate Avenue
Hershey PA, USA 17033
Tel: 717-533-8845
Fax: 717-533-8661
E-mail: cust@igi-global.com
Web site: http://www.igi-global.com

Library of Congress Cataloging-in-Publication Data

Names: Kharchenko, Valeriy, 1938- editor. | Vasant, Pandian, editor.
Title: Renewable energy and power supply challenges for rural regions / Valeriy Kharchenko and Pandian Vasant, editors.
Description: Hershey, PA : Engineering Science Reference, an imprint of IGI Global, [2019] | Includes bibliographical references.
Identifiers: LCCN 2018059000| ISBN 9781522591795 (hardcover) | ISBN 9781522591801 (softcover) | ISBN 9781522591818 (ebook)
Subjects: LCSH: Rural electrification. | Renewable energy sources. | Rural development projects.
Classification: LCC TK4018 .R35 2019 | DDC 621.04209173/4--dc23 LC record available at https://lccn.loc.gov/2018059000

This book is published in the IGI Global book series Advances in Computer and Electrical Engineering (ACEE) (ISSN: 2327-039X; eISSN: 2327-0403)

British Cataloguing in Publication Data
A Cataloguing in Publication record for this book is available from the British Library.

Advances in Computer and Electrical Engineering (ACEE) Book Series

Srikanta Patnaik
SOA University, India

ISSN:2327-039X
EISSN:2327-0403

Mission

The fields of computer engineering and electrical engineering encompass a broad range of interdisciplinary topics allowing for expansive research developments across multiple fields. Research in these areas continues to develop and become increasingly important as computer and electrical systems have become an integral part of everyday life.

The **Advances in Computer and Electrical Engineering (ACEE) Book Series** aims to publish research on diverse topics pertaining to computer engineering and electrical engineering. **ACEE** encourages scholarly discourse on the latest applications, tools, and methodologies being implemented in the field for the design and development of computer and electrical systems.

Coverage

- Electrical Power Conversion
- VLSI Fabrication
- Computer science
- Digital Electronics
- Computer Architecture
- Applied Electromagnetics
- Optical Electronics
- Circuit Analysis
- Sensor Technologies
- Qualitative Methods

IGI Global is currently accepting manuscripts for publication within this series. To submit a proposal for a volume in this series, please contact our Acquisition Editors at Acquisitions@igi-global.com or visit: http://www.igi-global.com/publish/.

The Advances in Computer and Electrical Engineering (ACEE) Book Series (ISSN 2327-039X) is published by IGI Global, 701 E. Chocolate Avenue, Hershey, PA 17033-1240, USA, www.igi-global.com. This series is composed of titles available for purchase individually; each title is edited to be contextually exclusive from any other title within the series. For pricing and ordering information please visit http://www.igi-global.com/book-series/advances-computer-electrical-engineering/73675. Postmaster: Send all address changes to above address.

Titles in this Series

For a list of additional titles in this series, please visit: www.igi-global.com/book-series

Advancing Consumer-Centric Fog Computing Architectures
Kashif Munir (University of Hafr Al-Batin, Saudi Arabia)
Engineering Science Reference • copyright 2019 • 217pp • H/C (ISBN: 9781522571490) • US $210.00 (our price)

New Perspectives on Information Systems Modeling and Design
António Miguel Rosado da Cruz (Polytechnic Institute of Viana do Castelo, Portugal) and Maria Estrela Ferreira da Cruz (Polytechnic Institute of Viana do Castelo, Portugal)
Engineering Science Reference • copyright 2019 • 332pp • H/C (ISBN: 9781522572718) • US $235.00 (our price)

Advanced Methodologies and Technologies in Network Architecture, Mobile Computing, and Data Analytics
Mehdi Khosrow-Pour, D.B.A. (Information Resources Management Association, USA)
Engineering Science Reference • copyright 2019 • 1857pp • H/C (ISBN: 9781522575986) • US $595.00 (our price)

Emerging Innovations in Microwave and Antenna Engineering
Jamal Zbitou (University of Hassan 1st, Morocco) and Ahmed Errkik (University of Hassan 1st, Morocco)
Engineering Science Reference • copyright 2019 • 437pp • H/C (ISBN: 9781522575399) • US $245.00 (our price)

Advanced Methodologies and Technologies in Artificial Intelligence, Computer Simulation, and Human-Computer Interaction
Mehdi Khosrow-Pour, D.B.A. (Information Resources Management Association, USA)
Engineering Science Reference • copyright 2019 • 1221pp • H/C (ISBN: 9781522573685) • US $545.00 (our price)

Optimal Power Flow Using Evolutionary Algorithms
Provas Kumar Roy (Kalyani Government Engineering College, India) and Susanta Dutta (Dr. B. C. Roy Engineering College, India)
Engineering Science Reference • copyright 2019 • 323pp • H/C (ISBN: 9781522569718) • US $195.00 (our price)

Advanced Condition Monitoring and Fault Diagnosis of Electric Machines
Muhammad Irfan (Najran University, Saudi Arabia)
Engineering Science Reference • copyright 2019 • 307pp • H/C (ISBN: 9781522569893) • US $225.00 (our price)

The Rise of Fog Computing in the Digital Era
K.G. Srinivasa (Chaudhary Brahm Prakash Government Engineering College, India) Pankaj Lathar (Chaudhary Brahm Prakash Government Engineering College, India) and G.M. Siddesh (Ramaiah Institute of Technology, India)
Engineering Science Reference • copyright 2019 • 286pp • H/C (ISBN: 9781522560708) • US $215.00 (our price)

IGI Global
DISSEMINATOR OF KNOWLEDGE

701 East Chocolate Avenue, Hershey, PA 17033, USA
Tel: 717-533-8845 x100 • Fax: 717-533-8661
E-Mail: cust@igi-global.com • www.igi-global.com

Editorial Advisory Board

Table of Contents

Detailed Table of Contents

Looking at the history of solar energy and renewable energy in general, the authorities and scientists have been paying much attention to the recent period, due to the depletion of fossil energy resources and the growing difficulties in solving environmental problems. The development of solar energy has led to the use of solar energy concentrators. Concentrators are used to concentrate sunlight onto PV cells. This allows for a reduction in the cell area required for producing a given amount of power. The goal is to significantly reduce the cost of electricity generated by replacing expensive PV converter area with less expensive optical material. In this chapter, the authors talk about concentrators in solar energy, especially about modules based on holographic films. Holographic solar panels (HSP) in recent decades have appeared in large-scale production and been actively used in solar energy. Evaluations of other types of existing concentrators are presented.

The chapter considers information systems that provide analysis of the RES resource base and support decision making. RES resources in the Republic of Kazakhstan are briefly analyzed. The methods and systems for evaluation of theoretical, technical, and economic potential of renewable energy sources are considered. Based on this overview, the authors propose six steps of estimation that should be realized in intellectual GIS. One of the major steps is the identification the factors that affect using of renewable energy sources. These factors can contribute to impeding the development of green energy. The simple taxonomy of these factors is proposed. The second important step is to aggregate estimations of the factors. As a result, the authors propose the aggregation method based on Bayes rule. Compared to other methods, this approach allows us to obtain two estimates: probabilities of positive and negative hypotheses. The methods discussed are at the heart of the information system supporting the development of renewable energy sources in the Republic of Kazakhstan.

At present, the production of electricity for agricultural consumers remote from the centralized electrical power grid is carried out using diesel-generator technology with a limited service life of engines and extremely low efficiency of the expensive fuel used. In this chapter, an innovative technology has been considered for the combined electrical and heat power production using the preliminary conversion of diesel fuel into synthesis gas with its subsequent supply to a high temperature electrochemical generator (ECG). Synthesis gas for the operation of the electrochemical generator was produced by air conversion of motor diesel fuels in a catalytic burner reactor. On the basis of heat balances of the burner, ECG and waste-heat boiler-utilizer, electrical efficiency of the solid oxide fuel cells' (SOFC) battery, chemical efficiency of the burner, the temperature at the SOFC anode, the EMF of the planar cell, a portion of hydrogen oxidized at the SOFC anode, specific consumption of diesel fuel for the production of electrical and heat power were calculated.

The presented research was carried out at existing solar power plants and renewable energy sources laboratories, whose purpose was to increase the energy efficiency of photovoltaic installations with parallel and mixed switching of photocells, operating under uneven illumination, parallel voltage arrays of photovoltaic modules due to voltage equalization. Experimental characteristics of photovoltaic installations with the developed device for coordinating the array of photoelectric modules (DCA), realizing a method of selection of electric energy and without its use are given. It is experimentally shown that the use of DCA increases the electrical power of the array with partial shading up to 2.6 times with partial shading. The results of the research can be used to design new photovoltaic installations and upgrade existing ones.

The chapter considers the factors that affect the incomplete use of power. The analysis and systematization of accumulating systems on the types of accumulated energy, on the processes occurring during storage, types of devices and types of generated energy is performed. Each of the possible ways of accumulating energy is analyzed in detail. Various variants of conformity of schedules of change of a consumed load within a year and a total resource of solar radiation are considered. The analysis of the parameters of heat-accumulating materials and their classification depending on the material class, the way of heat accumulation and return, on the cyclicity of work, etc. are provided. It is shown that the level of temperature, the scale of the storage unit, and the necessary duration of heat storage determine the requirements for the construction of batteries and the choice of heat-storage substances. The prospects of research on the search for new ways of accumulating energy and technical means for their implementation are considered.

The chapter focuses on different aspects and challenges of power supply for remote energy consumers in the Russian Arctic. The authors discuss the potential use of renewable energy, some specified technological features and risks related to the broader deployment of decentralized renewable energy systems in the Arctic region. Even though there is limited experience of renewable energy systems installation in the Russian Arctic and greater technological challenges, the authors see it as a potential opportunity to contribute to innovative and sustainable development of the region. The authors underline the potential synergistic effect of broader deployment of renewable energy systems in the Russian Arctic. The key argument is that a sustainable and efficient energy system will open new development opportunities and stimulate future socioeconomic development of the region through the use of local and renewable energy resources and the implementation of new power generation modes and technologies.

The prospects of increasing access to electricity for the population of rural areas of Africa are considered. The main international funds and organizations aimed at sustainable energy development in Africa are described. An analysis of the state and possible options for using renewable energy sources for this purpose in decentralized energy supply through the creation of mini-grids or stand-alone systems is given. The risks by developing renewable energy sources in rural areas and modern mechanisms for financing in solar energy are presented.

The object of research in the chapter is the solar power plant as the source of additional economically expedient power supply of the electrical energy consumer. The purpose of this research is to analyze the options for the layout of solar power plant, taking into account the solar energy potential of the district, the design features of the proposed location, the load curve of the consumer, and the cost of the generated electrical energy. The chapter presents the results of calculation and selection of the parameters of solar power plant elements on the roof of the consumer's production building. The chapter presents the results of research of the dependence of the cost of the electricity generated by the solar power plant on the number of installed panels, which in order to increase the realized solar energy potential of the district also allows adding photoelectric modules and accumulating devices in the layout of the operating station at tariff growth. The chapter presents the results of researching these areas, that are conducted by the authors and which are completely original.

The activity of farm enterprises is directly connected with the efficient use of resources such as electrical energy and water. Taking such conditions as factors of economic expediency and ecological safety into account, it is more reasonable to use wind stations in order to provide consumers with energy and water. The use of conventional wind and electrical stations is an easy and reliable solution. However, annual wind velocity on most settled territories does not exceed 6 m/s or even 4 m/s. It makes the efficient use of wind electrical stations more complicated. One of the solutions can probably be the use of wind and electrical stations on the basis of slow speed non-transmission arc-shaped-stator inductor-type generators with an integrated radial and ring-shaped rotor. Another efficient solution to provide areas with water and electrical energy is to use a combined wind station with a crank-and-rod mechanism and the rod of the driving mechanism of the back-and-forth motion of the core of a magnetic and electrical linear generator and the piston of a plunger pump.

The purpose of this chapter is to compare the design and operational parameters of energy efficiency (power generation, capacity utilization factor) of a wind power plant (WPP) in order to improve the accuracy of the design parameters estimation at the preliminary design stage. Recommendations have been given for accounting different factors influencing on energy efficiency and economic indicators of a WPP. In this chapter, the research is carried out on the example of existing wind-diesel complexes (WDC) in the settlement of Ust-Kamchatsk (Kamchatka Krai), Novikovo (Sakhalin Oblast), and projected WDC on Popova Island (Vladivostok city district).

Doubly fed induction generator (DFIG) is a widely spread technology in modern wind turbines (WT) due to its capability to operate with variable speed, partial scale power converter, and ability to control active and reactive power independently. The main drawback of DFIG is its complicated protection systems. In the chapter, several strategies for DFIG protection are reviewed, and the authors provide a conclusion about their advantages. Penetration of renewable energy sources (in particular, wind power) have a huge impact on power systems; thus, wind turbines should be considered as conventional generation units in terms of frequency and voltage regulation. Modern grid codes require WT stay connected during grid fault and be capable to provide appropriate grid support. Therefore, it is important to implement a DFIG protection system that could meet grid code requirements.

This chapter describes the advanced technologies of energy and water saving on the Uzbekistan largest pumping stations of Karshi Main Canal. The unique Karshi pumping stations stage consumes more than 4% of the total electricity generated in Uzbekistan. Such a significant consumption of electricity by irrigative pumping stations, gives the problem of energy conservation to them a status of special importance and relevance. Advanced technologies of energy saving and non-cavitation work are scientifically substantiated and developed. The results of carried out field tests are described. The scientific results of the research are recommended to allow efficient use of water and energy resources and to ensure reliable operation of the power equipment of pumping stations in the rural branches of economy.

Chapter 13
Improving the Power Quality of Rural Consumers by Means of Electricity Cost Adjustment 312
Alexander Vinogradov, Federal Scientific Agroengineering Centre VIM, Russia
Maxim Borodin, Orel State Agrarian University, Russia
Vadim Bolshev, Federal Scientific Agroengineering Centre VIM, Russia
Natalya Makhiyanova, Orel State Agrarian University, Russia
Nadezhda Hruntovich, Sukhoi State Technical University of Gomel, Belarus

Improving power quality of agricultural consumers by adjusting the cost of electricity is considered. The chapter lists the reasons for the output of power quality indicators beyond the standard values. There is the assessment of the quality of electricity in the Orel region. Distribution functions for distortion of power quality indicators are constructed. The relevance of the adjustment of the cost of electricity depending on its quality is justified. The mathematical model for determining the cost of consumed electric energy depending on its quality is developed in chapter. The chapter gives the method allowing to adjust the cost of consumed electric energy depending on its quality and source of distortion. In the chapter, there is the mechanism of carrying out calculations for the consumed and transferred electric power considering change of electric power cost depending on its quality. The feasibility study of the proposed methods and means of improving power quality is realized.

Chapter 14
Resonant Power System for Electric Transport 342
Dmitry Strebkov, Federal State Budgetary Scientific Institution, Russia
Alexey Nekrasov, Federal State Budgetary Scientific Institution, Russia
Anton Nekrasov, Federal State Budgetary Scientific Institution, Russia

Noncontact electric power supply with the use of cable line is an innovative trend in the development of electric transport. Work to develop such systems is carried out in the USA, South Korea, Germany, Israel, the Russian Federation. The VIESH researchers have developed and patented experimental models of a car and a tramway of the future without accumulators, which receive energy from external energy system through air gap from a single wire cable laid under the surface of road pavement. Noncontact systems for electric power supply to transport are developed at VIESH on the basis of resonant methods of electric power supply through a single-wire cable. This is a basic difference of the technology under consideration from similar ones. This technology makes it possible to realize at the new technological level the need for electric roads with minimal costs. Given that all the world's automakers are working on the production of an electric vehicle for mass production, the need for creating a motorway in the city and on inter-urban autobahns is even more topical.

Foreword

In order to meet these requirements of a steadily new involvement and economic circulation of new lands and territories – particularly for food supply, economic and managerial activities and their smart organization are continuously needed. Those territories and pieces of land are, by a rule, placed far away from the main populations and their activities whereas, at the same time, centers of energy supply are centralized. Agriculture is regarded as one of the most suitable activities within those territories. In order start a business there, it is easiest to get involved in the production of agricultural products through small-scale enterprises, with a small number of persons employed. Small-scale production permits to quickly deploy work, it ensures, facilitates and speeds up the sale of products. Such activities highly necessitate particular technical support measures for both the livelihood of people dwelling there and the organization a highly efficient and cost-effective production processes. Of course, in those territories it is most convenient and fast when deciding for and providing activities related to the production of agricultural goods. Such companies can more easily be provided with energy, especially, with renewable energy sources. However, existing technologies of energy supply, cropping and processing of crops are often hard to adapt to a use in small and medium-sized enterprises. Manufacturers and developers of technical facilities and technologies and scientific organizations engaged in the development of technologies and equipment for the power industry on the countryside, production processes of various types, and problems of housing and communal solutions, raise a great interest in new efficient techniques of conducting agricultural business, in particular, at small scale and for problems of the rural population.

The book describes a series of technological methods and various technical means allowing for an organization of production of different goods in landscapes and regions that have become newly allocated for economic activity, or cultivated again for those purposes. The needed and identified technologies and means are adapted to be used in production at small scale, and in companies of small and medium-sized business. Furthermore, the book also considers the design of technical solutions to guarantee a vital activity in new territories with a high degree of comfort and high standards of living conditions. In fact, this work pays a lot of attention to the organization of efficient production processes and to social comfort, even when facing scarce energy resources and the need to a rational use of them.

It is a great achievement of this book to having satisfied all the aforementioned needs, as not any similar work had been complied before for the sake of developmental issues of rural business designed in the intended mode as explained before. Herewith, this book should cause a high attraction and an attention by many specialists of rural agribusiness activities in remote areas, of a wide range of practical management projects, and by businessmen as well as researchers involved in the activities investigated and designed in this new work.

Modelling and solution of the global problem described above requires the creation of technological and organizational preconditions. Today's state-of-knowledge and -art in science and technology allows the offer of new techniques and equipment for necessary power supply of such territories and perspectives of electro-technologies for transportation of electric power and its effective utilization. This should enable manufacturing of agricultural commodities and permitting a comfortable living of the population. This book has been open to cover all problem complexities in the decision making on the formation of a distributed economy and the development of new territories as a core element of this problem.

In this compendium, innovative technologies which provide an effective production, transfer and use of electric energy, whose use can allow for a successful development of new territories for realization of the concept of a distributed economy, are described. A particular interest has been given by the fact that a considerable part of the material presented in the book was unpublished so far in Anglo-Saxon literature and, consequently, it was hardly available for a wide range of interested readers in many countries worldwide.

The readership of this work was targeted as a wide range of persons, such as students of agro-engineering and power specialists, experts and heads of municipal unions, managers of the ministries and other organizations responsible for development of new territories, the separate businessmen who are starting activities in new territories, and many others, as well as organizations such as large international consortia specialized in various businesses in rural areas, primarily related to -but not limited to- the production of agricultural products, including food. This very useful book will serve for small commodity producers on the countryside, various entrepreneurs of medium and small companies or projects as well as for inter-regional public organizations that assist in the enhancement of new technologies in the social sphere, regarding new environmentally friendly and sustainable development, self-determined and at the same time with responsibility for the whole and wider population. In general, this book intends to represent newest advanced achievements in agro-engineering and on rural-business energy related technical means which will be useful to a broad variety of readers and doubtlessly have a positive impact on a solution of sustainable development in new rural territories.

This work provides a reference of premium quality to graduate and postgraduate students, decision makers and investigators in private domains, universities, traditional and emerging industries - on agriculture, forestry, tourism and energy sector and beyond, for governmental and non-governmental organizations, in the areas of various Operational Research, Artificial Intelligence, geo- and earth-sciences, engineering, management, business and finance, where ever one has to represent and solve uncertainty-affected practical and real-world problems related to rural, remote, developing and emerging regions and territories. In the coming times, mathematicians, statisticians, computer scientists, economists, game theorists, physicists, chemists, geologists, representatives of environmental, civil, electrical and electrical engineering, but also biologists, neuroscientists, social scientists, scientists on natural resources, landscape architects, and representatives of the humanities, are cordially welcome to join this discourse and enter the collaboration for reaching improved and sustainable solutions. It has been well understood that predictability under uncertain environments and rapidly changing scenarios exercise a core request and a challenge in all fields of engineering, science and management. In this respect, this book is following a very modern perspective; finally, it promises to become very significant both in academia and in practice, and very fruitful and important for humankind.

The topics of this book's chapters were open, but not limited to:

- Solar thermal energy installations,
- Geothermal energy and Geo-Sciences,
- PV solar power plants for rural objects energy supply,
- Wind-based energy generation installations and systems,
- Small hydropower plants,
- Cogeneration PV Thermal modules fabrication and application,
- RES application best practices in agriculture,
- RES potential in remote territories of different countries,
- Specific RES potential evaluation methodologies in condition of meteo-data absence,
- Petroleum Engineering for local needs,
- Diagnostic and monitoring methods for energy equipment status in rural conditions,
- Providing parameters of solar power plant with given generation graph,
- Innovative compact controlled electricity transmission lines,
- Resonant Power Systems for Renewables based Electricity Grids,
- Resonant system for efficient electricity transfer,
- Energy saving system for illumination of greenhouse plants,
- Energy efficient electric equipment for heat supply in Agriculture,
- Scientific and methodological support of rural electric system operation,
- Solutions for diesel, wind and solar power plants joint use.

Let me please underline the high importance of this book for everyone related with it so far, and for all its forthcoming readers and appliers. Indeed, I trust that in the years to come, Operational research, optimization, and optimal control will remain in the center of giving Decision-making aid and support which this book targets at. In this context, I especially mention stochastic optimal control under impulses and regime switching of economic and cultural kinds. Here, my coauthors and I are working on scientifically, towards further inclusion of real-word situations and cases, with all their immense uncertainties and with the "human factors" even, into an optimal decision-making.

To all the authors of this valuable book, I convey my appreciation and thanks for having shared their expertise and idealism with the academic community and, eventually, with humankind. Moreover, I extend our gratitude to the two editors of this compendium, Prof. Dr. Valeriy Kharchenko and Prof. Dr. Pandian Vasant, for their dedicated hard-work and vision, of having gained an impressive variety of prosperous contributions. Finally, I am very thankful to the publishing house of IGI Global, for making possible and become reality a premium book of real-life significance and of potential high impact for the world of tomorrow, for our present generation and for the following generations.

Now, I wish us all a lot of pleasure and gain when browsing through and studying this interesting book, and I hope for a remarkable benefit going to be received from it scientifically, personally and societally.

Gerhard-Wilhelm Weber
Poznan University of Technology, Poland
November 2018

Gerhard-Wilhelm Weber *is a Professor at Poznan University of Technology, Poznan, Poland, at Faculty of Engineering Management, in the Chair of Marketing and Economic Engineering. His research is on mathematics, operational research), finance, economics, optimization and control, neuro-, bio- and earth-sciences, medicine, data mining, analytics, artificial intelligence, education and development; he is involved in the organization of scientific life internationally. He received Diploma and Doctorate in Mathematics, and Economics / Business Administration, at RWTH Aachen, and Habilitation at TU Darmstadt (Germany). He replaced Professorships at University of Cologne, and TU Chemnitz, Germany. At Institute of Applied Mathematics, Middle East Technical University, Ankara, Turkey, he was a Professor in both Financial Mathematics and Scientific Computing, and Assistant to the Director, and has been a member of five further graduate schools, institutes and departments of METU. G.-W. Weber has different affiliations at, e.g., Universities of Siegen (Germany), Federation University (Ballarat, Australia), University of Aveiro (Portugal), University of North Sumatra (Medan, Indonesia), Malaysia University of Technology, Chinese University of Hong Kong, and KTO Karatay University (Lean & Green SCM Lab; Konya, Turkey), at EURO (Association of European OR Societies) where he is "Advisor to EURO Conferences" and IFORS (International Federation of OR Societies), where he is member in many national OR societies and working groups, at Pacific Optimization Research Activity Group, etc. G.-W. Weber has (co-)supervised many MSc. and PhD. students, (co-)authored numerous books, special issues and articles, and given many presentations from a diversity of subject areas, on theory, methods and practice. He has been a member of many international editorial and award boards; he participated at and was in charge of numerous research projects; G.-W. Weber received various recognitions and awards by students, universities, conferences and scientific organizations, nationally and internationally.*

Preface

INTRODUCTION

Nowadays circumstances year by year puts forward new problems and challenges. Mankind food and other vital goods consumption steadily grows. To meet these requirements constant involving in economic circulation of new lands and territories, suitable for the economic activities organisation is required.

As a rule these territories and pieces of lands disposed far of main populations activity and in the same time from centralized centers of energy supply. The most suitable activity for these territories is agriculture

Start a business in such territories is easiest if you are engaged in the production of agricultural products on small-scale enterprises, with a small number of employed people. Small scale production makes it possible to quickly deploy work, easier and faster to ensure the sale of products.

Such activities require special technical support measures both for the livelihood of people living there and for organizing a highly efficient and cost-effective production process. Naturally, in such territories it is easiest and quickest to deploy activities related to the production of agricultural products. Such enterprises are easier to provide with energy resources, in particular, using renewable energy sources.

Existing technologies of energy supply, cropping and processing of crops cannot always be easily adapted to use in small and medium-sized enterprises.

There was such a situation that information of this kind is needed not only directly by the worker in the village. The manufacturers and developers of such technical facilities and technologies and scientific organizations that are engaged in the development of technologies and equipment for the power industry of the village, production processes of various types, as well as the problems of housing and communal solutions, are showing great interest in new efficient means of conducting agricultural business, especially on a small scale, problems of the rural population.

An important issue is the organization of reliable power supply in an autonomous mode. The use of renewable energy sources can play an important role in this issue. And this means that information on the possible technologies of autonomous generation of electric and thermal energy, its rational consumption through the use of energy-saving technologies will inevitably be in demand. Of great importance in the organization of active activities in rural areas can also be the use of intensive technologies for the cultivation of products and its processing on site with the least expenditure of energy, the amount of which at remote sites may be critically small.

OBJECTIVE OF THE BOOK

The book includes a description of a series of technological methods and a number of technical means that ensure the organization of production of various products in the territories newly allocated for economic activity or cultivated for these purposes. The specified technologies and technical means are adapted to application in production of small scale, at the enterprises of small and medium business.

Technical solutions designed to provide vital activity in new territories with a high level of comfort and high standards of living conditions are also considered.

Modern technologies of energy supply of life and production activity at remote, first-time territories are considered.

Much attention is paid to work aimed at organizing an efficient production process and social comfort in the face of scarcity of energy resources and the need to rationally use them.

It is an obvious fact that any similar published books complied altogether issues of rural business development concerned in the intended mode doesn't exist. No doubts that the book being published attract attention of a huge number specialists related to rural agribusiness activity on remote territories and looks to be very interesting for wide range of practical managements and businessman as well as scientists involved in activity connected with considered in the book technical field.

The particular interest to the book may be connected with the fact that the materials represented in the book, mainly was not published in English yet and consequently was not available for readers throughout the world.

TARGET AUDIENCE

The book will be useful to a wide range of persons, such as students of agro-engineering and power specialties, experts and heads of municipal unions, managers of the ministries and other organisations responsible for development of new territories, the separate large businessmen beginning business at new territories and many others as for organisations like large international consortia specializing in various business in rural areas, primarily related to but not limited to the production of agricultural products, including food. A very useful book will be for small commodity producers in the countryside, various entrepreneurs of medium and small businesses as well as for inter-regional public organizations for assistance to advancement of new technologies in social sphere in view of environmentally friendly sustainable development. In general the book is intended for representation of the wide public advanced achievements in the field of agro-engineering and related to the rural business energy technical means which will be useful to a wide range of readers and doubtless positive impact on a solution of a problem of a sustainable development of new rural territories.

Topics reflected in the book:

- Solar thermal energy installations
- PV solar power plants for rural objects energy supply
- Wind based energy generation installations and systems
- Cogeneration PV thermal modules fabrication and application
- RES application best practices in agriculture
- RES potential in remote territories of different countries

- Specific RES potential evaluation methodologies in condition of meteo data absence
- Providing parameters of solar power plant with given generation graph
- Innovative compact controlled electricity transmission lines
- Resonant Power Systems for Renewables based Electricity Grids;
- Resonant system for efficient electricity transfer
- Energy saving system for illumination of greenhouse plants
- Energy efficient electric equipment for heat supply in Agriculture
- Scientific and methodological support of rural electric system operation
- Solutions for diesel, wind and solar power plants joint use

The material presented in the book is devoted to the presentation of the results of theoretical and experimental research in the field of renewable energy sources. Devices, technologies and examples of practical use of systems based on renewable energy sources are considered.

Part of the book is devoted to the consideration of the issues of energy supply of agricultural production and household aspects of life in rural areas, including both heat and electricity supply.

It is necessary to note the fact that a significant amount of the material presented in the book is devoted to works made in different countries, but not published in English. This will allow the English-speaking reader to get to know them. The creators of the book hope that these sections of the book will be useful to English-speaking readers and will be in high demand.

The Chapter 1 dedicates to consideration of concentrators of solar energy with use concentrators based on holographic films. Looking at the history of solar energy and renewable energy in general, the authorities and scientists have been paying much attention to the recent period, due to the depletion of fossil energy resources and the growing difficulties in solving environmental problems. The development of solar energy technologies has led to the use of solar energy concentrators. Concentrators use to concentrate sunlight onto PV cells. This allows to provide for a reduction of the cell area required for producing a given amount of power. The goal is to significantly reduce the cost of electricity generated by replacing expensive PV converter area with less expensive optical material. In this chapter, the authors talk about concentrators of solar energy, especially about modules based on holographic films. Holographic solar panels (HSP) last decades appeared in large-scale production and actively used in solar energy. Evaluations of other types of existing concentrators are presented as well.

The Chapter 2 considers information systems that provide analysis of the RES resource base and support decision making. RES resources in the Republic of Kazakhstan are briefly analysed. The methods and systems for evaluation theoretical, technical and economic pothential of renewable energy sources are considered. Based on this overview we propose six steps of estimation that should be realized in intellectual GIS. One of the major steps is the identification the factors that affect using of renewable energy sources. These factors can contribute to and impeding the development of green energy. The simple taxonomy of these factors is proposed. The second important step is to aggregate estimations of the factors. As result we propose the aggregation method based on Bayes rule. Compared to other methods this approach allows us to obtain two estimates: probabilities of positive and negative hypotheses. The methods discussed are at the heart of the information system supporting the development of renewable energy sources in the Republic of Kazakhstan.

The Chapter 3 is devoted to consideration the processes of electrical and heat power production with use the products of air conversion of motor diesel fuel and electrochemical generator for agricultural consumer At present, the production of electricity for agricultural consumers remote from the centralized electrical power grid is carried out using diesel-generator technology with a limited service life of engines and extremely low efficiency of the expensive fuel used. In this paper, an innovative technology has been considered for the combined electrical and heat power production using the preliminary conversion of diesel fuel into synthesis gas with its subsequent supply to a high temperature electrochemical generator (ECG). Synthesis gas for the operation of the electrochemical generator was produced by air conversion of motor diesel fuels in a catalytic burner reactor. On the basis of heat balances of the burner, ECG and waste-heat boiler-utilizer, electrical efficiency of the solid oxide fuel cells' (SOFC) battery, chemical efficiency of the burner, the temperature at the SOFC anode, the EMF of the planar cell, a portion of hydrogen oxidized at the SOFC anode, specific consumption of diesel fuel for the production of electrical and heat power were calculated.

Presented in the Chapter 4 research carried out at existing solar power plants and laboratory renewable energy sources, whose purpose was to increase the energy efficiency of photovoltaic installations with parallel and mixed switching of photocells, operating under uneven illumination, parallel voltage arrays of photovoltaic modules due to voltage equalization. Experimental characteristics of photovoltaic installations with the developed "Device for coordinating the array of photoelectric modules" (DCA), realizing a method of selection of electric energy and without its use are given. It is experimentally shown that the use of DCA, allows to increase the electrical power of the array with partial shading up to 2.6 times with partial shading. The results of the research can be used to design new photovoltaic installations and upgrade existing ones.

The Chapter 5 considers the factors that affect the incomplete use of power. The analysis and systematization of accumulating systems on the types of accumulated energy, on the processes occurring during storage, types of devices and types of generated energy is performed. Each of the possible ways of accumulating energy is analyzed in detail. Various variants of conformity of schedules of change of a consumed load within a year and a total resource of solar radiation are considered. The analysis of the parameters of heat-accumulating materials and their classification depending on the material class, the way of heat accumulation and return, on the cyclicity of work, etc. are provided. It is shown that the level of temperature, the scale of the storage unit and the necessary duration of heat storage determine the requirements for the construction of batteries and the choice of heat-storage substances. The prospects of research on the search for new ways of accumulating energy and technical means for their implementation are considered.

The Chapter 6 focuses on different aspects and challenges of power supply for remote energy consumers in the Russian Arctic. Authors discuss the potential use of renewable energy, some specified technological features and risks related to the broader deployment of decentralized renewable energy systems in the Arctic region. Even though there is limited experience of renewable energy systems installation in the Russian Arctic and greater technological challenges, the authors see it as a potential opportunity to contribute to innovative and sustainable development of the region. The authors underline the potential synergistic effect of broader deployment of renewable energy systems in the Russian Arctic. The key argument is that sustainable and efficient energy system will open new development opportunities and stimulate future socioeconomic development of the region through the use of local and renewable energy resources, implementation of new power generation modes and technologies.

In the Chapter 7 renewable energy sources development in rural areas of African countries is considering. The prospects of increasing access to electricity for the population of rural areas of Africa are considered. The main international funds and organizations aimed at sustainable energy development in Africa are described. An analysis of the state and possible options for using renewable energy sources for this purpose in decentralized energy supply through the creation of mini-grids or stand-alone systems is given. The risks by developing renewable energy.

The object of research in the Chapter 8 is the solar power plant as the source of additional economically expedient power supply of the electrical energy consumer. The purpose of this research is to analyze the options for the layout of solar power plant, taking into account the solar energy potential of the district, the design features of the proposed location, the load curve of the consumer and the cost of the generated electrical energy. The chapter presents the results of calculation and selection of the parameters of solar power plant elements on the roof of the consumer's production building. The chapter presents the results of research of the dependence of the cost of the electricity generated by the solar power plant on the number of installed panels, which in order to increase the realized solar energy potential of the district also allows adding photoelectric modules and accumulating devices in the layout of the operating station at tariff growth. The chapter presents the results of researching these areas, that are conducted by the authors and which are completely original.

In the Chapter 9 the activity of farm enterprises which is directly connected with the efficient use of resources such as electrical energy and water is considering. Taking such conditions as factors of economic expediency and ecological safety into account, it is more reasonable to use wind stations in order to provide consumers with energy and water. The use of conventional wind and electrical stations is an easy and reliable solution. However, annual wind velocity on most settled territories does not exceed 6 m/s or even 4 m/s. It makes the efficient use of wind electrical stations more complicated. One of the solutions can probably be the use of wind and electrical stations on the basis of slow speed non-transmission arc-shaped-stator inductor-type generators with an integrated radial and ring-shaped rotor. Another efficient solution to provide areas with water and electrical energy is to use a combined wind station with a crank-and-rod mechanism and the rod of the driving mechanism of the back-and-forth motion of the core of a magnetic and electrical linear generator and the piston of a plunger pump.

The purpose of the Chapter 10 is to compare the design and operational parameters of energy efficiency (power generation, capacity utilization factor) of a wind power plant (WPP) in order to improve the accuracy of the design parameters estimation at the preliminary design stage. Recommendations have been given for accounting different factors influencing on energy efficiency and economic indicators of a WPP. In this article, the research carried out on the example of existing wind-diesel complexes (WDC) in the settlement of Ust-Kamchatsk (Kamchatka Krai), Novikovo (Sakhalin Oblast) and projected WDC on Popova Island (Vladivostok city district).

Considered in the Chapter 11 doubly Fed Induction Generator (DFIG) is widely spread technology in modern wind turbines (WT), due to its capability to operate with variable speed, partial scale power converter and ability to control active and reactive power independently. The main drawback of DFIG is its complicated protection systems. In the chapter, several strategies for DFIG protection are reviewed and provided a conclusion about their advantages. Penetration of renewable energy sources (in particular wind power) have a huge impact on power systems, thus, wind turbines should be considered as conventional generation units in terms of frequency and voltage regulation. Modern grid codes require WT stay connected during grid fault and be capable to provide appropriate grid support. Therefore, it is important to implement DFIG protection system, which could meet grid code requirements.

The Chapter 12 describes the advanced technologies of energy and water saving on the Uzbekistan largest pumping stations of Karshi Main Canal. The unique Karshi pumping stations stage consumes more than 4% of the total electricity generated in Uzbekistan. Such a significant consumption of electricity by irrigative pumping stations, gives the problem of energy conservation to them a status of special importance and relevance. Advanced technologies of energy saving, and non-cavitation work are scientifically substantiated and developed. The results of carried out field tests are described. The scientific results of the research are recommended to allow efficient use of water and energy resources and to ensure reliable operation of the power equipment of pumping stations in the rural branches of economy.

In the Chapter 13 relevant of improving power quality of agricultural consumers by adjusting the cost of electricity is considered. The chapter lists the reasons for the output of power quality indicators beyond the standard values. There is the assessment of the quality of electricity in the Orel region. Distribution functions for distortion of power quality indicators are constructed. The relevance of the adjustment of the cost of electricity depending on its quality is justified. The mathematical model for determining the cost of consumed electric energy depending on its quality is developed in paper. The paper gives the method allowing to adjust the cost of consumed electric energy depending on its quality and source of distortion. In the paper there is the mechanism of carrying out calculations for the consumed and transferred electric power considering change of electric power cost depending on its quality. The feasibility study of the proposed methods and means of improving power quality is realized.

In the Chapter 14 noncontact electric power supply with the use of cable line is considered. It is an innovative trend in the development of electric transport. Work to develop such systems is carried out in the USA, South Korea, Germany, Israel, the Russian Federation. The VIESH researchers have developed and patented experimental models of a car and a tramway of the future without accumulators, which receive energy from external energy system through air gap from a single wire cable laid under the surface of road pavement. Noncontact systems for electric power supply to transport are developed at VIESH on the basis of resonant methods of electric power supply through a single-wire cable. This is a basic difference of the technology under consideration from similar ones. This technology makes it possible to realize at the new technological level the need for electric roads with minimal costs. Given that all the world's automakers are working on the production of an electric vehicle for mass production, the need for creating a motorway in the city and on inter-urban autobahns is even more topical.

The Chapter 15 analyzes the power supply restoration time. The analysis shows that this time consists of the time for obtaining information about failure, the time of information recognition about failure, the damage repair time and the time of connection and disconnection harmonization. Each of them, in turn, is also considered in more detail that allows identifying reserves to reduce the power supply outage time. The paper proposes reducing the power supply restoration time by going to the monitoring system combination of both operating modes and technical condition of power networks. Based on the study results there are methods allowing to reduce the power supply outage time of rural consumers by means of automating rural electrical networks and improving the operating mode monitoring of power supply systems. The paper also suggests a set of devices that allows to implement the proposed methods.

Valeriy Kharchenko
Federal Scientific Agroengineering Center VIM, Russia

Pandian Vasant
Universiti Teknologi PETRONAS, Malaysia

Acknowledgment

The editors would like to acknowledge the efforts of all the people which were involved in the activity on creation of this book. Particularly they would like to thank each one of the authors for their contributions. Our sincere gratitude goes to the chapters' authors who contributed the time and expertise to this book. Special gratitude should be given to the people which beside the preparing their chapters took part in the reviewing process. Without invaluable support of all of them this book would not have become a reality. Among all these persons special gratitude should be given to such persons as A. Vasiliev, V. Bolshev, T. Gallyamova, D. Budnikov.

We give a lot of thanks to the editorial advisory board members which have made a lot for making the high quality of the book. We especially need to underline the essential role of Editorial Board member, professor Gerhard-Wilhelm Weber, for providing excellent reviews and invaluable assistance.

The editors are confident that this book will certainly be useful to readers, because, among other things, it gives them the opportunity to get acquainted with the results of research that has not been published in English-language literature so far. Therefore, it would be wrong not to mention the terrific role that IGI Global publishing house has played in order for this information to become published.

Finally, the editors would like to express their gratitude to the entire IGI Global publishing house staff, who helped the editors and authors in preparing the manuscript of this book in a professional and extremely friendly manner and have done a lot to make this a premium quality publication.

Valeriy Kharchenko
Federal Scientific Agroengineering Center VIM, Russia

Pandian Vasant
Universiti Teknologi PETRONAS, Malaysia

Chapter 1
Solar Concentrators:
Solar Modules Based on Holographic Films

Bekhruzi Talbi Shokhzoda
Tajik Technical University, Tajikistan

Mikhail Georgievich Tyagunov
MPEI, Russia

ABSTRACT

Looking at the history of solar energy and renewable energy in general, the authorities and scientists have been paying much attention to the recent period, due to the depletion of fossil energy resources and the growing difficulties in solving environmental problems. The development of solar energy has led to the use of solar energy concentrators. Concentrators are used to concentrate sunlight onto PV cells. This allows for a reduction in the cell area required for producing a given amount of power. The goal is to significantly reduce the cost of electricity generated by replacing expensive PV converter area with less expensive optical material. In this chapter, the authors talk about concentrators in solar energy, especially about modules based on holographic films. Holographic solar panels (HSP) in recent decades have appeared in large-scale production and been actively used in solar energy. Evaluations of other types of existing concentrators are presented.

INTRODUCTION

Energy is the largest field of the current world economy. It`s considered one of the main life-supporting sectors of the national economy, the level of development of which largely determines the economic power of the state and its geopolitical importance in the world society. The pace of Global Electricity development is commensurate with the growth of population and economy. As the growth rate of population, of course, different types of autonomous consumers are growing. Modern achievements of technology which can`t be imagined without electrical energy, provide a high level of life and comfort for man. Therefore, the task of ensuring the safety and reliability of their work is not easy and at the same time very important, since its solution depends not only uninterrupted supply of electricity but also often the lives of people. Scientifically technical progress, the emergence of new sectors of the economy,

DOI: 10.4018/978-1-5225-9179-5.ch001

improvement of technology, improving the quality and living conditions of people lead to the expansion of the use of electricity and increase the requirements for reliable and uninterrupted. Therefore, in order to avoid restrictions on electricity consumption, it is necessary to find new ways of using non-traditional types of energy and if possible find new types of energy sources.

One of the environmentally safe ways to provide the population with energy (electrical and thermal) is solar energy. Since conventional solar modules have low efficiency now, many talks about the inefficiency of the use and even the development of this industry. Of course, it is not so. There are several simple reasons why solar energy should develop:

- Exhaustion of Fossil Fuels;
- Clean and Safe;
- Decentralized Electricity Source;
- Space Exploration;
- Enough Territory with Good Solar Radiation;
- Energy Independence;
- New Workplace.

All the above-mentioned advantages are only a superficial assessment. It is possible to itemize the advantages of solar energy in terms of electricity and heat. There are also disadvantages to the use of solar energy and technologies that in this industry:

- Uneven Distribution of Solar Energy;
- Low Density of Solar Energy;
- Instability of Solar Energy;
- Low Efficiency of Solar Cells;
- High Cost of Solar Cells;
- Sensitive to Temperature Changes.

Conversion method of solar energy into electrical energy by means of semiconductor solar cells is currently the most developed in scientific and practical terms. It is widely used in power supply systems of spacecraft and is increasingly using in ground conditions to provide electricity of autonomous consumers (portable equipment, beacons, automatic meteorological stations etc.).

The main problem of using solar energy is the high cost of crystalline silicon, which forms the basis of solar power plants, as well as the low efficiency of modern solar batteries, due to the fact that from the entire solar spectrum only a narrow part is used to convert into electricity. Taking into account these problems, solar energy is developing in several directions; some are trying to save on the production of solar cells or looking for a new method of production to reduce their cost. Others are trying to build hybrid systems or new designs to use solar energy. There are those who are working on the creation of new elements or new technology.

BACKGROUND

The authors of the chapter review the use of different types of solar concentrators of old and new generations. Nevertheless, focuses on a new design with the planar holographic solar concentrators and provides research data. First, it`s talking about the relevance of the theme, the history of the use of solar energy and give some historical facts that are important in the history of the development of solar energy. Different types of solar concentrators are considered, and their positive and negative sides are evaluated. A new design of a hybrid solar installation using solar concentrators and its operation mode for certain terrain conditions is considered. A few words about future research and what directions could be a continuation of this study. At the end of the chapter, a conclusion is made about the study of a hybrid solar installation and an additional list of references is given that could be useful.

HISTORICAL FACT

Since ancient times in Central Asia, sunlight and clean air have been used to gain strength and improve health. Children were allowed to walk naked in the sunlight, which is of great importance for the normal growth and development of the child's body. For storing the fruits, they were made dry with the help of sunlight.

The real "solar boom" began in the XVIII century, when science, liberated from the shackles of religious superstitions, went forward by leaps and bounds. The first solar heaters appeared in France. Natural scientist J. Buffon created a large concave mirror that focused reflected sunlight in one point. This mirror was capable of on a clear day, quickly igniting the dry wood at a distance of 68 meters. Soon after the Swedish scientist, Saussure N. built the first water heater. It was just a wooden box with a glass lid (cover), but the water poured into the no-wonder device was heated by the sun to 88 °C. In 1774, the French scientist Lavoisier used lenses for the first time for the concentration of thermal energy. Soon in England, polished a large biconvex glass that melting cast iron in three seconds and granite - per minute (Tikhonov et al., 2009).

Historical dates of solar energy development:

- 700 BC – Sunlit Fires. Humans figured out how to make fires by concentrating the sunlight with a magnifying glass.
- 214–212 B.C. – Archimedes' Heat Ray. The focusing properties of curves of solid surfaces are known since the time of Archimedes, who according to legend set fire to enemy ships, focusing on them the sun's rays from polished to shine copper shields of soldiers lined up in a semicircle along the shore (Figure 1).
- 1767 – The First Solar Oven. A solar oven, or solar cooker, uses sunlight to heat meals or drinks. Already in 1767, the first solar oven was invented. It is the credit goes to a Swiss physicist Horace de Saussure.
- 1816 - On September 27, 1816, Robert Stirling applied for a patent for his economizer at the Chancery in Edinburgh, Scotland. By trade, Robert Stirling was actually a minister in the Church of Scotland and he continued to give services until he was eighty-six years old! But, in his spare time, he built heat engines in his home workshop. Lord Kelvin used one of the working models

during some of his university classes. This engine was later used in the dish/Stirling system, a solar thermal electric technology that concentrates the sun's thermal energy in order to produce power.

- 1839 – Photovoltaic Effect. This is a big year in the history of the solar electricity because Alexandre Edmond Becquerel, a French physicist, discovered (photovoltaic effect) that there is a creation of voltage when a material is exposed to light.
- 1873 - Willoughby Smith discovered the photoconductivity of selenium.
- 1876 – Electricity from Light. Building on Smith's discovery three years before, professor William Grylls Adams, accompanied by his student, Richard Evans Day, were the first to observe an electrical current when a material was exposed to light. They used two electrodes onto a plate of selenium, and observed a tiny amount of electricity when the plate was exposed to light.
- 1880 - Samuel P. Langley, invents the bolometer, which is used to measure light from the faintest stars and the sun's heat rays. It consists of a fine wire connected to an electric circuit. When radiation falls on the wire, it becomes very slightly warmer. This increases the electrical resistance of the wire.
- 1883 - The First Design of a Photovoltaic Cell. An American inventor, Charles Frits creates the first solar cell of selenium, covered with a very thin layer of gold.
- 1905 - Einstein and the Photoelectric effect. Albert Einstein published a paper advancing the hypothesis that light energy is carried in discrete quantized packets to explain experimental data from the photoelectric effect. This model contributed to the development of quantum mechanics (The Nobel Prize, n.d.).
- 1908 - William J. Bailley of the Carnegie Steel Company invents a solar collector with copper coils and an insulated box—roughly, it's present design.
- 1918 – Single-Crystal Silicon. Jan Czochralski, a Polish scientist, figured out a method to grow single-crystal silicon. His discoveries laid the foundation for solar cells based on silicon. The method invented by Czochralski produces all of today's silicon, from which diodes, transistors, chips and the most common solar panels are made.
- 1932 - Audobert and Stora discover the photovoltaic effect in cadmium sulfide (CdS).
- 1938 - A. F. Ioffe introduced for the first time the government of the USSR program of energy use of solar photovoltaic roofs. He was one of the first who put the problem of converting light energy into electrical energy with the help of semiconductors.
- 1947 - Passive solar buildings in the United States were in such demand, as a result of scarce energy during the prolonged W.W.II, that Libbey-Owens-Ford Glass Company published a book entitled Your Solar House, which profiled forty-nine of the nation's greatest solar architects.
- 1953 - Gerald Pierson, an American physicist conducting experiments at Bell Laboratories, found that silicon, with certain impurities, is much more sensitive to sunlight than selenium.
- 1954 - Production of a solar photocell with a practically significant efficiency of about 5% (J. Pearson, D. Chapin., K. Fuller, Bell Laboratories, USA).
- Mid 1950s - Architect Frank Bridgers designed the world's first commercial office building using solar water heating and passive design. This solar system has been continuously operating since that time and the Bridgers-Paxton Building, is now in the National Historic Register as the world's first solar heated office building.
- 1955 - In Phoenix, Arizona (USA), the Association for the use of solar energy was founded. At the same time, an international Symposium and the first exhibition of solar energy technology were

organized. Thousands of scientists from 36 countries took part in this event, about 80 inventions were exhibited.

- 1958 - The first satellites with solar batteries were launched (USA – Avangard 1, USSR - Sputnik-3). The cost of one watt-hour of energy was $ 500. The first Soviet solar battery, manufactured by May 1958, with a power of 2.0 watts, was installed on the Sputnik-3, launched on May 15, 1958.
- 1960 - Hoffman Electronics achieves 14% efficient photovoltaic cells. Silicon Sensors, Inc., of Dodgeville, Wisconsin, is founded. It starts producing selenium and silicon photovoltaic cells.
- 1962 - Bell Telephone Laboratories launches the first telecommunications satellite, the Telstar (initial power 14 watts).
- 1964 - In the Karakum desert near Ashgabat in Turkmenistan, a solar battery with 0.25 kW concentrators for water lifting was tested.
- 1965 - The first practical application of advanced gallium arsenide solar cells for energy purposes was even more exotic than in the case of silicon batteries. They provided electricity Soviet spacecraft operating in the vicinity of the planet Venus.
- 1967 - USSR scientists A.P. Landsman, N.S. Lidorenko and D.S. Strebkov developed a new class of photoconverters - multi-transition and solar high-voltage elements made of silicon.
- 1970 - Ion implantation technology has been used in the production of solar cells. USSR scientists have developed a technology of photovoltaic cells with bilateral sensitivity. The cost of electricity per watt-hour was reduced to $ 100. USA. At that time, all satellites were equipped with solar panels made of silicon. Efficiency at this point reached 10% and about two decades kept at this point.
- 1973 - The cost per watt-hour was reduced to us $ 50 due to the use of cheaper silicon boards. Funding for many studies in solar energy was curtailed, as the price of silicon at that time was an unacceptable luxury compared to oil prices.
- 1980 - The technology of solar multi-transition elements based on GaAIAs-GaAs was developed.
- 1984 - Charles Fritts installed the world's first rooftop photovoltaic solar array, using 1%-efficient selenium cells, on a New York City roof.
- 1985 - The USSR first industrial solar power plant (SPP) was built in the Crimea, near the town of Shelkino. SPP-5 had a peak power of 5 MW. At that time, about 30,000 payphones were working on solar energy throughout the California highway. Street solar lights installed during the day they accumulated energy, and at night lit the streets. With the help of solar energy, bus stops were also illuminated.
- 1988 - Using a new mirror-based technique, Roland Winston and his team set a new record for concentration of solar energy, concentrating sunlight to more than 60,000 times its normal intensity (Wilford, 1988).
- 1993 - Efficiency was achieved in 30% for solar cascade solar cells based on GaAIAs-GaAs heterostructure on germanium substrate. New classes of holographic, prismatic, parabolic concentrators and optical systems based on them were developed.
- 1995 - The cost per one Wh (watt hour) of solar electricity fell to $ 15. In all developed countries, intensified developments in the field of solar energy began. Governments have tried to reduce the cost of solar energy as quickly as possible to the level of conventional electricity. By this time, the efficiency of solar photovoltaics was raised to 15%.
- 1999 - A low-temperature chlorine-free technology for obtaining solar silicon has been developed.

- 2004 - This year, the generation reaches 1 million kilowatts of electricity. Solar revenues are close to $ 6.5 billion. USA. The first places in the "solar race" are Japan, Germany and the United States of America.
- 2014 - The 392 MW Ivanpah installation is the largest concentrating solar power plant in the world now, located in the Mojave Desert of California (Blood & Skolof, 2014). In the Netherlands, the world's first bicycle track from solar panels was opened.
- 2015 - Google has invested $ 300 million in the project of SolarCity Corp for the installation of 25 thousand solar panels throughout America. Thus, the Internet giant has made the largest investment in alternative energy in history.
- 2018 - Apple said 100% of the electricity used in our company's premises, now comes from renewable sources. We include offices, retail stores and data centres in 43 countries around the world.

TYPES OF SOLAR ENERGY CONCENTRATORS

Solar concentrators are optical devices that increase the solar flux density. Concentrators in the solar energy field using to generate thermal and electrical power. For solar power plants with thermal cycles of energy conversion, concentrators allow you to create the temperatures necessary to produce steam with certain parameters. For PV they are solved the following challenges:

- Increases the conversion efficiency of solar energy into electrical energy;
- Makes it possible to achieve the greatest conversion efficiency of solar energy into electrical energy;
- Reduce the cost of the output electric power unit.

Using solar concentrators is one way to reduce the cost of solar energy. The types of solar energy concentrators that currently exist in the world are below (Strebkov & Tveryanovich, 2007):

Figure 1. Wall painting in the Uffizi gallery in Florence, created by the painter Giulio Parigi in 1599-1600

- Paraboloid solar concentrators;
- Fresnel Lens Solar Concentrator;
- Prismatic concentrators;
- Compound cone concentrators (focon, Winston concentrator);
- Stationary cylindrical concentrators with U-shaped receiver;
- Cylindrical parabolic concentrator (S and W shaped);
- Louvered heliostats.
- Holographic and luminescent concentrators;

Paraboloid Solar Concentrators

The shape of the reflecting surface of the paraboloid formed by the rotation of the parabola around the axis of symmetry (see Figure 2). One of the properties of the parabola is the convergence of all light rays falling parallel to the main optical axis (the symmetry axis) to the focus F. The paraboloid creates an image of the remote object in the focal plane. Parabola belongs to the category of high-potential concentrators that radiation concentration can exceed hundreds and thousands of times the solar constant. The paraboloid configuration has a clear advantage over other forms in terms of the size of the concentrating ability. Therefore, it is widespread in solar systems. The optimal opening angle of real paraboloid concentrators, in contrast to the angle of an ideal paraboloid concentrator (45°), is close to 60° (Li et al., 2013).

Another type of parabolic concentrator is solar dish concentrator that resembles a plate (see Figure 3). Solar radiation is concentrated on the receiver that "Stirling engine" connected on it. An engine can convert thermal energy directly into mechanical work or electricity. Such systems can achieve efficiency

Figure 2. Operational scheme of parabolic concentrator

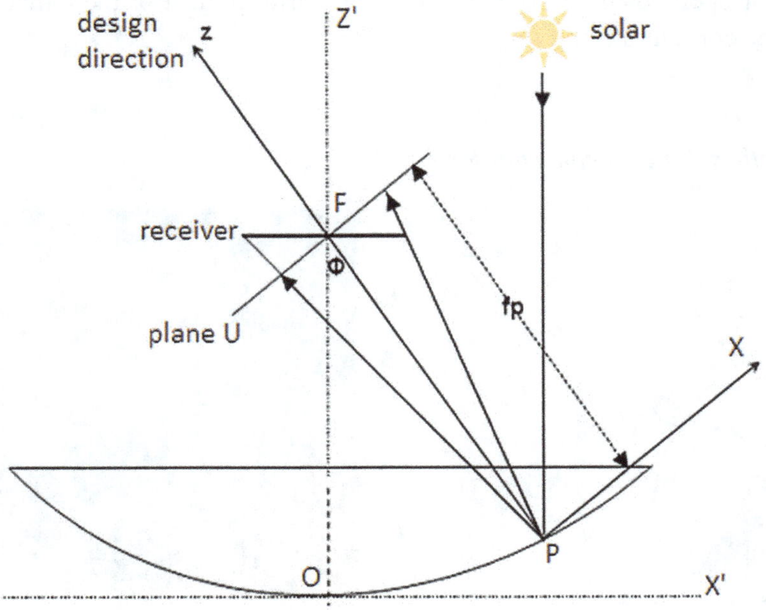

of more than 30%. They designed to operate in stand-alone mode also have the ability to combine multiple systems into one solar power plant.

Solar Concentrator Based on Fresnel Lens

Refractive systems (different types of lenses) have one vital advantage over the reflecting concentrators in that the radiation receiver is placed on the rear side of the receiving solar radiation surface, that is, it is not obscured by the receiver and its supports. The following devices can be attributed to lens concentrators (see Figure 4):

- Concentric Fresnel lenses that made in the form of a set of concentric refractive elements with straight lines or curves forming the working surfaces;
- Linear Fresnel lenses with refractive elements in the form of triangular prisms;
- Dispersion Fresnel lenses, the main principle of which is the decomposition of light into the spectrum which is selected photovoltaic radiation receivers with appropriate spectral sensitivity;
- Holographic Fresnel lenses.

The main feature of the lens concentrating systems is the presence of chromatic aberrations. In lenses with a high concentration of chromatic aberration the degree of concentration can reduce, but in the calculation of the thermal solar systems of lenses with a concentration of about 100 it can be neglected. For dispersion lenses, the dispersion of light is increased and strictly controlled.

The advantages of Fresnel lenses are as follows (Strebkov & Tveryanovich, 2007)

- High technological ability to manufacture Fresnel lenses (pressing, rolling) and relatively low cost;
- High constructive qualities of lenses as compact, flat objects, conveniently mounted in load-bearing structures with a high degree of occupancy;
- Convenient for operational purposes is the mutual arrangement of the radiation receiver on the rear side of the concentrator;

Figure 3. A parabolic solar dish concentrator

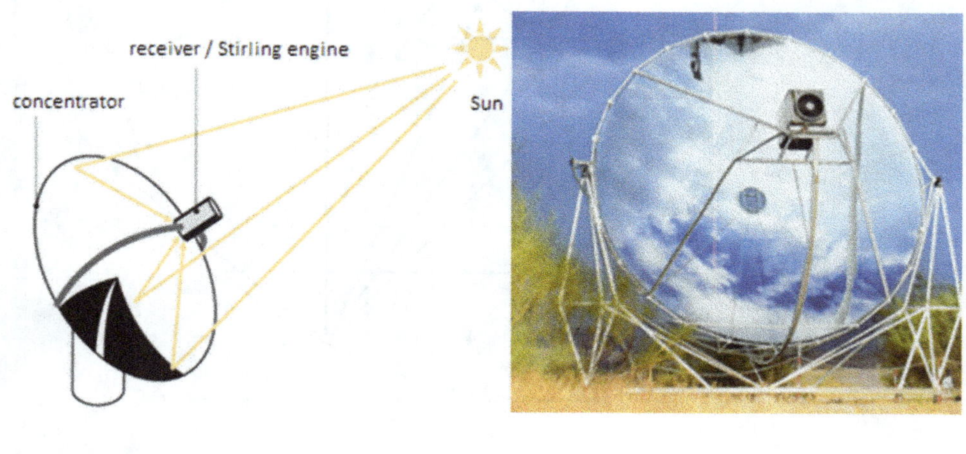

Figure 4. Principles of formation of the working profile of Fresnel lenses to obtain the necessary irradiation in the focal point

- The possibility of forming the necessary irradiance of solar energy in the focal spot due to the corresponding profile of the working layer of the lens.
- No need for precise tracking than mirror concentrators.

Prismatic Concentrators

Prismatic concentrators work as follows: radiation falls on the front face at an angle "i" refracts. After reflects from the back face and comes to the front face at an angle greater than the angle of total internal reflection. To date, many varieties of prismatic concentrators have been developed and are considered for both electricity and heat generation and as well as secondary concentrators (see Figure 5).

Major light losses in the prismatic concentrators:

- The loss of the absorption in the material prismatic (glass, organic glass);
- Light loss when reflected from the rear reflective surface;
- Light loss at total internal reflection;
- Output Losses.

Compound Cone Concentrators (Focon and Winston Cone Concentrator)

Compound cone concentrators - fiber-optics element consisting of a beam of conical fibers. The stationary concentrator was recognized the compound parabolic concentrator that invented independently by Prof. Winston in the USA (Winston, 1970) and Prof. Baranov in Russia (Baranov, 1976) (Figure 6). This concentrator has the field of view defined by its parametric angle and is able to focus radiation, which is falling on it with different angles less than the parametric. All parameters of this concentrator are directly dependent on its parametric angle (Formula 1). Significant differences of focons and Winston cone concentrators from other type's reflective lenses and lenses concentrators are:

$$d = D * sin\alpha \ , \quad H = \frac{\left(D + d\right)}{2} * ctg\alpha \qquad (1)$$

Figure 5. Main types of prismatic concentrators; a) prismatic on the basis of the oblique triangular prism, b) vertical triangular, c) on the basis of a trapezium, d) compound prism, e) with a central deflecting prism, f) for thermal collectors, g) prismatic as a secondary concentrator

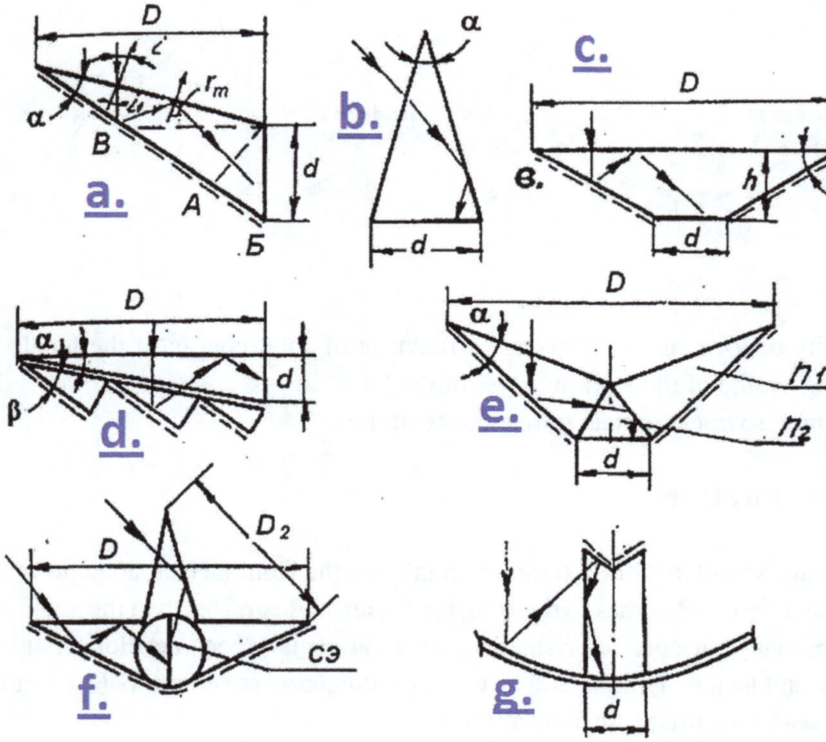

- Long-term operation within the parametric angle of the concentrator without tracking the sun position (see Figure 6);
- The possibility of using some part of the scattered radiation around the solar disk. This factor should be taken into account when choosing a specific type of concentrator for a solar power plant. Since total power generation of concentrators like focons and Winston cone concentrators increases significantly compared to concentrators that concentrating only direct solar radiation (for example parabolic).

Such type concentrators can be effectively used as secondary reflectors, increasing the concentration of light stream from the primary mirror (Strebkov & Tveryanovich, 2007).

One of the disadvantage of focons as concentrator are relatively large values of the depth 'l' that making them difficult to manufacture and increases the materials consumption. Another advantage of focons is that with a small loss of concentration can significantly shorten the depth 'l' (see Figures 6-7).

Since the focons are the mirror concentrators, the main requirements for the development of the technology of their production are ensuring a high quality reflective surface and protection of the mirror layer. In this task, the following ways have been implemented:

Figure 6. Compound cone concentrator; l- the height of the reflector; D, d - hole of entry and exit

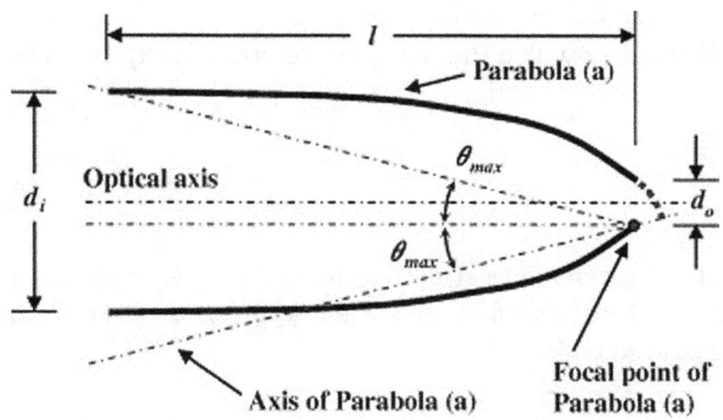

- The development of the glass focons;
- The development of the metal focons with a possible hermetic sealing of the inner reflective surface;
- Aluminum electroplating.

Stationary Cylindrical Concentrators With U-Shaped Receiver

The ideal profile of a U-shaped symmetric stationary concentrator with a continuous reflective coating is formed by two branches of the parabola AB and CD, deployed relative to the General focus F at the

Figure 7. The field of view of focon with a parametric angle α

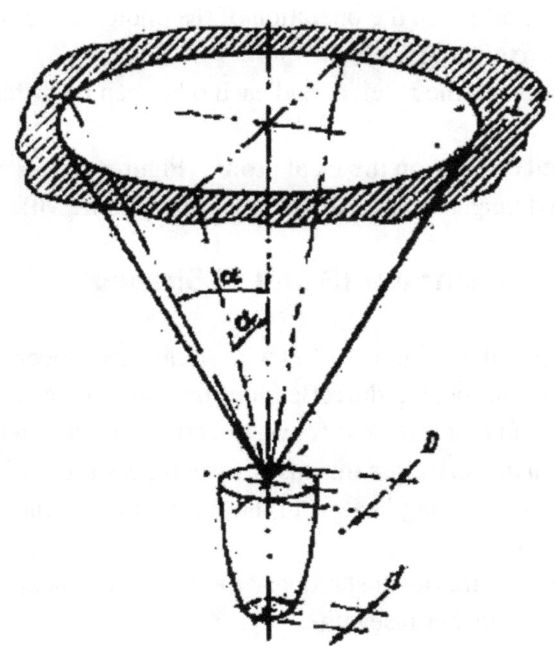

angles α_1 and α_2 respectively, at the aperture angle of the stationary concentrator (α_1 and α_2) and connected along the circumference BC (Figure 8).

The power of the solar module with a stationary concentrator is expressed by the formula:

$$P = E_o * K_g * s * \eta * \tau, \qquad (2)$$

E_o - the arrival of direct solar radiation per unit surface, W/m^2; K_g - geometric concentration of the stationary concentrator; s - effective surface area of the receiver; η - the efficiency of the receiver; τ - optical efficiency of the concentrator.

The geometric concentration is declination angle function of the Sun δ and the hour angle ψ (Formula 3). Formula of changing the angle of the Sun δ depending on the day number beginning of the year is given below (Formula 4).

$$K_g = K_{go} * \cos\delta * cos\psi, \qquad (3)$$

$$\delta = 23.45 * \sin\left(360 * \frac{284 + n}{365}\right) \qquad (4)$$

The design profile of a stationary concentrator can be used to obtain a facet stationary concentrator. The calculation of facet systems has a number of specific features, in contrast to systems with a solid reflective coating. When using facets, the distribution of concentrated solar energy on the receiver is more evenly, this has a positive effect on the operation of the photovoltaic cells and the facets are more durable than the solid reflective coating.

The location of facets relative to the receiver and each other can be different (Figure 9):

- When the facet is located directly on the ideal profile (Figure 9a);
- When the facet is located tangentially to the ideal profile (Figure 9b).

Cylindrical Parabolic Concentrator (S and W Shaped)

Using of the presented solar modules (Figure 10-11) with stationary concentrators allows increasing the coefficient of geometric concentration, but the reflections increase; there are problems with shading the secondary mirrors of the incident and reflected flows, the effect on the shading of the brackets holding the secondary reflectors. When the reflection increases, the temperature surface of the receiver increases which may have to think about removing heat from the receiver (example if the receiver is solar cell) (Strebkov & Tveryanovich, 2007).

However, these types of concentrators in stationary mode give a fundamentally high degree of concentration and are of interest for further research.

Figure 8. The geometry of the ideal profile of the U-shaped concentrator: a) asymmetric profile; b) symmetrical profile

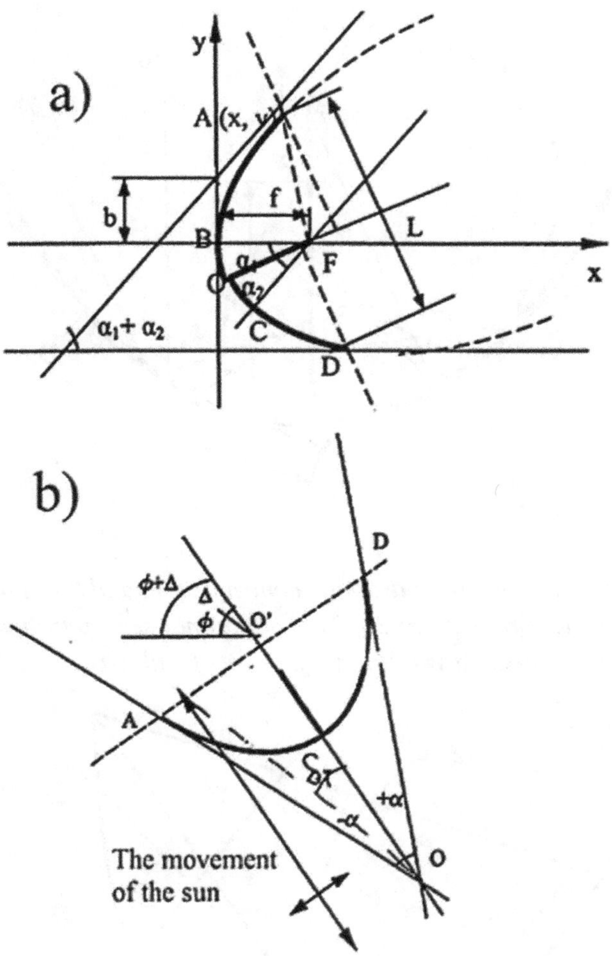

Louvered Heliostats

Louvered heliostats concentrators (LHC) are parallel rows of synchronous mirrors located horizontally or at an angle to the horizon. If the mirrors, after reflection, preserve the parallelism of the sun's rays then they work in heliostat mode. If the reflected rays produce a convergent light beam, the system operates in the concentrator mode (Strebkov & Tveryanovich, 2007). In this case, one movement tracking is carried out by turning the mirrors around parallel axes (tracking the height of the Sun), the other - by turning around the vertical axis (tracking the hour angle) (Figure 12-13) (Yellowhair et al., 2015).

As compared to traditional heliostats, louvered heliostats concentrators have a number of distinctive features:

- The possibility of placing the radiation receiver in a plane located below the level of the mirrors;
- The possibility of placing in a reflected flow of a concentrator to obtain acceptable values of the irradiation of the receiver, while the entire system will be more compact than for a whole heliostat;

Figure 9. Facet surface of U-shaped parabolic concentrator: a) according to the ideal profile; b) the tangent to the ideal profile (calculated) profile

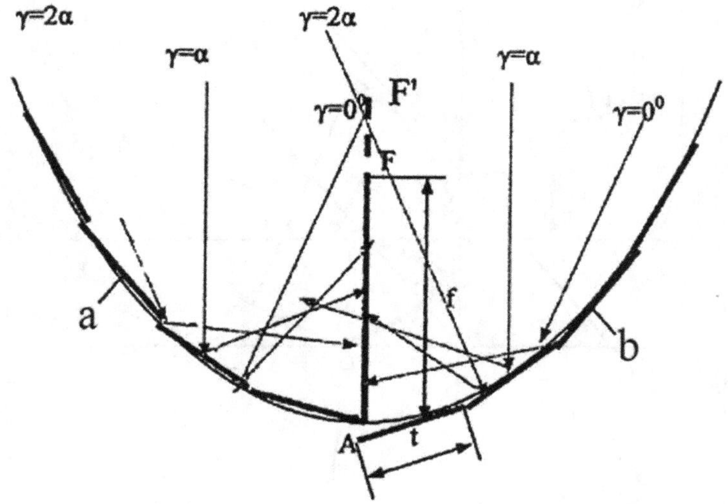

Figure 10. Solar module with a symmetric parabolic concentrators and S - shaped secondary reflectors. 1, 2 - semi parabolic cylindrical reflectors; 20, 21 – additional semi parabolic cylindrical reflectors; 3- radiation receiver with double-sided working surface; 4 - focal axis

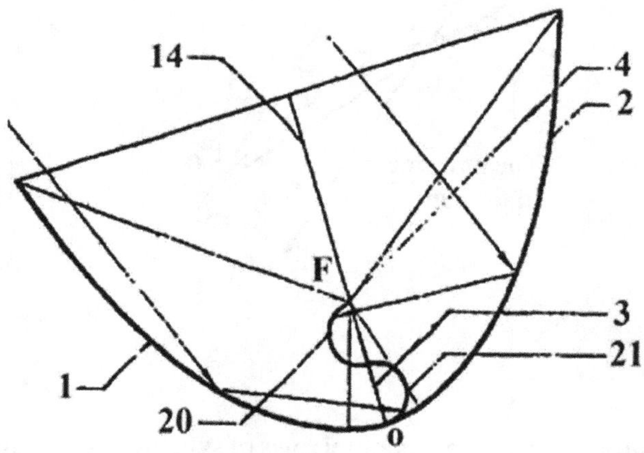

- Significant reduction in wind loads;
- Reduction in the device metal consumption;
- Systems "closed type": The possibility of the whole solar system placed under a transparent protective coating.

The acceptability of such systems requires close analysis, taking into account the possible automatic averaging of energy production throughout the year, which may be of interest to countries with hot climates.

Figure 11. A photovoltaic module based on w shaped concentrator

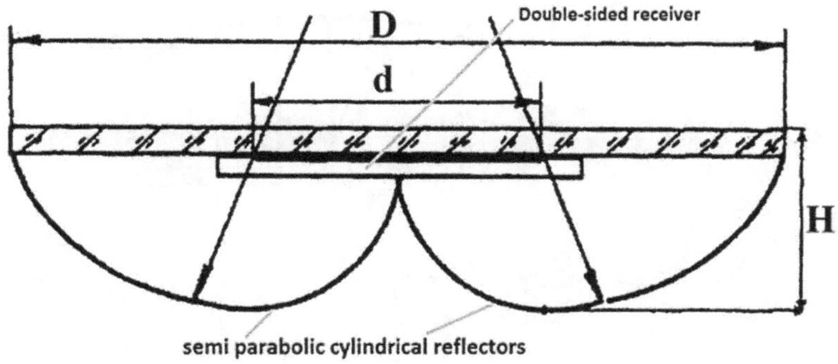

Holographic Concentrator

Holographic optical elements (HOE) - a generic name various optical elements, which influence on the wave front based on the diffraction of light. Micro-optical hologram element – it is a class of holographic optical elements (HOE), the sizes and structure (period, the relief height) that does not exceed micrometer. Micro-optical hologram (MOH) element is most often represents a planar glass substrate, on one side of which is formed diffractive microstructure, which makes the phase delay due to modulation of coefficient of the transmittance or the height of the microrelief. The simplest examples of micro-holographic elements (MHE) are the diffraction lattice and a Fresnel zone plate.

Micro-optical hologram element (as a hologram optical element) is performed by recording the specific type of volume hologram photosensitive material. Beam incident such a hologram element is diffracted through modulation the refractive index within the manifested photosensitive layers.

Figure 12. Example 1: method of using heliostats

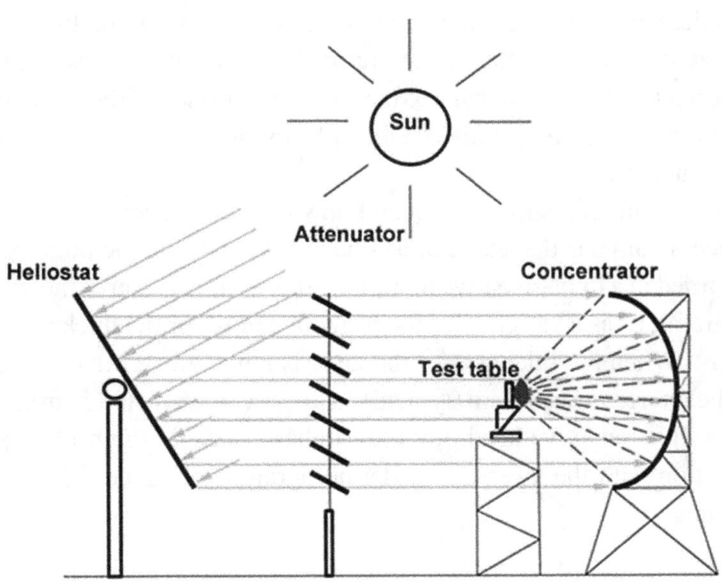

Figure 13. Example 2: method of using heliostats

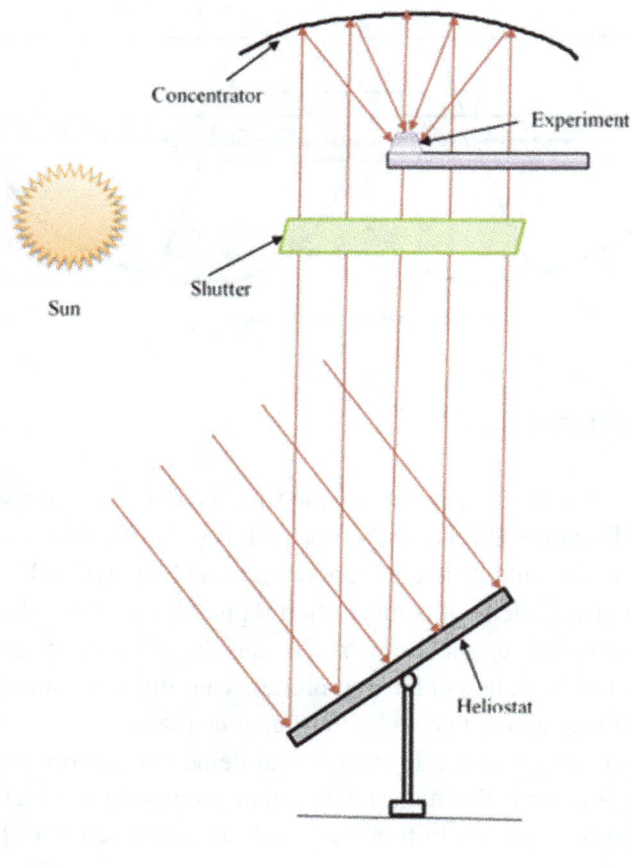

The most suitable realization of a microoptics holographic element for elements of solar batteries is the multilevel diffraction lattice. On the basis of a multilevel diffraction grating (DG) was developed the schematic diagram of collimator system, comprising on the basis of a planar light guide plate with holographic micro-optical elements for input and output radiation from the light-guide fiber.

For the manufacture of a holographic concentrator (holocon), it is advisable to take the reference beam collimated and direct it along the normal to the receiving surface of the light-sensitive layer thereby it is likened to the geometry of solar radiation. Object beam depending on the desired type of holocon may have a different structure.

Figure 14 shows a diagram of a simple holocon known in holography as a hologram of a point light source. Here the subject beam has the shape of a cone with a vertex at the point of future focus.

A hologram is recorded in a light-sensitive material either as a variation of the transmittance or reflection (amplitude hologram), or as a variation of the refractive index or the thickness of the light-sensitive layer (phase hologram). When the thickness of the layer is much greater than the distance between the isophase surfaces of the interference pattern (typical thickness values of 10-15 cm), it turns out the volume hologram. With a thin light-sensitive layer, not the isophase surfaces themselves are recorded, but the traces of their intersection with the layer, and a flat hologram is obtained. There are also transmissive and reflective holograms.

1 – laser, 2 - beam splitter, 3 - support beam expander, 4 - reference beam, 5 -mirror, 6 - lens, 7 - object beam (signal beam), 8 - light-sensitive layer, 9 - glass photographic plates.

An important energy parameter of the hologram is its diffraction efficiency, which is equal to the ratio of the diffracted radiation intensity to the incident radiation intensity. For amplitude holograms this value is only a few percent and they are hardly interesting for use in solar engineering (heliotechnics). For phase volumetric transmission and reflection holograms, the theoretical value of the diffraction efficiency is 100%, for the phase plane transmission hologram is 33.9%, and the experimentally obtained values are close to the theoretical ones.

When considering diffraction on volumetric transmission holograms, it is necessary to take into account that each beam is consistently scattered from a large number of periodically located surfaces of density maxima. To maximize the amplitude of the resulting diffracted wave, the waves scattered by successive layers must be in phase. This condition is satisfied if Bragg's law is observed (Strebkov & Tveryanovich, 2007):

$$2dsin\Theta = \lambda_0 * n \quad (5)$$

Where, d - period of the holographic grating, Θ - grazing angle - additional angle to the angle of incidence. It's the angle between the object beam and the midstation plane, λ - wave length, n - average refractive index of the photosensitive layer.

The diffraction efficiency of 100% for phase volumetric holograms can be obtained only if Bragg's law is observed. Deviation from the Bragg angle reduces the diffraction efficiency. Following this law, the developers come up with various applications of holographic optical elements. These developments relate to energy supply (electricity and heat) and lighting of buildings by solar energy (Wagner, 2004).

Daylighting design interacts with many design issues of a building, hence it is important to follow an integrated approach. Daylighting in buildings is not only a matter of energy saving and performance but also of comfort and appearance. A conventional approach considers building issues like daylighting, solar shading and glare separately. Mostly, a conventional approach leads to high energy consumption of a building, e.g. a conventional shading device reduces solar heat gains, but also affects the daylighting by excluding the diffuse light.

Figure 14. Principle of formation of holographic lenses

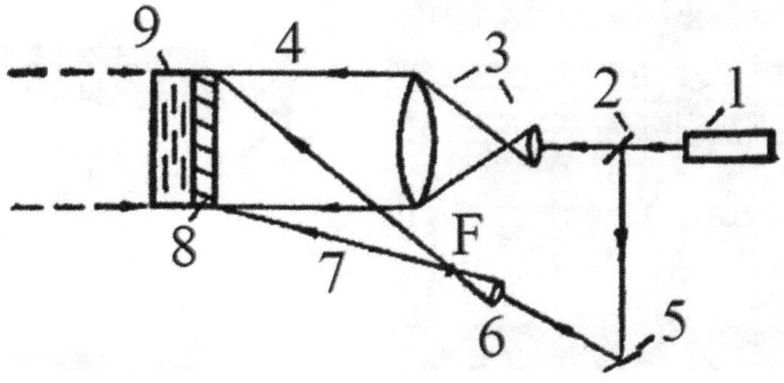

Integration of holographic optical elements (HOE) into the building envelope of a built environment makes solar radiation as a renewable source of power accessible. Holographic optical elements (HOEs) have useful properties for diffuse light transmission and radiation control.

As the investigation and experiments show the holographic systems for illumination and solar control are applicable in existing and new buildings. They contribute to the decrease of energy demand and CO2 emissions, depending on local conditions as well as to the increase of comfort for the user (Wagner, 2004).

Solar Panels Based on Holographic Concentrators

Solar panels using holographic concentrators are already being produced. Since 2005, Prism Solar Technologies company designs and developments holographic planar concentrator (HPC) that incorporates proprietary holographic optical film technology to achieve low module manufacturing costs (Prism Solar Technologies, n.d.). Company assert that Prism's bifacial modules harvest sunlight on both faces, making brilliant use of the sun by generating significantly more energy (kWh) per watt of power than traditional modules throughout the changing seasons of the year. Current-voltage characteristic of one module is shown in figure 15. As it is seen in the graph, bifacial cells producing up to 27% more energy (kWh) in standard installations as compared to conventional modules with the same nameplate (STC) rating. But do not forget that they are bifacial modules and the number of solar cells more than standard installations. As shown by the temperature characteristics of the module with the increased surface temperature of the module, the efficiency is reduced to 30 per cent of the nominal value (Prism Solar Technologies, n.d.). Keeping the efficiency at increasing the concentration and raising the ambient temperature is also one of the issues of using concentrators for the production of electrical energy in solar energy.

Three holographic panels (one's reference coupon) were assembled for the temperature analysis of the HPCs (Holographic Planar Concentrator) with the same type of diced bifacial solar cells (Castillo-Aguilella, 2012). All of the coupons had the same glass surface area, with dimensions of 304 mm x 406 mm. The reference coupon consisted of nine 125 mm x 25 mm cells, with the cells arranged in three rows of three, spaced 2 mm in either direction. The 1 to 1 coupon consisted of nine 125 mm x 25 mm cells, with the cells arranged in three rows of three, spaced by the holographic film width of 25 mm, and the cell columns separated by 2 mm. The 2 to 1 coupon consisted of fifteen 12.5 mm x 12.5 mm cells,

Figure 15. Light directing system with conventional shading device (Wagner, 2004)

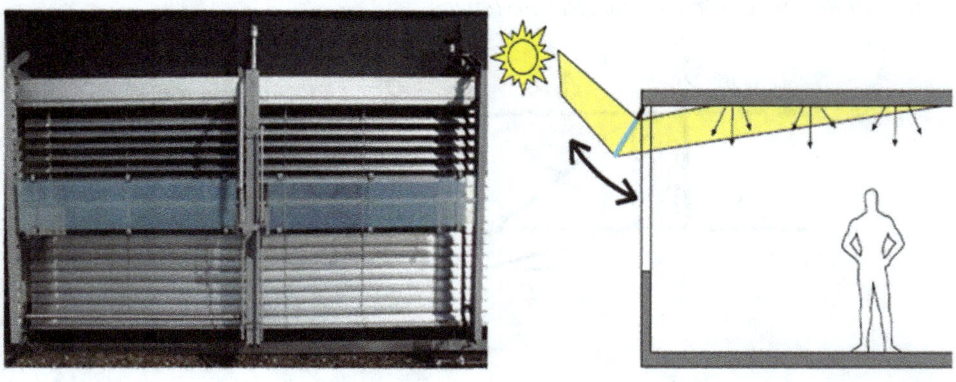

Figure 16. Irradiance dependence Bi60-368BSTC (Prism Solar Technologies Company, n.d.)

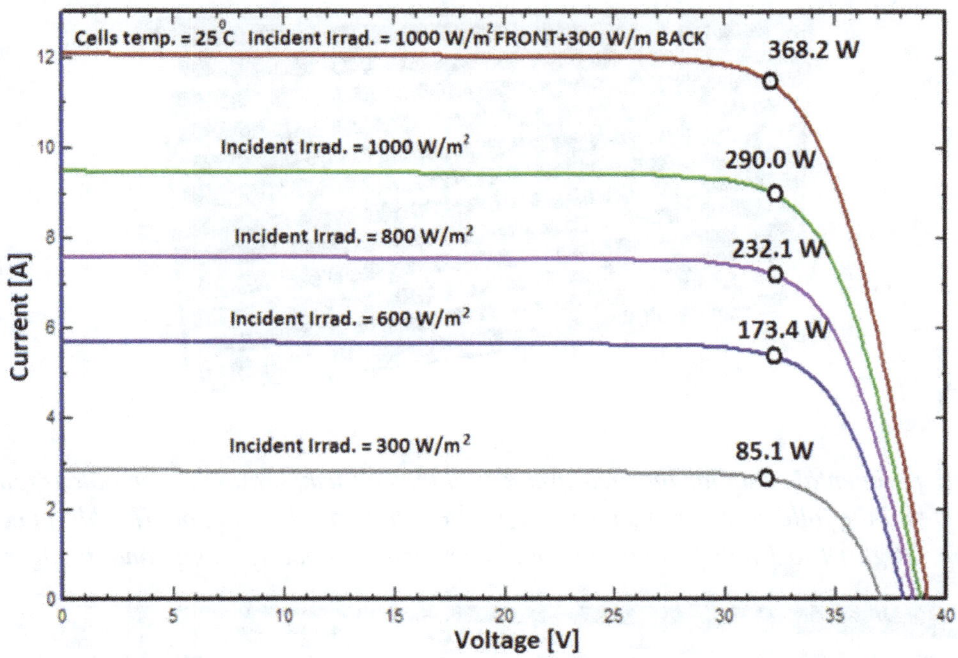

arranged in five rows of three cells, with the cell rows spaced by the holographic film width of 25 mm, and the cell columns separated by 2 mm.

The temperature of the cells was measured with an imbedded thermocouple placed behind the center cell with the test coupons operating under no electrical load condition. Several assumptions were used while developing the model which are given in the literature 16.

Full analysis of the article can be found in the literature 16. As experiments and analyses show, one of the good methods to reduce the temperature of the PV cells is to arrange the elements (PV cells and hologram) in rows (Figures 18-22). This method does not lead to additional costs, but there is an increase in efficiency. Thanks to the alternating system up to 40% can reduce the use of photovoltaic cells by 1 m^2.

Bifacial HPC produced by Prism Solar Technologies Company and actively used in solar projects. These modules are also under study by various researchers and research centres to improve and find new methods of application (Singh et al., 2012).

Utility Model Using Holographic Concentrators

In the patent N°149536 RU U1 "Solar photovoltaic battery," (Figure 23) authors used a holographic to concentrate the solar radiation at a certain point. According to the authors of the patent, in the construction of the model considered the main disadvantages of classical solar panels:

- High cost due to the high cost of the initial silicon;
- Decreased power generation in the absence of tracking the position of the sun;
- Decreased efficiency due to heating of the photocell.

Figure 17. Bifacial module of Prism Solar Technologies Company (n.d.)

Figure 18. Experimental setup for the measurement of the cell temperature of the holographic panels. The coupons had the following arrangement (Left) 2 to 1 holographic coupon, (Middle) packed reference coupon, (Right) 1 to 1 holographic coupon. The system was pointed south and tilted for latitude.

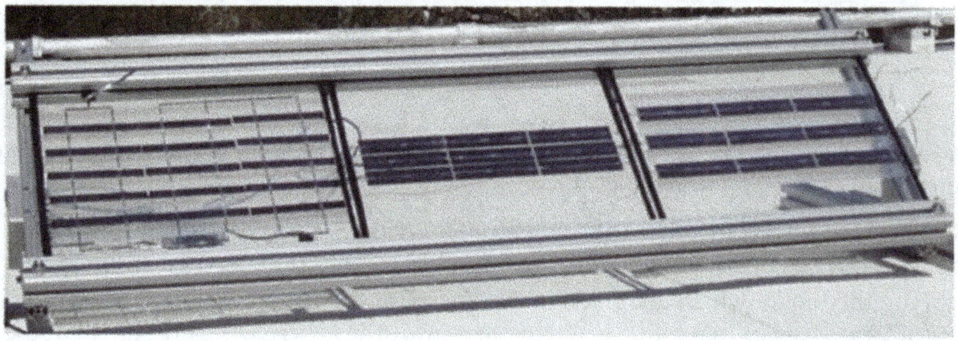

Figure 19. Ross coefficient [Km^2W^{-1}] for the measured data from the test coupons

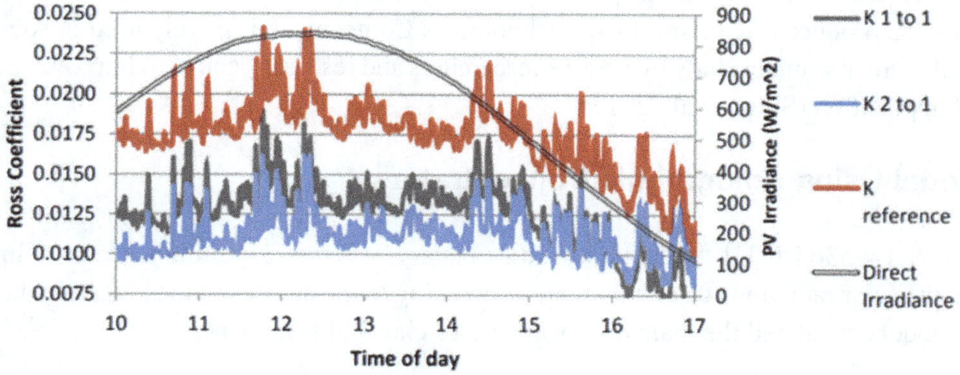

Figure 20. Experimental temperature data for the HPC coupon modules for a day in late May, 2010. (Upper) temperature for the three tested coupons. (Lower) wind data for the experiment site (N=0, E=90)

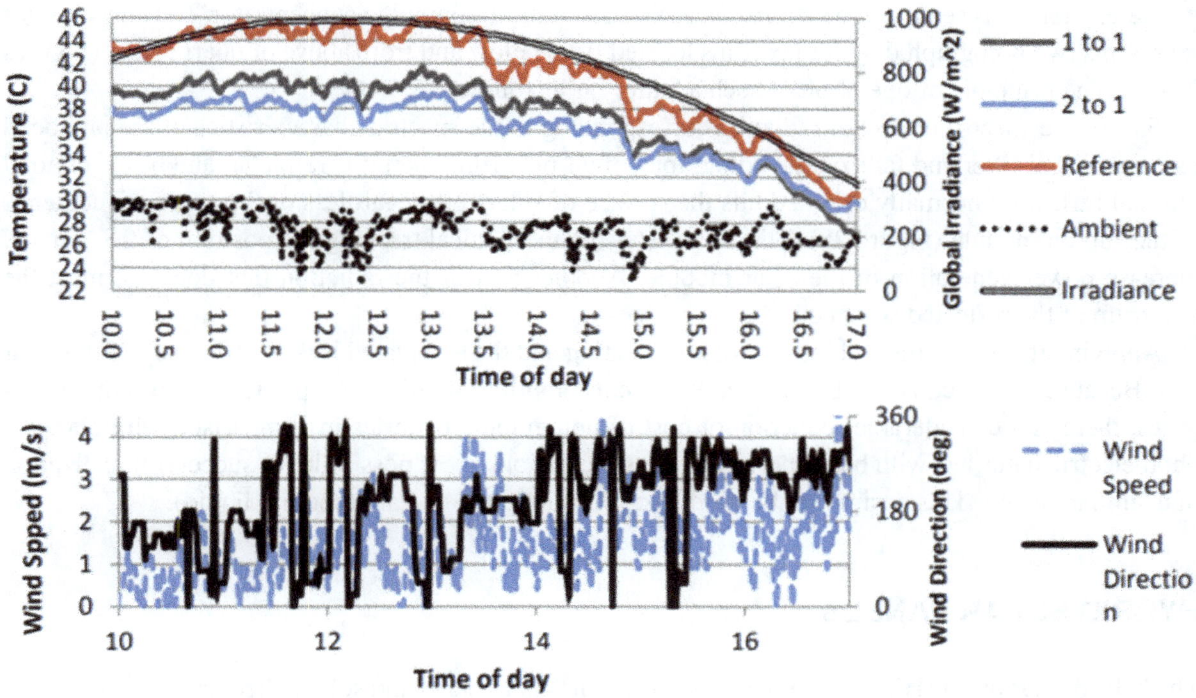

Figure 21. Model derived heat maps at the solar cell layer for an EPV of 800 Wm^{-2} and a heat transfer of 15 Wm^{-2} for the glass surfaces for the test coupon configurations. The color scales varies from 30C (light blue) to 56C (bright red).

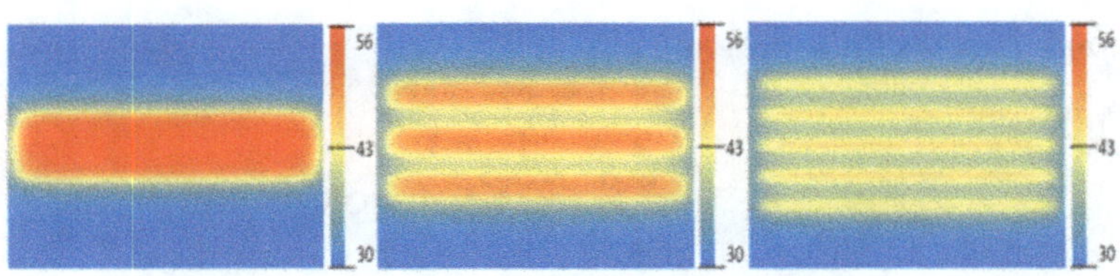

Figure 22. Model derived heat maps at the solar cell layer for an EPV of 800 Wm-2 and a heat transfer of 30 Wm-2 for the glass surfaces for the test coupon configurations. The color scales varies from 30C (light blue) to 56C (bright red).

In fact, all developments in the creation of new solar panels take into account all these shortcomings but each solves the problem differently as already mentioned.

Here, photocells settle down on three face surfaces of the holographic concentrator. The concentrator represents two holographic multiplex films located from below and from above of quartz glass (Figures 23-24). The principle of operation of such a battery is as follows:

Part of the spectrum of solar radiation corresponding to the spectral characteristics of the photocell passes spectral filter and falls on the holographic multiplex film, then it is refracted and based on total internal reflection normally oriented hits the surface of solar cell, regardless of the angle of incidence of the sun on module (Figure 25). This eliminates the mechanical tracking the position of the sun and increase power generation of the solar module. Another part of the radiation that does not meet the spectrum of the reflected solar cell.

Applying this structure enables to avoid the heating and the associated loss of efficiency of the solar cell. Because of the considerable decrease in a consumption of silicon on production of similar batteries, there is a considerable reduction of cost of holographic batteries in comparison with classical photoelectric batteries, with better performance. Calculations show possibility of successful realization of a similar design. This design can be used also for heating the coolant (cooling liquid).

HYBRID SOLAR PANELS

The hybrid solar panel (HSP) combines the abilities of a solar panel and solar collector. This means that it can heat water and produce electrical power at the same time. This combination gives a higher total effect of output per panel. Several companies produce hybrid solar panels without the use of holographic concentrators (Figure 26) (DanSolar, n.d.; Solimpeks, n.d.; DualSun, n.d.).

Figure 23. Holographic photovoltaic battery

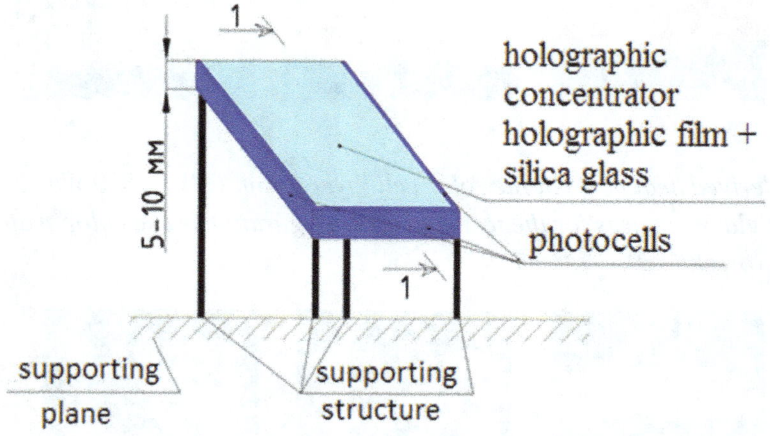

Figure 24. The holographic photovoltaic battery in a section

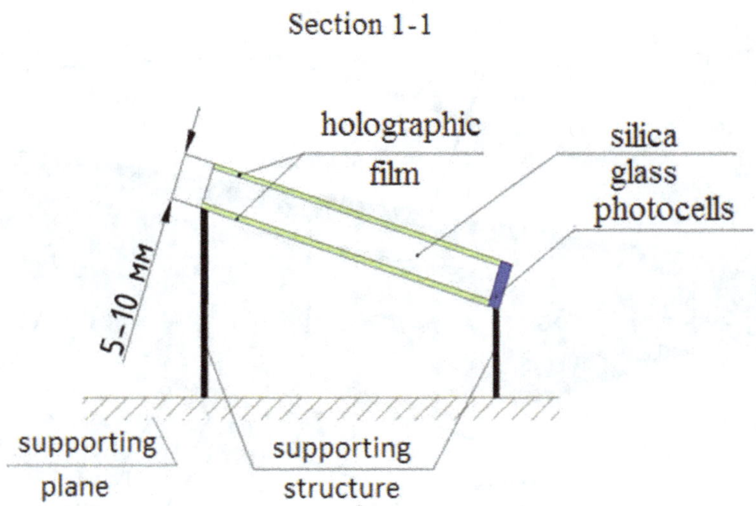

Figure 25. Scheme of operation of the holographic battery

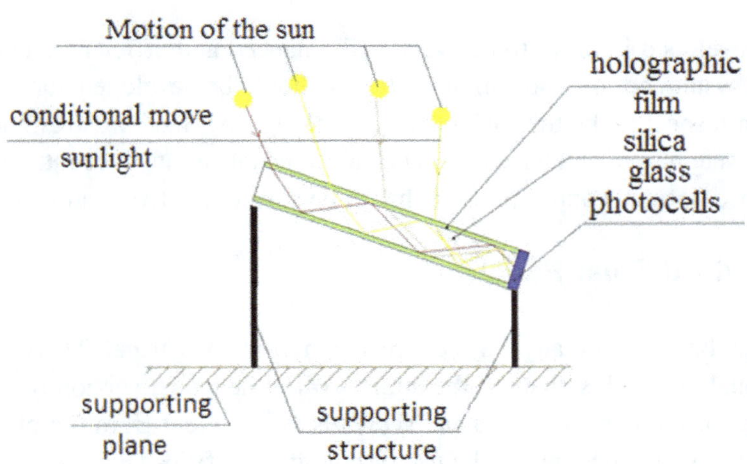

Hybrid Solar Panels With HOE

Authors propose to solve the above-listed disadvantages of existing solar panel problems with the new solar hybrid system based on holographic concentrators and removing heat for the simultaneous production of the thermal energy. Authors consider the option of using holographic concentrators to improve the efficiency and reduce the cost of electricity and heat, which are produced by solar energy. In the simplified design variants of solar modules for power generation only part of the incoming solar energy is used, the remaining part is dissipated as heat energy in the surrounding space. In the chapter task is a more efficient use of solar energy arriving at the surface of the solar module by integrating photovoltaic panels using holographic concentrators and solar collectors in one technological device, i.e. the design of an integrated solar system. Here, due to the increased complexity of the design space, an optimal design

Figure 26. HSP – Power Therm-cross-section (Solimpeks, n.d.)

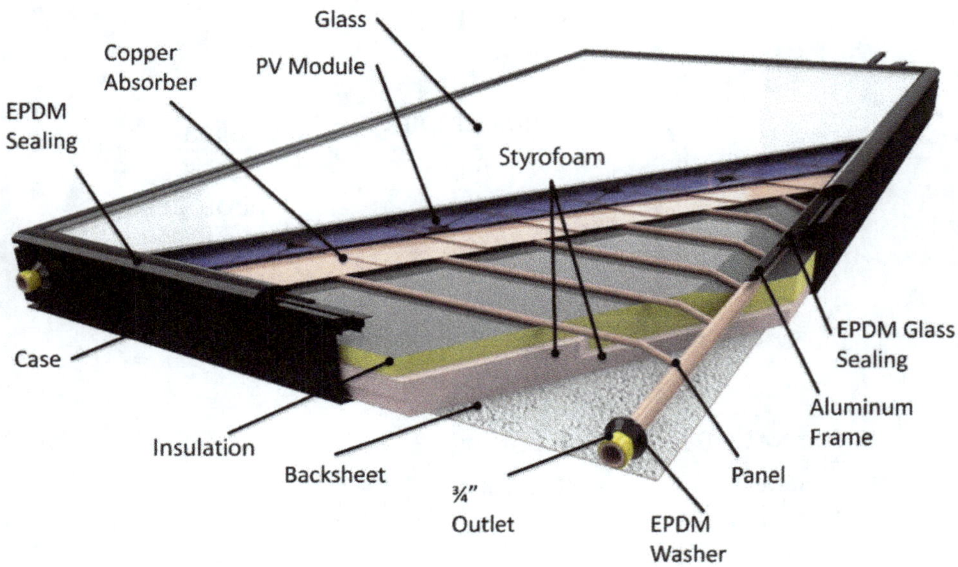

strategy should be developed in order to derive a well balanced and efficient overall system. For the new integrated solar system optimal operation strategies should be developed guaranteeing an optimal system behavior with respect to the thermal and electrical regime. Here, weather and demand forecast information and the system state and its constraints have to be taken into account. As these vary during operation the described optimal control problem has to be solved iteratively and online.

Description of Hybrid Solar Panel

Holographic films can be used in many ways to create a hybrid solar panel. There are already options for a hybrid solar panel but in this chapter, the authors offer their own version of the installation [9]. Practically the design of the hybrid solar panel (HSP) is no different from the classical solar panels with water-cooling. The difference just in the use of holographic films (Fig. 28). In the Figure 28-29, you can see the cut view of the HSP. As we using one series put the solar cell other a glass with total internal reflection and HOE. In direct contact solar radiation (SR) with the surface of a solar cell a solar cell generates electric energy, but if SR gets a different angle and not directly (at any angle) that solar radiation is refracted and under the effects of total internal reflection falls on the surface of the solar cell. The special proposed design of the panel shown Figures 28-29.

Part of the spectrum of solar radiation corresponding to the spectral characteristics of the photocell passes spectral filter and falls on the holographic multiplex film, then it is refracted and based on total internal reflection normally oriented hits the surface of solar cell, regardless of the angle of incidence of the sun on module. This eliminates the mechanical tracking the position of the sun and increase power generation of the solar module. Another part of the radiation that does not meet the spectrum of the solar cell is ignored and falls on the copper pipe. Applying this structure enables to avoid the heating and the associated loss of efficiency of the solar cell. The other solar spectrum shown in Figure 28-29 with other

Figure 27. Hybrid solar system 1. Polycarbonate monolith, 2. concentrator of solar radiation – holographic film, 3. Photovoltaic cells, 4. heat conducting grease, 5. absorber, 6. heat transfer fluid, 7. insulation, 8. body.

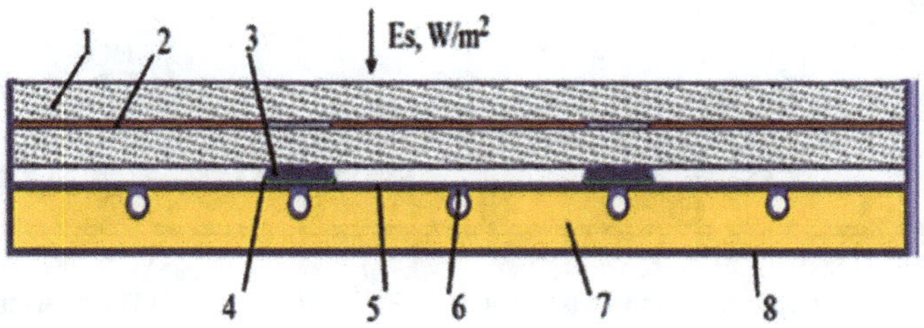

Figure 28. The cut view of the HSP (section - A – see Figure 29)

Figure 29. The cut view of the HSP section A (see Figure 3). Where 1- a glass with total internal reflection, 2 – HOE, 3 – solar cell, 4 - copper plate, 5 - copper pipe, 6 - heat-transport fluid (for example, water), 7 - reflective film, 8 – frame.

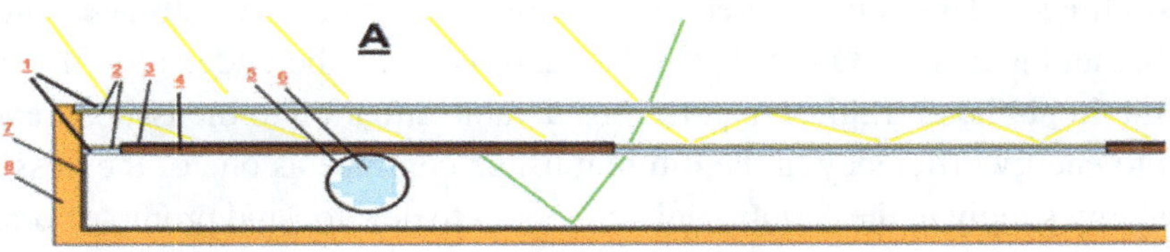

color. Simplified design of solar modules for power generation is used only part of the incoming solar energy, the remainder is dissipated as thermal energy in the surrounding space.

Because of considerable decrease in a consumption of silicon on production of similar batteries, there is a considerable reduction of cost of holographic batteries in comparison with classical photoelectric batteries, with better performance. Thanks to the spectral filter and a holographic film the rest of the spectrum that is not identical with the spectrum of a solar cell we can use as thermal energy.

The second variant of the hybrid solar module construction without the law of total internal reflection is proposed (Figure 30). The diffraction grating of holographic films is designed so that when sunlight hits their surface they are refracted and fall on the surface of solar cells. In this design, it is better to choose a width of holographic films less than solar cells, as this may depend on the height of the holographic films over solar cells. This can lead to an increase in the size of the solar installation. The calculation of diffraction gratings also depends on the height of the films above the solar cells.

Figure 30. Design of hybrid solar panel without the use of the law of total internal reflection

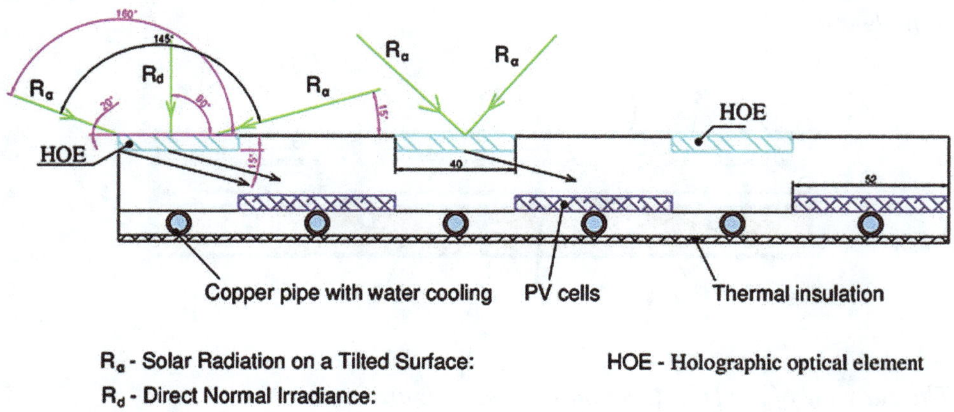

Installation Condition

For the following system development for HOE applications, it was necessary to take into consideration different geographical and climate situations as well as the market situation for HOEs where will be applied. For calculating invited to consider the consumer that located in the Republic of Tajikistan in the region of Dusti (37,5° N, 68.5° E). Therefore, climate and market investigations for Tajikistan have been undertaken (Shokhzoda & Tyagunov, 2016; Singh et al., 2012). Based on this information, the application possibilities of HOEs in architecture have been examined. To estimate solar radiation used database of NASA and METEONORM. From the analysis, it can be understood that recent years the Government of the Republic of Tajikistan paid considerable attention to the issue of energy efficiency. Paying attention to energy efficiency in the future must be continue as one of the possible solutions to the problem of energy supply in the country, it is important to develop and promote companies involved in the supply of equipment of all form of RES. There are various reasons for the development of solar energy in Tajikistan. There are the main reasons for the development of heating and hot water supply based on solar energy:

- Huge solar energy potential;
- The annual increase in electricity tariffs;
- High efficiency of solar collectors;
- Ecological compatibility.

In addition to taking into account the geographical and climatic conditions the place of installation to increase the efficiency of the module`s is necessary to select energy consumers. The criterion for choosing a consumer is the coincidence of the load schedule of the consumer with the arrival of solar radiation (Figure 31). There are a lot of such energy consumers. For example, kindergartens, schools, universities, administrative buildings of large. The peak power of the load may coincide with the peak arrival of solar radiation. This approach will give us the most efficient use of solar energy.

Figure 31. Diagram of a hypothetical daily load shape of school

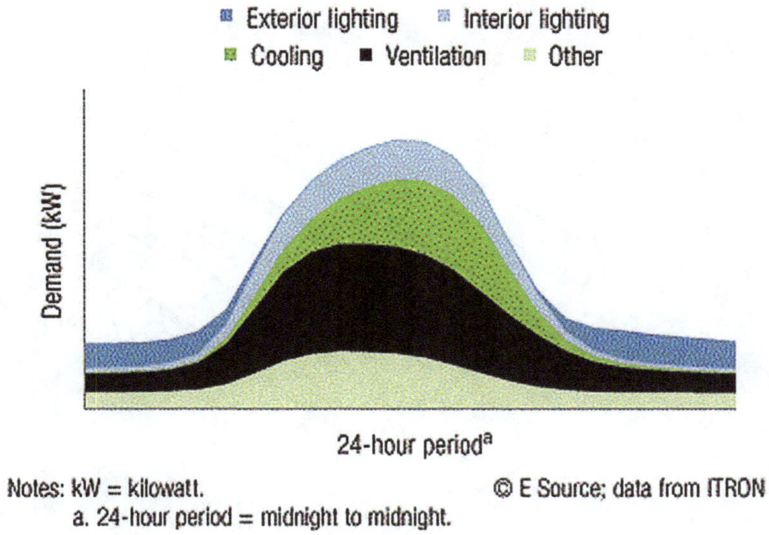

The Simulation of Hybrid Solar Panel

Modern computer-aided design and simulation of heterogeneous physical processes were used (Simulink, n.d.). In the thesis of Je Vin "The study of the efficiency of solar energy use for autonomous power supply systems in the Republic of the Union of Myanmar" (Je Vin, 2013), devoted to the design of solar panels and heat exchangers, a comparative analysis of a number of factors that have the greatest impact on the efficiency of the thermal cooling system. In this case, the following parameters are subjected to a comparative analysis (Shokhzoda & Kokin, 2016):

- Ambient temperature;
- Coolant pipe length;
- Mass flow of coolant;
- Diameter of coolant pipes;
- The arrival of solar radiation.

Among the listed parameters, it is advisable to create a system whose efficiency is based on the variation of the mass flow of the coolant, and to estimate the temperature of the coolant at the exit from the cooling system and the mean temperatures of the elements of the solar panel. The data obtained in this thesis allow us to choose the parameters of the heat supply system for the solar panels in a valid way. As a result, the analysis will be subjected to the following possible design of the solar panel:

General characteristics of the system: 30 solar modules 100 mm*100 mm combined into 3 rows of 10 modules. Through each number is 5 coolant channels of 2 mm diameter each. The total area of the irradiated surface will be 0.3 m^2 (Figure 32).

Figure 32. Schematic depiction of the cell of the solar cell under study

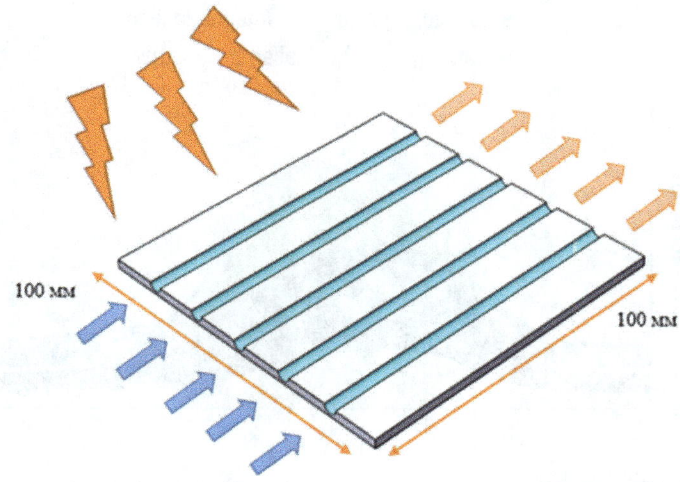

An important step in the efficiency analysis was the assessment of water heating during the passage of the solar battery cooling system and the determination of the created head in the system during the operation of the pump TL-B04/S/PV. For this purpose the program of modeling of hydrodynamic processes of ASONIKA-P was applied (Figure 33) (Volovikov, 2002).

The model of the processes of coolant pumping through the solar cells is a source of pressure depending on the flow rate in the system and describing the operation of the pump. A further part of the hydraulic

Figure 33. An example of a graphical representation of the model of hydrodynamic processes in the program ASONIKA-P

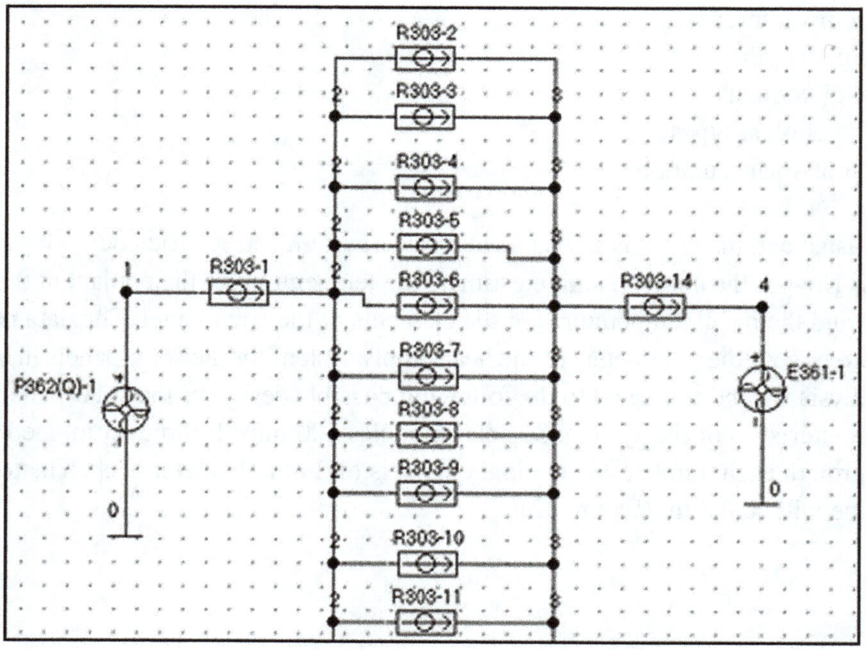

Figure 34. Graph reduction of the coolant flow rate in the system with the increase in the height of the solar panel lift relative to the pump obtained in the simulation

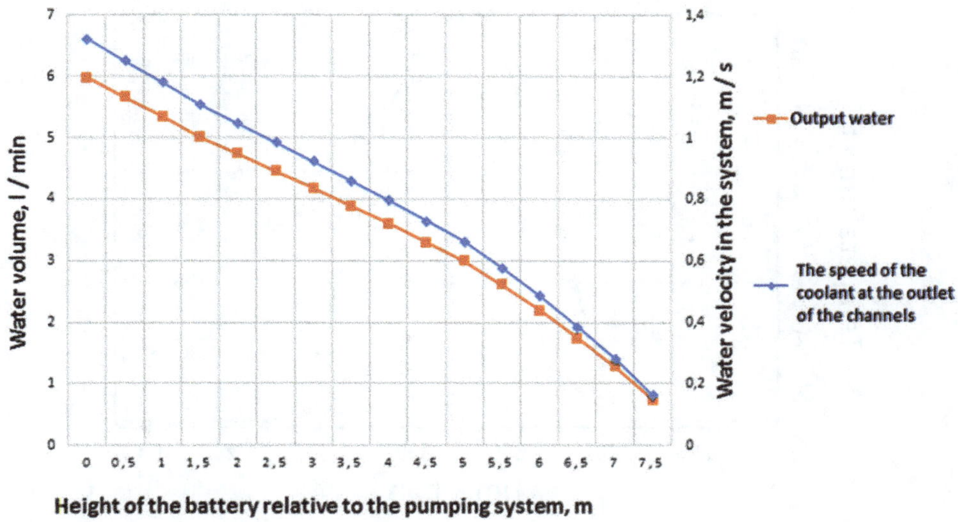

resistance of the channels that occurs when pumping through them coolant. Also added to the system is a pressure describing the effect of the height difference between the solar panel and the water pump.

The obtained data were used to create a model of thermal processes that describe the process of distribution of thermal energy entering the solar panel modules during the daily course of temperature.

The set of heat energy distribution processes for the solar module can be represented as follows:

- Solar battery heating due to solar radiation;
- Radiation of thermal energy from the heated surface of the solar module into the surrounding space;
- The transition of thermal energy from solar modules due to forced convection to the coolant;
- Convective heat transfer between the surface of the solar battery with the environment.

The maximum power of solar radiation falling on each module 100 mm*100 mm is 9.8 V. The results of modeling of thermal processes can be seen in Table 1.

Provided that the efficiency of the solar module (SM) falls by no more than 1%, it can be said that in the course of the day it is possible to obtain at least 43.2 liters of water at a temperature of 42°C, or more if the design changes with an orientation to a longer path of passage of the coolant through the system, or the concentration of solar energy from a larger area. As shown in Figure 36-37 with high solar activity and high ambient temperature, the efficiency of CM is maintained while maintaining electricity generation of more than 22%.

Under the specified operating conditions, several benefits can be obtained simultaneously;

- Maximum output;
- Hot water;
- Maintain the service life of solar modules;
- Reduce emissions into the atmosphere.

Figure 35. Temperature dynamics of the heated and coldest solar panel modules per day at an input water temperature of 25 ° C

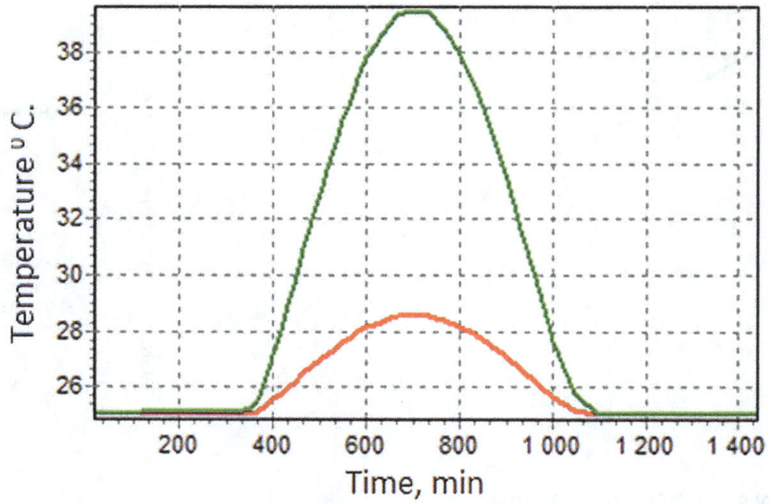

The Dependence of the Efficiency of Solar Cells From the Temperature and Solar Radiation

The energy produced by a photovoltaic generation is estimated from the data the global irradiation on inclined plane, the ambient temperature data manufacture for the PV module used. The power output of PV array can be calculate from the following equation (Figure 36) (Singh et al., 2012):

$$P_{pv} = G * A_{pv} * N_m * \eta_{pv} = G * A_{pv} * N_m * \eta_{pc} * \eta_{pc} \left(1 - \alpha\left(T_j - T_{jref}\right)\right) \tag{6}$$

Where, A_{pv} - is the total area of the photovoltaic generator; G - is a solar radiation on tilted plane module; η_r - is the reference efficiency of the photovoltaic generator; η_{pc} - is the power conditioning efficiency which is equal to 1 if a perfect maximum power tracker (MPPT) is used; α_{sc} - is the temperature coefficient of short-current ($/^0K$) as found on the datasheet; T_j - is cell temperature and T_{jref} is the reference cell temperature (25^0C).

Model (Simulink) is associated with the production of thermal energy. In the model allowed a number of assumptions:

- The temperature on the all surface of the copper plate the same;
- For heat sink take water;
- The water temperature at the inlet to the copper tube of the same;
- The model was used ready-made blocks from the Simulink program (Rekioua & Matagne, 2012; Simulink, n.d.).

The model is made so that you can connect the pipe between or increase the overall dimensions. This type of models allows to assemble a large system and analyze the operation of the system.

Table 1. Temperature dependence of the solar module elements on the volume of the pumped coolant (l/min). Where A - air temperature, B - the temperature of the solar module without cooling, C - the temperature of the solar cell (minimum), D - the temperature of the solar cell (max)

Time hour	A	B	5.971 l/min, 1.2 m/s		2.99 l/min, 0.75 m/s		0.721 l/min, 0.1 m/s	
t			C	D	C	D	C	D
0	14	14	14	14	14	14	14	14
1	14	14	14	14	14	14	14	14
2	14	14	14	14	14	14	14	14
3	14	14	14	14	14	14	14	14
4	14	14	14	14	14	14	14	14
5	14	14	14	14	14	14	14	14
6	14	14	14	14	14	14	14	14
7	14	23	14.02	14	14	14	14	14
8	18	33	14.16	14.36	14.14	14.64	14.43	16.71
9	22	46	14.3	14.71	14.29	15.29	14.86	19.43
10	26	55	14.44	15.07	14.43	15.93	15.29	22.14
11	30	61	14.58	15.43	14.57	16.57	15.71	24.86
12	34	65	14.72	15.79	14.71	17.21	16.14	27.57
13	38	66	14.86	16.14	14.86	17.86	16.57	30.29
14	42	67	15	16.50	15	18.50	17.00	33.00
15	38	66	14.86	16.14	14.86	17.86	16.57	30.29
16	34	61	14.72	15.79	14.71	17.21	16.14	27.57
17	30	55	14.58	15.43	14.57	16.57	15.71	24.86
18	26	46	14.44	15.07	14.43	15.93	15.29	22.14
19	22	33	14.3	14.71	14.29	15.29	14.86	19.43
20	18	23	14	14	14	14	14	14
21	14	14	14	14	14	14	14	14
22	14	14	14	14	14	14	14	14
23	14	14	14	14	14	14	14	14

After analyzing the results of both methods of modeling, it can be seen that the operating temperature of photovoltaic cells at a good arrival of solar radiation and high ambient temperature can reach 70 ^{0}C, which leads to a decrease in the efficiency of solar cells more than 20% of the maximum value. When using the cooling system, as can be seen from figure 42 and table 1, the difference between the input and output temperature reaches 10 ^{0}C. When regulating the water supply to the tubes of the hybrid solar panel on each 1 m^2 of the installation, up to 50 liters of water with a temperature of up to 45 0C can be obtained. The efficiency of hot water preparation can be increased if a solar collector is connected to the outlet of the hybrid solar panel tube. This design increases the output water temperature to 30 0C, and the overall efficiency of hot water by 3%. Thus, the use of a cooling system allows you to simultaneously save the life of solar cells and maintain their production at the maximum level, and provide the

Figure 36. Changing the efficiency of the solar module with and without cooling

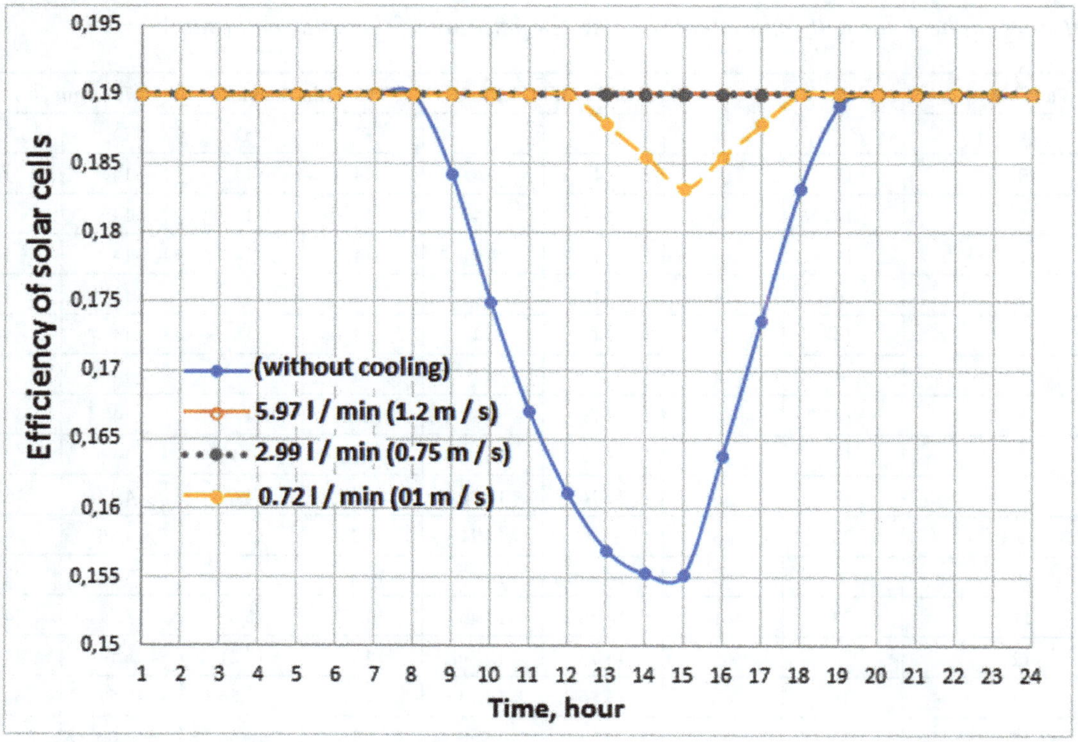

Figure 37. The variation of PV output power by temperature

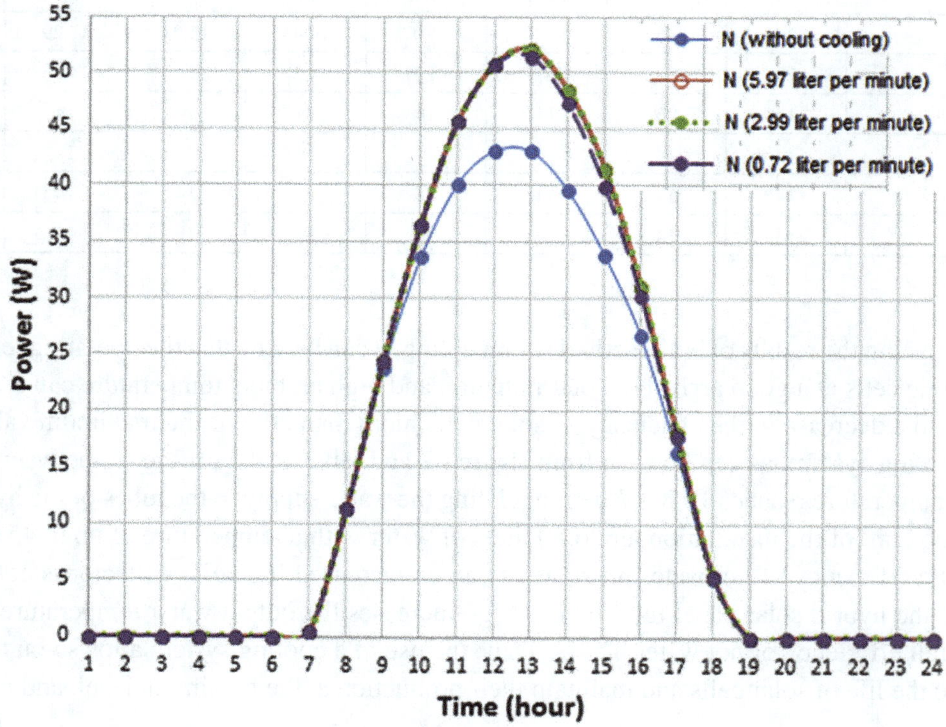

Figure 38. The model diagram. 1 - the source data (ambient temper-ature, solar radiation), 2 - the point of interaction laws of CvHT and RHT, 3 - heat transfer by conduction through a layer of material (from copper plate to pipe see Figure 34).

Figure 39. The cut view of the HSP section, which include solar cell, copper plate and pipe

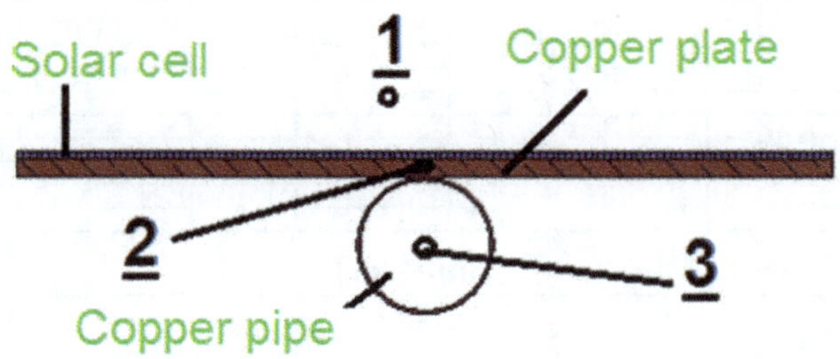

Figure 40. Simulate of input data in Simulate

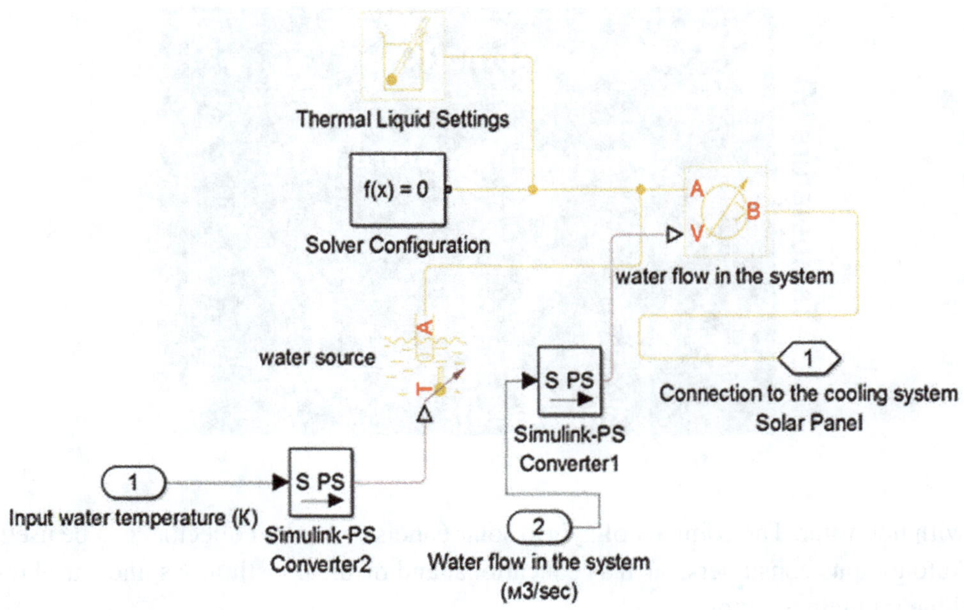

Figure 41. The variation of the solar radiation (W/m²) and ambient temperature (K) for a typical day on January in Tajikistan (Li et al., 2013).

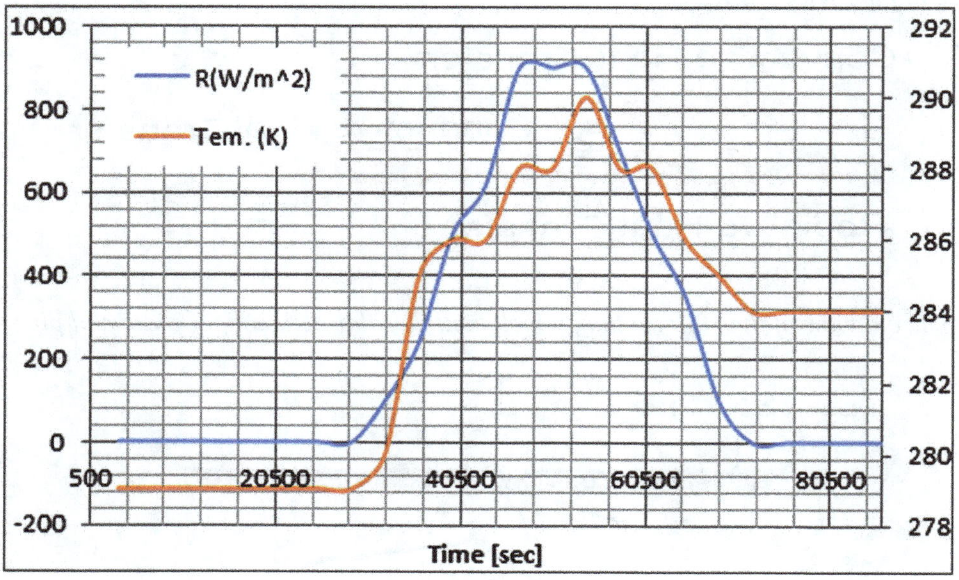

Figure 42. The variation of the output ant input temperature result from Simulink (the length of the copper tube 1 meter)

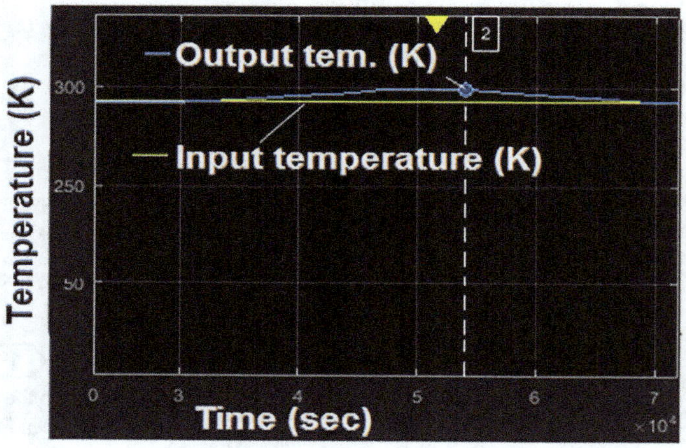

consumer with hot water. The complex of hybrid solar panels and solar collectors can be used to supply energy to Autonomous consumers, such as educational and medical institutions, industrial (small businesses) and the municipal sector.

The technical problem of the present model took place in increasing the efficiency of classic solar cells increases the efficiency of conversion of solar energy into electrical energy due to an increase in the density of solar energy and the maintenance of the efficiency of the photoconverter due to the dissipation of heat to obtain thermal energy.

Holographic concentrators with diffraction grating were used to increase the density of solar radiation coming to the surface of the photoconverter. The diffraction gratings of the holographic multiplex film are calculated in such a way that each layer effectively refracts the solar radiation incident at a certain angle to its surface. This solution allows you to:

- Refuse mechanical tracking of the solar module for the position of the Sun;
- Avoid reducing the efficiency of the solar module associated with heating through the use of spectral filters;
- Increase the power generation of the solar module in relation to the stationary solar module;
- Reduce the cost of energy production by reducing the use of solar cells.

The task of maintaining the efficiency of the photoconverter at a high temperature is performed by heat removal. This solution allows you to:

- Avoid reducing the efficiency of solar photovoltaic cells by removing heat;
- Avoid reducing the service life of solar cells ahead of time;
- Producing thermal energy using special regimes of the heat supply;
- Reduce the area used to generate electricity and heat.

In General, according to the results of the research, it can be argued that there is the feasibility and economic efficiency of using this model for Autonomous consumers.

FUTURE RESEARCH DIRECTIONS

Currently, solar panels as power sources are widely used both in stationary and mobile facilities. It is obvious that this type of power supply is the most promising and will be actively used in the future. The scope of their use is expanding and the cost decline. But at the same time, all the above disadvantages of solar panels remain unresolved. An obvious solution to this problem is the use of various means of changing the direction of the beam, the position of the batteries, reflectors or solar energy concentrators. The main advantages of holographic concentrators from other concentrators is their weight and low cost in large-scale production. Therefore, the future research directions are the use of holographic solar concentrators to generate electricity and heat. An interesting part of this directions are the selection of modes and determining the optimal parameters of the hybrid system. The parameters that you need to select to get the maximum effect from the system are the following:

- The ratio of the surface area of solar cells and holograms;
- Pipe diameter;
- Speed of water (or another coolant) through the pipe;
- Mode and algorithm of water supply in the pipe.

CONCLUSION

In this Chapter, the authors have considered the concentrators of the old and completely new generation. The use of solar radiation concentrators allows several times to reduce the required area for obtaining a given electric power, increase their resistance to external factors and, as a result, significantly reduce the cost of electricity. The appearance on the market of the first generation of holographic concentrators (Prism Solar Technologies, n.d.), the so-called flat holographic concentrator has become this reason. Holographic films are much cheaper than large mirrors or lenses. Prospects of using solar holographic concentrators are great.

As shown by the thermal characteristics of the hybrid solar panel, the operating temperature of photovoltaic cells at a good solar radiation and high ambient temperature can reach 70 degrees Celsius, which leads to a decrease in the efficiency of solar cells by more than 20% of the maximum value. When using the cooling system as shown in Fig. 35, 42 the difference between the input and output temperature values reaches 10 degrees Celsius. When regulating the water supply to the tubes of the hybrid solar panel on each 1 m^2 of the installation, up to 50 liters of water with a temperature of up to 45 degrees Celsius can be obtained. The efficiency of hot water preparation can be increased if a solar collector is connected to the outlet of the hybrid solar panel tube. This design increases the output water temperature to 30 degrees Celsius, and the overall efficiency of the hot water supply by 3%. Thus, the use of the cooling system of the hybrid solar panel allows you to simultaneously save the life of solar cells and maintain their production at the maximum level, and provide the consumer with hot water. The complex of hybrid solar panel and solar collector can be used for power supply of Autonomous consumers, such as educational and medical institutions, industrial (small enterprises) and municipal sector.

All of the above and the importance of the energy problem will lead to the fact that this direction will soon become one of the topical issues of scientists and departments of non-traditional and renewable energy sources.

I hope that the material included in the work, as well as the sequence (logical order) and style of its presentation is completely corresponding with the purposes of writing this chapter.

ACKNOWLEDGMENT

I Shokhzoda B. T. would like to express my appreciation to Doctor of Technical Sciences, Professor Tyagunov M. G. for this during the term of completion of the article. Without his valuable assistance, this work would not have been completed. Finally, we would like thank the large number of open source and related projects that critically facilitated this work.

REFERENCES

Baranov V. K. (1976). *Methods of Calculation for the compound parabolic concentrator profiles, Applied Solar Energy*. Moscow: Publishing house of Nauka.

Blood & Skolof. (2014). *Huge thermal plant opens as solar industry grows*. Associated Press.

Castillo-Aguilella. (2012). *Non imaging applications of volume diffractive optics.* Department of electrical and computer engineering, the University of Arizona.

DanSolar. (n.d.). *Danish solar energy supplier.* Retrieved from http://dansolar.dk/en/

DualSun. (n.d.). Retrieved from https://dualsun.com/en/

Iurevych, Gubin, & Dudeck. (2013). Modelling of a Hybrid Solar Panel with Solar Concentration. *Electrical Engineering Research, 1*(2).

Iurevych, O., Gubin, S., & Dudeck, M. (2012). Combined Receiver of Solar Radiation with Holographic Planar Concentrator. *IOP Conference Series: Materials Science and Engineering Physics*, 29. 10.1088/1757-899X/29/1/012016

Li, H., Huang, W., Huang, F., Hu, P., & Chen, Z. (2013). *Optical analysis and optimization of parabolic dish solar concentrator with a cavity receiver*. Elsevier.

Prism Solar Technologies. (n.d.). *Prism Solar Technologies*. Retrieved from https://www.prismsolar.com/

Rekioua, D., & Matagne, E. (2012). *Optimization of Photovoltaic Power Systems. In Green Energy and Technology, Modelization, Simulation and Control.* London: Springer-Verlag. doi:10.1007/978-1-4471-2403-0

Shokhzoda & Tyagunov. (2016). Assessment of solar energy potential on the territory of the Republic of Tajikistan. *Twenty-second international scientific and technical conference of students and postgraduates "Radio electronics, electrical technology and power engineering* (vol. 3). Moscow publishing house MPEI.

Shokhzoda & Tyagunov. (2017). Heat supply based on Solar Energy in Tajikistan. *Conference on Sustainable Energy Supply and Energy Storage Systems*, 181-186.

Shokhzoda, B. T. (2016). *Solar mobile power station", International scientific and technical conference.* "Energy supply and energy saving in agriculture", Russia, Moscow.

Shokhzoda B. T., & Kokin N. N. (2016). *Evaluation of hybrid systems electrical and heat supply based on solar energy*. Moscow: Publisher MPEI Vestnik.

Shokhzoda, B. T., & Kuznetsova, V. A. (2013). Development of power supply system for residential house in Tajikistan. *Nineteenth international scientific and technical conference of students and postgraduates "Radio electronics, electrical technology and power engineering* (vol. 4). Moscow publishing house MPEI (in Russian).

Simulink. (n.d.). *The MathWorks, Inc. Version 8.7 (R2016a).* retrieved from https://www.mathworks.com/products/simulink.html

Singh, Saw, & Kumar. (2012). Technical viability of holographic film on solar panels for optimal power generation. *International Journal of Advances in Engineering and Technology.*

Solimpeks. (n.d.). *Solimpeks Solar Energy corp*. Retrieved from http://www.solimpeks.com/

Strebkov & Tveryanovich. (2007). *Solar Radiation Concentrator*. VIESH.

Swanson. (n.d.). *Photovoltaic Concentrators*. Sun Power Corporation.

The Nobel Prize. (n.d.). *The Nobel Prize in Physics 1921*. Retrieved 2013-03-16 from https://www.nobelprize.org/prizes/physics/1921/summary/

Tikhonov, A., Tugov, I. I., & Yuferev, L. Y. (2009). Technologies of renewable (solar) energy. Moscow: Publisher Research and Manufacturing Association "SODIS".

Vin, J. (2013). *Study of solar energy efficiency for Autonomous power supply systems in the Republic of the Union of Myanmar* (Thesis). Moscow. MPEI.

Vissarionov, Deryugina, Kuznetsova, & Malinin. (1998). *Calculation of solar energy resources, a textbook for courses "Theoretical foundations of energy renewable resources" "Design and operation of solar and wind farms"*. Moscow: Publishing house of MPEI.

Volovikov, V. V. (2002). *Use of mathematical modeling in professional work in radio electronics: proceedings / Information technologies in design, production and education*. KGTA.

Wilford. (1988). *Scientists, Using Mirrors, Match Energy Intensity of Sun's Surface*. Retrieved from https://www.nytimes.com/1988/03/11/us/scientists-using-mirrors-match-energy-intensity-of-sun-s-surface.html

Winston, R. (1970). Light Collection Within the Framework of Geometrical Optics. *Journal of the Optical Society of America, 60*(2), 245–247. doi:10.1364/JOSA.60.000245

Yellowhair, J., Yellowhair, J., Ho, C. K., Ho, C. K., Ortega, J. D., Ortega, J. D., ... Andraka, C. E. (2015). Testing and optical modeling of novel concentrating solar receiver geometries to increase light trapping and effective solar absorptance. In *Proc. SPIE 9559. High and Low Concentrator Systems for Solar Energy Applications.*

Chapter 2
Tools and Methods of Information Support for the Development of Renewable Energy

Ravil I. Mukhamediyev
ISMA University, Latvia

Sophia Kiseleva
Lomonosov Moscow State University, Russia

Kirill Yakunin
Institute of Information and Computational Technologies, Kazakhstan

Renat Mustakayev
Institute of Information and Computational Technologies, Kazakhstan

Jelena Muhamedijeva
Institute of Information and Computational Technologies, Kazakhstan

ABSTRACT

The chapter considers information systems that provide analysis of the RES resource base and support decision making. RES resources in the Republic of Kazakhstan are briefly analyzed. The methods and systems for evaluation of theoretical, technical, and economic potential of renewable energy sources are considered. Based on this overview, the authors propose six steps of estimation that should be realized in intellectual GIS. One of the major steps is the identification the factors that affect using of renewable energy sources. These factors can contribute to impeding the development of green energy. The simple taxonomy of these factors is proposed. The second important step is to aggregate estimations of the factors. As a result, the authors propose the aggregation method based on Bayes rule. Compared to other methods, this approach allows us to obtain two estimates: probabilities of positive and negative hypotheses. The methods discussed are at the heart of the information system supporting the development of renewable energy sources in the Republic of Kazakhstan.

DOI: 10.4018/978-1-5225-9179-5.ch002

INTRODUCTION

Use of renewable energy sources (RES) is a modern powerful trend in energy development. "Green energy" technologies (technologies of gathering energy from renewable sources) are actively developed and will allow reduce significantly the use of fossil resources (oil, gas, coal, peat), reduce the ecological impact of energy plants, improve the ecology around populated areas, reduce the cost of energy generation, increase the autonomy of life support systems and energy security of the country (International Energy Agency 2014, Environmental Consulting, 2010, Mustafina, 2010) in the future.

According to experts, there is no reasonable alternative to development of renewable energy sources and energy efficiency, as decarbonisation of the energy sector will allow avoid the catastrophic consequences of global warming. Under the best case scenario of development it is planned to reduce the demand for oil (30%), coal and gas with overall growth of energy supply by 60%. By the year 2050 the share of fossi00l fuels will be slightly more than 40% of total energy.

The prospects of RES (primarily solar, wind, bio-sources and geothermal plants) use in Kazakhstan are very high (Antonov, 2014). The part of renewable energy (production) in Kazakhstan should be 50% by 2050 according to the "The Concept of transition to green economy" accepted in 2013.

Evaluation of technical and economic efficiency of new technologies requires special approaches (Sysoeva, Pahamov, 2011) and those must be adapted to the conditions of Kazakhstan.

The chapter considers of the following topics

- Information systems applying in the field of renewable energy (RE);
- Kazakhstan's renewable energy resources;
- Inhibitors and accelerators of renewable energy utilization;
- Methods of heterogeneous factors aggregation;
- In conclusion, we summarize the information from the presented parts of the paper.

BACKGROUND

Despite large potential of renewable energy sources (RES) it might be economically indefensible to harness them in full. Consequently it is necessary to select the locations in the territory where use of RES would be most useful. Although such a work has been performed in a range of projects mentioned below, purely engineering considerations are not sufficient for detailed analysis of the specified territories, as deployment of such facilities is influenced by a variety of different factors, which should be evaluated and consolidated in a generalized estimate. Such factors encompass geographical (environmental, geomorphological, location (Arán, Espín, Aznar, 2008), ecological, technical, economical, social factors. Particularly, recent research show that there should be taken into account the problem of generator recycling (Bodrova, Solomin, 2016) (Gorbunova, Gorbunov, 2016) landscape and aesthetic limiting criteria, emerging in recreational area locations (Gorbunova, Gorbunov, 2016) etc.

RES potential evaluation approaches are considered in different research papers. The paper (Angelis-Dimakis, 2011) defines basic stages of that process (Figure 1). It shows that the total scope of existing potential of a certain renewable (theoretical RES potential) is evaluated at the first stage. Technological potential, depending on parameters of the environment, generator effectiveness, service lines, etc., is evaluated at the second stage. Economic strength is evaluated at the third stage, based on greatest possible number of factors. As soon as RES depend mostly on geographic conditions, the paper (Solovjev, 2016) suggests an additional step – evaluation of geographic RES potential. Geographic potential is determined as a part of technological potential, being geographically available and necessary in a certain region.

Different methods are used for evaluation of theoretical RES potential. For instance, the atlas (Kiseleva, Ermolenko & Popel,2015) describes techniques and results of potential estimates for natural resources and solar and wind energy, minor streams, peat, agricultural waste biomass, forestry and wood industry waste in the territory of Russia. The detailed calculation of RES potential for one of Spanish regions is described in the paper (Ruiz-Arias, 2012) . The paper might be considered as a case study for the certain region. The region counts the population of about 669 thousands and occupies the area of 13500 sq.km. The territory is specific with the mountain system and the valley in the Upper Guadalquivir riverbed. The major regional production is olive oil, which made at the time of writing 20% and 50% of production in the world and in Spain respectively. The region produced less power than consumed. Along with that significant RES resources appeared to be unused. The paper evaluated energy potential of three RES types: bioenergy, wind energy and solar radiation. Olive oil production waste constitutes the main source in bioenergy production. The evaluation methodology is based on the following prerequisites.

Residual plant growing index was used in evaluation of bioenergy potential, which relates residual (secondary) production to plant growing products. The regional database information on olive tree (the major regional olive oil producing culture) planting lands is used. The minimum delivery cost map was produced for estimation of biomass supply cost (depending on road quality rate). Furthermore, the map of power stations distribution was produced (based on existing capacities) and five locations for electric power plant deployment were determined, to minimize the cost of biomass delivery. Such an optimal variant allows using of 712366 t biomass (93%) and generation of 763 GW (25% consumed in the region).

European Solar Radiation Atlas (Scharmer, Greif, 2000) was used for estimation of solar energy potential. The Atlas is often applied for estimation of Solar Radiation, with use of images from space satellites. This data was used for production of monthly maps with spatial interpolation using 90-metre grid surfaces. Interpolation and total solar irradiance maps at temporal resolution 12 minutes and spatial resolution 90 meters were produced using a numerical weather prediction model. There were calculated daily average totals including effects of eclipse. According to the technique, stipulated in paper (Blaesser, Munro, 1995) the authors calculated the collected energy for the case of fixed south-facing 3- degree tilted PV panels. Ultimately, based on calculations, installation of 1 mw plant will require the area of 2 ha. Considering all the limitations, the calculations did not include residential areas, industrial zones, recreational and commercial zones, irrigated and protected lands, surrounding territories of water reservoirs, providing minimum 200 m from highways and 100 m from railways. Landscapes featuring slopes

Figure 1. Stages of RES potential evaluation

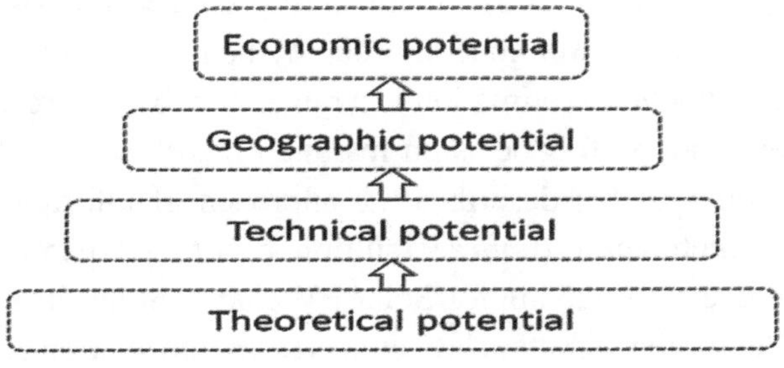

over 10 degrees and located farther than 1 km from transmission lines were excluded (which allows to avoid large investments into installation of power line). In such conditions, the available area is reduced to 7622 ha with production of 5935 GWh. Selection of the zones with the area of at least 70 ha, will allow to install 12 solar power stations with 420 MW capacity, which will generate 656 GWh of electric power (21% of consumed in the region).

In estimation of wind energy potential numerical weather prediction models are used widely, using principles of physics (so called 'dynamical downscaling'), as solely surface measurements do not suppose measurements at heights of wind turbines (60-80 m and higher). These models allow to generate high-resolution spatial and temporal data. The paper applied Weather Research and Forecasting (WRF) model (Skamarock, 2008). The atmosphere is described with 27 layers of uneven thickness (closer to Earth's surface) with account of turbulent transfer processes. Due to the fact that most of existing wind turbines is installed at the height of 80 metres, wind speed was interpolated to this height, where it varies from 4 to 7.5 m/s depending on geographic location (it is higher in mountainous regions). Furthermore, with help of regularizable splines, spatial interpolation to 1 km was made. At the last stage calculation of wind energy potential was based on power capacity curves of two types of Gamesa Corporation turbines. The time of turbine operation during at least 1500 hours a year, Turbine breakaway rate, distance between turbines (at least 7 rotor diameters) and deployment of wind farms with surface area of at least 110 ha were taken into account. These predicates allow to deploy 253 wind turbines, which will generate about 825 GWh (27% of annual consumption).

Forest biomass potential of Italy was studied in detail in the paper (Sacchelli, De Meo, Paletto, 2013), where all possible functions and scenarios of forest utilization were considered, also from the point of energy production. Italian forests are mainly multifunctional (Farrell, Fuhrer, Ryan, 2000). Forest policy should follow Eurostandards oriented at multifunctionality (MCPFE, 2002). Important aspects of forest utilization are prevention of soil erosion and regulation of water cycle, as well as recreation and tourism. The paper suggests the model that takes into account territorial differences if forest regions. It addresses the limitations related to forest biomass production: slope gradient, surface stability, soil structure and distance to roads. Compromise Programming (CP) methodology (Malczewski, 1999) was applied for evaluation of factors for different forest utilization scenarios. CP methodology allows determining the optimal volume of produced biomass in compliance with all forest functions.

The paper (Hoesen, Letendre, 2010) represents an example of RES (solar and wind) potential estimation for rural regions. It shows that for remote regions the structure of household costs supposes that about 20% is spent on energy, heating, transportation, household appliances. In the framework of a local project with use of mainly local sources of energy it is possible to reduce these costs sufficiently. Differential energy data integration is made with help of GIS.

A technically related approach was applied in the thesis paper (Rafikova, 2015), where combination of factors, impacting deployment of power generators was classified by their significance for decision-making. Based on the suggested classification there were produced GIS-assisted regional maps with specification of favorable zones for power generator deployment.

Certain techniques of RES potential estimation in a range of papers were summarized with regard to establish information systems, supporting decision-making processes.

The paper (Sánchez-Lozano, 2013) describes the approach of information system establishment based on GIS and MCDM combination (*abbrev.* multiple-criteria decision making). The information system provides support of decision making for Solar PV plants (Solar farms). GIS is used for factors and limitations mapping. The source of factors lays comes from information about territory development planning, closed territories, road networks, water sources, mountains, etc. The factors can be estimated for definition of potential Solar farm deployment. Analysis and calculation of those factors was made using Analytic Hierarchy Process (AHP) (Saaty, 1989). Estimation of alternative options according to their degree of adequacy was made using TOPSIS (Technique for Order Preference by Similarity to Ideal Solution). GIS gvSIG was applied as a visualisation technique. The software allows processing a wide variety of alternatives and producing maps of most favorable power plant deployment zones.

MCDM method is also used in the work (Lee, Chen, Kang, 2009). Analytical hierarchical process (AHP) is applied in the MCDM model. This process is associated with benefits, opportunities, cost and risk (BOCR). AHP methodology supplemented by BOCR helps in the selection of a suitable wind power plant project.

Consider the previous experience we can proposed to develop a system which can integrate the described stages of assessing the potential of any RES kind. For this purpose the system can include methods of aggregating heterogeneous factors, depending on types of energy, technologies and equipment used. For example it would be analytical hierarchy process (Díaz-Cuevas, et al, 2018), Bayesian (Mukhamediev, Mustakaev, 2016), Weighted Sum Model, TOPSIS, VIKOR, etc. (Kumar, Sah, et al, 2017). The latest results in machine learning (Muhamedyev, 2015) make it possible to use them with promising results (windcat.ch). The system should support advanced visualization tools based on GIS (Muhamedyev, Kiseleva, et al, 2016).

Summarizing, the following steps in the process of estimating the potential of RES, which can be implemented in the intellectual GIS are:

1. The estimation of theoretical RES potential
2. Identification the factors that affect using of renewable energy sources
3. Usage the widest range of territory and RES databases with global and local nature.
4. Estimation of factors affecting using different methodologies (AHR, etc.)
5. Application of intellectual methods generalizing the influence of various factors
6. Using GIS for result visualization.

KAZAKHSTAN'S RENEWABLE ENERGY RESOURCES

At present the Republic of Kazakhstan (the RK) is one of large energy producers among the countries of Central Asia, due to its major deposits of coal, oil, natural gas, uranium, and economic stability of the state is provided also on account of fuel energy resources exports. Consequently the basis of the energy sector in the Republic is formed with fossil (hydrocarbon) resources, which impacts as the structure of the energy sector in whole (Figure 2), as the power generation sector in particular (Table 1)

Against the background of the structure of electric power generation and network system the country may be divided into 3 zones (Figure 3):

- Northern – Eastern Kazakhstan, Pavlodar, Akmolinsk, Karaganda, Northern Kazakhstan and Kostanai regions;
- Western – Western Kazakhstan, Atirau, Aktyubinsk and Mangistau regions;
- Southern - Almaty, Zhambyl, Kizilorda and Southern Kazakhstan regions.

Transmission and distribution of electric power among three zones is complex owing to the fact that Kazakhstan has an underdeveloped trunkline grid with voltage in the range of 220-500-1150 kV and suffers significant wear of trunk and distribution lines, which results in severe losses of electric power (trunk lines – around 20%, but distribution lines – 25-50%). Both this fact and developed agricultural production, implying seasonal or annual off-line consumers of heat and electric power, actualize development of renewable energy projects in the territory of the Republic.

In this regard the priority task is to estimate, to rate and to analyze distribution of different renewable energy sources in the territory of the country.

Table 1. Classification of power plants in Kazakhstan by destination

Power	Characteristics	Examples
National	Large-scale CHPs and hydropower plants with large capacities, providing generation and sales of electric power to consumers on the wholesale market of electric power	- Ekibastuzsk regional power station-1; - JSC "Ekibastuzskaya GRES-2»; - Aksussk regional power plant; - Karaganda regional power plant -2; - Zhambyl regional power plant; - Ust-Kamenogorsk hydropower plant; - Shuljbinsk hydropower plant.
Industrial	CHPs with combined cycle of heat and electric power production, providing power and heat supply to large industrial manufacturers and neighbouring communities	- GTPP PLC "Tengizshevroil"; - CHP-3 "Karaganda-Zhylu"; - CHP-2 JSC "Arselor Mittal Temirtau"; - Rudnensk CHP; - Balhash CHP; - Zhezkazgan CHP of "Kazahmis" corporation; - Pavlodarsk CHP-1 JSC "Aluminy Kazahstana"; - Shimkentsk CHP-1,2 (JSC Yuzhpolimetal") and others
Regional	CHPs, selling electric power via networks of regional electric supply companies and power transmission organizations, as well as heat supply of neighbouring municipalities	- Almaty CHP-1; - Semipalatinsk CHP-1,2 and others

Source: Power Sector Development Program in the Republic of Kazakhstan for the period 2010 - 2014.

Figure 2. Structure of the energy sector of the RK (electric power generation segment) by types of generating capacity

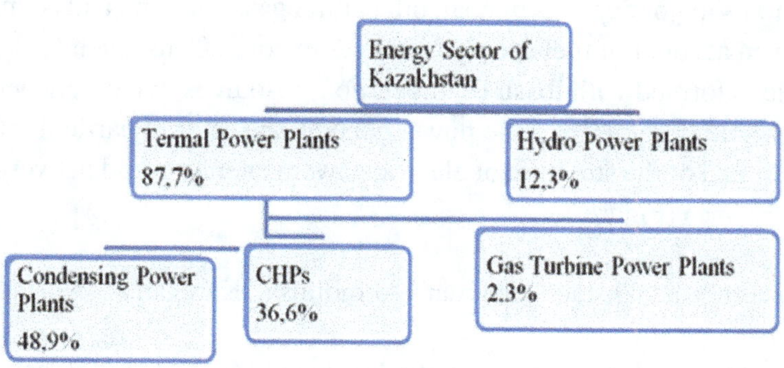

Wind Energy

Total Wind Energy Potential Estimates in the Territory of Kazakhstan

According to the United Nations Development Programme (UNDP) experts wind energy potential of Kazakhstan is estimated as 0,929 - 1,820 mlrd. kWh per year. Whereby over the area of around 50 thous. km^2 the annual average wind speed exceeds 6 m/sec. This makes them attractive for development of wind energy. Wind energy resources in Jungar Gates and Shelek Corridor are considered to be the most significant.

Data Sources and Methods of Wind Energy Potential Estimation

Primary estimation of spatial wind resources may be based on the data from weather stations (initial or processed). The data needed for long-term estimation of wind speed distribution by grades is contained in USSR Climate Reference Books, issued in series in 1950's, 60's, and 90's (Scientific and Applied Guide of the USSR, 1966-1969). They present data on wind patterns for more than 1000 USSR weather stations by 15 parameters. The territory of Kazakh Soviet Socialist Republic features 131 weather stations (Figure 3).

At the time being the weather station network in Kazakhstan remains distributed. Despite the fact that the number of automatic weather stations is gradually increasing, their number is still very small for such a large territory (Figure 4). Note that the paper (Solar radiation, radiation balance and sunshine,1967) shows location of meteorological stations in the Kazakh SSR.

The next task for acquisition of adequate wind energy estimates is to select (or to develop) calculation techniques. Development of wind resource estimation methods and basically the USSR resource estimation, including Kazakhstan, has a rather long history. In 1935 Krasovsky N.V. estimated wind energy resources of the USSR based on the map of average monthly and annual wind speeds, with adjustments by a typical wind generator capacity (Krasovsky, 1935).

Figure 3. Wind speed distribution over the territory of Kazakhstan (height of measurements – 80 m) (The Kazakhstan an Electricity Association Committee on Renewable Energy Sources)

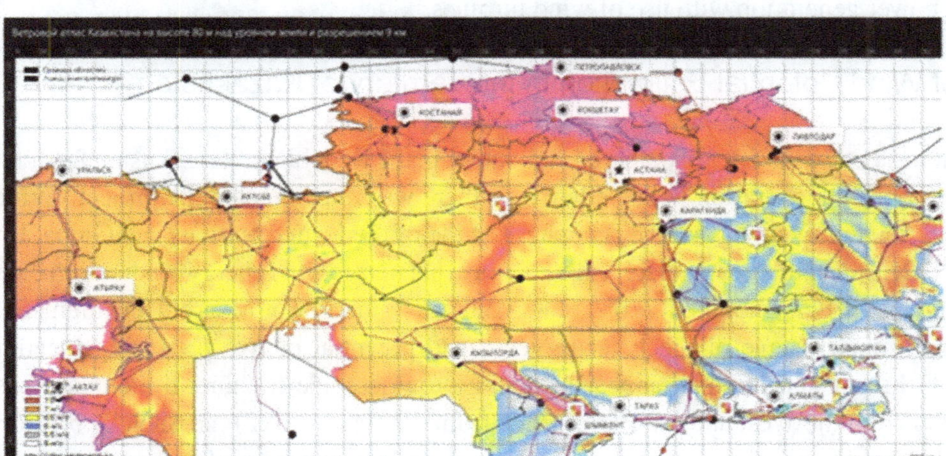

Figure 4. Automatic weather stations in the territory of Kazakhstan (Sister, and others, 2001)

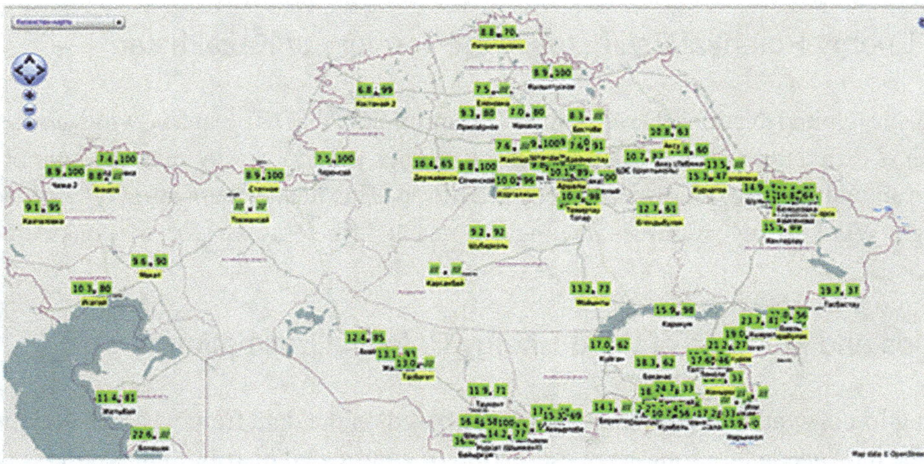

In 50's-60's of the 20th century the meteorologist Grinevich G.A. suggested a new method of wind energy potential estimation for Kazakhstan and Central Asia, which was applied for development of wind energy inventory (Grinevich, 1952).

Furthermore a range of the USSR wind energy potential estimation (including, Kazakhstan) were made with use of the method, suggested by A. I. Voeikov Main Geophysical Observatory and и NPO (R&D institution) "Vetroen" (Recommendations for the determination of climatic characteristics, 1989) according to USSR climate reference book data for the period from 1936 till 1960-1965. Weather station measurements for that period were recognized most credible because they had been made using a unified technique at the same hours of local solar time with help of calibrated devices. Figure 6 shows the map of average annual wind speeds, from National Atlas of the Republic of Kazakhstan, based exactly on long-term meteorological data.

It is apparent from the map in Figure 3 that most part of the territory of Central Kazakhstan, Peri-Caspian Lowland, as well as southern regions feature rather high average annual wind speeds (5-6 m/sec), which would allow developing wind energy in these regions. As outlined above, wind climate in certain regions – Jungar gates and Shelek corridor (Alamty region) are also potentially very favourable for electric power generation with use of wind turbines.

Results of Wind Potential Estimation in the Territory of Kazakhstan

To carry out in-depth wind energy potential estimation in the Republic of Kazakhstan from 1997 to 1998 in the framework of cooperation with Netherlands there was implemented the project "Wind energy in Kazakhstan". That project became the first phase of the of the United Nations and Global Environment Facility (UN-GEF) development project "Elimination of barriers to wind energy production in Kazakhstan", and then the UN-GEF development project "Kazakhstan-Wind Power Market Development Initiative".

At the preliminary stage of the latter project wind potential of two promising locations was estimated – (1) Jungar gates, which appear to be a narrow tectonic fracture, a mountain pass between Jungar Alatau from the West and the Barlik ridge from the East; and (2) the Shelek corridor in Almaty region. With regard to acquire wind energy potential data in central part of Jungar gates four 32 m high

Figure 5. Average annual wind speeds in the territory of the Republic of Kazakhstan (National Atlas of the Republic of Kazakhstan, 2006)

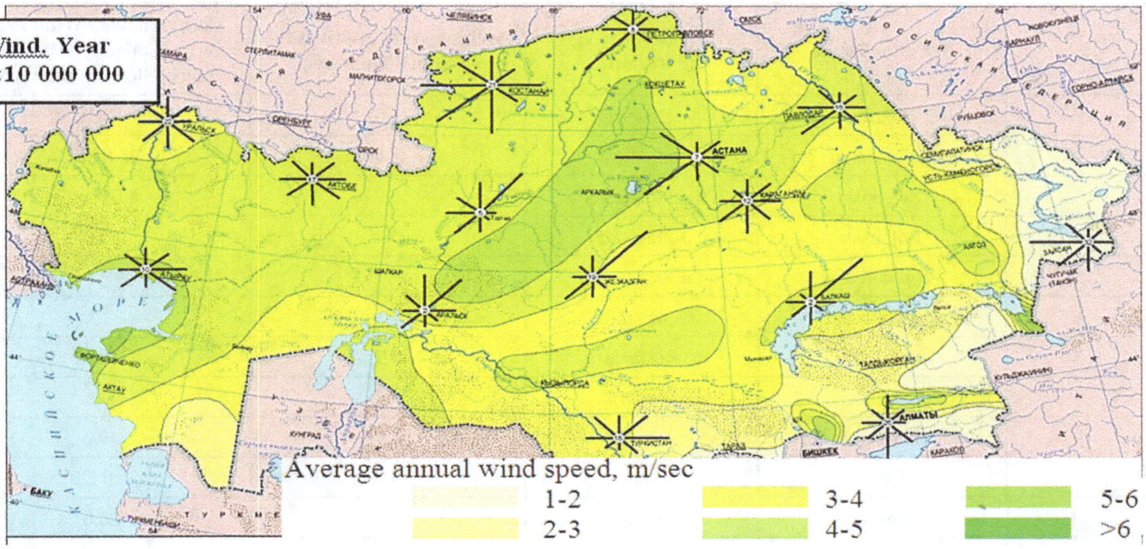

wind-monitoring poles were installed. In the Shelek corridor one 30 m high pole was functioning. The wind monitoring results in the framework of UNDP project, confirmed high potential of Akmolinsk and Karaganda regions, Mangistau and Atirau regions (UNDP/GEF project,2010).

Spot wind monitoring in the framework of UNDP Project allowed the project executors give some estimates of total wind potential of the territory of the Republic: it was estimated, as outlined above, as 0,989 – 1,82 mlrd. kWh per year, or 354 GW of installed capacity, which exceeds the installed capacity of Kazakhstan power plants by 18 times at present. A range of regions, with total area of 50 thous. km^2 have good wind resources (7-8 m/sec), in some areas of Almaty region the average annual wind speed is very high (8-9 m/sec), and extremely high (over 9 m/sec). Most significant wind resources belong to the region of Jungar gates. As the research show, the average annual wind speed there is 9,7 m/sec at the height of 50 m, but the wind flow energy density is around 1050 W/m^2. One more location, featuring good wind potential is the Shelek corridor, situated between the mountain ridge Zhetigey and Zayliysky Alatau 150 km far from Almaty. Average annual wind speeds there are 7,8 m/sec at the height of 50 m, wind flow energy density is around 510 W/m^2.

Based on analysis of meteorological data and with reference to the following factors:

1. accessibility of transmission lines and substations for power delivery,
2. topographic expression and elevation above sea level,
3. availability of transportation lines,
4. presence of energy consumers,
5. opportunity of a wind farm deployment
6. existing preliminary engineering of a wind farm deployment

There were determined favourable locations for deployment of wind farms (Table 2).

Table 2. List of favourable locations for deployment of wind farms. The UNDP and the government of Kazakhstan Project "Kazakhstan - Wind Power Market Development Initiative"

Nr	Location Name	Kazakhstan Region	Average Wind Speed at the Height 10 m,m/sec	Recommended Wind Farm Capacity,MW (First Phase in Brackets)
1.	Zharma	Eastern	5,6	40,0
2	Yerementau	Akmolinsk	5,4	35,0 (20)
3	Seletinsk	Akmolinsk	5,9	40,0 (20)
4	Balkhash	Karaganda	4,4	10,0
5	Yeghindibulak	Karaganda		
6	Karkaral	Karaganda	4,3	10
7	Arkalik	Kustanai	5,7	10,0
8	Sakril	Western	5,2	10,0
9	Atirau - Karabotan	Atirauss	4,4	40,0
10	Akkistau	Atirauss	5,5	50,0 (20)
11	Inder	Atirauss	5,4	20,0
12	Prorva	Atirauss	6,2	40,0 (20)
13	Fort Shevchenko	Mangistau	6,0	40 (20)
14	Sai-Utyos	Mangistau		
15	Cordai	Zhambil	5,1	20,0 (10)
16	Aralsk	Kizilorda	4,9	10,0
17	Karmakchin	Kizilorda	5,5	20,0 (5)
18	Chayan	Southern	5,0	40,0 (20)

Solar Energy

The initial data on the fluxes of solar radiation and total incident energy on an area of different orientations at a given geographic point on Earth's surface can be obtained in several ways:

- Analytical, when necessary parameters for a specific geographic point are determined by calculation;
- From long-term terrestrial actinometric measurements;
- Mathematical modeling of solar energy distribution in the atmosphere based on satellite monitoring and terrestrial data.

Analytical methods for determining the values of solar radiation rely on basic positions of solar geometry. The amount of solar energy incident on the earth's surface is affected by presence of Earth's atmosphere. While travelling through the atmosphere, the intensity of solar radiation gradually decreases because light is scattered by gas molecules, dust particles, aerosols, etc. At the same time, scattered radiation appears in the atmosphere, part of which goes back to outer space. In addition, there occurs absorption of solar radiation by ozone (in the upper layers of the atmosphere), water vapor, carbon dioxide and oxygen (in the lower layers). So, evaluation methods are currently used only for preliminary calculations of solar energy potential in individual regions, they are complex and their accuracy is not high (Vissarionov, Deriugina, Krivenkova,1998).

Measurements of solar radiation flux density directly at the place of the intended deployment are associated with certain organizational difficulties. Moreover, for justifiable estimates, such measurements should be perennial. Therefore, the main sources of initial actinometric information are climate directories and databases created by specialized organizations based on processing of the results of long-term actinometric observations at weather stations. For the territory of the former USSR, the main sources of actinometric data can serve as multivolume directories on the climate of the USSR, which contain the results of regional and central meteorological organizations (Reference book on the climate of the USSR, 1967).

In the absence of actinometric measurements, one of the characteristics in evaluation of the potential resources of solar energy is sunshine duration. As a result of the recalculation of sunshine duration according to Angstrem formula, there is an opportunity to obtain data of arrival of solar energy (in energy units).

Traditionally, in winter time, solar energy is more efficiently collected by vertically located receiving devices (photoelectric energy converters) (in moderate latitudes), and in summer - by horizontal ones. In connection with this, for the fullest use of solar energy, either a fixed inclined orientation of receiving surfaces (for example, the angle is equal to the latitude of the space) is often used, or the angle of the oblique receiver.

In Figure 7 characteristics of incident solar energy in terms of solar luminosity, are showed, namely the average duration of sunshine in hours per year in the Republic of Kazakhstan. It can be seen that the number of sunshine hours per year reaches 3000 or more in Kazakhstan.

From the standpoint of energy indicators, the solar energy resource of the territory of Kazakhstan is estimated at 2.5 billion kWh per year (Energy-efficient Russia, 2016). This is almost three orders of magnitude smaller than the wind potential, howeverthis indicator is very significant. The average annual daily sum of total solar radiation is 3.8 - 5.2 kWh / m2. This is one of the best world indicators. Some parameters of distribution of incident solar energy in the territory of the Republic of Kazakhstan are presented in Table 3.

As can be seen from the table, the greatest amount of solar energy enters the southern regions of the Republic, where the maximum efficiency of solar energy application can be expected. Figures 8, 9 present data on monthly total incident solar radiation on horizontal plane surface and direct solar radiation on the surface normal to the ray in 3 regions (Fort Shevchenko, Almaty, Aral Sea).

As can be seen from the figure, the values of total and direct solar radiation for the selected regions are practically equal. However, the maximum values of these indicators are observed in the Aral Sea regions. The total solar radiation is important in design of solar collectors and photoelectroconverters (PECs) without concentrators and a solar tracking system. For tracking systems or concentrators, it is important to know direct solar radiation on the surface normal to the beam of surface.

Due to the fact that the number of meteorological (and especially actinometric) quantity of stations in the territory of Kazakhstan is very little, there is a necessity in additional sources of data on incident solar radiation. To solve the problem of insufficient ground-based data on solar energy resources, mathematical modeling has been increasingly used in recent decades. As such, one can consider, first, reanalysis, which is widely used in meteorology, climatology, oceanology and other fields. At present, both in the NCEP reanalysis data (The NCEP/NCAR Reanalysis Project, 2016) and in the ERA-reanalysis (ERA

Table 3. Total incident solar radiation on horizontal plane surface for some regions of the Republic of Kazakhstan (EnergyPartner LLP, 2016)

Region	KWh / m2 per Year	KWh / m2 per Day		
		Average	June	December
Shymkent	1780	4,88	7,95	1,65
Aktau	1442	3,95	6,71	0,98
Astana	1297	3,55	6,47	0,83
Semei	1441	3,95	6,74	1,20
Taldykorgan	1703	4,67	7,40	1,58
Fort Shevchenko	1503			
Aral Sea	1703			
Almaty	1461			

Figure 6. Duration of sunshine per year for the territory of Kazakhstan (National Atlas of the Republic of Kazakhstan, 2006)

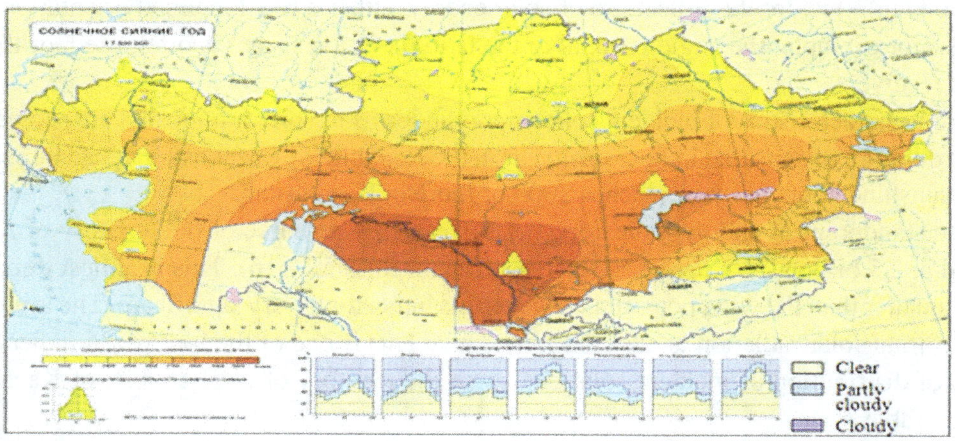

Interim reanalysis system) presents data on high resolution incident solar radiation . So, in case of the ERA, monthly average values of the following values pertaining to estimates of solar energy resources are available for each month of the year in the interval (1979 - present) on a grid with a step of 0.25° (the last 20 years - with a step of 0.125°):

- Sunshine duration;
- Surface net solar radiation;
- Surface solar radiation downwards;
- TOA incident solar radiation и др.
- Open archives of NCEP reanalysis contain similar data:
- Clear sky downward solar flux;
- Downward solar radiation flux;

Figure 7. Average monthly daily sum of total solar radiation incident on horizontal plane surface

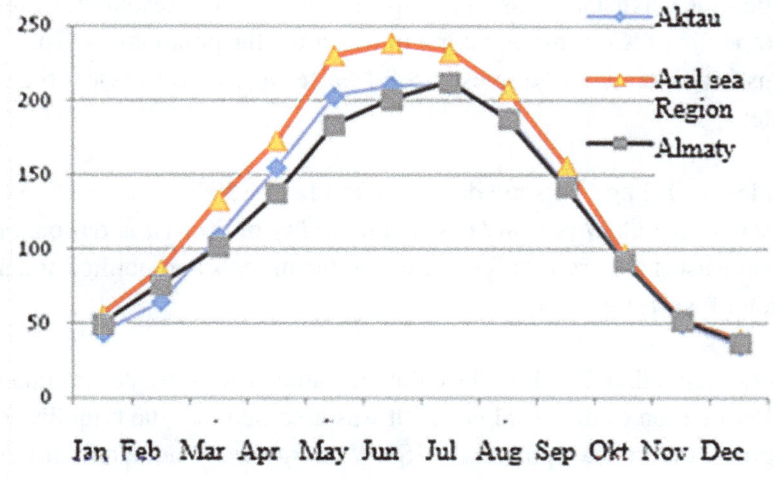

Figure 8. Average monthly daily sum of direct solar radiation on the surface normal to the beam of surface, kWh / m2 (European Bank for Reconstruction and Development. 2016)

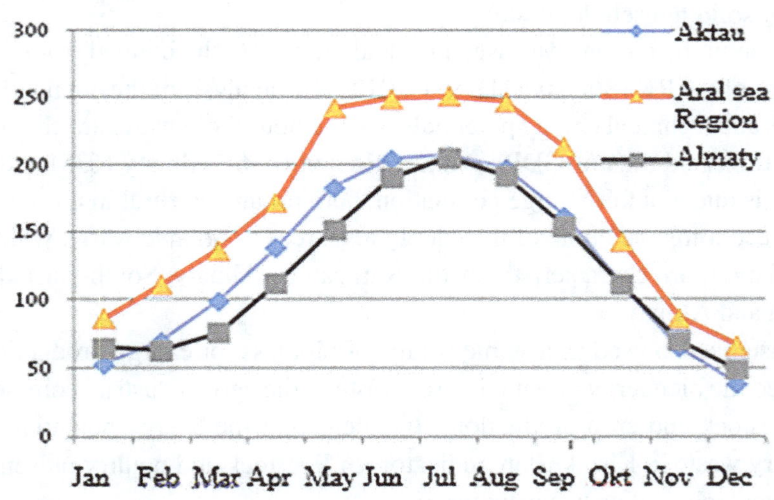

- Near IR beam downward solar flux, near IR diffuse downward solar flux.

The question of the applicability of specified arrays for the purposes of solar energy for authors is still open and requires further study and verification.

Energy of Biomass

The Republic of Kazakhstan, being an agro-industrial country, has a great potential for bioenergy development due to significance and diversity of biomass resources. According to (Energy efficiency and Renewable Energy Consulting, 2016), the processing potential of agricultural waste is 35 billion kWh of electric power in the state, and 44 Gcal of heat energy per year.

Data on amount of domestic solid waste (DSW) and sewage sludge (SALT) was analyzed in the regions of the Republic of Kazakhstan, to assess the potential of energy resources of organic wastes from settlements of the Republic of Kazakhstan, taking into account the population (Sister, and others, 2001). The following norms for generation of domestic solid waste were used to assess the energy potential of domestic solid waste:

- For urban residents - 1.2 kg / person / day at a humidity of 50%;
- For rural residents - 0.52 kg / person / day at a humidity of 50% (it is reasonably assumed that in rural areas, food wastes are used for feed of domestic animals and poultry and are not included in the composition of waste)

According to (Sister and others, 2001), The calorific value of solid waste was taken equal to 0.2 tons of fuel equivalent. Per one ton of dry solid waste. It was assumed that the humidity of solid household waste is 50%. The gross (theoretical) potential of SHW energy was calculated for the entire population of the Republic and separately for the regions. The technical energy potential was calculated taking into account the population of cities and the number of rural residents in the region (assuming that residents of villages use food waste for animal feed): The economic potential was calculated on the basis of the generated amount of solid household waste for the urban population (it is assumed that in the next few years, processing of solid household waste)

Data of population in the regions was used in calculations, which obtained from the materials of the Agency for Statistics of the RK (Statistical Digest, 2012). The analysis of the obtained results shows that the maximum gross and technical energy potentials of solid household waste are characteristic for South Kazakhstan (1148 thousand tons of GDP), Almaty Region (836), Almaty (635) and East Kazakhstan Oblast (611)). This is due to a fairly large population, both urban and rural in these regions. The maximum values of the economic potential of the energy resources of organic waste, which directly depend on the urban population, are characteristic of the Karaganda, Almaty, South-Kazakhstan oblasts, and the cities of Astana and Almaty.

Agricultural waste is also used as a stable source of biomass for energy production in Kazakhstan. The authors assessed the bioenergy potential of the waste of the agro-industrial complex of the Republic of Kazakhstan (livestock and crop production). In calculating the energy potential of the biomass of livestock and poultry waste in Kazakhstan, indicators of livestock and poultry presented by (Statistical Digest, 2012) and zootechnical norms were used.

The gross potential of energy resources of organic wastes of livestock and poultry was calculated for all available livestock and poultry for all categories of farms, including personal subsidiary plots of the population. It is assumed that the technical potential of waste is equal to the gross one, since all wastes can technically be processed into energy. As the economic potential of the energy of livestock and poultry wastes, the amount of waste and their energy resources that are generated at large livestock and poultry enterprises where these wastes are accumulated continuously and easily collected.

The energy potential of crop wastes was calculated by the main types of agricultural crops in the Republic of Kazakhstan: cereals, potatoes, sugar beets, vegetables and sunflower.

When calculating the gross potential, waste generation norms were used for all types of agricultural crops grown in the RK on an industrial scale, their humidity and energy content. Gross energy potential of straw was calculated for all categories of farms using the formula:

The technical potential was assumed equal to the gross. The economic potential is half the gross and technical, since up to 50% of the straw and stems produced are used in farms for other economic needs: as bedding material, in construction, they return to arable land, etc. The summarized results of estimations of plant waste in the RK are presented in Table 4.

It should be noticed that timber has an insignificant part in the balance of bioenergy resources. In Kazakhstan, forests occupy an area of more than 10 million hectares, which is only 4% of the total territory of the country. The volume of wood waste on logging sites and woodworking in woodworking enterprises, as well as wood, which is used as firewood, is almost 1.3 million m3 or 1 million tons. Thus, the energy potential of wood waste is no more than 200 thousand tons of fuel equivalent (AD).

In summary, annual yield of livestock and poultry waste on a dry weight basis is 22.1 million tons, or 8.6 billion cubic meters. (Cattle - 13 million tons, sheep - 6.2 million tons, horses - 1 million tons), plant residues - 17.7 million tons (wheat - 12 million tons, barley - 6 million tons) or 8.9 billion cubic meters), which is equal to 14-15 million tons of equivalent fuel, or 12.4 million tons of fuel oil, or more than half of the oil produced (The Kazakhstan Electricity Association, 2016)

According to the obtained results, the total gross potential of livestock wastes of agriculture and solid household waste in the Republic of Kazakhstan is 48 963 652 tons of fuel equivalent or 398.6 billion kWh.

AN INHIBITORS AND ACCELERATORS OF RENEWABLE ENERGY UTILIZATION

At the initial stage, as indicated above, it is necessary to identify a set of factors on which the location of the generator will depend. The set of these factors depends on type of energy used and type of generator. In the above-mentioned work (Lee, Chen, Kang, 2009), in order to implement multi-criteria support for decision-making in the field of wind energy, the following 4 categories were considered: Benefits, Opportunities, Costs, Risks.

Categories include criteria and sub-criteria as listed below.

Benefits:

1. Wind availability
    a. Geographical distribution of wind speed frequency
    b. Mean wind power density
    c. Annual mean wind speed
2. Site advantage
    a. Influence of selected height of installation

Table 4. Regions of the maximum gross and economic biomass potential of crop wastes

Agricultural Crops	Cereals	Potatoes	Sugar Beet, Vegetables	Sunflower
Region of the Republic of Kazakhstan	Kostanay, North-Kazakhstan	Almaty, North and East Kazakhstan	Dzhambul, Almaty	Pavlodar, Kostanay
The Republic of Kazakhstan as a whole, Tpa / year	11 458 000	92 283	126 423	350 657

   b. Effect of wind gusting
   c. Micro-siting of WEGs
3. WEG functions
   a. Real and technical availability
   b. Affordable, reliable, and maintenance free
   c. Power factor, capacity factor

Opportunities:

1. Financial schemes
   a. Switchable tariff
   b. Discount of tax rate and duty rate
   c. Other investment and production incentives
2. Policy support
   a. Wind power concession program
   b. Clean development mechanisms program
   c. Other policy supports
3. Advanced technologies
   a. Computerized supervisory
   b. Variable speed wind power generation
   c. Swept area of a turbine rotor
   d. Static reactive power compensator, etc.

Costs:

1. Wind turbine
   a. Design and development
   b. Manufacturing
   c. Installation, maintenance
2. Connection
   a. Electric connection
   b. Grid connection
3. Foundation
   a. Main construction
   b. Peripheral construction

Risks:

1. Concept conflict Entrepreneurs, policy makers, residents
2. Technical risks Technical complexity and difficulties
3. Uncertainty of land Loyalty or lease agreement, geology suitability, etc.

Paper (Sánchez-Lozano, 2013) considers set of factors in order to select the optimal locations for solar stations. The hierarchical set includes the following elements:

1. Environmental
2. Agrological capacity
3. Geomorphological
    a. Slope
    b. Orientation
    c. Area
4. Location
    a. Distance to road
    b. Distance to power lines
    c. Distance to villages
    d. Distance to Substantions
5. Climatic
    a. Solar irradiation potential
    b. Average temperature

A set of interdependent factors (ecological, economic, political) that affect the extraction of forest biomass energy are considered in the work (Sacchelli, DeMeo, Paletto, 2013). Some of considered factors are: steepness of the slopes, reliability of surface (it is necessary to take into account when using mechanical means for forest harvesting), soil structure, distance from road etc. The dependence of factors is determined in particular by the fact that if many forest biomass are extracted, the forests can degrade due to the lack of nutrients available in the "dead" trees. On the other hand, if you didn't remove the fallen trees, risk of fires increases.

In the work (Ruiz-Arias, 2012) the main factor influencing the production of bioenergy is the volume of olive fruits used for oil production. The other factors are distance from roads (for example, the collection of biomass at distances more than 50 km from the power plant is not profitable) and steepness of the slopes. For solar power plants, along with the assessment of the potential, an important factor is availability of free area (for example, the installation of 1 megawatt station will require about 2 hectares of area). The important factor for wind power plants are parameters of wind turbines.

In the work (Rafikova, 2015), the factors affecting the installation of RES are divided into natural, environmental, technical, economical and social.

These factors and criteria can both promote the use of renewable energy sources (catalysts) and prevent their using (inhibitors).

TAXONOMY OF FACTORS AFFECTING THE INSTALLATION OF POWER GENERATORS

As result of the analysis, it is proposed to divide all factors into the following groups (Figure 9). Table 5 represents factors categories in more details, extending categorization from introduction above.

Depending on the type of RES generator, each of the parameter can be either an inhibitor or catalyzer. Some of the inhibitors absolutely prohibit installation of generators in the area and are called hard inhibitors – they are not taken into account in further ranking of factors, and if the hard inhibitor is present in the area, the further calculations are terminated. Example of such hard inhibitor is national parks.

Figure 9. Groups of factors affecting the installation of power generators

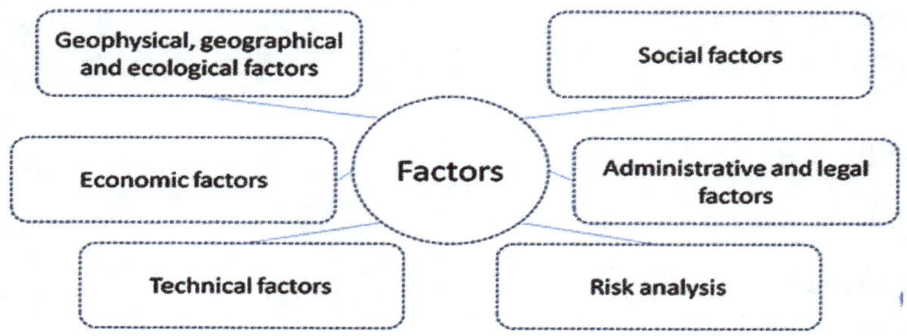

Since the number of parameters is substantial, an issue of ranking (weighting) them arises. Analytical Hierarchy Process (AHP), developed by Thomas Saaty could used to solve this problem. It allows inferring weight of factors based on pairwise comparison of factors by experts in the field. However, since AHP method requires pairwise comparisons, in case of large number of factors – 65 in this case – the number of pairwise comparisons is unjustifiably high – 65^2/2 = 2112 comparisons. For that reason categories of factors, marked as "AHP" in the Table 5 are firstly pairwise compared, ranked and weighted, and then pairwise comparisons inside each category is performed. It significantly reduces number of required comparisons.

After comparative evaluation of the factors we need to obtain aggregate assessment related to particular territory.

AGGREGATION METHODS OF HETEROGENEOUS FACTOR

Methods of aggregation of factors influencing the assessment of suitable territory for application of RES are considered in the works (Ruiz-Arias, 2012) (Sacchelli, De Meo, Paletto, 2013) (Hoesen, Letendre, 2010) (Rafikova, 2015) (Sánchez-Lozano, 2013) (Lee, Chen, Kang, 2009). Two aggregation models are considered in the paper (Mukhamediev, Mustakaev, 2016): multiplicative and Bayesian models. For multiplicative aggregation inhibitors with weight value 0 exclude the use of territory. The Bayesian model uses subjective probabilities.

The last model seems to be promising in the case of expert assessments since it allows obtain the probabilities of both positive and negative solution. However, the need for direct calculations of the energy potential or the probabilities of abnormal weather phenomena can complicate the model.

As a result of analyzing models of aggregation of data, a hybrid of multiplicative and Bayesian models was developed. Those categories of factors, which can be hierarchal compared using AHP method are aggregated using Bayesian inference, then the results of the inferences is aggregated with other factors (statistically or computationally evaluated) using multiplicative model.

The process of decision making using AHP consists of three steps:

1. Evaluating vector of factors weights
2. Computing matrix of option scores
3. Ranking the options

Table 5. Factors affecting the installation of power generators. The Inhibitor and catalyst are marked as "-" and "+" respectively

N	The Categories(Method of evaluation) and the Factors	Inhibitorcatalyst
G	Geophysical, geographic and environmental factors	
GN	Natural potential (Computed evaluation)	
	Available potential of this energy type	
1	Potential of solar energy	+
2	Potential of wind energy	+
3	Hydropower potential	+
4	Potential of bio energy	+
5	The potential of geothermal energy (availability of geothermal sources in the considered area)	+
GC	Consistency (stability) of energy source (Part is hard-threshold inhibitor, the rest – AHP)	
6	Variability of solar radiation	-
7	Variability of wind force	-
8	Variability of hydropower	-
9	Variability of bio resources for energy production	-
GG	Geographical and geophysical factors(Statistical evaluation)	
10	Specially Protected Natural Areas	-
11	Non-linear hydrological objects	-
12	Linear hydrological objects	-
13	Ornithological territories and migration routes of birds	-
14	Logistic accessibility of sources (livestock farms) for bioenergy stations	+
15	Territory of livestock farms	-+
16	The territory of the forest fund land	-
17	Distance to the road line	+
18	Distance to railway line	+
19	The territory roadways and railways, other areas occupied by structures	-
20	The territory for Ministry of Defense objects	-
21	Territory of settlements	-+
22	Distance to the built-up area	_+
23	The presence of forests, pastures and pastures, gardens near the installation site	-+
24	Distance to the nearest power line and substation	+
GGE	Areas with unfavorable geotechnical conditions (Computed, hard-threshold inhibitors)	
25	Waste storage areas	-+
26	The karst area	-
27	Permafrost propagation zones	-
28	The territory with sharp slopes	-
29	Territory of sand	-
GGW	Abnormal weather phenomena (Statistical evaluation):	
30	Abnormally high temperature	-
31	Abnormally low temperature	-

continued on following page

Table 5. Continued

N	The Categories(Method of evaluation) and the Factors	Inhibitorcatalyst
32	Abnormal wind gusts (hurricanes, typhoons, tornadoes, etc.)	-
	Dust storms	
GGN	Abnormal natural phenomena (Statistical evaluation)	-
33	The possibility of an earthquake	-
34	Flood potential	-
35	The possibility of volcanic eruptions and other volcanic manifestations	-
36	Possibility of landslides	-
37	The possibility of avalanches, rockfall, etc.	-
38	Possibility of floods	
E	Economical factors (AHP)	
39	Price competitiveness or lack thereof, payback, service cost	-+
40	Independence from import of energy carriers	-
41	Reduction of the length of power transmission lines	+
42	Commercial effect from attraction of “clean” technologies in housing construction, hotel business, on recreational territories	+
43	Structure of energy consumption in the region	-+
44	Investment incentives	+
45	Economic incentives from the state, for example, guaranteed purchase of energy at stable prices	+
T	Technical factors (AHP)	
46	Technical complexity of project implementation	-
47	Constructive limitations	-
48	The novelty of technology	-
49	Production incentives	+
S	Social factors(AHP)	
50	mproving the living conditions. Compliance with smart city criteria	+
51	Guaranteed power supply in rural and remote areas	+
52	Employment creation.	+
53	Raising the level of education	+
54	Presence of the personnel providing the operation of the generator	+
A	Administrative and legal factors	
55	Availability of construction plans	-
56	Private area	-+
57	A large number of permissive documents (the number of instances that need to be passed for obtaining a building or installation permit)	-
58	Presence of non-legal relations and corruption	-
59	Presence of laws and government resolutions that promote the using of renewable energy	+
F	Risk factors (Statistical evaluation)	
60	The attitude of entrepreneurs, officials and the public	-+
61	Increase in the risk of fires in case of inadequate cleaning of the forest fund	-+
62	Decrease or increase of tourist and aesthetic appeal	-+

In the case of the system, mostly the first step is used on order to rank the factors and then apply the developed Bayesian aggregation model.

Let's describe the process of evaluating vector of factors weights:

A pairwise comparison between all of the factors is conducted. For each pair of i-s and j-s factors an a_{ij} is evaluated by expert, which can take one of the following values, if j is more important, than i:

$a_{ij} = 1$ – j and i are equally important
$a_{ij} = 3$ – j is slightly more important than i
$a_{ij} = 5$ – j is more important than i
$a_{ij} = 7$ – j is strongly more important than i
$a_{ij} = 9$ – j is absolutely more important than i

It should be noted that $a_{ij} * a_{ji} = 1$, which means that if j is more important then i, than i is proportionally less important than j, and matrix A of a_{ij} is symmetrical relative to the main diagonal.

Than a normalized pairwise comparison matrix A_{norm} is calculated, so that sum of each column is equal to 1:

$$a_{jinorm} = a_{ji} / \sum a_{jk}$$

Finally, the factor weight vector w is calculated by averaging the entries of each row:

$$w_j = \sum a_{jk} / m$$

This vector represents weights of each of the factors based on pairwise comparison. Than this vector is applied to a list of alternatives and they are ranked. In the case of the system the w vector is than used as a parameter to the developed Bayesian aggregation model.

Bayesian model is based on Bayesian formula (1):

$$p(h \mid e) = \frac{p(e \mid h) \times p(h)}{p(e \mid h) \times p(h) + p(e \mid \infty h) \times p(\infty h)} \tag{1}$$

Where p(e|h) – conditional probability of event e, given h

P(h) – prior probability of hypothesis h

P(e|~h) – conditional probability of e, given absence of h

P(~h) – prior probability of h being false.

In developed version of applying Bayesian theorem to decision making, we propose to hypothesis:

H1 = 0.1 – possibility of feasibility of installing RES generator in the area

H2 = 0.2 – possibility of installing RES generator to be non-feasible

Then, we consider a list of factors, for example:

The issue of evaluating p(e|h1) is resolved using Analystical Hierarchy Process method. Inside each of the categories the method is applied, then the resulting weights (in range [0; 1]) are renormalized according to inhibitory/catalyzing properties of the factor. For inhibitors is is normalized to range [0; 0.5], for catalyzers – [0.5;1]. As simple example, Table 6 shows some factors' subjective probabilities.

Let it be known, that on the given territory, factors e3 and e4 are present. Than using formula (4.1):

$$p\left(h1 \mid e4\right) = \frac{p\left(e4 \mid h1\right) \times p\left(h1\right)}{p\left(e4 \mid h1\right) \times p\left(h1\right) + p\left(e4 \mid \infty h1\right) \times p\left(\infty h1\right)} = \frac{0.4 \times 0.1}{0.4 \times 0.1 + 0.5 \times 0.9} = 0.08163; \text{p}$$

$$\left(h2 \mid e4\right) = \frac{p\left(e4 \mid h2\right) \times p\left(h2\right)}{p\left(e4 \mid h2\right) \times p\left(h2\right) + p\left(e4 \mid \infty h2\right) \times p\left(\infty h2\right)} = \frac{0.6 \times 0.1}{0.6 \times 0.1 + 0.5 \times 0.9} = 0.11765$$

After applying e4 factor to the prior probability of h1 and h2 hypothesizes, the posterior probability becomes prior probability for the next iteration of calculations. So, prior probabilities are basically re-calculated using each known factor:

$$p\left(h1 \mid e3\right) = \frac{p\left(e3 \mid h1\right) \times p\left(h1\right)}{p\left(e3 \mid h1\right) \times p\left(h1\right) + p\left(e3 \mid \infty h1\right) \times p\left(\infty h1\right)} = \frac{0.9 \times 0.08163}{0.9 \times 0.08163 + 0.5 \times \left(1 - 0.08163\right)}$$

=0.13793

$$p\left(h2 \mid e4\right) = \frac{p\left(e3 \mid h2\right) \times p\left(h2\right)}{p\left(e3 \mid h2\right) \times p\left(h2\right) + p\left(e3 \mid \infty h2\right) \times p\left(\infty h2\right)} = \frac{0.1 \times 0.11765}{0.1 \times 0.1 + 0.5 \times \left(1 - 0.11765\right)}$$

=0.02579

Table 6. Examples of factors' subjective probabilities

ID	Factor	p(e|h1)	p(e|~h1)	p(e|h2)	p(e|~h2)
e1	Roads	0	0.5	1	0.5
e2	Railroads	0	0.5	1	0.5
e3	Favorable geological conditions	0.9	0.5	0.1	0.5
e4	Pastures, hayfields	0.4	0.5	0.6	0.5
e5	Populated locality	0.2	0.5	0.8	0.5
e6	Forests, gardens	0.2	0.5	0.8	0.5

Such approach solves the problem of unavailable data – data is heterogeneous and incomplete, so conventional multiplicative model, normally applied to multivariate factor analysis is not applicable in the given circumstances.

As the result, the all of the aggregated results for each categories are multiplied in order to get a final evaluation.

FUTURE RESEARCH DIRECTIONS

As the future research direction the prospects for the working out of multi criteria decision support system for support decision making in the field of RES is considered. One of the main task of the future system is to assess the territory in terms of installing power generators. The system should support methods of heterogenius data generalization and capabilities of horizontal and vertical scaling.

CONCLUSION

The present analysis shows us developed methods for assessing theoretical and technical potential of RES. Estimations of geographical and economic potential are very similar although they vary for different territories in particular. They are based on the analysis of factors, techniques for a weighted impact assessment of these factors and assessment of some integrated territorial attractiveness. Further maps of favorable zones are created manually or in an automated mode depending on applied software. Examples of these systems for some type of energy and different areas are described. All this allows us to say that now there appear prerequisites for development of a software system performing assessment of capacity by stages and decision-making process support in the field of renewable energy sources exploitation. Operation of the system can be based on data collection and its detailed analysis with regard to identify accelerating factors; ese of impact evaluation methods for the mentioned factors and territorial assessment methods for suitability and necessity of RES. Visualization of results should be performed in GIS environment.

Kazakhstan is one of the richest countries in the world in terms of renewable energy resources. Favorable places for development of wind energy are located in the South (Jungar Gates and the Shelek corridor), in the Caspian region, as well as in the center of the Republic. Favorable conditions for development of solar energy exist in southern Kazakhstan (South Kazakhstan, Zhambyl, Almaty region), for hydropower engineering in eastern Kazakhstan (East Kazakhstan region), for bioenergy in South Kazakhstan, Almaty, and Kostanay regions.

The present assessment of the economic renewable energy potential is energy balance analysis of the regions and identification of those regions of Kazakhstan where energy consumption exceeds production. Energy deficient regions of Kazakhstan include Akmola, Aktobe, Dzhambul, Karaganda, Kostanay, Kyzylorda, South Kazakhstan. The greatest electricity shortage is observed in Kostanai (-3215 million kWh / year), Dzhambul regions (-2171) and Almaty (-3549).

For detailed analysis of the territory of the Republic of Kazakhstan, the main factors influencing assessment of the territory from the point of RES application have been determined. The main categories include: Geophysical, geographical and ecological factors, Economic factors, Technical factors, Social factors, Administrative/ legal factors and risks.

Each category counts approximately 5-10 factors, the overall number of currently taken into account factors is 65. Some of the data could be parsed from open sources, such as NASA SSE, RETScreen, OpenStreetMap datasets, WDPA datasets, etc.

Detailed analyses need to take into account local data sources. Factors differ in representation, scope of influence, RES type and technical parameters of generators.

There are numberical methods used to assess these factors. Calculation of those factors was performed with help of Analytic Hierarchy Process, Technique for Order Preference by Similarity to Ideal Solution, Benefits-Opportunities-Costs-Risk model. Analytical Hierarchy Process (AHP), developed by Thomas Saaty allows inferring weight of factors based on pairwise comparison of factors by experts in the field.

The weight of the factors should be aggregated in order to obtain an estimate of the territory from the point of application of various RES. We use Bayesian model to aggregate weights of factors. Bayesian model allows obtaining two assessments in form of probabilities. We can calculate probabilities of two hypotheses, where H1 is a possibility of installing a RES generator in the area and H2 is an impossibility of installing a RES generator.

Based on described approaches a prototype of multi-criteria decision support system is developing.

ACKNOWLEDGMENT

The paper was funded by grants No. 0168/GF4, BR05236839 of the Ministry of Education and Science of the Republic of Kazakhstan. The materials provided by Regina Kireeva in terms of RES resources of the Republic of Kazakhstan were used in this work

REFERENCES

Angelis-Dimakis, A., Biberacher, M., Dominguez, J., Fiorese, G., Gadocha, S., Gnansounou, E., ... Robba, M. (2011). Methods and tools to evaluate the availability of renewable energy souces. *Renewable & Sustainable Energy Reviews*, *15*(2), 1182–1200. doi:10.1016/j.rser.2010.09.049

Antonov, O. (2014). *Green Energy in Kazakhstan in 21st Century: Myths, Realities and Prospects.* Retrieved from http://www.kursiv.kz/news/details/kompanii/Zelenaya-energetika-Kazahstana-v-21-veke-mify-realnost

Arán Carrión, J., Espín Estrella, A., Aznar Dols, F., Zamorano Toro, M., Rodríguez, M., & Ramos Ridao, A. (2008). Environmental decision-support systems for evaluating the carrying capacity of land areas: Optimal site selection for grid-connected photovoltaic power plants. *Renewable & Sustainable Energy Reviews*, *12*(9), 2358–2380. doi:10.1016/j.rser.2007.06.011

Blaesser, G., Munro, D. (1995). *Guidelines for the assessment of PV plants. Document B: Analysis and presentation of monitoring data.* Joint Research Centre of the European Commission: Ispra, Varesse, Italy. CL-NA-16-339-EN-C

Bodrova, E., & Solomin, E. (2016). Overview of the problem of wind power plants utilization. *Proceedings of the Russian Scientific Conference with International Participation and the X Scientific Youth School. Renewable Energy Sources*, 16-21.

Díaz-Cuevas, P., Camarillo-Naranjo, J., & Pérez-Alcántara, J. (2018). Relational spatial database and multi-criteria decision methods for selecting optimum locations for photovoltaic power plants in the province of Seville (southern Spain). *Clean Technologies and Environmental Policy*, 1–14.

Dubrovsky, V., & Viestur, U. (1988). *Methane Fermentation of Agricultural Wastes*. Riga: Zinatne. (in Russian)

Earth System Research Labarotory. (2016). *The NCEP / NCAR Reanalysis Project*. Retrieved from http://www.cdc.noaa.gov/cdc/reanalysis

Energy Efficiency and Renewable Energy Consulting. (2016). *Bioenergetics*. Retrieved from http://www.energypartner.kz

Energy-efficient Russia. (2016). *Kazakhstan alternative*. Retrieved from http://energohelp.net

Energy Strategy of Russia for the period up to 2030. (2010). *Environmental Consulting*. Author.

EnergyPartner LLP. (2016). *Solar energy*. Retrieved from http://www.energypartner.kz

ERA Interim Reanalysis System. (n.d.). Retrieved from http://www.era

Farrell, E., Fuhrer, E., Ryan, D., Andersson, F., Huttl, R., & Piussi, P. (2000). European forest ecosystems: Building the future on the legacy of the past. *Forest Ecology and Management*, *132*(1), 5–20. doi:10.1016/S0378-1127(00)00375-3

Gorbunova, T., & Gorbunov, R. (2016). Study of resource opportunities and land-use restrictions on the use of solar energy in the example of the southeastern Crimea. *Proceedings of the Russian Scientific Conference with International Participation and the X Scientific Youth School.*

Grinevich, G. (1952). *Experience in developing elements of a small wind energy cadastre in Central Asia and Kazakhstan*. Publishing House of Academy of Sciences.

Hoesen, J., & Letendre, S. (2010). Evaluating potential renewable energy resources in Poultney, Vermont: A GIS-based approach to supporting rural community energy planning. *Renewable Energy*, *35*(9), 2114–2122.

International Energy Agency. (2014). *Energy Technology Perspectives. Executive Summary*. Retrieved from http://www.iea.org/publications/freepublications/publication/name-51003-en.html

Kiseleva, S., Ermolenko, G., & Popel, O. (2015). *Atlas of renewable energy resources in Russia*. Moscow: D.I. Mendeleyev RCTU.

Krasovsky, N. (1935). Wind power resources of the USSR and prospects of their use. Wind Power Resources, 1(3).

Kumar, A., Sah, B., Singh, A. R., Deng, Y., He, X., Kumar, P., & Bansal, R. C. (2017). A review of multi criteria decision making (MCDM) towards sustainable renewable energy development. *Renewable & Sustainable Energy Reviews*, *69*, 596–609. doi:10.1016/j.rser.2016.11.191

Lee, A. H. I., Chen, H. H., & Kang, H. Y. (2009). Multi-criteria decision making on strategic selection of wind farms. *Renewable Energy*, *34*(1), 120–126. doi:10.1016/j.renene.2008.04.013

Malczewski, J. (1999). *GIS and multicriteria decision analysis*. New York: John Wiley & Sons Inc.

MCPFE. (2002). *Improved Pan-European Indicators for Sustainable ForestManagement*. Paper presented at the MCPFE Expert Level Meeting, Vienna, Austria.

Muhamedyev, R. (2015). Machine learning methods: An overview. *Computer Modelling and Network Technologies, 19*, 14–29.

Muhamedyev, R., Kiseleva, S., Gopejenko, V.I., Amirgaliyev, Y., Muhamedyeva, E., Gopejenko, A.V., & Abdoldina, F. (n.d.). Visualization of the renewable energy resources. *Lecture Notes in Computer Science, 9768*, 218.

Mukhamediev, R., & Mustakaev, R. (2016). *Aggregation of heterogenius data in the decision support system for the use of renewable energy sources*. International Scientific Conference "Informatics and Applied Mathematics", Almaty, Kazakhstan.

Mustafina, R. (2010). The issue of energy security in Kazakhstan regions. *Vestnik PSU, 2*, 110–116.

Petersen. (2011). *The UNDP and the Government of Kazakhstan Project "Elimination of barriers to development of wind energy in Kazakhstan. Wind energy potential of Jungar gates and the Shelek corridor in Kazakhstan. Abstract*. Astana European Bank for Reconstruction and Development. Retrieved from http://ebrd.com

Petersen. (1999). *UNDP and the Government of the Republic of Kazakhstan project "Removing Barriers to Wind Energy Development in Kazakhstan. Wind energy potential of the Dzungar gate and Shelek corridor in Kazakhstan. Brief annotation.* National laboratory RISO.

Rafikova, Yu. (2015). *Geoinformational mapping of renewable energy resources (on the example of the South of Russia)* (PhD diss.). Lomonosov Moscow State University.

Recommendations for the determination of climatic characteristics of wind energy resources. GGO. NGO "Vetroen". (1989). Leningrad: Gidrometeoizdat.

Ruiz-Arias, J., Terrados, J., Pérez-Higueras, P., Pozo-Vázquez, D., & Almonacid, G. (2012). Assessment of the renewable energies potential for intensive electricity production in the province of Jaén, southern Spain. *Renewable & Sustainable Energy Reviews, 16*(5), 2994–3001. doi:10.1016/j.rser.2012.02.006

Saaty, T. (1989). *Group decision making and the AHP*. New York: Springer Verlag. doi:10.1007/978-3-642-50244-6_4

Sacchelli, S., De Meo, I., & Paletto, A. (2013). Bioenergy production and forest multifunctionality: A trade-off analysis using multiscale GIS model in a case study in Italy. *Applied Energy, 104*, 10–20. doi:10.1016/j.apenergy.2012.11.038

Sánchez-Lozano, J., Teruel-Solano, J., Soto-Elvira, P. L., & Socorro García-Cascales, M. (2013). Geographical Information Systems (GIS) and Multi-Criteria Decision Making (MCDM) methods for the evaluation of solar farms locations: Case study in south. *Renewable & Sustainable Energy Reviews, 24*, 544–556. doi:10.1016/j.rser.2013.03.019

Scharmer, K., & Greif, J. (2000). *The European solar radiation atlas.* Les Presses Mines ParisTech.

Sister, V. (2001). *Solid household waste.* Publisher of the ACH.

ADDITIONAL READING

Díaz-Cuevas, P., Camarillo-Naranjo, J. M., & Pérez-Alcántara, J. P. (2018). Relational spatial database and multi-criteria decision methods for selecting optimum locations for photovoltaic power plants in the province of Seville (southern Spain). *Clean Technologies and Environmental Policy*, 1–14.

Kazak, J., van Hoof, J., & Szewranski, S. (2017). Challenges in the wind turbines location process in Central Europe–The use of spatial decision support systems. *Renewable & Sustainable Energy Reviews*, *76*, 425–433. doi:10.1016/j.rser.2017.03.039

Latinopoulos, D., & Kechagia, K. (2015). A GIS-based multi-criteria evaluation for wind farm site selection. A regional scale application in Greece. *Renewable Energy*, *78*, 550–560. doi:10.1016/j.renene.2015.01.041

Rahman, M. M., Paatero, J. V., Lahdelma, R., & Wahid, M. A. (2016). Multicriteria-based decision aiding technique for assessing energy policy elements-demonstration to a case in Bangladesh. *Applied Energy*, *164*, 237–244. doi:10.1016/j.apenergy.2015.11.091

Vasileiou, M., Loukogeorgaki, E., & Vagiona, D. G. (2017). GIS-based multi-criteria decision analysis for site selection of hybrid offshore wind and wave energy systems in Greece. *Renewable & Sustainable Energy Reviews*, *73*, 745–757. doi:10.1016/j.rser.2017.01.161

KEY TERMS AND DEFINITIONS

Aggregation Methods: Method of heterogeneous data generalization to obtain single value that assess the suitability of territory.

Bayesian Aggregation Model: Method of generalization based on Bayes law, proposed in the work.

Decarbonization: Process of excluding carbon-based energy sources and using green, RES energy instead.

Inhibitors and Accelerators: Negative and positive factors respectively, affecting the target function.

MCDM: Multi-criteria decision making algorithms, used for solving multi-criteria aggregation problems. Includes AHP, multiplication method, PROMETHEE, and others.

RES Potential Evaluation: Multi-stage process to assess territories' suitability for renewable energy source generators installation.

Chapter 3
Electrical and Heat Power Production Using the Products of Air Conversion of Motor Diesel Fuel and Electrochemical Generator for Agricultural Consumers

Sergey Evgenevich Shcheklein
Ural Federal University, Russia

Alexei Mihailovich Dubinin
Ural Federal University, Russia

ABSTRACT

At present, the production of electricity for agricultural consumers remote from the centralized electrical power grid is carried out using diesel-generator technology with a limited service life of engines and extremely low efficiency of the expensive fuel used. In this chapter, an innovative technology has been considered for the combined electrical and heat power production using the preliminary conversion of diesel fuel into synthesis gas with its subsequent supply to a high temperature electrochemical generator (ECG). Synthesis gas for the operation of the electrochemical generator was produced by air conversion of motor diesel fuels in a catalytic burner reactor. On the basis of heat balances of the burner, ECG and waste-heat boiler-utilizer, electrical efficiency of the solid oxide fuel cells' (SOFC) battery, chemical efficiency of the burner, the temperature at the SOFC anode, the EMF of the planar cell, a portion of hydrogen oxidized at the SOFC anode, specific consumption of diesel fuel for the production of electrical and heat power were calculated.

DOI: 10.4018/978-1-5225-9179-5.ch003

INTRODUCTION

At present, the production of electric energy consumer remote from agro-based centralized networks is done using diesel-generator technology with limited service life of the engine and the extremely low efficiency in the use of expensive fuel. In this paper, is considered an innovative technology of combined production of electricity and heat using a preliminary conversion of diesel fuel in the synthesis gas and then serving it at high temperature electrochemical generator. Essential progress in considerate use of organic fuels with simultaneous simplification of the power plant design can be achieved by using technologies for the direct electrical power production with the help of electrochemical generators or their combination with GTP and STP (Shpil'rain et al., Malyshenko, & Kuleshov, 1984; Lisbona & Romeo, 2008; Promes et al., 2015).

Recently, the test study of the electrical power installation of the electrochemical generator based on the SOFC with the power of 5 kW were performed on the natural gas steam conversion products (Munts et al., 2015) and calculations were carried out on the products of steam gasification of coal of the Borodinskoye coal deposit (Dubinin & Shcheklein, 2015). This study presents a mini-CHPP, where direct conversion of the energy of the chemical reaction of hydrogen oxidation from synthesis gas obtained by air conversion of motor diesel fuel is carried out. This project allows prediction of the required ECG parameters: electrical efficiency of the SOFC battery, chemical efficiency of the burner, the temperature at the SOFC anode, EMF of the planar cell, a portion of hydrogen oxidized at the SOFC anode, specific consumption of diesel fuel for the production of electrical and heat power.

MODELLING OF PHYSICOCHEMICAL PROCESSES IN THE UNITS OF THE MINI-CHPP BASED ON THE CATALYTIC BURNER FOR AIR CONVERSION OF MOTOR DIESEL FUEL AND ELECTROCHEMICAL GENERATOR

Principle and Technological Model of the Process

In (Peters et al., 2013) more than 220 variants of the SOFC operation with recirculation of anode gases and without it have been calculated. It has been concluded that the choice of the scheme has little effect on the process efficiency, thus it is advisable to choose the simplest scheme. In the scheme proposed the anode gases recirculation is not considered. Planar fuel cells with direct flow of fuel and oxidizer are used. Catalysts at the anode and cathode are made of Ni-YSZ and $(ZrO_2)_{0.9}$ $(Sc_2O_3)_{0.1}$ is used as an electrolyte.

Figure 1 presents a schematic diagram of conversion of the motor diesel fuel into synthesis gas by the process of air conversion in the catalytic burner in order to ensure the operation of the electrochemical generator. The main units of the installation for the synthesis gas production are a catalytic burner for conversion of the diesel fuel vapor - 1, a blower - 3, a synthesis gas cooler - 4, an electrical customer - 5, an electrochemical generator based on SOFC - 6, a waste-heat boiler-utilizer (WHB) - 7, a diesel fuel filter - 9, a fuel pump - 10, a network heater – 12 and thermal insulation - 13.

Motor diesel fuel is supplied to WHB for heating to boiling and evaporation and enters the catalytic burner in a vaporized form. There also comes the air heated in WHB. When the air flow rate $\alpha = 0.4$ diesel conversion occurs with the formation of synthesis gas. Then the synthesis gas is cooled from 1284 °C to 800 °C in the cooler by the air supplied to the cathode channel.

Figure 1. Schematic diagram of mini-CHPP on products of air conversion of motor diesel fuel and electrochemical generator:1 - catalytic burner; 2 - air pipe; 3 - compressed-air blower; 4 - synthesis gas cooler; 5 – electrical customer; 6 - electrochemical generator based on SOFC; 7 - waste-heat boiler-utilizer; 8 - motor diesel fuel manifold; 9 - fuel filter; 10 - fuel pump; 11 – flow-sensing element; 12 – network heater; 13 - thermal insulation I- air input; II - motor diesel fuel input; III - output of combustion products; IV- electrical power output.

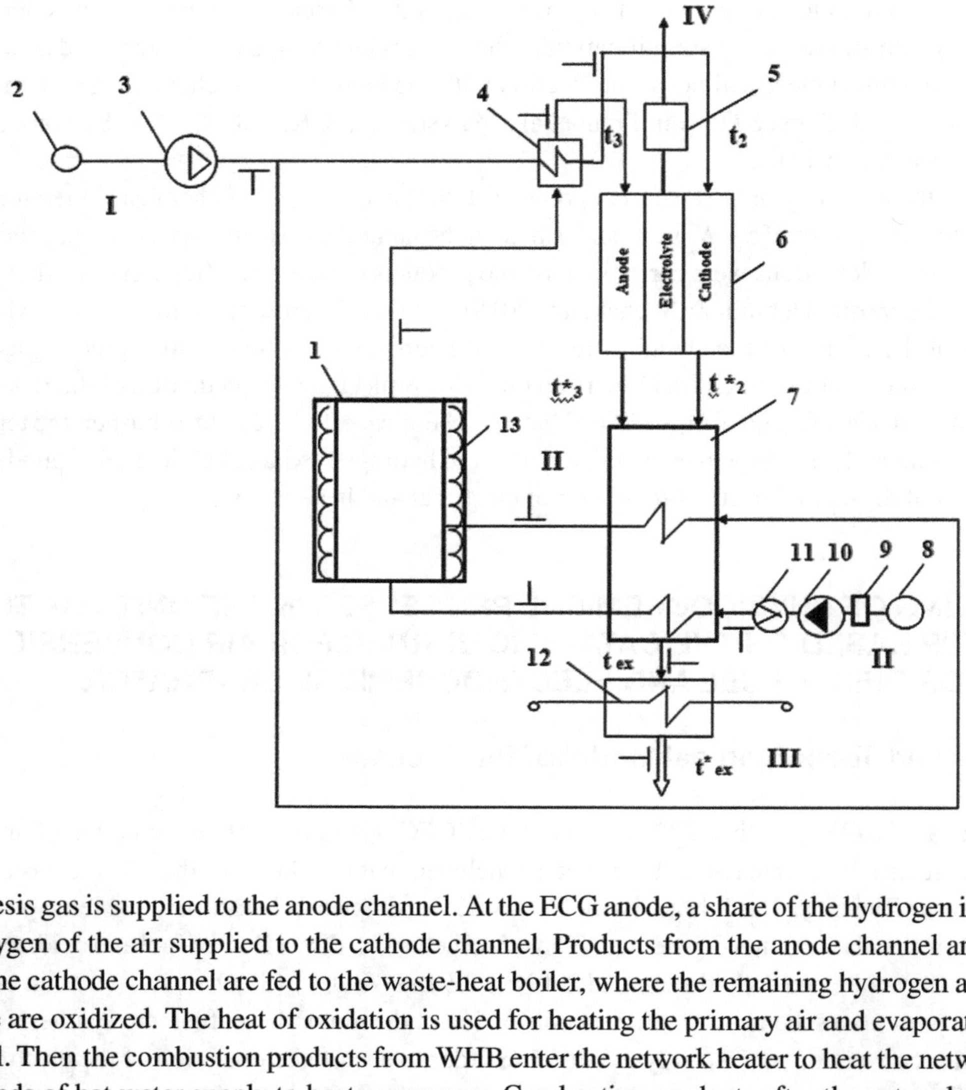

Synthesis gas is supplied to the anode channel. At the ECG anode, a share of the hydrogen is oxidized by the oxygen of the air supplied to the cathode channel. Products from the anode channel and vitiated air from the cathode channel are fed to the waste-heat boiler, where the remaining hydrogen and carbon monoxide are oxidized. The heat of oxidation is used for heating the primary air and evaporation of the diesel fuel. Then the combustion products from WHB enter the network heater to heat the network water for the needs of hot water supply to heat consumers. Combustion products after the network heater are removed through the stove chimney at a temperature of 120 °C.

Modelling of the Processes in the Technological Scheme Units

Modelling of the Processes in the Catalytic Burner

It is known (Orlin & Kruglov, 1983) that motor diesel fuel contains (in weight %) carbon - 87, hydrogen - 12.6 and oxygen - 0.5. For further calculations let us define the diesel fuel composition in atomic fractions of the elements contained in it (Orlin & Kruglov, 1983; Baskakov, 1974). It contains $n = 87/12 = 8$ of carbon, $2n + 2 = 18$ of hydrogen and $l = 0.5/16 = 0.03$ of oxygen. Since the oxygen content is low, it can be neglected for further calculations. In this case, the diesel fuel formula can be written as C_8H_{18}. The minimum coefficient of air consumption at which the emission of soot in the products of conversion is thermodynamically impossible:

$$\alpha_c = (n - 1) / (2n + 0.5m - 1). \quad (1)$$

We take -the coefficient of air flow $\alpha > \alpha_c$ ($\alpha = 0.4$). -when in the reaction volume of the catalytic burner the air flow rate is equal to 0.4, synthesis gas is produced according to the stoichiometric equation:

$$C_8H_{18} + 5(O_2 + 3.76N_2) = 7.938H_2 + 7.056CO + 0.089CO_2 + 1.062H_2O + 0.994CO_2 + 18.8N_2. \quad (2)$$

The composition of the synthesis gas produced (in volume %): H_2 – 22.1; CO – 19.68; H_2O – 2.96; CO_2 – 2.77; N_2 – 52.49.

The heat of combustion of the synthesis gas is determined according to the stoichiometric equation of complete oxidation of synthesis gas:

$$7.938H_2+7.056CO+1.062H_2O+0.994CO_2+18.8N_2+7.497O_2+28.18N_2 = 8CO_2+9H_2O+47N_2. \quad (3)$$

q_{chem} = - 4872 kJ/(kg of synthesis gas)

Complete combustion of the diesel fuel at the air flow rate factor α equal to 1 is described by the stoichiometric equation:

$$C_8H_{18} + 12.5(O_2 + 3.76N_2) = 8CO_2 + 9\,H_2O + 0.994CO_2 + 47N_2. \quad (4)$$

The number of kilomoles of air at the flow rate factor $\alpha = 0.4$ is equal to $12.5 \bullet 0.4 = 5$ (see Equation (1)). The combustion value of the diesel fuel is Q^p_{low} = 41800 kJ/(kg of diesel fuel).

The boiling temperature is t_1 = 380 °C.

According to preliminary calculations for the ECG to achieve the electrical power level of 10 kW the synthesis gas is required with the heat of combustion q_{chem} = 4872 kJ/(kg of synthesis gas).

$$B_{sg} = Q_e/(q_{chem}\eta^*), \quad (5)$$

where CE of the SOFC battery is tentatively assumed to be equal to $\eta^* = 0.386\left(38,6\%\right)$ according to the data presented in (Korovin, 2005).

For the production of B_{sg} it is necessary to introduce into the burner the following amount of diesel fuel, kg/s, given the 20% heat loss.

$$G_{df} = 1{,}2 \cdot B_{sg} \cdot 114/800 \quad (6)$$

and primary air, kg/s

$$G_{airl} = G_{df}\, 686.4/114. \tag{7}$$

Balance Equation for the Catalytic Burner

$$Q_{low}^{p}\, a114 / 800\left(1 - q_3 - q_5\right) + q_{air1} + q_v = C_c t \tag{8}$$

the left side of Equation (8) is the heat input from incomplete combustion with taking into account losses with chemical underburning and external heat losses through thermal insulation $Q_{low}^{p} \cdot a \cdot 114 / 800 \cdot \left(1 - q_3 - q_5\right)$, from primary air heated in WHB - q_{air1} and steam - q_v ; on the right side – heat consumption for heating the synthesis gas conversion products.

The coefficient 114/800, which converts $Q_{low}^{p}\, kJ / (\text{kg of diesel fuel})$ to kJ/(kg of synthesis gas), has been found from Equation (2).

The heat introduced by primary air into the burner heated to $t_1 = 380$ °C

$Q_{\text{air1}} = C_{\text{air}} \cdot t_1/800 = 686 \cdot 1.056 \cdot 380/800 = 344$ kJ/(kg of synthesis gas).

The heat introduced by vaporized diesel fuel

$q_v = C_v \cdot t_1 \cdot 114/800 = 3.5 \cdot 380 \cdot 114/380 \cdot 189$ kJ/(kg of synthesis gas),

where C_{air}, C_v, true specific isobaric heat capacities of air and vaporized diesel fuel at 380 °C, are equal to 1.056 and 3.5 kJ/(K· kg), respectively.

True specific isobaric heat capacity of synthesis gas $C_{sg} = 1.62$ kJ/(K· kg of synthesis gas).

From the heat balance equation, the temperature in the catalytic burner can be determined as:

$$t = [Q_{low}^{p}\, a\, \left(1 - q_3 - q_5\right) 114 / 800 + q_{air1} + q_v] / C_{sg} \tag{9}$$

Chemical efficiency of the catalytic burner can be written as:

$$\eta_{chem} = B_{sg} \cdot q_{chem} / (Gdt \cdot Q_{low}^{p}). \tag{10}$$

Synthetic Gas in the Electrochemical Generator: Physicochemical Processes

Synthesis gas cooled in the heat exchanger (4) down to $t_3 = 800$ °C is fed to the anode channel, while the air heated up to $t_2 = 620$ °C in the heat exchanger (4) is fed to the cathode channel of the E. According to (Korovin, 2005), contribution to the EMF value from the CO oxidation reaction athe SOFC anode is less than 1% of that from hydrogen oxidation; therefore, it is neglected in the calculations, and assumed that only hydrogen is oxidized at the SOFC anode.

The water-gas reaction $CO + H_2O \underset{<-}{->} CO_2 + H_2$ cannot be a donor of hydrogen, since this reaction proceeds only on the iron-chromium catalyst (Baskakov, Dubinin, & Tuponogov, 2008), which is absent in the anode channel.

At the SOFC anode, the oxidation of hydrogen from the synthesis gas by oxygen coming from the electrolyte is described by a stoichiometric equation assuming all the hydrogen to be completely oxidized:

$$7.938H_2+7.056CO+1.062H_2O+0.994CO_2+18.8N_2+3.969O_2=9H_2O+7.056CO+0.994CO_2+18.8N_2. \quad (11)$$

The exothermic effect of reaction (11) under standard conditions ΔH^0_1 is equal to 2400 kJ/(kg of synthesis gas), while the change of Gibbs free energy for the reaction under standard conditions $\Delta G^0 =$ -2863.59 kJ/(kg of synthesis gas).

According to (Baskakov, Dubinin, & Tuponogov, 2008), the temperature of the hydrogen oxidation products at the anode is

$$T_3^* = (\Delta H^0_1 - \Delta G^0) / \Delta S^0 \quad (12)$$

The temperature level is in accordance with (Munts et al., 2015). The EMF of a planar cell at the temperature of T_3^* is equal to

$$E_e = \Delta G/(2F), \quad (13)$$

where $F = 9.648 \cdot 10^4$ C/mol is the Faraday constant; ΔG kJ/(kmol of water) is determined at T_3^*($\Delta G = 190 \cdot 10^3$).

Fuel cells usually operate under isobaric-isothermal conditions. The work in the isobaric-isothermal process is equal to the negative change in of the free Gibbs energy (Baskakov & Volkova, 2013) and the electrical power is equal to the negative change in the free Gibbs energy ΔG^*, kJ/(kg of synthesis gas) at T_e^*, multiplied by the synthesis gas consumption B_{sg}, kg/s

$$Q_e = B_{sg} \Delta G^*, \quad (14)$$

where

$$\Delta G^* = 7,938 \cdot 190 \cdot 10^3 / 800 = 1884 \text{kJ} / (\text{kg of synthesis gas}).$$

On the other hand, the electrical power is (Munts et al., 2015):

$$Q_e = B_{sg} q_{chem} \eta \quad (15)$$

Equating these expressions, the electrical efficiency of the SOFC battery is determined and then compared with (Korovin, 2005; Zhao et al., 2011).

Heat Balance Equation for the Electrochemical Unit

The power released during the oxidation of hydrogen at the SOFC anode B_{sg} φ_T $\Delta H^0{}_1$, brought into by synthesis gas from the cooler (see Fig.), $B_{sg}C_{sg}$ t_3 and air coming from the boiler-utilizer G_rC_{air} t_2 is equal to the electrical power Q_e, removed from the ECG, by the synthesis gas, partly oxidized, outgoing from the anode channel $B^*{}_{sg}C^*{}_{sg}$ $t^*{}_3$ and air, outgoing from the cathode channel $G^*{}_{air}$ $C^*{}_{air}$ $t^*{}_2$, where φ_T is a portion of hydrogen oxidized at the anode.

$$B_{sg}\varphi_T\Delta H_1 + B_{sg}C_{sg}t_3 + G_{air}C_{air}t_2 = Q_e + C^*_{sg}B^*_{sg}t^*_3 + C^*_{air}G^*_{air}t^*_2. \tag{16}$$

The temperature t_3 at the inlet of the synthesis gas to the anode channel is assumed to be 800°C. The specific true isobaric heat capacity of the synthesis gas at 800°C is determined from Equation (2) as C^*_{sg} =.5 kJ/(K · kg of synthesis gas).

The air temperature at the outlet from the synthesis gas cooler (at the inlet to the cathode channel) t_2 is determined from the heat balance equation of the synthesis gas cooler and equal to 620°C. The specific true isobaric heat capacity of the air at the temperature of 620°C C_{air} = 1.115 kJ/K · kg (found from Equation (3)). The temperature t^*_3 is equal to 780°C (it was defined earlier, see Form. (12)).

The specific true isobaric heat capacity of the oxidized synthesis gas at the outlet from the anode at 7°C C^*_{sg} = 1.161 kJ/(K•kg of synthesis gas) is determined from equation (11). The air temperature at the outlet from the cathode channel is assumed to be by 100°C below t^*_3, i.e. 780 - 100 = 680°C.

The specific true isobaric heat capacity of the air depleted in oxygen at the outlet from the cathode channel is (7.497 - 3.969). O_2 +28.18N_2 at 680°C,C^*_{air} = 1.166 kJ/K · kg.

The air flow rate into the cathode channel necessary for the complete oxidation of synthesis gas can be calculated using Equation (3) as:

$$G_{air} = (240 + 789)\, B_{sg} / 800 = 1.285 \cdot 5.31 \cdot 10^{-3} = 6.83 \cdot 10^{-3}\text{kg} / \text{s}$$

T flow rate of vitiated air at the outlet of the cathode channel can be determined from Equation (3) and (11):

$$G^*_{air} = G_{air} - 3.969 \cdot 32 \cdot B_{sg} / 800 = (6.83 - 0.158 \cdot 5.31) \cdot 10^{-3} = 5.99 \cdot 10^{-3}\text{kg} / \text{s}$$

The consumption of synthesis gas at the outlet of the anode channel can also be determined from Eq. (11):

$$B^*_{sg} = \left(9 \cdot 18 + 7.656 \cdot 28 + 0.994 \cdot 44 + 18.8 \cdot 28\right) B_{sg} / 800 =$$
$$1.16 \cdot 5.31 \cdot 10^{-3} = 6.15 \cdot 10^{-3}\text{kg} / \text{s}$$

The materials balance at the ECG input and output has converged.

From the heat balance equation (16), a portion of hydrogen oxidized at the anode can be calculated (all the parameters included in the heat balance equation are presented in Table 1):

$$\varphi_T = \left(Q_e + C^*_{sg} B^*_{sg} t^*_3 + C^*_{air} G^*_{air} t^*_2 - B_{sg} C_{sg} t_3 - G_{air} C_{air} t_2\right) / B_{sg} \Delta H_1. \tag{17}$$

The last equation of the parameters presented in Table 1 gives the value of φ_T, which agrees with the data of (Baskakov, Volkova, Plotnikov, 2014).

Modelling of the Processes in the Waste-Heat Boiler-Utilizer: Physicochemical Processes

The rest hydrogen, not oxidized at the anode, is oxidized is WHB with power release:

$$(1 - \varphi_T) B_{sg} \Delta H_1^0 .$$

Carbon oxide, coming from the anode channel, is oxidized by oxygen-depleted air (7.497 - 3.969) O_2 + 28.18N_2, outgoing from the cathode channel, according to the stoichiometric equation:

$$9H_2O + 7.056CO + 0.994CO_2 + 18.82N_2 + 3.528O_2 + 28.18N_2 = 9H_2O + 8CO_2 + 47N_2 . \tag{18}$$

The exothermic effect of reaction (18) is $\Delta H_2^0 = -2472$ kJ/(kg of synthesis gas).

Power Balance for the Waste-Heat Boiler-Utilizer

The power coming with products from the anode channel $C^*_{sg} B^*_{sg} t^*_3$, with air from the cathode channel $C^*_{air} G^*_{air} t^*_2$, released from the oxidizing the rest hydrogen, non-oxidized at the anode $(1 - \varphi_T) \cdot B_{sg} \Delta H_1^0$. and carbon oxide $\Delta H_2^0 B_{sg}$, with primary air, inflowing to the catalytic burner $G_{air1} C_{air} t_{air}$, with diesel fuel $G_{df} C_{df} t_{df}$ is spent on heating the primary air $G_{air1} C_{air} t_1$, heating to boiling and evaporating the

Table 1. Values of the parameters in Equation (17)

Designation	B_{sg}	B^*_{sg}	C_{sg}	C^*_{sg}	C^*_{air}	C_{air}	G_{air}	G^*_{air}	t_2	t^*_2	t_3	t^*_3	ΔH_1	Q_e
Dimension	kg/s	kg/s	kJ/K•kg	kJ/K•kg	kJ/K•kg	kJ/K•kg	kg/s	kg/s	°C	°C	°C	°C	kJ/K•kg	kW
Value	5.31•10^{-3}	6.15•10^{-3}	1.5	1.61	1.166	1.115	6.83•10^{-3}	5.99•10^{-3}	620	680	800	780	2400	10

diesel fuel $G_{df}[C_{df}(t_1 - t_{df}) + r_{df})]$ and heating exhausted combustion products from the boiler-utilizer $G_{ex}C_{ex}t_{ex}$.

$$C^*_{sg}B^*_{sg}t^*_3 + C^*_{air}G^*_{air}t^*_2 + (1-\varphi_T)B_{sg}\Delta H^0_1 + \Delta H^0_2 B_{sg} =$$

$$= G_{air1}C_{air}(t_1 - t_{air}) + G_{df}[C_{df}(t_1 - t_{df}) + r_{df})] + \; G_{ex}C_{ex}t_{ex}. \tag{19}$$

For the heat balance equation, the temperature of the gases outgoing from the WHB can be determined (all the parameters in Equation (19) are presented in Table 2).

The specific isobaric heat capacity of diesel fuel is $C_{df} = 1.84\,\text{kJ} / \text{K} \cdot \text{kg}$.

$$G_{ex} = B^*_{sg} + G^*_{air} = B_{sg} + G_{air} = \left(6.15 + 5.99\right) \cdot 10^{-3} = \left(5.31 + 6.83\right) \cdot 10^{-3} = 12.14 \cdot 10^{-3}\text{kg} / \text{s}$$

r_{df} – the heat of evaporation of diesel fuel kJ / kg.

$$\begin{aligned} t_{ex} = {} & C^*_{sg}B^*_{sg}t^*_3 + C^*_{air}G^*_{air}t^*_2 + (1-\varphi_T)B_{sg}\Delta H^0_1 + \Delta H^0_2 B_{sg} - G_{air1}C_{air}(t_1 - t_{air}) - \\ & -G_{df}[C_{df}(t_1 - t_{df}) + r_{df})] / (G_{ex}C_{ex}) \end{aligned} \tag{20}$$

It is assumed that the SOFC battery and WHB is ideally insulated, the chemical underburning in WHB is equal to zero. The results of the calculations are summarized in Table 3.

Power supplied to the network heater to heat the network water for hot water needs of heat consumers can be calculated as

$$Q_{nh} = G_{ex}C_{ex}(t_{ex} - t^*_{ex}), \tag{21}$$

where t^*_{ex} is the temperature after the network heater, which is assumed to be equal to 120 °C.

Table 2. Values of the parameters in Equation (20)

Designation	B_{sg}	B^*_{sg}	C^*_{sg}	C^*_{air}	C_{air}	C_{ex}	G_{air1}	G^*_{air}	C_{df}	G_{df}	G_{ex}	t_1	t_{air}	t^*_2	t^*_3	t_{df}	ΔH_1	ΔH_2	φ_B	r_{df}
Dimension	kg/s	kg/s	kJ/ K•kg	kJ/ K•kg	kJ/ K•kg	kJ/ K•kg	kg/s	kg/s	kJ/ K•kg	kg/s	kg/s	°C	°C	°C	°C	°C	kJ/kg	kJ/kg	Rel. unit	kJ/ kg
Value	$5.31 \cdot 10^{-3}$	$6.15 \cdot 10^{-3}$	1.61	1.166	1.056	1.41	$5.467 \cdot 10^{-3}$	$4.62 \cdot 10^{-3}$	$1.84 \cdot 10^{-3}$	$0.906 \cdot 10^{-3}$	$12.14 \cdot 10^{-3}$	380	20	680	780	20	2400	2472	0,899	3454

Table 3. Results of calculations for the mini-CHPP typical operation

Parameter designation	Designation	Dimension	Equation number	Value
A flow rate at the sooting formation point	α_A	dim. less	(1)	0.32
Synthesis gas flow rate (specific consumption)	B_{sg}	kg/s	(5)	$5.31 \cdot 10^{-3}$
Diesel fuel flow rate (specific consumption)	G_{df}	kg/s	(6)	$0.906 \cdot 10^{-3}$
Primary air flow rate to the burner	G_{air1}	kg/s	(7)	$5.467 \cdot 10^{-3}$
Burner temperature	t	°C	(9)	1284
Burner chemical efficiency	η_{chem}	%	(10)	68.2
Temperature at the SOFC anode	T_3^* t_3^*	K °C	(12)	1053 (780)
EMF of the single planar cell	E_e	V	(13)	0.985
ECG electrical power	Q_e	kW	(14)	10
ECG electrical efficiency (gross)	η .	(%)	(15)	0.386
Volume fraction of hydrogen oxidized at the anode	φ_T	%	(17)	89.9
Temperature of gases outgoing from the boiler-utilizer	t_{ex}	°C	(20)	928
Network water heater capacity	Q_{nh}	kW	(21)	14.5
Coefficient of fuel utilization in CHPP	η_{ut}	dim. less	(22)	0.637
Absolute consumption of diesel fuel for the production of electrical power	G_e	kg/h	(23)	1.35
Specific consumption of diesel fuel for the production of electrical power	b_e	kg/kW•h kg r.f./kW•h	(24)	0.135 0.192
Absolute consumption of diesel fuel for the production of heat power	G_T	kg/h	(25)	1.91
Specific consumption of diesel fuel for the production of heat power	b_T	kg/GJ kg r.f./GJ	(26)	37.4 53.4
Difference in diesel fuel flow rates supplied to ECG and used beneficially in power plant	ΔG_{df}	kg/s	(27)	0
Reagents input flow rate	G_{in}	kg/s	(28)	$13.1 \cdot 10^{-3}$
Reagents output flow rate	G_{out}	kg/s	(29)	$12.14 \cdot 10^{-3}$

continued on following page

Table 3. Continued

Parameter designation	Designation	Dimension	Equation number	Value
Flow rate imbalance	ΔG	kg/s	(30)	$0.6 \cdot 10^{-3}$
Power input	Q_{in}	kg/s	(31)	25.83
Power output	Q_{out}	kg/s	(32)	25.88
Power imbalance	ΔQ	kg/s	(33)	0.05

Coefficient of fuel utilization in the mini-CHPP can be calculated as follows:

$$\eta_{ut} = \left(Q_e + Q_{sg}\right) / \, (Q^p_{low} G_{df}) \tag{22}$$

Energy Performance of the Mini-CHPP: Differentiation of the Diesel Fuel Consumption Between the Production of Electrical and Heat Energy and the Specific Fuel Consumption

Absolute consumption of diesel fuel for the production of electrical power, kg/h

$$G_e = Q_e \cdot 1 \cdot 3600 \, / \, (Q^p_{low} \, \eta_{ut}). \tag{23}$$

Specific consumption of diesel fuel for the production of electrical power, kg/kW·h

$$b_e = G_e \, / \left(Q_e \cdot 1\right). \tag{24}$$

Absolute consumption of diesel fuel for the production of heat power, kg/h

$$G_T = Q_{nh} \cdot 1 \cdot 3600 \, / \, (Q^p_{low} \, \eta_{ut}) \, . \tag{25}$$

Specific consumption of diesel fuel for the production of heat power, kg/kW·h

$$b_T = G_T \cdot 10^6 \, / \left(Q_{\mathrm{nh}} \cdot 1 \cdot 3600\right). \tag{26}$$

Imbalance of the diesel fuel utilization, kg/h

$$\Delta G_{df} = G_{df} - \left(G_e + G_T\right) / \, 3600 \, . \tag{27}$$

The levels of diesel fuel specific consumption correspond to a high-efficient heat power-station (Sokolov, 2001).

Material Balance of the Electric Power Plant

Input flow rate

$$G_{in} = G_{air} + G_{air1} + G_{df}. \tag{28}$$

Output flow rate

$$G_{ex} = B^{*}_{sg} + G^{*}_{air}. \tag{29}$$

Mass flow rate imbalance

$$\Delta G = G_{in} - G_{out}. \tag{30}$$

Heat Balance of the Electric Power Plant

Input power

$$Q_{in} = Q^{p}_{low} G_{df} \eta_{chem} \tag{31}$$

Output power

$$Q_{out} = Q_{e} + G_{ex} C_{ex} t_{ex}. \tag{32}$$

Power imbalance

$$\Delta Q = Q_{in} - Q_{out}. \tag{33}$$

INVESTIGATION OF THE INFLUENCE OF FUEL TYPE ON ENERGY INDICATORS OF AN ELECTROCHEMICAL GENERATOR AS A PART OF A COGENERATION UNIT

The use of other fuels is based on a process of prior converting them into synthesis gas. Show how to work on an example of the installation of ethanol.

Modelling of Processes in an Electrochemical Generator

As an example of the modelling of the power unit operation the process of direct conversion of ethanol into electrical and heat power has been considered in detail.

Ethanol C_2H_5OH with molecular weight of μ_{eth} has been taken as a fuel and air – as an oxidant.

The stoichiometric equation for complete oxidation of ethanol by air can be written as follows:

$$C_2H_5OH + 3.5(O_2+3.76N_2) = 2CO_2 + 3H_2O + 13.16N_2. \tag{34}$$

The heat of oxidation of ethanol:

$$Q^{p}_{low} = \left[M_{CO2}\left(-\Delta H_{CO2}\right) + M_{H2O}\left(-\Delta H_{H2O}\right) - M_{eth}\left(-\Delta H_{eth}\right)\right] / \mu_{eth} =$$

$$= \left[2\left(-393.8 \cdot 10^{3}\right) + 3\left(-242.02 \cdot 10^{3}\right) - 1\left(-235.46 \cdot 10^{3}\right)\right] / 46 = 27785 \text{kJ / kg of ethanol.}$$

The coefficient of air consumption, below which free carbon (soot) appears in the system, has been determined by the method described in (Baskakov, 1974):

$$\alpha_A = \left(n - l\right) / \left(0.5m + 2n - l\right),$$

where the atomic fractions of C, H and O in the formula of ethanol $n = 2$; $m = 6$; $l = 1$.

$$\alpha_A = \left(2 - 1\right) / \left(0.5 \cdot 6 + 2 \cdot 2 - 1\right) = 0.166.$$

The value of α for the subsequent calculations with a certain margin was taken equal to 0.2857. Therefore, the coefficient of air in the stoichiometric equation of partial oxidation of ethanol is:

$$M_{air} = 3.5 \cdot \alpha = 3.5 \cdot 0.2857 = 1.$$

The stoichiometric equation for ethanol partial oxidation at $\alpha = 0.2857$:

$$C_2H_5OH+(O_2+3.76N_2)=2CO+2H_2+H_2O+3.76N_2 \tag{35}$$

Weight of synthesis gas in Equation (35) can be calculated as:

$$G_{sg} = \mu_{eth} + \mu_{O2} + 3.76\mu_{N2} = 46 + 32 + 3.76 \cdot 28 = 183.28\, kg.$$

Number of kilomoles of synthesis gas in Equation (35):

$$\sum_{1}^{4} M_i = M_{CO2} + M_{H2} + M_{H2O} + M_{N2} = 2 + 2 + 1 + 3.76 = 8.76.$$

Density of synthesis gas under normal conditions:

$$\rho_{sg} = G_{sg} / \left(\sum_{1}^{4} M_i \cdot 22.4 \right) = 183.28 / \left(8.76 \cdot 22.4\right) = 0.934\,kg / m^3 .$$

Synthesis gas composition (vol. %): $r_{CO} = 22.83; r_{H_2} = 22.83; r_{H_2O} = 11,4; r_{N_2} = 42.94.$

The heat of oxidation of synthesis gas can be determined according to the formula, proposed by D.I. Mendeleev (Karapetyants, 1975):

$$q_{chem} = \left(124.6 r_{CO} + 108 r_{H_2}\right) / \rho_{sg} =$$
$$= \left(124.6 \cdot 22.83 + 108 \cdot 22.83\right) / 0.934 = 5685\,kJ / \left(kg\ \ of\ \ synthesis\ \ gas\right).$$

The contribution to the EMF value from the CO oxidation reaction at the SOFC anode is less than 1% of that from the H_2oxidation, and it is usually neglected in calculations. It is assumed that only the oxidation of H_2 takes place at the SOFC anode (Korovin, 2005; Vagraftic, 1956).

The stoichiometric equation of the oxidation reaction of H_2, containing in synthesis gas, at the SOFC anode can be written as follows:

$$2CO+2H_2+H_2O+3.76N_2+O_2=2CO+3H_2O+3.76N_2. \quad (36)$$

The heat of oxidation of H_2 at standard parameters (Dubinin, Shcheklein, Tuponogov, & Ershov, 2017):

$$\Delta H_1^0 = \left[M_{H_2O} \left(-\Delta H_{H_2O}^0\right) - M_{H_2O}^{*} \left(-\Delta H_{H_2O}^0\right) \right] / G_{sg} =$$

$$= [3\left(-242.02 \cdot 10^3\right) - 1(-242.02 \cdot 10^3)]/183.28 =$$

$$= -2641 \text{ kJ} / \left(\text{kg of synthesis gas}\right).$$

A change in free Gibbs energy for reaction (3) at standard parameters (Dubinin et al., 2017)

$$\Delta G^0 = \left[M_{H_2O}\left(-\Delta G^0_{H_2O}\right) - M^*_{H_2O}\left(-\Delta G^0_{H_2O}\right)\right] / G_{sg} =$$
$$= \left[3\left(-288.74\cdot 10^3\right) + 1\left(-288.74\cdot 10^3\right)\right] / 183.28 = -3151 kJ / kg \ of \ synthesis \ gas$$

An entropy change for reaction (3) at standard parameters (Dubinin et al., 2017).

$$\Delta S^0 = \left[M_{H_2O}\left(S^0_{H_2O}\right) - M_{H_2}\left(S^0_{H_2}\right) - M^*_{H_2O}\left(S^0_{H_2O}\right) - M_{O_2}\left(S^0_{O_2}\right)\right] / G_{sg} =$$
$$= \left[3\cdot 188.85 - 2\cdot 130.67 - 1\cdot 188.85 - 1\cdot 205.16\right] / 183.28 = \quad .$$
$$= -0.482 kJ / K\cdot kg \ of \ synthesis \ gas.$$

The temperature of products of the H_2 oxidation at the anode (Baskakov & Volkova, 2013).

$$T_1 = (\Delta H^0_1 - \Delta G^0)/(\Delta S^0) =$$

$$= \left(-2641 + 3151\right) / \left|-0.482\right| = 1058\,K \ (785^{\circ}\mathrm{C})$$

The change in free Gibbs energy at the temperature $T_1 = 1058\mathrm{K}\ \mathrm{kJ} / \left(\mathrm{kg\ of\ synthesis\ gas}\right)$.

$$\Delta G^{**}_{1058\,K} = \mathrm{M}_{\mathrm{H}_2}\Delta G_{1058\,K} / G_{\mathrm{sg}},$$

where $\Delta G_{1058\,K} = 189\cdot 10^3$ kJ/kmol of water at the temperature of $1058\,K$.

$$\Delta G^{**}_{1058\,K} = 2\cdot 189\cdot 10^3 / 183.28 = 2062\ \mathrm{kJ} / \left(\mathrm{kg\ of\ synthesis\ gas}\right),$$

where $M_{H_2} = 2$ – the stoichiometric coefficient of hydrogen in Equation (36).

The synthesis gas flow rate to obtain the electrical power of the SOFC battery Qe = 10 kW (Dubinin & Shckeklein, 2015):

$$B_{\mathrm{sg}} = \mathrm{Qe} / \Delta G^{**}_{1058\,K} = 10/2062 = 4.848\cdot 10^{-3}\,\mathrm{kg/s}.$$

To produce B_{sg} it is necessary to supply ethanol into the catalytic burner, with taking into account heat losses through the thermal insulation $q_5 = 0.2$, with the flow rate:

$$G_{\mathrm{eth}} = B_{\mathrm{sg}}\left(1 + q_5\right)\mu_{\mathrm{eth}} / G_{\mathrm{sg}} = 4.848\cdot 10^{-3}\cdot 1.2\cdot 46 / 183.28 =$$
$$= 1.459\cdot 10^{-3}\mathrm{kg/s}\ (5.252\mathrm{kg/h}).$$

The power released with exhaust gases from the boiler-utilizer to the network heater for heating water for hot water supply:

$$Q_2 = B_{sg} q_{chem} - Q_e - i'' G_{eth} = 4.848 \cdot 10^{-3} \cdot 5685 - 10 - 2308 \cdot 1.459 \cdot 10^{-3} = 14.2 kW,$$

where *i"* is the enthalpy of dry saturated vapour, kJ/kg.

The fuel utilization ratio (Baskakov, 1974):

$$\eta_{ut} = (Q_2 + Q_э) / (Q^p_{low} \cdot G_{eth}) = \left(14.2 + 10\right) / (1.459 \cdot 10^{-3} \cdot 27785\right) = 0.596\left(59.6\%\right).$$

The chemical efficiency of the catalytic burner:

$$\eta_{chem} = B_{sg} \cdot q_{chem} / G_{eth} Q^p_{low} = \left(4848 \cdot 10^{-3} \cdot 5685\right) / (1.459 \cdot 10^{-3} 27785) = \\ = 0.6798\left(67.98\%\right).$$

The electrical efficiency of the SOFC battery:

$$\eta^* = Q_e / B_{sg} \cdot q_{chem} = 10 / (4.848 \cdot 10^{-3} \cdot 5685) = 0.363\left(36.3\%\right).$$

The efficiency of the ECG (gross):

$$\eta_e = Q_e / (G_{eth} \cdot Q^p_{low}) = 10 / \left(1.459 \cdot 10^{-3} \cdot 27785\right) = 0.247\left(27.7\%\right).$$

The consumption of ethanol for the production of electrical power:

$$G_{eth} = Q_e \cdot 3600 / (\eta_{ut} \cdot Q^p_{low}) = 10 \cdot 3600 / (0.596 \cdot 27785) = 2.17 \text{kg} / \text{h}.$$

The consumption of ethanol for the production of heat power for the heat networks:

$$G_h = G_{eth} - G_e = 5.252 - 2.17 = 3.082 \text{kg} / \text{h}.$$

The specific consumption of ethanol for the production of 1 kWh of electricity:

$$b_e = G_e / Q_e = 2.17 / 10 = 0.217 \text{kg n.f.} / \text{kW} \cdot \text{h} (0.205 \text{kg r.f.} / \text{kW} \cdot \text{h}).$$

The specific consumption of ethanol for the production of 1 GJ of heat power for heat networks:

$$b_h = G_h \cdot 10^6 / (Q_2 \cdot 3600) = 3.082 \cdot 10^6 / \left(14.2 \cdot 3600\right) =$$
$$= 65.6\,\text{kg n.f.} / \text{GJ}(62.1\,\text{kg r.f} / \text{GJ}.$$

The analysis of the dependence of the energy indicators for other types of fuels mentioned above was carried out using the developed methodology. The results are presented in Table 4, from which it can be seen that hydrogen is the most preferred fuel in terms of energy indicators. Hydrogen is followed by methane, motor diesel, ethanol, motor gasoline and methanol. The fuel utilization ratio for these types of fuels differs by 2.1 times. According to the value of fuel utilization ratio as well as specific fuel consumption for generating electricity and heat for the purpose of heat supply to consumers, the most undesirable fuel for a power plant based on SOFC is methanol.

Discussion of the Results of Calculation Studies

The calculated values of the fuel utilization ratio when using methane as a fuel and air as an oxidant of 75.5% and the specific fuel consumption for generating electrical power of 0.162 kg r.f. /kWh and heat power of 45 kg r.f./GJ are in the range given in (Peters et al., 2013), which confirms the correctness of the calculation methodology proposed.

Figure 2 and Figure 3 a,b depict, respectively, the fuel utilization ratio and the specific fuel consumption for the production of electrical and heat power for the fuels studied.

Obviously, the higher the heat of oxidation of the fuel, the greater is the fuel utilization factor and the lower is the specific fuel consumption for the production of electricity and heat for heat supply purposes.

The overall level of fuel efficiency for a cogeneration-type power plant based on SOFC with hydrogen fuel and methane exceeds the level of modern combined-cycle power plants, and when using diesel, ethanol, gasoline, and methanol, it exceeds the level of cogeneration of thermal power plants based on internal combustion engines.

*Table 4. Dependence of energy indicators on the type of fuel and oxidant***

Nº	Fuel	Energy indicators					
		η_{ut} rel.u.	b_e $\frac{\text{kg r.f.}}{\text{kW} \cdot \text{h}}$	b_h $\frac{\text{kg r.f.}}{\text{G}}$	η_{chem} rel.u.	η^* rel.u.	η_e rel.u.
1	Hydrogen	1	0.122	34	-	0.785	0.785
2	Methane	0.755	0.162	45	0.756	0.49	0.374
3	Diesel	0.637	0.192	53.4	0.682	0.387	0.264
4	Ethanol	0.596	0.205	62.1	0.679	0.363	0.247
5	Gasoline	0.556	0.22	61.3	0.598	0.299	0.179
6	Methanol	0.359	0.475	83.7	0.447	0.786	0.352

(**) with all fuels, air was used as an oxidant.

Figure 2. Diagram of the fuel utilization ratio of various types of fuels

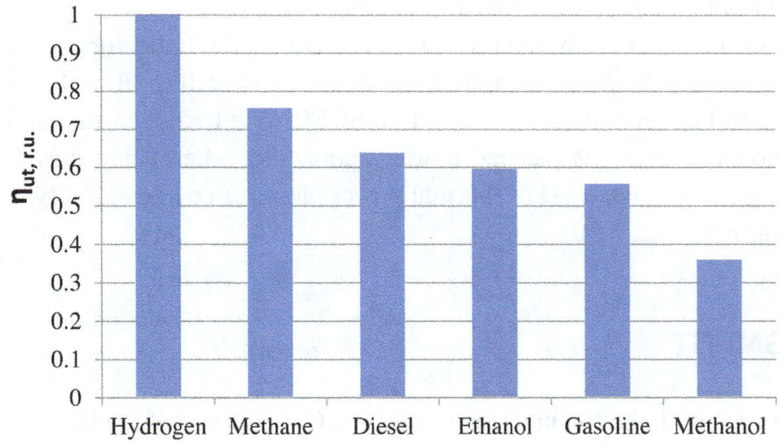

Figure 3. Diagram of the specific fuel consumption for the production of electricity (a) and heat for heat supply purposes (b) for various types of fuels

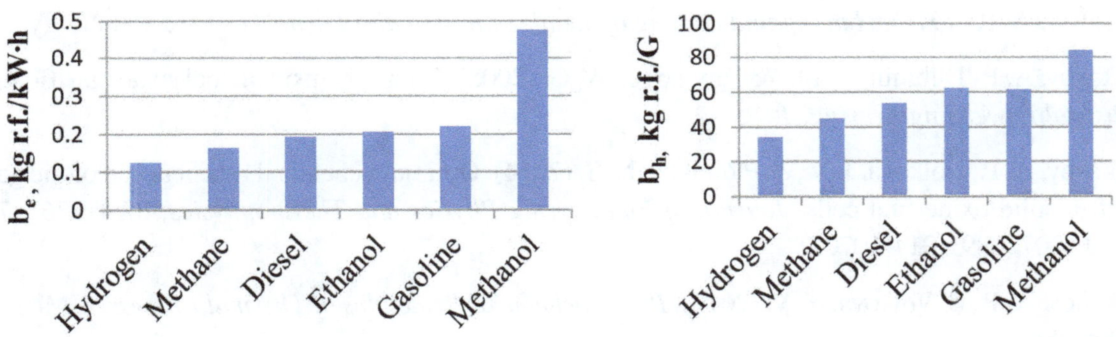

a **b**

CONCLUSION

1. The principal possibility for using the products of the motor diesel fuel air conversion for the operation of the SOFC-based electrochemical generator has been demonstrated.
2. Calculation studies have shown that the heat of combustion of the products of the diesel fuel air conversion (at $\alpha = 0.4$) is equal to 4872 kJ /(kg of synthesis gas), the fuel cell EMF is 0.985 V, the portion of hydrogen oxidized at the anode is 0.899, the temperature of hydrogen oxidation at the SOFC anode is 780 °C. The results of calculations carried out for the ECG are in reasonable agreement with the literature data obtained for other liquid hydrocarbon fuels.
3. Comparison with modern power sources has shown that both the specific consumption of diesel fuel for the production of electric power equal to 135 g/kWh (192 g r.f./kW·h) and heat power equal to 37.49 kg/GJ (54.3 kg r.f./GJ, 227.5 kg r.f./ Gcal) corresponds to a high-efficient heat power-station and more than 3 times lower than in modern diesel-electric stations of equal power

4. It is noted that thermal energy for heating and hot water supply in this cogeneration technology is an additional product that reduces overall consumer costs.
5. Investigations of the influence of the type of fuel on the energy indicators of electrochemical generators based on SOFC have revealed the following: The overall level of the fuel utilization ratio of a power plant of cogeneration type based on SOFC with hydrogen fuel and methane exceeds the level of modern steam-gas thermal power plants; and, when using diesel, ethanol, gasoline and methanol, it exceeds the level of thermal power plants of cogeneration type based on internal combustion engines.

ACKNOWLEDGMENT

The article was prepared with the financial support of the Government of the Russian Federation (Contract N°02.A03.21.0006).

REFERENCES

Baskakov, A. P. (1974). *Heating and cooling of metals in the fluidized bed*. Moscow: Metallurgy.

Baskakov, A. P., Dubinin, A. M., & Tuponogov, V. G. (2008). On mechanism of coal steam gasification. *Industrial Power Engineering*, *4*, 40–42.

Baskakov, A. P., Volkova, J. V., & Plotnikov, N. S. (2014). Optimum chemical regeneration of the gases burnt in solid oxide fuel cells. *Journal of Engineering Physics and Thermophysics*, *87*(4), 763–778. doi:10.100710891-014-1070-9

Baskakov, A. P., & Volkova, Y. V. (2013). *Physicochemical Principles of Thermal Processes*. Moscow: Teplotekhnik.

Dubinin, A. M., & Shcheklein, S. E. (2015). Coal mini CHPP plants on the basis of a gas generator and electrochemical generator. *Alternative Energy and Ecology*, *7-9*, 60–74.

Dubinin, A.M., Shcheklein, S.E., Tuponogov, V.G., & Ershov, M.I. (2017). Mini CHP based on the electrochemical generator and impeded fluidized bed reactor for methane steam reforming. *Alternative Energy and Ecology*, (19-21), 95-105.

Halinen, M., Thomann, O., & Kiviaho, J. (2012). Effect of anode off-gas recycling on reforming of natural gas for solid oxide fuel cell system. *Fuel Cells (Weinheim)*, *12*(5), 754–760. doi:10.1002/fuce.201200047

Karapetyants, M. K. (1975). *Chemical thermodynamics* (3rd ed.). Moscow: Khimiya.

Korovin, N. A. (2005). *Fuel Cells and Electrochemical Power Installations*. Moscow: MPEI.

Lisbona, P., & Romeo, L. M. (2008). Enhanced coal gasification heated by unmixed combustion integrated with an hybrid system of SOFC/GT. *International Journal of Hydrogen Energy*, *33*(20), 5755–5764. doi:10.1016/j.ijhydene.2008.06.031

Moscow, V. N. B. (Ed.). (1956). *Thermal-physical properties of substances*. Gosenergoizdat.

Munts, V. A., Volkova, Y. V., Plotnikov, N. S., Dubinin, A. M., Tuponogov, V. G., & Chernishev, V. A. (2015). Studying the characteristics of a 5 kW power installation on solid-oxide fuel cells with steam reforming of natural gas. *Thermal*, *62*(11), 779–784. doi:10.1134/S004060151511004X

Orlin, A. S., & Kruglov, M. G. (1983). *Internal Combustion Engines: Theory of Reciprocating and Compound Engines*. Moscow: Mashinostroenie.

Peters, R., Deja, R., Blum, L., Pennanen, J., Kibiaho, J., & Hakala, T. (2013). Analysis of solid oxide fuel cell system concepts with anode recycling. *International Journal of Hydrogen Energy*, *38*(16), 6809–6820. doi:10.1016/j.ijhydene.2013.03.110

Promes, E. J. O., Woudstra, T., Schoenmakers, L., Oldenbroek, V., Thattai, A. T., & Aravind, P. V. (2015). Thermodynamic evaluation and experimental validation of 253 MW Integrated Coal Gasification Combined Cycle power plant in Buggenum, Netherlands. *Applied Energy*, *155*, 181–194. doi:10.1016/j.apenergy.2015.05.006

Shpil'rain, E. E., Malyshenko, S. P., & Kuleshov, G. G. (1984). *An Introduction to Hydrogen Power Engineering*. Moscow: Energoatomizdat.

Sobyanin, V. A. (2003). High-temperature solid oxide fuel cells and methane conversion. *Russian Chemical Journal*, *47*(60), 62–70.

Sokolov, E. Ya. (2001). *Heating and heating networks*. Moscow: MPEI.

Yakovlev, B. V. (2008). Increase in efficiency of central heating and heat supply systems. Moscow: Novosti teplosnabzheniya.

Zhao, Y., Sadhukhan, J., Lanzini, A., Brandon, N., & Shah, N. (2011). Optimal integration strategies for a syngas fuelled SOFC and gas turbine hybrid. *Journal of Power Sources*, *196*(22), 9516–9527. doi:10.1016/j.jpowsour.2011.07.044

APPENDIX

Table 5. List of notations

Greek Alphabet Letters	
α	Excess air coefficient (flow rate)
η	Efficiency (Coefficient of efficiency)
η_{ut}	Coefficient of fuel utilization in the CHPP
φ_T	Volume fraction of hydrogen oxidized at the anode
Latin Alphabet Letters	
B	Synthesis gas flow rate (specific consumption)
b	Diesel fuel flow rate (specific consumption)
c	Specific true mass isobaric heat capacity kJ/K kg
F	Faraday constant
G	Specific consumption of reagents
ΔG^o	Change in Gibbs free energy under standard conditions for the hydrogen oxidation reaction, kJ/kg (kJ/kmol of water).
ΔH_1^0	Heat of oxidation of hydrogen in synthesis gas, kJ/kg of synthesis gas
ΔH_2^0	Heat of oxidation of carbon monoxide and methane in synthesis gas, kJ/kg of synthesis gas.
Q_e	Electrical power, kW
q_{chem}	Synthesis gas combustion heat, kJ/kg.
q_3	Heat loss with chemical underburning, kJ/kJ.
q_5	External heat loss, kJ/kJ.
r_{df}	Diesel fuel evaporation heat, kJ/kg.
ΔS^o	Change in entropy of the hydrogen oxidation reaction under standard conditions, kJ/ K kg.
T	Temperature, K.
t	Temperature, °C.
Units of Measurements	
kg n.f.	Kilogram of natural fuel
kg r.f.	Kilogram of reference fuel
kJ/K •kg	Kilojoule per Kelvin and kilogram
kg n.f./GJ	Kilogram of natural fuel per gigajoule

continued on following page

Table 5. Continued

Upper Indexes	
*	At the outlet of the anode and cathode channels.
o	Under standard conditions.
p	Per unit mass.
Lower Indexes	
air	Air
v	Vapor
in	Inlet of reagents
out	Outlet of reagents
low	Low
c	Combustion products
sg	Synthesis gas
nh	Network heater
ex	Exhausted gases
e	Electrical power
chem	Chemical
Abbreviations	
CE	Coefficient of efficiency
SG	Synthesis gas
WHB	Waste-heat boiler-utilizer
SOFC	Solid Oxide Fuel Cell
CHPP	Combined heat and power plant
EMF	Electromotive force
ECG	Electrochemical generator
GTP	Gas-turbine plant
STP	Steam-turbine plant

Chapter 4
Increasing the Energy Efficiency of Photovoltaic Systems Operating Under Conditions of Uneven Illumination:

Leonid Yuferev
Federal State Budgetary Scientific Institution, Russia

Pavel Nikolaevich Kuznetsov
Federal State Budgetary Scientific Institution, Russia

ABSTRACT

The presented research was carried out at existing solar power plants and renewable energy sources laboratories, whose purpose was to increase the energy efficiency of photovoltaic installations with parallel and mixed switching of photocells, operating under uneven illumination, parallel voltage arrays of photovoltaic modules due to voltage equalization. Experimental characteristics of photovoltaic installations with the developed device for coordinating the array of photoelectric modules (DCA), realizing a method of selection of electric energy and without its use are given. It is experimentally shown that the use of DCA increases the electrical power of the array with partial shading up to 2.6 times with partial shading. The results of the research can be used to design new photovoltaic installations and upgrade existing ones.

BACKGROUND

Wider implementation of solar power plants prevents a number of reasons, the main ones are: high cost, low efficiency, imperfection of energy storage technologies. Specialists from various countries, including Russia, are working to reduce the impact of the factors outlined. A great contribution in this direction was made by A.F. Ioffe, J.I. Alferov, D.S. Strebkov, G. Rauschenbach, M. Prince, A.P. Landsman, J. Loferensky, V.V. Kharchenko, V.A. Mayorov and others (Kuznetsov P.N., Avdeev D.S. 2017).

DOI: 10.4018/978-1-5225-9179-5.ch004

Figure 1 shows a graph showing the dynamics of the unit cost of solar installations and modules with silicon photocells, from which it can be seen that the cost reduction of both plants and modules is rapid due to their technological and technical improvement, which is confirmed by the values of the modules' efficiency made in different years (Renewable Energy Technologies: Cost Analysis Series. 2012).

One of the main trends in the development of photovoltaics in the world is the creation of new technologies for the manufacture of photovoltaic cells and modules, focused on reducing costs and increasing efficiency.

In the first half of the 20th century, the efficiency of photocells was no more than 1%. Currently, the average efficiency of the most common silicon elements is 14-18%. Elements made based on cascade heterostructures have an efficiency of up to 36.9%, and with the use of arsenide of gallium (GaAs) up to 47.5%. The theoretical efficiency of terrestrial cascade solar cells is 49% (Alferov Zh.I., Andreev V.M., & Rumyantsev V.D. 2004).

At the same time, it should be noted that the cost of such installations is still high, which calls for continuing work in this direction. Moreover, not only photovoltaic modules need to be improved, but also all the additional elements included in the installation - conversion devices, supporting structures, energy storage systems and many others that significantly affect its cost in general. The data of the International Renewable Energy Agency (IRENA) show that the cost of a solar installation can be up to four to five times the cost of the modules.

The improvement of additional elements of photovoltaic plants, in addition to reducing costs, should be aimed at improving the reliability and stability of energy output in the face of changing environmental parameters, or the effects of external factors (shading, pollution, etc.), and achieving maximum energy efficiency. To this end, in recent years a number of new devices and technologies have been created, the main of which are (Kitaeva, 2015; Gevorkian, 2016; Strebkov & Tver'yanovich, 2007):

Figure 1. Dynamics of unit cost of solar installations (blue) and modules with silicon photocells (orange)

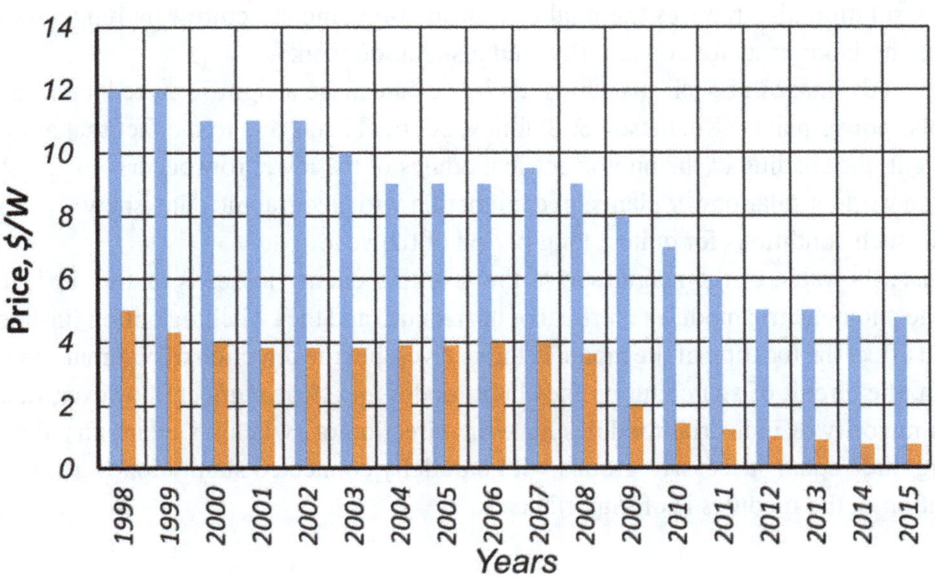

- Application of solar concentrators;
- The use of active systems for tracking the sun;
- The use of multi-row designs.

Solar concentrators allow increasing the intensity of radiation coming to photoelectric converters, increasing the amount of electric energy received from them. In addition, concentrated radiation increases the efficiency of conversion of solar radiation.

However, despite all the apparent attractiveness, the use of solar concentrators is not widely used in the industrial generation of electricity by photovoltaic installations. This is due to the fact that their use requires considerable complication of the design and its operation, an increase in the material consumption and the cost of manufacturing concentrators (especially those with complex geometry), the availability of additional means of heat removal, the need for a rather precise orientation to the sun, and so on. In addition, the most common silicon photocells are not designed for operation under conditions of concentrated radiation, which requires the use of significantly more expensive photoelectric converters.

The use of active solar tracking systems also allows the generation of more electrical energy from photovoltaic installations by providing more solar radiation to the surface of solar cells. In addition, such systems increase the daily interval of energy production.

The main disadvantages of using tracking systems are their high cost (in most cases, two or more times the cost of modules), the need for regular, expensive service (usually at least every two years), and a decrease in reliability. Also, the wide use of these systems, in the industrial generation of electrical energy by photoelectric converters, is hindered by the fact that their use leads to mutual shading, with relatively close transverse placement of installations. A consequence of this is a significant reduction in the output of the entire solar power plant, as will be shown in the following sections. Rare station location is also an inefficient solution due to reduced specific power, which leads to increased area and additional electrical losses.

Another way to increase the efficiency of photovoltaic systems is to increase their specific power. For this, multi-row designs are used, which allow placing photoelectric modules in several rows on one support. This solution also reduces the total cost of load-bearing structures, by reducing the material consumption and labor costs for construction and installation work.

Despite the advantages of such structures, their use can cause a significant reduction in the energy output of solar power plants (Kuznetsov & Safonov, 2016). This is due to the fact that at low values of the sun's height, the shading of the photoelectric modules of the lower row occurs (Figure 2). Research projects of the world's solar power plants, in conjunction with geographic data, show that about 70% of them work in such conditions for quite a long period of the year.

In this case, the value of the magnitude of the resulting energy losses is mainly influenced by the location of the photoelectric modules (vertical or horizontal) and the switching option (in-line, or block). This does not affect the location of the modules themselves, but the direction of commutation of the solar cells in them, the circuit of switching on the shunt diodes and their number. However, due to the fact that the vast majority of industrial modules have the direction of switching solar cells along the larger side, forming three equal successive groups with parallelly connected shunt diodes, it can be assumed that the location of the modules is of importance.

Figure 2. Shading of photovoltaic modules in solar power plants

The effect of the location of the modules on the value of energy losses is because with partial shading of any group of solar cells, it is shunted by a diode. Thus, with the horizontal arrangement of the modules, in the case of shading of the elements of the lower rows, there is no shunting of the remaining (unshaded) groups of modules (Figure 3b), and for the vertical - all groups are shunted (shunted), since each of them includes shaded solar elements of the lower row (Figure 3a). However, it should be noted that not all photovoltaic module manufacturers allow their mounting in the horizontal position.

Different options for switching photovoltaic modules also significantly affect the value of energy losses. The main switching options used in industrial solar power plants are "block" and "line" (Figure 4). In case of block switching, fewer wires are used than in case of in-line switching, which positively affects the efficiency of electric power transmission from modules to inverters. However, in the case of shading, all arrays of modules (except the first row) operate under partial shading conditions. Moreover, the input voltage range of the inverter can be higher than the voltage range near the maximum power point of such arrays, as a result of which the selection of energy from them will be performed in a non-

Figure 3. Disconnection of photovoltaic module solar cells by shunt diodes
a - for block switching, b - for line switching

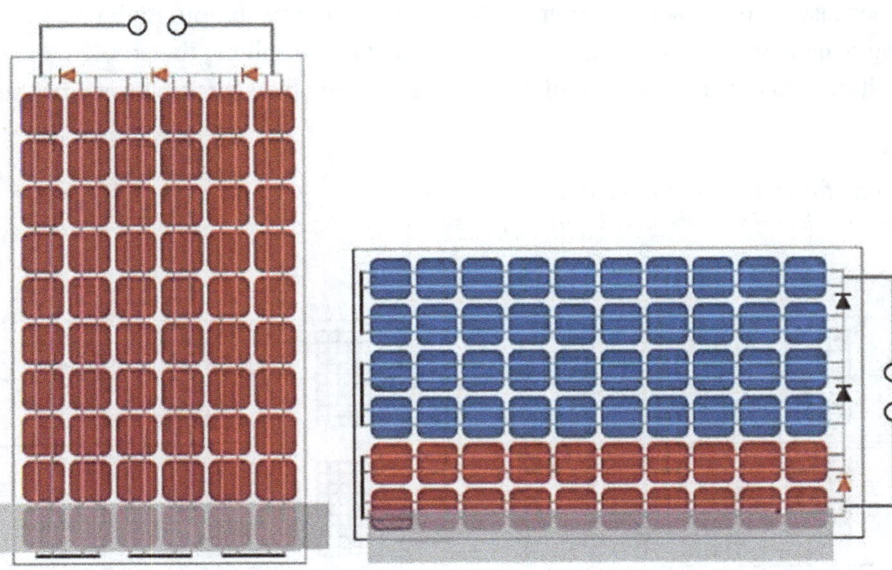

optimal mode. An example of such a case is the reduction in the power output of the Sevastopol solar power plant, where the change in the switching of modules from the block vertical to the line-type allowed to increase the daily power generation of the station in winter time by more than 2 times (Kuznetsov & Safonov, 2016). The use of in-line switching avoids the operation of the upper array of modules under partial shading conditions, increasing the efficiency of the entire station, while the lower (shaded) arrays are significantly limited in power transmission.

An analysis of these methods of increasing efficiency shows that their use is mainly hampered by problems associated with uneven illumination of solar cells. In this regard, the solution of this problem is an important task for the further development of industrial photovoltaics.

The existing methods of selecting electrical energy from solar photovoltaic installation operating under conditions of uneven illumination are in some cases not effective or economically justified solutions. Shunting diodes disconnect (shunt) groups solar cell that have partial shading and lead to a voltage mismatch in the maximum power point (MPP) parallel array of photoelectric modules. Individual matching converters significantly increase the cost of installations and also lead to voltage mismatch in the parallel array of modules. Micro-inverters significantly increase the cost of installation, reduce their efficiency in conditions of uniform illumination and cannot be used on existing solar power plants with a central inverter. Tunable switching systems are an expensive and inefficient solution with a small number of shadowed modules. Differential systems for selecting maximum power are also an expensive, energy-inefficient solution.

The analysis shows that currently there is a sufficiently large number of technical solutions that allow to partially level the effect of uneven illumination on the energy production of photovoltaic modules. Moreover, each of them has its own peculiarities that limit their widespread use in solar power plants that have a wide variety of design solutions, including various versions of supporting structures, switching solar cells, orientation systems, lighting patterns, etc. In this regard, to evaluate the effectiveness of their implementation, it is necessary to create mathematical models, the result of which are the values of the energy losses of solar installations with various options for switching solar cells and methods of selecting electrical energy.

At the same time, some of the methods initially do not seem promising for implementation in industrial power plants because of the practical complexity of their implementation, high cost or inability to be built in existing plants. Such methods include methods that allow solving the problem of voltage matching of photovoltaic array arrays with parallel commutation-tunable switching systems, micro invertors

Figure 4. Options for switching photovoltaic modules
a - row; b – block; (1 - photoelectric module, 2 - array of modules)

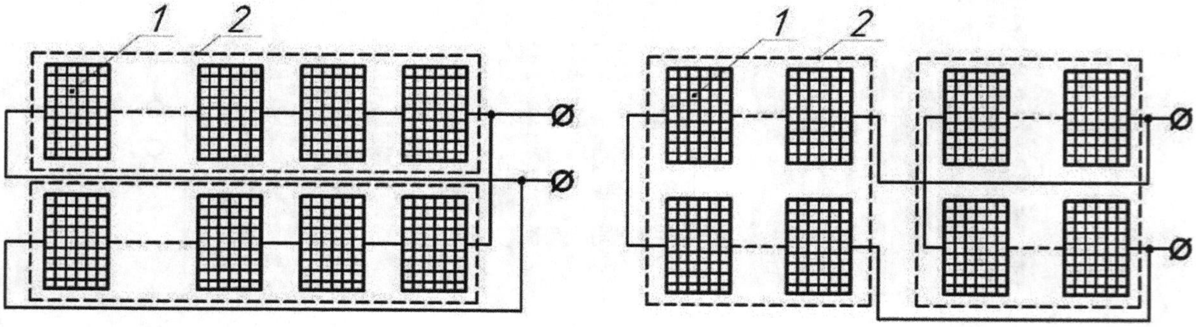

and differentiated systems for selecting maximum power. Because of this, one of the most urgent tasks is the creation of an economically justified method that allows the selection of electrical energy from parallel groups of photovoltaic modules.

SOLUTIONS AND RECOMMENDATIONS

Thus, at the present time, devices capable of solving the voltage matching problem for parallel array of photoelectric modules are not represented. The existing methods of harmonization do not appear to be expedient due to low efficiency and high cost, as evidenced by their use only in special installations used in the space industry (Chernyshev, Yu, Shinyakov, Gordeev, 1991; Kudryashov, Elman, Nesterishin, 2009).

To solve the problem of voltage matching of parallel arrays of photoelectric modules, the author of the article invented the method of " Receiving of electric energy from batteries of the photovoltaic converters" (Kuznetsov & Borisov, 2017) and its practical implementation in the form of "Device for coordinating the array of photoelectric modules" (DCA). The technical result of the method is to increase the efficiency of electric energy extraction from arrays connected in parallel with one or several other arrays, with a different nominal voltage in the maximum power point (MPP), or by shunting the part of the modules with diodes due to shading, contamination, and failure. Particular attention in the development of the device is given to its technical and economic component, since the issue of payback period is one of the most important for its implementation on existing ground-based solar power plants. And the ways to reduce the cost of the device were not related to the simplification of its design, or the use of inexpensive less qualitative materials, but in the application of more technological, modern engineering and scientific solutions.

The essence of the method is explained in Figure 5 and consists in the fact that for the selection of the maximum power from the arrays of photoelectric modules (1), the array is reconciled by including an additional power element (2) in each of them (in need of coordination). The voltage of each additional power element is regulated by means of an microcontroller device (5), up to a value at which the total power of the photovoltaic installation is maximized. Electric energy in the additional power element is supplied from the same photovoltaic cells, providing galvanic isolation, or from any other source of electrical energy. At the same time, the point of maximum power of the photoelectric module arrays is monitored and optimized using the maximum power take-off (3). After that, the electrical energy is transferred directly to the load (4), or through a converter, or accumulated.

The essence of the proposed method is also explained in Fig. 6, which depicts the current-voltage characteristic of arrays of photovoltaic modules with a large nominal voltage (all converters are included in the work) and an array with a lower rated voltage, or having a shaded, contaminated, faulty one or several solar cells (some of the elements are shunted by diodes). It can be seen from the graph that the maximum power points of the photoelectric module arrays (*P1* and *P3*) having different nominal voltages do not coincide. In this regard, the selection of electrical energy from the array having a lower rated voltage is produced at the non-optimal point *P2*, as a result, it does not produce the maximum electrical power, as shown in the second section.

In order to select the electrical energy at the maximum power point (P_3) from the array of solar cells with a lower rated voltage, it is proposed to include in it an additional battery element having a nominal voltage equal to ΔU and capable of producing a current not less than the current at the maximum

Figure 5. Structural diagram of the method for selecting maximum power from arrays of photovoltaic modules

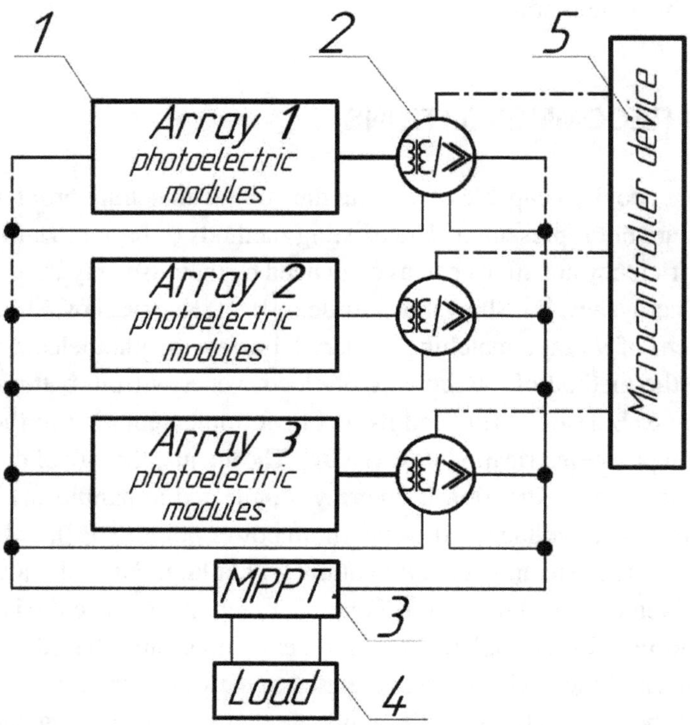

power point of the array not having shading I_1. At the same time, the value of its electrical power will be (1):

$$P' = \Delta U \cdot I_1, \tag{1}$$

where *P'* – power of the additional power element (W); ΔU – voltage of the additional power element, (V); I_1 – current at the MPP of the array without shading, (A).

Figure 14 shows that the area of the figure bounded by the points *U1, U2, P3, P1* is smaller than the figures *P1, P2, I2, I1*, because the voltage *U1* is much larger than ΔU, and the current I_1 is much larger than I_2. As a consequence, the electrical power of the supplementary battery *P'* is less than the power ΔP lost in the array with a lower rated voltage. From this it follows that when selecting electric energy in this way it is possible to get more of it in the case of shunting the part of modules by diodes due to their shading, pollution, failure of one or more solar cells, or using arrays of photovoltaic modules with different nominal voltages.

On the basis of the developed method, the functional scheme of the DCA, shown in Figure 7.

The principle of operation is as follows: the electric energy from the array of photovoltaic modules passes through the series-connected output of an adjustable DC-DC converter (up or down depending on the operating mode set by the contactor groups s1 and s2) and the current sensor (I_{count}). Information from the current sensor is fed to the analog-digital input of an microcontroller device (MCD) in which

Figure 6. Volt-ampere characteristics of arrays of photovoltaic modules with greater and lesser-rated voltage

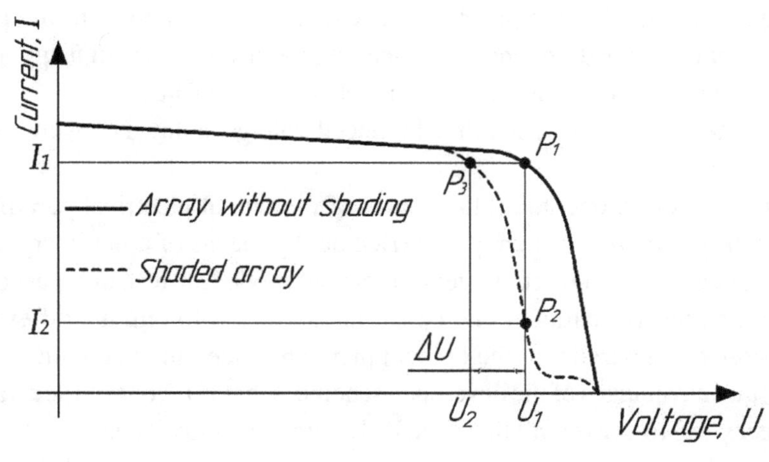

Figure 7. Functional diagram of the DCA; 1 - high-frequency converter with galvanic isolation; 2 - DC/DC buck (power); 3 - DC/DC boost (power); 4 - control device for DC/DC converters; 5 - display; 6 - microcontroller device; 7 - relay; 8 - DC/DC buck (low-current); 9 - voltage sensor 10 - current sensor.

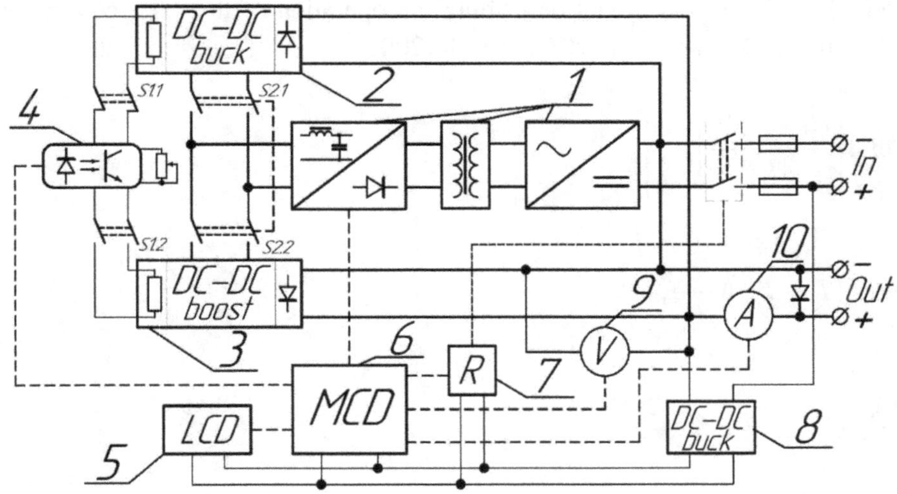

the value of the array current is processed and recorded. Then, the MCD, through the optocoupler, sends a control signal to the DC-DC converter to increase its output voltage by one quantum. In this case, the MCD monitors the change in the current value. If an increase in current occurs, it again sends a control signal to the DC-DC converter to increase the voltage by one more quantum. This sequence of operations is repeated until the increase in the DC-DC voltage of the converter ceases to cause an increase in the amperage of the array of photovoltaic modules, or the DC-DC voltage of the converter reaches its maximum value. After setting the optimal mode of operation, the device enters the monitoring mode, during which the parameters of the array are monitored according to the current values. In the event of a change in these parameters, the voltage of the DC-DC converter is reset and the adjustment starts again.

Since the electrical energy in the DC-DC converter comes from the same array, it is necessary to provide a galvanic isolation at its input. For this, a high-frequency voltage converter with a pulse transformer is used. This element consists of a computer unit that sends control signals to the inverter driver and provides short-circuit and overload protection by opening the inverter circuit using an electromagnetic relay (R), a high-frequency pulse transformer and a rectifier with a filter.

In order to provide the MCD with a stabilized reduced voltage, a down converter of a constant voltage is provided in the circuit.

The indication of the device operation, the values of the output electrical parameters, the selected operating mode and the operating efficiency is carried out by means of a liquid crystal display (LCD).

To regulate the voltage of DC-DC converters in manual mode (for research purposes), or to use the device in the case of parallel switching of photoelectric modules with initially different voltage values, a potentiometer R_v connected parallel to the optocoupler is provided in the circuit.

One of the important advantages of the developed scheme is that to reconcile the arrays of photovoltaic modules, it is not necessary to convert all the electrical energy, but only its parts and the fact that all converters have a sufficiently high efficiency, because they use the principles of high-frequency conversion.

For a theoretical analysis of the processes occurring in photovoltaic system with different variants of commutation of solar cells operating in conditions of uneven illumination, it is necessary to compile a mathematical model of the solar cell, with accuracy sufficient for engineering calculations - not exceeding the limits of constructive tolerance.

Thus, the initial mathematical model describing the operation of the solar cell model chosen (2) (Tian, H., 2012, Petreuş, D., Fărcaş, C., & Ciocan, I. (2008):

$$I = I_{ph} - I_0 \left[\exp\left(\frac{q(U + IR_s)}{AkT} \right) - 1 \right],$$

$$I_D = \left[I_{sc(stc)} + K_i \left(T - T_{stc} \right) \right] \frac{\beta}{\beta_{stc}},$$

$$T = T_{at} + K_t \beta,$$

$$I_0 = \frac{I_D}{\exp\left[\frac{q\left[U_{oc} + K_u \left(T - T_{stc} \right) \right]}{AkT} \right] - 1},$$

where I - load current, (A); I_{ph} - photocurrent, (A); I_d - the current flowing through the diode, (A); I_0 - reverse saturation current, (A); q - is the electron charge, ($1.602 \cdot 10^{-19}$ C); U - output voltage, (V); k - the Boltzmann constant, (1.381 · 10-23 J/K); T - the absolute temperature of the solar cell, (K); R_s - series resistance of the solar cell, (Ohm); A - coefficient of ideality, depending on the thickness of the p-n-junction and the material, takes values for silicon PEC from 1.2 to 5; $I_{sc(stc)}$ - the SC current in standard test conditions (STC); K_u – temperature coefficient of no-load voltage, (V / K); T_{at} – ambient temperature, (K); U_{oc} - open circuit voltage, (V). K_t – temperature coefficient of temperature increase, ((K·m2)/W); β - intensity of solar radiation, (W/m^2); β_{stc} - intensity of solar radiation at stc, (W/m^2).

Parallel switching of the photovoltaic cell (PEC) into a solar photoelectric device shown in Fig. 8:

The current-voltage characteristic of the uniformly illuminated battery in the first quad-welt, with an accuracy sufficient for engineering calculations described analytically by the equation (3). This equation is derived according to the Kirchhoff law and the Shockley equation written for a "simplified" equivalent circuit for replacing the PEC, taking into account that the current at the output of the unit will be equal to the sum of the currents of each PEC ($I = \sum_{i=1}^{n} I_i$) and the voltage will be equal to the voltage at the output of each PEC ($U = U_1 = U_2 = \ldots = U_n = const$):

$$I = n_p \left[I_{ph} - I_0 \left[\exp\left(\frac{q(U + IR_s)}{AkT} \right) - 1 \right] \right],$$

where n_p is the number of parallel connected PEC in the installation.

It can be assumed that in case of uneven illumination of the PEC with the same temperature, the maximum power of the installation will decrease by the amount of decrease in the maximum power of the individual PECs, due to their shading. However, this happens somewhat differently because the voltage of the PEC at the points of maximum power (PMP1 and PMP2) is not the same for different values of the illumination, as shown in Figure 9:

Due to the different voltage, it is not possible to obtain electrical energy from the installation at the maximum power point of each PEC. The value of this difference can be found from equation (4), solving it with respect to the voltage for each of the points:

$$\Delta U = U_{max1} - U_{max2} =$$

Figure 8. Installation with parallel solar cells in the battery (1 - PEC; 2 - solar photovoltaic installation; 3 - the device for finding the maximum power point and optimizing the output power (MPPT))

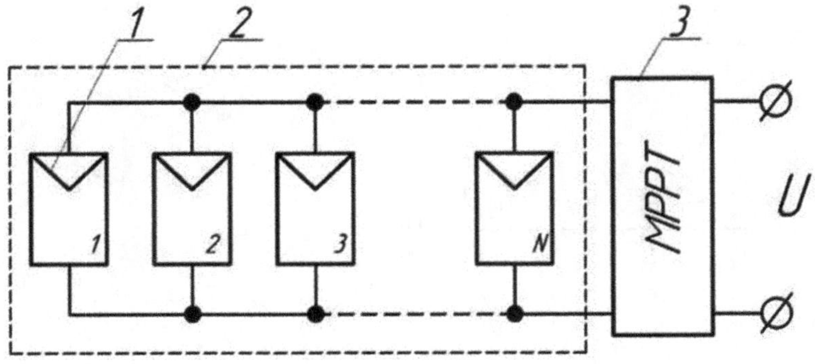

Figure 9. Current-voltage (a) and power (b) characteristics at different illumination PEC

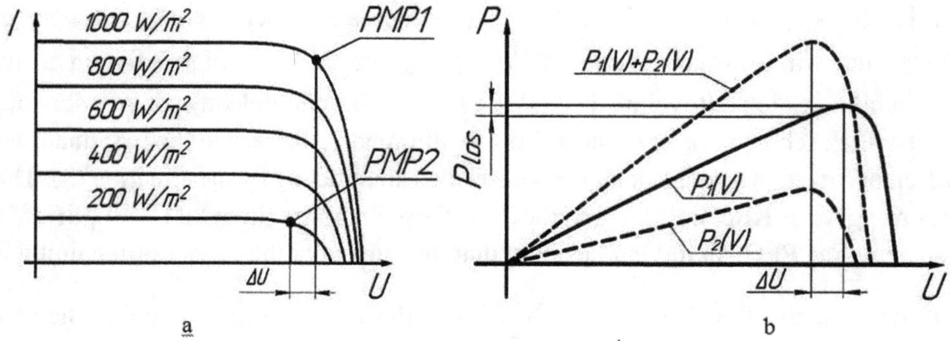

$$= \frac{AkT}{q}\ln\left(\frac{I_{ph1}}{I_{01}} - \frac{I_{max1}}{I_{01}} + 1\right) - I_{max1}R_s - \frac{AkT}{q}\ln\left(\frac{I_{ph2}}{I_{02}} - \frac{I_{max2}}{I_{02}} + 1\right) - I_{max2}R_s \approx$$

$$\approx \frac{AkT}{q}\ln\left(\frac{\left(I_{ph1} - I_{max1}\right)I_{02}}{\left(I_{ph2} - I_{max2}\right)I_{01}}\right) - R_s\left(I_{max1} + I_{max2}\right),$$

where U_{max1}, U_{max2} - the voltage at the point of maximum power of PEC1 and PEC2, (B); I_{ph1}, I_{ph2} - photocurrent of PEC1 and PEC2, (A); $I_{max1,}$ I_{max2} - load currents PEC1 and PEC2 at the points of maximum power, (A); I_{01}, I_{02} - reverse saturation currents of PEC1 and PEC2, (A).

An analysis of the characteristic of such a facility shows that the voltage at the absolute maximum point (AMP) will with some assumption correspond to the voltage at the point of maximum power of the shaded PEC in the range of the radiation intensity divergences from 10 to 100%. From this it follows that in such a case, PEC having a capacity greater brightness value will be limited to less illuminated PEC. Then the power losses will be (5):

$$P_{loss} = P_{max1} - P_1 = I_{max1}U_{max1} - I_1U_{max2} =$$

$$= \left[I_{ph1} - I_{01}\left[\exp\left(\frac{q\left(U_{\max 1} + I_{\max 1}R_s\right)}{AkT}\right) - 1\right]\right]U_{\max 1} - \quad (5)$$

$$-\left[I_{ph1} - I_{01}\left[\exp\left(\frac{q\left(\left(U_{\max 1} - \Delta U\right) + I_{\max 1}R_s\right)}{AkT}\right) - 1\right]\right]U_{\max 2}.$$

For experimental confirmation of these events was assembled installation consisting of two PEC. As a PEC in the experiment, two silicon single crystal plates produced by the company "KVAZAR" with the dimensions of 100x100 mm, pseudo square filters and two commercially produced PS-250 photovoltaic panels from Progeny Solar, USA, were used in the experiment. Their use presupposes a mixed connection of the PEC, but since in the experiment the panels are evenly illuminated, then their use in the experiment is permissible and gives an opportunity to obtain results that are more reliable. The intensity of radiation was PEC $IR = 1000$ Вт/м^2, and the other *(0,1 – 1)IR*. The illumination was measured with two luxmeters UT381 and U116. The experimental results are presented in Figure 10:

It can be seen from the graphs that in such a case the PEC having normal illumination loses up to 7.4% of the power, depending on the illumination of the shaded PEC, which can be significant in photovoltaic systems with mixed commutation, as will be shown below.

With mixed switching of the PEC, the installation current is composed of the currents of each parallel PEC array ($I = \sum_{i=1}^{n} I_i$), and the voltage is equal to the voltage at the output of each such array, which in turn will be composed of voltages each PEC included in this array ($U = \sum_{i=1}^{n} U_i$). Thus, the current-voltage characteristic, with a uniform illumination will be described by equation (6).

$$I = n_p \left[I_{ph} - I_0 \left[\exp\left(\frac{q\left(U + IR_s\right)}{n_s \left(AkT\right)} \right) - 1 \right] \right], \tag{6}$$

where n_s - the number of sequentially connected PEC in a parallel array.

In case of uneven illumination of a photovoltaic plant with a mixed PEC connection, the reduction in energy production occurs for the reasons that arise during serial and parallel commutation, but with some peculiarities. These features are related to the fact that when changing the PMP of any of the parallel connected arrays, it becomes impossible to select the maximum power from each group due to the voltage difference (ΔU), which leads to a decrease in the battery power by an amount exceeding the reduction in the maximum power of individual groups, due to their shading. At the same time, power

Figure 10. Plots of power loss (a) and voltage difference (b) PEC of radiation intensity

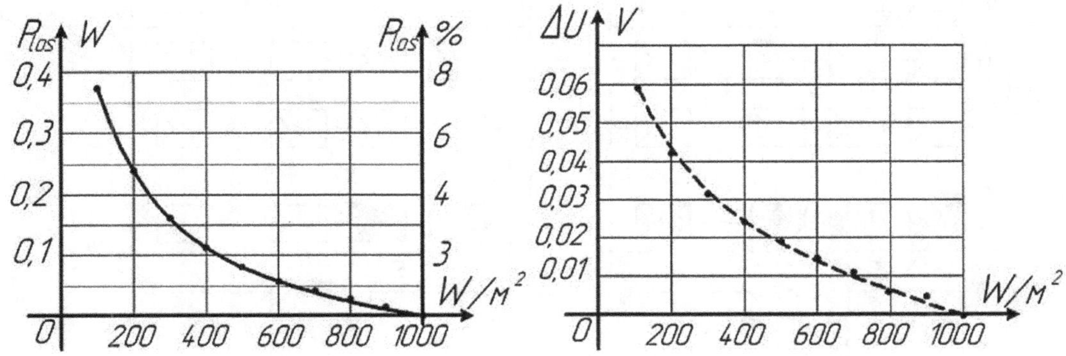

reduction occurs in different ways depending on the nature of the shading - uniform and non-uniform (Figure 11).

With a uniform shade band solar cell, power reduction of the entire battery (P_{los}) will consist of the power loss caused by a decrease in shaded illumination groups (P_{sh}), and the power loss caused by the difference in voltages MPP lighted and shaded arrays ($P_{\Delta U}$) according to the equation (7):

$$P_{los} = P_{sh} + P_{\Delta U} = n_{p\,sh}(P_{mpp1} - P_{mpp2}) + P_{\Delta U}, \tag{7}$$

where P_{mpp1} – maximum power of a normally illuminated array, (W); P_{mpp2} – maximum power of the photoelectric unit with uniform partial shading, (W).

The value of the power losses caused by the difference in voltages can be calculated from the relation (8):

$$P_{\Delta U} = U_{1max}\sum_{i=1}^{n_p} I_{1max\,i} - U_{2max}\sum_{i=1}^{n_p} I_{2max\,i}, \tag{8}$$

where n_p – number of parallel arrays; U_{1max}, U_{2max} – voltage in the MPP of parallel connected arrays operating without shading and with shading, (V); I_{1max}, I_{2max} – currents in the MPP arrays without shading and with shading, (A).

The value of the sum of currents for arrays without shading can be determined from the relation (6), substituting the voltage value corresponding to the MPP of the photovoltaic installations (U_{1max}). This voltage value, taking into account that it characterizes the extremum point of the battery power curve, can be calculated from the equation:

Figure 11. Photovoltaic installation mixed switched PEC: a - uniform shading; b - non-uniform shading (1 - normally illuminated PEC; 2 - array of PEC; 3 - solar installation; 4 - MPPT; 5 - shaded PEC)

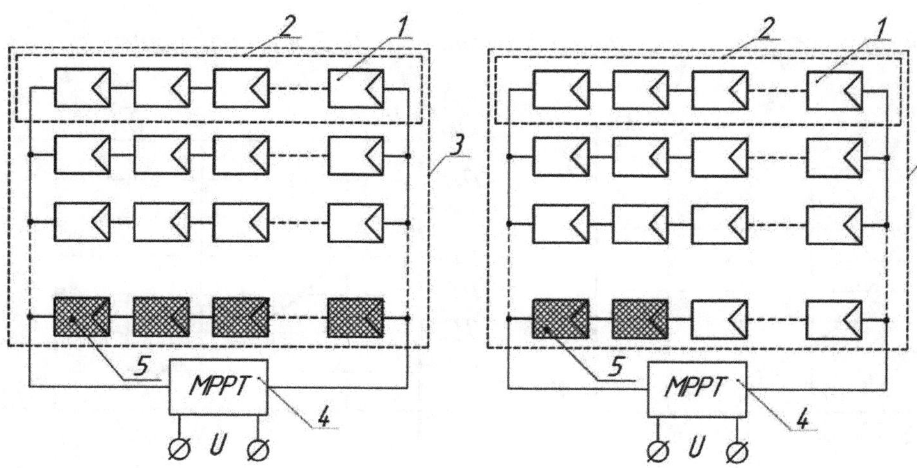

$$\frac{dP_{1max}}{dU_{1max}} = I_{1max} + U_{1max}\frac{dI_{1max}}{dU_{1max}} =$$

$$= n_p\left[I_{ph} - I_0\left[\exp\left(\frac{q\left(U_{1\max} + I_{1\max}R_s\right)}{n_s\left(AkT\right)}\right) - 1\right]\right] -$$

$$-\frac{n_p q\left(U_{1\max} + I_{1\max}R_s\right) I_0 \exp\left(\frac{q\left(U_{1\max} + I_{1\max}R_s\right)}{n_s\left(AkT\right)}\right)}{n_s\left(AkT\right)} = 0,$$

or:

$$\frac{q\left(U_{1\max} + I_{1\max}R_s\right)}{n_s\left(AkT\right)} - \ln\frac{I_0 + I_{ph}}{I_0\left(1 + \frac{q\left(U_{1\max} + I_{1\max}R_s\right)}{n_s\left(AkT\right)}\right)} = 0. \tag{9}$$

where U_{1max}, U_{2max} - the values voltage in the MPP of parallel arrays without shading and with shading, (V); I_{1max} и I_{2max} - values of currents in MPP without shading and with shading, (A).

The value of the voltage of an array having uniform partial shading in the case of MPPT setting of energy extraction at the global maximum point will be determined by the point of the global extremum of the power curve - equation (10), and in the case of setting MPPT to the local maximum point by equation (9).

$$\frac{dP}{dU_{2\max}} \approx n_{p\,sh}I_{ph\,sh} + n_{p\,il}I_{0\,il} - \exp\left(\frac{q\left(U_{2\max} + I_{1\max}R_s\right)}{n_s\left(AkT\right)}\right)\times$$

$$\times\left(\frac{q}{n_s\left(AkT\right)}\left(n_{p\,sh}I_{0\,sh}\left(U_{2\max} + I_{2\max}R_s\right) + n_{p\,il}I_{0\,il}\left(U_{2\max} + I_{1\max}R_s\right)\right)\right) = 0, \tag{10}$$

where $n_{p\,sh}$ и $n_{p\,il}$ - the quantity of arrays that have shading and normal illumination; n_p - number of parallel connected arrays; $I_{ph\,sh}$ и $I_{ph\,il}$ - photocurrent of the shaded and illuminated array, (A).

Thus, the total power loss of the photovoltaic installation at uniform shading groups of solar cells and the right determination of the maximum power point characterized by a system of equations (11):

$$P_{los} = U_{1\max} n_p \left[I_{ph>} - I_0 \left[\exp\left(\frac{q\left(U_{1\max} + I_{1\max} R_s\right)}{n_s\left(AkT\right)} \right) - 1 \right] \right] - U_{2\max}$$

$$\left[n_{p\,sh} \left[I_{ph\,sh} - \quad -I_0 \left[\exp\left(\frac{q\left(U_{2\max} + I_{2\max} R_s\right)}{n_s\left(AkT\right)} \right) - 1 \right] \right] + n_{p>} \left[I_{ph\,il} - I_0 \left[\exp\left(\frac{q\left(U_{1\max} + I_{1\max} R_s\right)}{n_s\left(AkT\right)} \right) - 1 \right] \right] \right],$$

$$\frac{q\left(U_{1\max} + I_{1\max} R_s\right)}{n_s\left(AkT\right)} - \ln \frac{I_0 + I_{ph}}{I_0 \left(1 + \frac{q\left(U_{1\max} + I_{1\max} R_s\right)}{n_s\left(AkT\right)} \right)} = 0, \tag{11}$$

$$n_{p\,sh} I_{ph\,sh} + n_{p\,il} I_{0\,il} - \exp\left(\frac{q\left(U_{2\max} + I_{1\max} R_s\right)}{n_s\left(AkT\right)} \right) \times$$

$$\times \left(\frac{q}{n_d\left(AkT\right)} \left(n_{p\,sh} I_{0\,sh} \left(U_{2\max} + I_{2\max} R_s\right) + n_{p\,il} I_{0\,il} \left(U_{2\max} + I_{1\max} R_s\right)\right)\right) = 0.$$

The value of the mismatch voltage, which leads to losses, can be calculated from the relation (12):

$$\Delta U = U_{1max} - U_{2max}, \tag{12}$$

The results of modeling the losses of a solar installation consisting of three PV-250 photovoltaic modules, with a uniformly shaded one module, are shown in Figure 12:

Figure 13 shows the results of simulation of the power loss and voltage mismatch same photovoltaic system, but when installing MPPT in the local maximum point:

In the case of uneven shading of a group of solar cells, a decrease in the power of the entire battery (P_{loss}), as well as with uniform shading, will consist of a power loss caused by a decrease in the illumination of the shaded groups (P_{sh}) and a power loss caused by the difference in voltage values in the TMM of the illuminated and shaded groups $P_{\Delta U}$). In this case, the difference in voltage values will be more significant and, accordingly, the losses caused by this difference will be greater than in the case of uniform shading. This is due to the shutdown of a part of the solar cells by shunting diodes, as was shown earlier.

Thus, the mismatch voltage, taking into account can be calculated from the equation (13):

$$\Delta U = U_{1max} - U'_{2max} = p\left(U_n + U_D\right), \tag{13}$$

Figure 12. Graphs of the dependence of losses of a photoelectric installation with mixed connection PEC and uniform partial shading on the radiation intensity in the shadow and setting the MPPT to the absolute maximum point for different illumination: a - loss from inconsistency in voltage, b - mismatch voltage

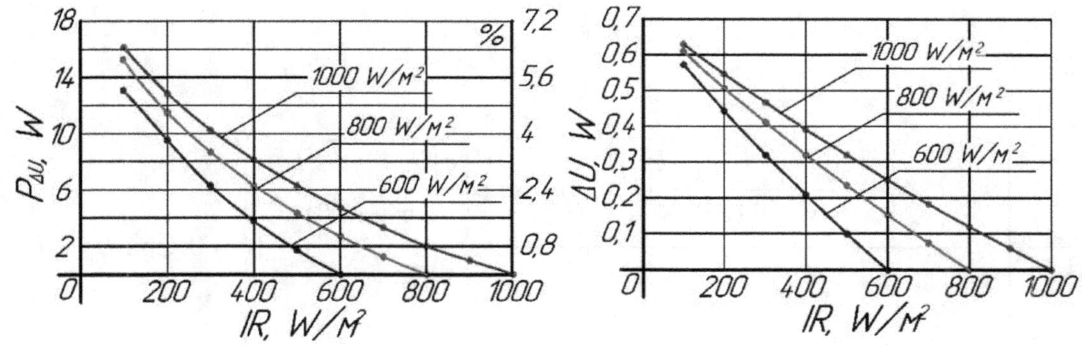

Figure 13. Graphs of the dependence of losses of a photoelectric installation with mixed connection PEC and uniform partial shading on the radiation intensity in the shadow and setting the MPPT to the local maximum point for different illumination: a - loss from inconsistency in voltage, b - mismatch voltage

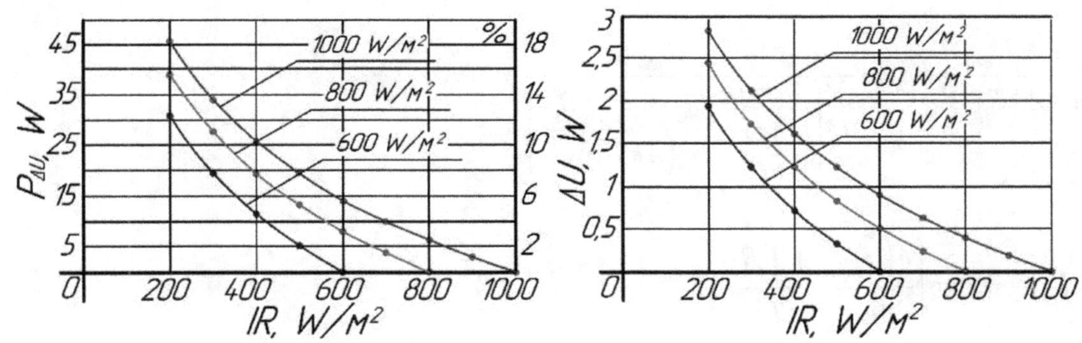

where U_n – voltage of a group of solar cells with a shunt diode connected in parallel under normal illumination; p - number of groups of solar cells shunted by diodes; U_D - total voltage drop across shunt diodes of shaded groups.

Thus was composed mathematical model (14) allows us to calculate the total loss of power photovoltaic installation in uneven shading groups of solar cells and setting MPPT to the point of absolute maximum:

$$P_{loss} = U_{1max} n_p I_{1max} - U'_{2max}\left(n_{p\,il} I'_1 + n_{p\,sh} I'_2\right),$$

$$P_{\Delta U} = P_{loss} - pP,$$

$$P = U_n I_{1max},$$

$$I_{1\max} = I_{ph} - I_0 \left[\exp\left(\frac{q\left(U_{1\max} + I_{1\max} R_s\right)}{n_{s\,il}\left(AkT\right)} \right) - 1 \right],$$

$$I_{ph}\left(n_{p\,sh} + n_{p\,il}\right) -$$

$$-\frac{U'_{2\max} q I_0}{AkT} \left(\frac{n_{p\,sh} \exp\left(\frac{q\left(U'_{2\max} + I'_2 R_s + U_D\right)}{n_{s\,sh}\left(AkT\right)} \right)}{n_{s\,sh}} + \frac{n_{p\,il} \exp\left(\frac{q\left(U'_{2\max} + I'_1 R_s\right)}{n_s\left(AkT\right)} \right)}{n_s} \right) = 0, \text{ при } U'_2 < U_{2max},$$

$$U'_{2max} = U_{1max}, \text{ при } U'_2 \geq U_{2max}, \tag{14}$$

$$\frac{q\left(U_{1\max} + I_{1\max} R_s\right)}{n_s\left(AkT\right)} - \ln \frac{I_0 + I_{ph}}{I_0 \left(1 + \frac{q\left(U_{1\max} + I_{1\max} R_s\right)}{n_s\left(AkT\right)}\right)} = 0,$$

$$I'_1 = I_{ph} - I_0 \exp\left(\frac{q\left(U'_{2\max} + I'_1 R_s\right)}{n_{s\,il}\left(AkT\right)} \right),$$

$$I'_2 = n_{p\,sh} \left(I_{ph} - I_0 \exp\left(\frac{q\left(U'_{2\max} + I'_2 R_s + U_D\right)}{n_{s\,sh}\left(AkT\right)} \right) \right), \text{ при } U'_2 < U_{2max},$$

$$I'_2 = n_{p\,sh} \left(I'_{ph} - I'_0 \exp\left(\frac{q\left(U'_{2\max} + I'_2 R_s + U_D\right)}{n_{s\,il}\left(AkT\right)} \right) \right), \text{ при } U'_2 \geq U_{2max}.$$

where $I'_1 8 I'_2$ – currents in parallel connected groups without shading and with shading under voltage U'_{2max}, (A); I'_{ph} – photocurrent of a group of solar cells with illumination equal to the illumination in the region of partial shading, (A); I'_o – the inverse saturation current of a group of solar cells with an illumination equal to the illumination in the region of partial shading, (A).

Figure 14 shows the power loss curves ($P_{?1}$, $8\, P_{\Delta U}$) and the output voltage (U'_{2max}) of a 30 kW photoelectric installation consisting of 12 parallel connected arrays consisting of ten connected PS-250 modules in series sequential partial shading of the modules in one of the arrays, when the MPPT set at the global maximum point, obtained as a result of simulation under standard conditions:

The graphs show that with uneven partial shading of the photovoltaic modules, the power of the array decreases significantly more from the voltage mismatch than from the direct reduction of the illumination of the shaded modules. Using the example of a photovoltaic system with a power of 30 kW, it can be seen that partial shading of one module results in a reduction in the array power by almost 40%, and shading more than one practically does not allow the generation of electrical energy to the array having shading.

In case of MPPT installation of energy extraction from the photoelectric installation at the first point of the local maximum (from the point of idling), the voltage U'_{2max} will be equal to the voltage U_{1max} ($U'_{2max} = U_{1max}$). These voltages will also be equal in the case when the voltage of the parallel array with shunted solar cells groups is less than the voltage in the MPP of the array having no shading ($U'_2 < U_{2max}$).

Thus, the total losses of the solar installation power in the case of uneven shadowing of the solar cells groups and the energy extraction at the local maximum point are determined by the system of equations (15):

$$P_{loss} = U_{1max} n_p I_{1max} - U_{1max} \left(n_{p\,il} I'_1 + n_{p\,sh} I'_2 \right),$$

$$P_{\Delta U} = P_{loss} - pP,$$

$$P = U_n I_{1max},$$

$$I_{1\max} = I_{ph} - I_0 \left[\exp\left(\frac{q\left(U_{1\max} + I_{1\max} R_s\right)}{n_{s\,il}\left(AkT\right)} \right) - 1 \right],$$

Figure 14. Graphs of the dependence of power losses and output voltage of SFU with a power of 30 kW from partial shading of photoelectric modules of one of the arrays

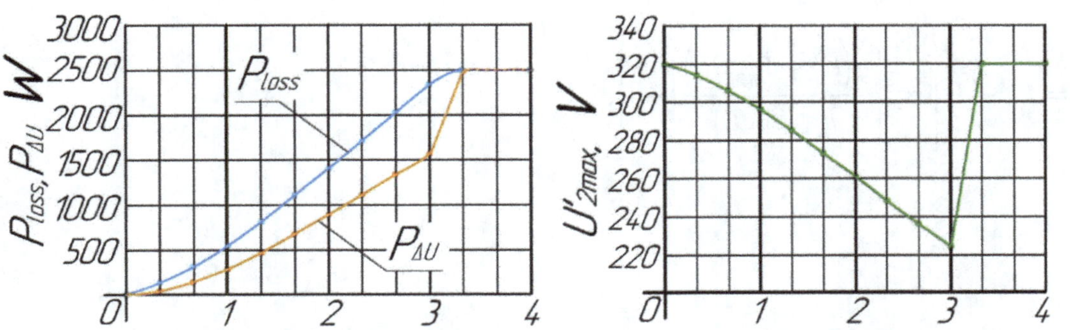

$$\frac{q\left(U_{1\max}+I_{1\max}R_s\right)}{n_s\left(AkT\right)}-\ln\frac{I_0+I_{ph}}{I_0\left(1+\frac{q\left(U_{1\max}+I_{1\max}R_s\right)}{n_s\left(AkT\right)}\right)}=0, \tag{15}$$

$$I'_1=I_{ph}-I_0\exp\left(\frac{q\left(U'_{2\max}+I'_1R_s\right)}{n_{?Ail}\left(AkT\right)}\right),$$

$$I'_2=n_{p\,sh}\left(I_{ph}-I_0\exp\left(\frac{q\left(U'_{2\max}+I'_2R_s+U_D\right)}{n_{p\,sh}\left(AkT\right)}\right)\right), \text{ при } U'_2<U_{2max},$$

$$I'_2=n_{p\,sh}\left(I'_{ph}-I'_0\exp\left(\frac{q\left(U'_{2\max}+I'_2R_s+U_D\right)}{n_{s\,il}\left(AkT\right)}\right)\right), \text{ при } U'_2\geq U_{2max}.$$

The use of DCA allows to obtain electrical energy from e mismatched array of solar cells in the MPP. This occurs as a result of the fact that the load on the output of such an array corresponds to the MPP. In this case, the voltage at its output is equal to the voltage of the other parallel arrays.

The loss of battery power with «in-line» switching of the DCA can be calculated from the system of equations (16):

$$P_{loss}=n_pP-\left(n_{pil}P_{il}+n_{psh}P_{sh}\right)=n_{psh}\left(P_{il}-P_{sh}\right),$$

$$P_{il}=I_{mpp1}U_1,$$

$$P_{sh}=I_{mpp1}U_2-\frac{\Delta UI_{mpp1}}{U_1}U_2\left(1-\eta_{dca}\right),$$

$$I_{mpp1}=I_{ph}-I_0\left[\exp\left(\frac{q\left(U_1+I_{mpp1}R_s\right)}{n_{s\,il}\left(AkT\right)}\right)-1\right],$$

$$\frac{q\left(U_1 + I_{mpp1}R_s\right)}{n_s\left(AkT\right)} - \ln\frac{I_0 + I_{ph}}{I_0\left(1 + \frac{q\left(U_1 + I_{mpp1}R_s\right)}{n_s\left(AkT\right)}\right)} = 0, \tag{16}$$

$$\frac{q\left(U_2 + I_{mpp1}R_s + U_D\right)}{\left(n_{s\,sh}\left(AkT\right)\right.} - \ln\frac{I_0 + I_{ph}}{I_0\left(1 + \frac{q\left(U_2 + I_{mpp1}R_s + U_D\right)}{n_{s\,sh}\left(AkT\right)}\right)} = 0.$$

where η_{dcs} is the DCA efficiency.

The results of the calculations show that the installation of the DCA significantly reduces the energy losses of the photoelectric battery with partial shading of the solar cells, which significantly increases the energy production of the battery operating under such conditions. A particularly significant effect from the use of DCA is achieved when used in the photovoltaic system, consisting of a large number of parallel arrays of solar cells, or the use of MPPT with algorithms that establish a mode of selecting electrical energy not in global maximum point. In such cases, partial shading of more than one panel in a parallel group consisting of ten panels practically does not allow the generation of electrical energy of such a group without the DCA.

To check the adequacy of the developed mathematical models and confirmation of the results of theoretical research was an experimental study.

Experimental studies were carried out at the existing solar power plants and installations in Sevastopol (LLC «S.Energy-Sevastopol») with an installed capacity of 2.99 MW, the Republic of Crimea (LLC «SPF Spetsproekt»), and an installed capacity of 9 kW, Samara (LLC «Eco Energy» and in the laboratory “Renewable Energy Sources” in “Sevastopol State University”.

Measurement of the current-voltage characteristics was carried out simultaneously with the measurement of the intensity of solar illumination on the surface of solar cells and the temperatures of the environment and solar cells. The measurements were carried out under natural sunlight on a clear, sunny day according to (GOST 28977 - 91 2004, GOST 51597 2000) at latitude 44 ° 37.0012 ‘N (the city of Sevastopol).

Measurement of the I-V characteristics in a classical way, using an ammeter, a voltmeter and a rheostat as a load, takes quite a long time. Because of this, the accuracy of the received data in the field is reduced. This is due to the changing parameters of solar cells in the process of measurement as a result of their heating from the effects of solar energy, or cooling, due to convective heat transfer by wind, etc. Also such measurements are not very convenient, since they require simultaneous manual load variation and data recording with instruments.

In connection with the foregoing, an information and measurement complex (IMC) was developed to remove I-V characteristic from laboratory photovoltaic systems, which allows automatic change in the load with measurement, recording and display of voltage and current values graphically. In addition, this complex makes it possible to measure and record the temperature of the environment and the surface of solar cells using a pyrometric method, as well as the intensity of solar radiation.

The structure of the IMC is shown in Figure 15:

The principle of the IMC is that the microcontroller (MC) (6) performs an automatic step change in the load of the solar cell (1) by means of a special adjustable load (4). Each change in load is accompanied by a cycle of current and voltage measurements by means of voltage sensors (3) and current (2) transmitting information to the A/D converter. The number of measurements in one cycle is 500. In addition, the MC receives information about the ambient temperature, the temperature of the solar cells and the illumination, through temperature sensors (5) and illumination (11). The MC converts the measuring signals, saves them to the electronic drive (7), displays on the graphical touch screen (8) in graphic and text form. The display also displays information on the power, voltage and current values in the absolute maximum point, determined by the MPPT algorithm in the MC, the current and voltage values at the points OC and SC, the series resistance of the photovoltaic system. The values of the ambient temperature and the surface of the solar cells are displayed in a graphic and text form in a separate window on the display screen, which is switched by touch control on the display. The illumination values on the surface are also displayed in a separate display window. To analyze and store the measurement information on a personal computer (PC), an interface MC (9) is provided in the complex, which transmits data packets to the PC via a serial port (10).

As the current sensor, the integrated current sensor from Allegro - ACS712 is used. Voltage measurement is made by voltage divider, calculated on the basis of the maximum voltage of the photovoltaic system of a laboratory installation with a margin of 25% - 140 V and a maximum input voltage of the MK - 5 V ADC.

To measure the temperature of solar cells and the environment, use the digital infrared temperature sensor MLX90615. A special feature of the sensor is that it uses a low noise amplifier, a 16-bit ADC and a high-performance signal processor. Due to this, high accuracy of temperature measurement (up to 0.5 °C) and resolution (0.02 °C) is achieved. The advantages of this sensor can also be attributed to the fact that it comes factory-calibrated in the temperature range: -40 °C to + 85 °C - for ambient temperature, and -40 °C to +115 °C for the temperature of the measured object . The temperature table is stored in the sensor ROM with a resolution of 0.02 °C. These values are read by means of a two-wire bus compatible

Figure 15. Structure of the IMC (1 - photovoltaic system; 2 - current sensor; 3 - voltage sensor; 4 - special adjustable load; 5 - solar cell temperature sensor and OS; 6 - microcontroller; 7 - electronic storage device; 8 - LCD TFT Touch Screen; 9 - interface microcontroller; 10 - personal computer; 11 - light sensor)

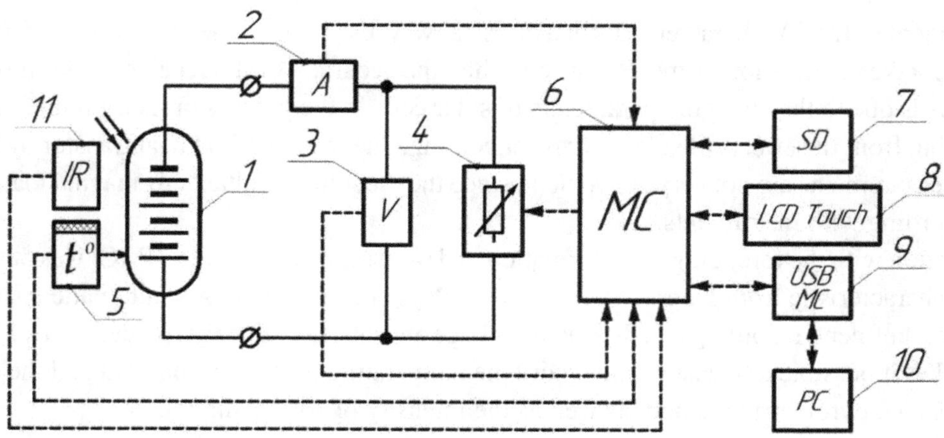

with SMbus. To exclude the influence of illumination on the sensor readings, the housing contains an optical filter with a bandwidth of 5.5 ... 14 μm.

The light sensor is a digital light sensor MAX44009. The sensor is made using BiCMOS technology, which uses the integration of two optical sensors: one for working in the infrared spectrum, and the second - in the visible. This allows digital signal processing algorithms to take into account the characteristics of both sensors and more accurately determine the level of illumination in a wide 22-bit dynamic range from 0.045 to 188,000 lux in the operating temperature range from -40 to + 85 °C. In addition, the sensor contains ADCs and other functional elements that allow you to read the measured values calibrated by the manufacturer by the I^2C interface (Guba A. 2006).

One measurement of illumination and temperature consists of a measurement cycle consisting of 500 automatic measurements with an average value.

As a special adjustable load, a boost converter was used, widely used in devices for searching for the maximum power point and optimizing the output power (Russkin V.A., Semenov S.M., Dickson R.K. 2016, Sridhar, T. and J. Anish kumar. 2012). To the output of the converter is connected a load of 0.1 Ohm with passive cooling, capable of withstanding current up to 15 A for 50 seconds and 8 A for 140 seconds, which is sufficient due to the fact that one measurement (with 50 points of I-V characteristic and the number of measurements at each point equal to 500), does not exceed 4.2 seconds. The load is controlled by the PWM.

Experimental characteristics (Figure 16) and graphics daily burnup (Figure 35b) array of 18 photovoltaic modules KV-260M, having uniform shading the solar cells (Figure 20a) were obtained in the 2nd and 3rd stages of the experiment included parallel 20 - to the same arrays that do not have shading, with DCA (curve 3) and without DCA (curve 2). The experiment was carried out at the Sevastopol solar power plant, which initially had block switching modules. In this connection, work was done on re-commutating the module with a block-by-line module (Kuznetsov P.N., Yuferev L.Yu. 2017).

During the research, the DCA was located in the power plant's connecting panels. The presented characteristics were obtained with a solar radiation intensity of 750 W/m² on the module surfaces without a shadow and 100 W/m² on the shaded modules. The area of shading was 10% of the module area. The value of the temperature of the solar cells of the shaded modules was + 13 °C, and the normally

Figure 16. Experimental characteristics of an array of photovoltaic modules (1 - without shading, 1 - with partial shading and in-line vertical switching, 3 - with partial shading, in-line vertical switching with DCA)

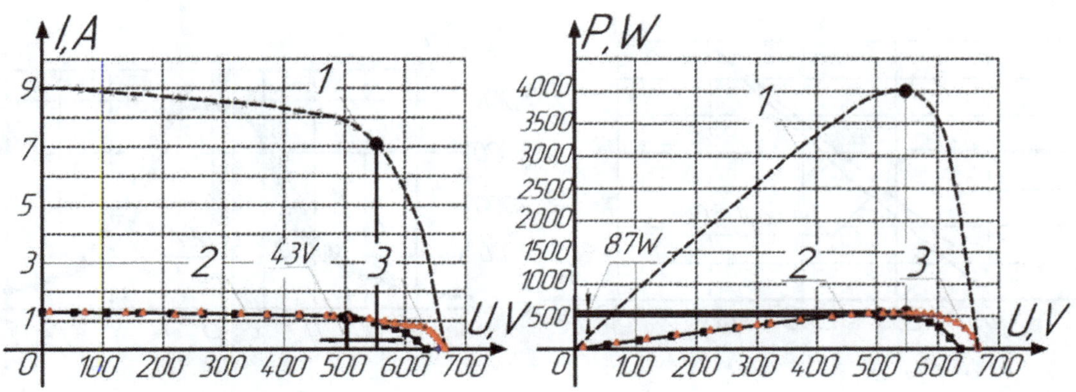

illuminated temperature was + 33 °C. It can be seen from them that when the voltage difference in the TMM is 43V, the power loss without using the DCA is 87W, which corresponds to 17.9% of the capacity.

The structural scheme of the inclusion of DCA in the investigated array is shown in Figure 17:

Research of daytime energy production showed that using the DCA with this kind of shading and switching makes it possible to increase the output capacity of the array to 22%, and the daily energy output by 8.7%. To assess the effect of radiation intensity in the shadow on the output power of the array, an appropriate experiment was also performed during this stage, the results of which are shown in Fig. 18a. The carried out experiment confirms the results obtained in mathematical modeling with an error not exceeding 8.5%.

The 4th and 5th stages of the experiment were carried out similarly to the second and third stages with the difference that the investigated array of photovoltaic modules worked under conditions of uneven shading (Figure 18b), created by means of an opaque rectangular screen located at a distance of 3 meters from the shaded modules. The displacement of the screen changed the area of shading of the PV modules, corresponding to the shading character for horizontal and vertical block as well as horizontal line commutation of the modules. To carry out the experiment with the horizontal arrangement of the modules at the Sevastopol power plant, modules were installed with the appropriate orientation in the

Figure 17. Structural diagram of the inclusion of the DCA (1 - solar module, 2 - shadow, 3 - DCA, 4 - voltmeter, 5 - ammeter)

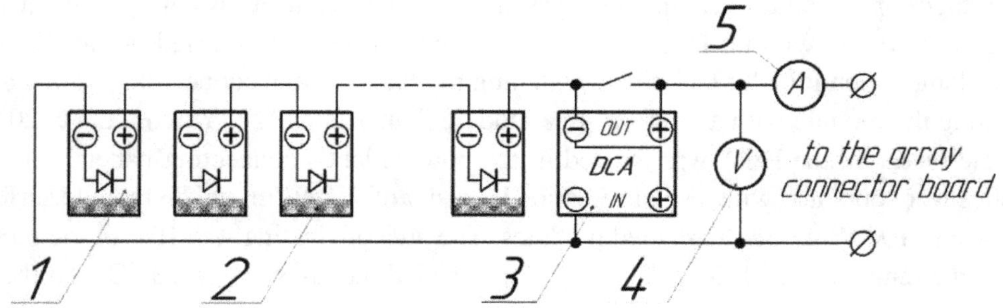

Figure 18. Output power of a uniformly shaded array of modules (a - on the intensity of radiation in the shadow, b - during the light day)

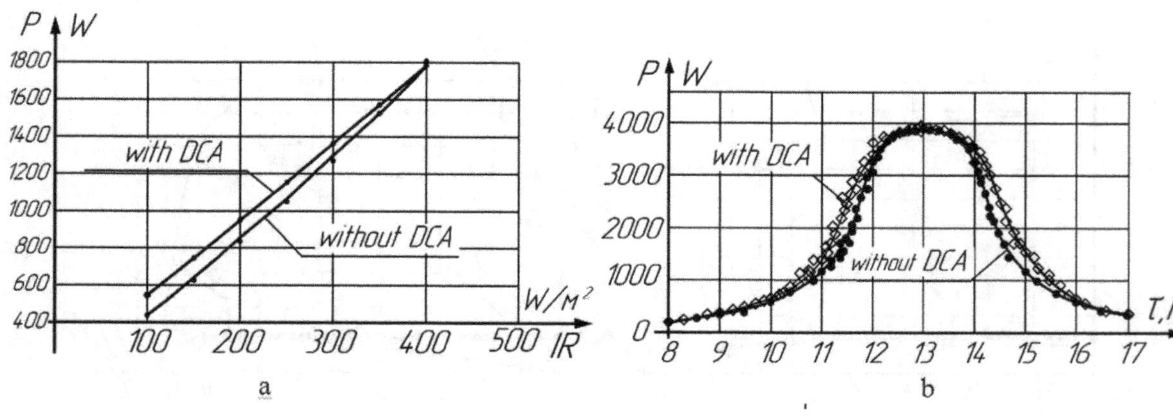

unfinished section of the station. Switching of such array was made in a junction box with twelve arrays, working without shading. Since in the experiment with block vertical and horizontal horizontal commutation the mismatch voltage was more than 90V, two DCA with series-connected outputs were installed in the module.

In Figure 19, the obtained I-V characteristic and W-V are shown, from which it is evident that the use of the DCA makes it possible to extract electric energy from the array of modules in the MPP over a wide range of voltages. Due to this, with their uniform partial shading (area 10-30%), the output power is increased by 78% (2092 W) with block vertical or horizontal switching and by 22.2% (787 W) with block horizontal switching.

Also in these stages, a study was made of the dependence of the energy losses of the array of modules, depending on the number of shunted shaded modules. This experiment was conducted on the industrial network solar power plant "Sevastopolskaya" LLC "S. Energi-Sevastopol" and LLC "NPF "Energospetsprokt". At the first power plant, parallel-connected array of modules consist of 18 serial-connected modules, and on the second - of 10. In the experiment, a consecutive increase in the area of the shadow was made for one group due to the fact that each module is divided by shunt diodes into 3 consecutive groups, which are switched off separately when the solar cells of the corresponding group are shaded. The results of this research are shown in Figure 20.

From the obtained graphs it follows that in the case of shading 2 panels, the use of DCA allows 2.6 times (from 34% to 13%) to reduce the power loss of an array consisting of 18 modules, and with a larger shading without DCA, such an array is practically incapable of producing electric energy, the value of which determines, mainly the intensity of radiation in the shade. In the case of shading of the 1st module in an array consisting of 10 modules, due to the use of DCA, the power losses are reduced 3.4 times (from 34% to 10%), and with shading of 2 or more power generation of the array as well as in the previous case, it is practically not produced.

Figure 19. Experimental characteristics of an array of photovoltaic modules (1 - without shading, 2 - with partial shading and block vertical switching, 3 - with partial shading and block horizontal switching, 4 - with partial shading and in-line horizontal switching, 5 - with partial shading, horizontal block switching and DCA, 6 - with partial shading, horizontal in-line, or block vertical switching with DCA)

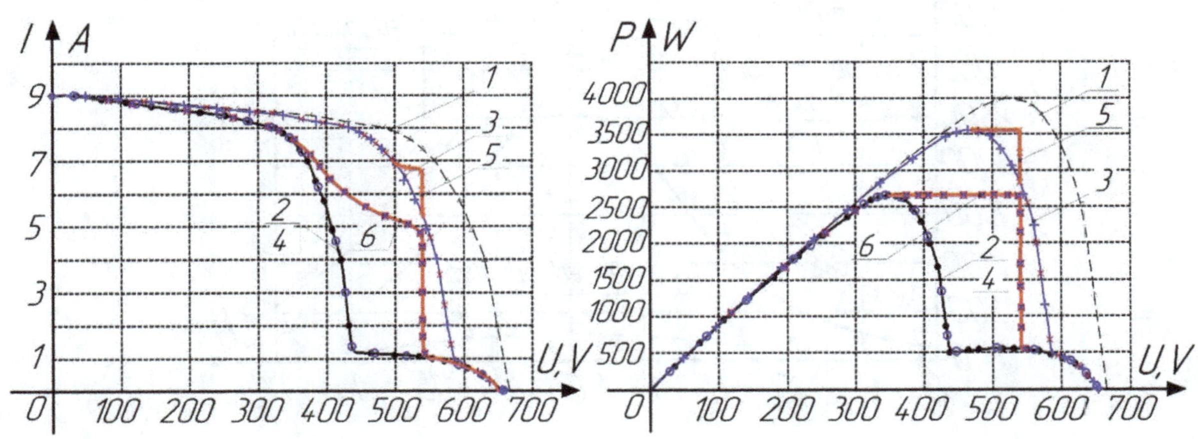

Figure 20. Graphs of power loss of the array of modules from partial shading of modules (a - at the "Sevastopol" solar power plant, b - at the station LLC "NPF" Energospetsproekt")

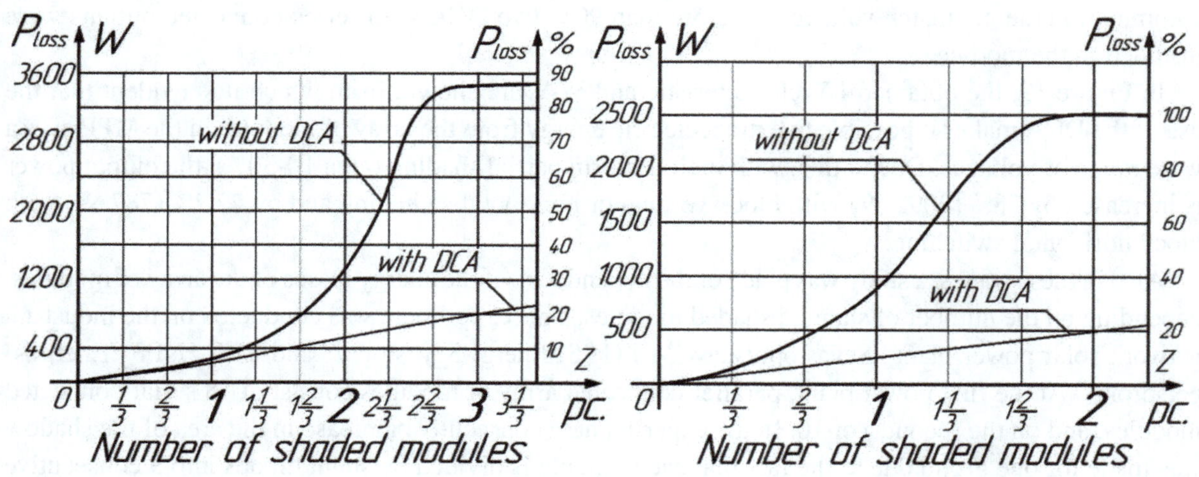

To obtain experimental data on the daily energy production of the array with partial uneven shadowing of modules and DCA, within the framework of these stages, a study was conducted during which during the daylight the artificial shading of the modules of one array of the Sevastopol solar power plant was performed alternately. After each shading, its output power was measured. The results of the study are shown in Figure 21, from which it can be seen that with the shading of 2 modules out of 18, the use of DCA allows to increase the daily power output of the array by 43.7%, and in the case of shading of the first one - by 19.7%.

Figure 21. Graphs of the daytime power generation of the array of modules

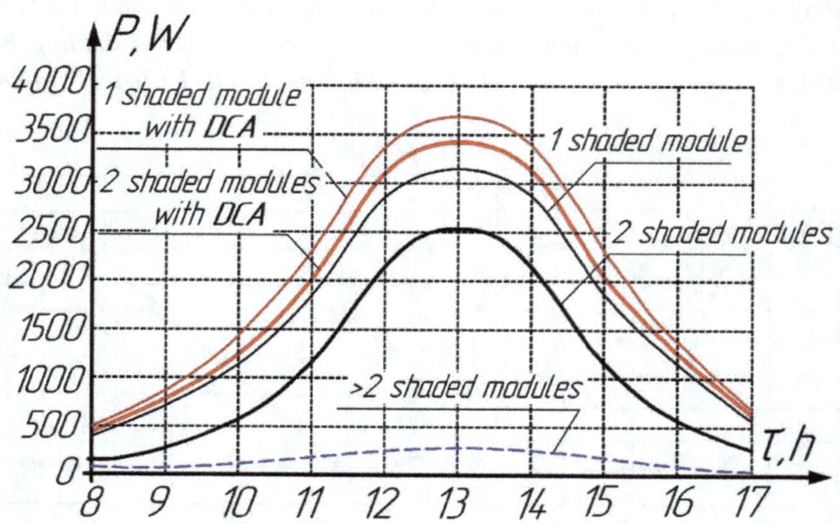

SOLUTIONS AND RECOMMENDATIONS

The research materials can be used in the design of solar photovoltaic power plants and installations that assume work in conditions of uneven lighting, and also for the introduction of existing solar power plants operating in such conditions to increase their energy efficiency.

The developed systems of equations make it possible to carry out a theoretical analysis of the energy and electrical characteristics of the photovoltaic system with parallel and mixed switching of solar cells and methods of extracting electrical energy from them under various illumination conditions, which allows choosing the most rational technical solution from the view of obtaining maximum energy efficiency for specific conditions.

The implementation of the developed photovoltaic system on the operating photovoltaic systems, which realize the method of selecting electric energy from parallel array of photovoltaic modules with different illumination, can significantly increase their energy production, especially at low values of the sun's height (in winter), at financial costs, not exceeding 1.2% of the installation cost.

Studies have shown that the use of DCA can significantly increase the energy production of the array of modules operating under partial shading conditions. So, with partial shading of 2 modules out of 18, the average daily power losses are more than 40%, and with greater shading, the modules are practically incapable of working without a device. DCA also increase the output capacity of the array to 22%, and daily energy production by 8.7% with uniform shading.

The developed and manufactured information and measuring system allows, both in field and laboratory conditions without the use of additional equipment, to perform with high accuracy in automatic mode the measurements and recording of various electrical and power characteristics of the photomultiplier, as well as the characteristics of environmental conditions with their reflection in the graphic and numeric form.

REFERENCES

Kuznetsov, P.N., & Safonov, V.A. (2016). Increase of the efficiency of the photovoltaic station. *Energy Security and Energy Saving, 3*, 26-30.

Kuznetsov, P.N., Yuferev, L.Y. (2017). Increase of efficiency of work of photoelectric converters at consecutive connection. *Bulletin of Agrarian Science of Don, 37*, 15-25.

Kuznetsov, P. N., & Avdeev, D. S. (2017) Development of a laboratory model of the device for coordinating photovoltaic modules. *Abstracts of the International Scientific and Technical Conference "Electrotechnical Complexes and Systems"*, 45-46.

Alferov, Zh. I., Andreev, V. M., & Rumyantsev, V. D. (2004). Tendencies and prospects for the development of solar photovoltaics. *Physics and Technology of Semiconductors, 38*(8), 937–948.

Kitaeva, M. V. (2015). Solar tracking systems for solar energy. *Collection XV International Scientific and Practical Conference "Modern technology and technology"*, 54-55.

Gevorkian, P. (2016). *Solar power generation Problems, Solutions, and Monitoring*. Cambridge University Press. doi:10.1017/CBO9781316341414

Strebkov, D. S., & Tver'yanovich, E. V. (2007). *Concentrators of solar radiation*. Moscow: Academic Press.

Chernyshev, Y.A., Shinyakov, A.I., & Gordeev, K.G. (1991). *Autonomous power supply system*. Patent of the Russian Federation 2035109 IPC H02J 7/35.

Kudryashov, V.S., Elman, V.O., & Nesterishin, M.V. (2009). *The power supply system of the spacecraft*. Patent of the Russian Federation 2396666 H02J7/35.

Kuznetsov, P.N., & Borisov, A.A. (2017). *A method for selecting electric energy from photovoltaic converter batteries*. Patent of the Russian Federation 2634590.

Tian, H., Mancilla-david, F., Ellis, K., Jenkins, P., & Muljadi, E. (2012). *A Detailed Performance Model for Photovoltaic Systems Preprint. Solar Energy Journal*.

Petreuş, D., Fărcaş, C., & Ciocan, I. (2008). Modelling and simulation of photovoltaic cells. *Electronics (Basel)*, *49*(1), 42–47.

GOST 28977 - 91 (IEC 904-187). (2004). *Photovoltaic devices. Part 1. Measurements of photoelectric volt-ampere characteristics*. IPK Publishing Standards.

GOST 51597 - 2000. (2000). *Unconventional power engineering. Modules solar photovoltaic. Types and basic parameters*. IPK Publishing Standards.

Guba, A. (2006). Digital sensors of illumination MAX44007 and MAX44009 of Maxim Company. *Electronic Components*, *6*, 11–13.

Russkin, V.A., Semenov, S.M., & Dickson, R.K. (2016). Investigation of the search algorithms for the point of maximum power for the boost converter of the solar inverter voltage. *Izvestiya Tomsk Polytechnic University. Engineering of Geo-Resources, 4*, 78-87.

Sridhar & Kumar. (2012). Development of Solar MPPT System Using Boost Converter with Micro-controller. *International Journal of Advanced Research in Electrical. Electronics and Instrumentation Engineering*, *1*(4), 334–340.

Chapter 5
Materials and Methods of Thermal Energy Storage in Power Supply Systems

Baba Dzhabrailovich Babaev
Dagestan State University, Russia

Valeriy Kharchenko
Federal Scientific Agroengineering Center VIM, Russia

Vladimir Panchenko
Russian University of Transport, Russia

Pandian Vasant
Universiti Teknologi PETRONAS, Malaysia

ABSTRACT

The chapter considers the factors that affect the incomplete use of power. The analysis and systematization of accumulating systems on the types of accumulated energy, on the processes occurring during storage, types of devices and types of generated energy is performed. Each of the possible ways of accumulating energy is analyzed in detail. Various variants of conformity of schedules of change of a consumed load within a year and a total resource of solar radiation are considered. The analysis of the parameters of heat-accumulating materials and their classification depending on the material class, the way of heat accumulation and return, on the cyclicity of work, etc. are provided. It is shown that the level of temperature, the scale of the storage unit, and the necessary duration of heat storage determine the requirements for the construction of batteries and the choice of heat-storage substances. The prospects of research on the search for new ways of accumulating energy and technical means for their implementation are considered.

DOI: 10.4018/978-1-5225-9179-5.ch005

INTRODUCTION

Under normal conditions energy production must correspond to the regime of its total consumption. The power of the power plants should be expected to satisfy the maximum possible load regardless of its duration. At the same time, the power system needs a sufficient reserve of power to ensure the continuity of energy supply to consumers in the event of a reduction in the capacity of individual power generating units of the system during a normal production process or due to sudden emergency or breakdowns.

Uneven energy consumption leads to the fact that the power of power plants is most fully used only at the time of maximum load. Incomplete use of the power system capacity can also be caused by:

- The presence in its composition of power plants with variable available capacity, for example, thermal power plants;
- Insufficient regulation of hydroelectric power stations;
- Lack of maneuverability of power regulation of nuclear power plants;
- Uneven output of solar and wind power plants due to diurnal and seasonal fluctuations in solar radiation intensity and wind speed;
- Other factors.

Increasing the use of the system's capacity and the creation of energy reserves are major problems in the design and operation of power systems (Popel & Fortov, 2015, Sorensen, 2011, Vasilev, Elistratov, Mukhammadiev & Pretro, 1995).

Needs in accumulation increases at the operation of power plants based on renewable energy sources (RES) – the energy of the sun, wind, etc.

In this case the inclusion of accumulating units in the power system improves the reliability and stability of the power supply to consumers, makes it possible to regulate the operation mode of power plants without strict dependence on the energy consumption mode by cutting the peak of the graph by means of discharging the battery and filling the graphs with charge of the battery.

BACKGROUND

There are different kinds of energy accumulators. They can be classified according to the types of accumulating energy, the processes occurring during its storage, the types of devices and the types of energy given out (figure 1).

The following species are among the most interesting and applicable accumulation means (Babaev & Danilin, 2002):

- Heat (cold) accumulators are installations in which the energy is stored as heat (cold) due to heat capacity, heats of phase transitions or endothermic (exothermic) reactions;
- Accumulators of electrochemical energy are installations in which the electric power is saved and given out as a result of chemical reactions;
- Mechanical accumulators are installations in which the energy is stored in the form of the potential and kinetic energy of physical bodies;

Figure 1. Types of energy accumulators

Main type of accumulated energy	Accumulation due to:	Types of energy storage batteries			Kind of energy released by the battery
		Chemical current sources (CCS)	Acidic	$Pb\text{-}PbO_2$	
Electrical	Electrochemical reactions		Alkaline	Ni-Cd; Fe-Ni; Ag-Zn; Li-ion, etc.	Electrical
		Thermal chemical sources of current (TCSC)	Electrolyte from molten salts (eutectic)	Na-S etc.	
				Fuel cell	
		Electrolysis	The accumulators H_2	Internal combustion engines	
				Burning	Thermal
			Storage of H_2, CH_4 and other combustible gases		
Thermal, electrical and light energy	Heat capacity, phase transitions, adsorption, photo- and thermochemical reactions	Photochemical and thermochemical	Tanks for adsorption, reversible photo-, endo- and exothermic reactions		
				Thermochemical (combustion)	
			Biomass (photosynthesis)	Biochemical (production of liquid fuels, biogas)	Thermal
				Agrochemical (extraction of fuels)	
		Heat-capacitive	Containers with solid, liquid and gaseous materials		
			"Sunny" pond		
		Phase Transition	Containers with phase-transition materials (phase transition - solid-solid, solid-liquid, solid-gas, liquid-vapor)		
	Kinetic energy	Flywheels (rotation)	Disc, stem and tipping		
Mechanical and electrical energy		Torsion bars (torsion)			Electrical
	Potential energy	Lifted cargo	Pumped storage power station, etc.		
		Compressed gas	Containers or underground cavity		
		Compressed spring, curved tape, rod, etc.			
Electric Energy	Electric and magnetic fields	Capacitive	Capacitors		Electrical
		Inductive	Inductors		

- Electric accumulators are installations in which the energy is stored in the form of electrical or electromagnetic energy.

Classification of Materials for Heat Accumulation

In some cases, the accumulation of heat is desirable and economical, especially when using renewable energy sources that are impermanent in time because of their very nature. For example, the use of solar energy for heating purposes is almost impossible without the use of heat accumulation – the battery warms up when solar radiation is present, and gives off heat when it is not there.

Figure 2 shows, for example, the graphs of changes in the load during the year and the total solar radiation (SR) resource from 1 km^2 in the corresponding zones of the Republic of Dagestan. The graphs illustrate the great unevenness throughout the year.

As can be seen from the figure 2, the heat load of heating is maximal in December-January and minimal in June-July, while solar energy supply, on the contrary, is minimal in winter, and in summer is maximum. For a stable supply of heat, it is necessary to capture solar energy more than is required at the moment and accumulate its excess in a heat accumulator. The energy reserve in the battery can be calculated as for several months – with seasonal accumulation as well for several hours or days with short-term accumulation.

In the summer months, a large amount of electricity is spent on air conditioning and cooling. Therefore, the development of new efficient refrigeration facilities designed for cooling of drinking water and beverages, storage of medical products, agricultural products, indoor air conditioning and other practical applications remains an urgent task. The accumulation of cold is a fundamentally important problem

Figure 2. Average monthly electricity consumption and SR resource from 1 km² in the zones of the Republic of Dagestan in kWh: a) foothill; b) the plain; c) mountain (calculation of SR by zones (Babaev, 2015)

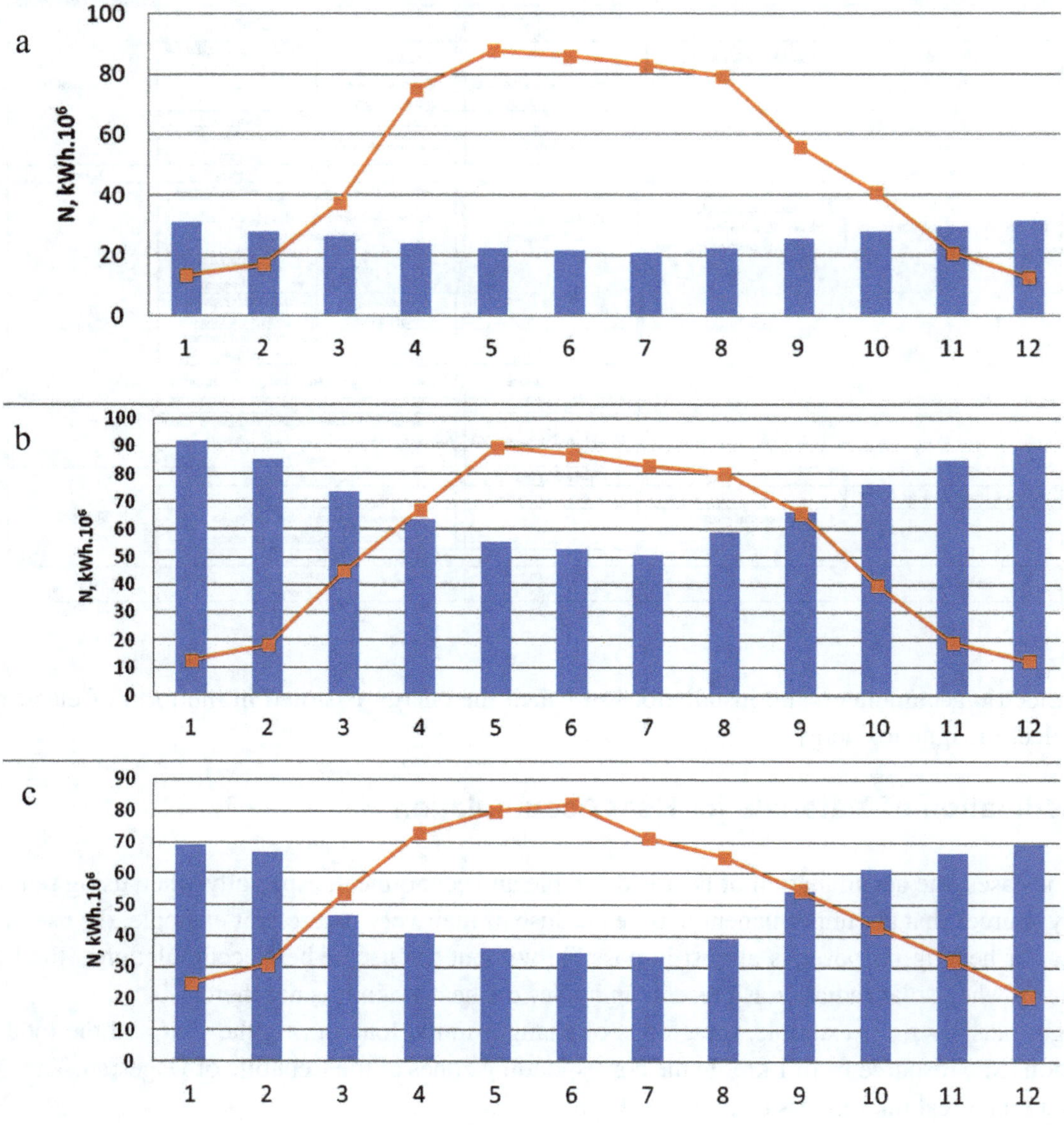

for many countries located in areas with a hot climate. It is of great practical interest for consumers who have seasonal needs in the cold (Popel & Fortov, 2015).

Today, the cold is obtained mainly by means of vapor compression refrigerating units, in which the drive of the compressor is provided by electricity use. However, there are numerous of consumers who do not have reliable power supply or use autonomous power plants operating on expensive imported fuel for electricity production. In this situation, it is of interest to search for effective technical solutions for inexpensive production of cold from local energy sources, primarily due to universally available solar energy, the use of which is particularly attractive for refrigeration due to the fact that the graphs of the demand for cold and solar energy intake correlate friendly.

In addition, the accumulation of heat (cold) can be considered as a means of equalizing the energy generation schedule. This is especially important for nuclear power plants for the operation of the nuclear reactor in the mode of nominal design load. With a decrease in power consumption, the excess heat released by the reactor can be stored to give it away when the power consumption again increases.

The creation of heat (cold) accumulators depends on the temperature level, the scale of the installation and the duration of the accumulation.

Classification of types and ways of heat (cold) accumulation is shown at fig. 3. According to the accumulation temperature level, heat and cold accumulators are divided into three groups: low-temperature (35 °C < t < 100 °C) medium temperature (100 °C < t < 500 °C); high-temperature (t > 500 °C) (Sorensen, 2011); accumulation cold storage (t < 15 °C). In terms of the use of heat and cold, they can be classified as small-scale (for decentralized consumers) and large-scale (for large centralized systems). For the duration of storage of heat or cold batteries are divided into: short-term (1 – 2 days); medium-term (up to 1 month); off-season (up to six months) (Duffie & Beckmann, 2013; Twidell & Weir, 1990). According to the type of material used, the accumulators are divided into capacitor-type accumulators (due to heat capacity), phase transition (mainly due to enthalpy change) and thermochemical (mainly due to chemical reactions) (figure 3).

Heat storage materials (HSM) may be classified according to the material class, the way of accumulation and heat release as well as from the cyclic operation (Babaev, 2014; Babaev & Danilin, 2002) (Figure 4).

According to (Babayev, 2014; Babaev & Danilin, 2002), the classification of HSM is based on the material belonging to a certain class, the way of heat accumulation and recycling, the cyclic nature of the material, and also preparation in the form of a continuous phase or in the form of various phase-transition materials.

The properties of the classes of substances suitable for the manufacture of heat-accumulating materials determine the properties of the materials themselves. Table 1 shows these classes and basic properties. They vary considerably within the class as well as between classes, which allows creating materials with different properties.

Review of the literature (Bekirov, 2011; Adadurov, 2005; Alekseev, 1991; Vasilev, Elistratov, Mukhammadiev & Pretro, 1995; Grigorev, 2003; Da Rosa, 2010; Babaev, 2016b; Beckmann & Gilli, 1987; Duffy & Beckman, 2013; Popel & Fortov; 2015; Reznitsky, 1998; Reznitsky, 1996; Sorensen, 2011; Twidell & Weir 1990) shows that currently among the main types of HSM used for thermal storage may be listed such as stones, water, gases and eutectic salts.

The level of temperature, the size of the storage unit and the necessary duration of heat storage determine the requirements for the design of batteries and the choice of heat storage materials.

Moreover, the higher the storage temperature, the more difficult it is to ensure a long accumulation time due to existing heat loss. But the improvement of thermal insulation reduces the intensity of heat losses, and therefore increases the possible duration of storage of stored energy. Therefore, the type, design and cost of heat accumulators depend on the desired duration of heat storage.

Recently, more and more attention has been paid to thermochemical batteries, which ensure the storage of energy as a result of the passage of a certain reversible chemical reaction. The accumulation of heat in this case occurs when a chemical reaction proceeds to the side in which heat absorption take place, and the heat return to the consumer is due to the reverse reaction (Babaev, 2014; Babaev, 2016a; Babaev & Danilin, 2002; Duffy & Beckman, 2013; Popel & Fortov, 2015).

Figure 3. Types and methods of heat (cold) storage

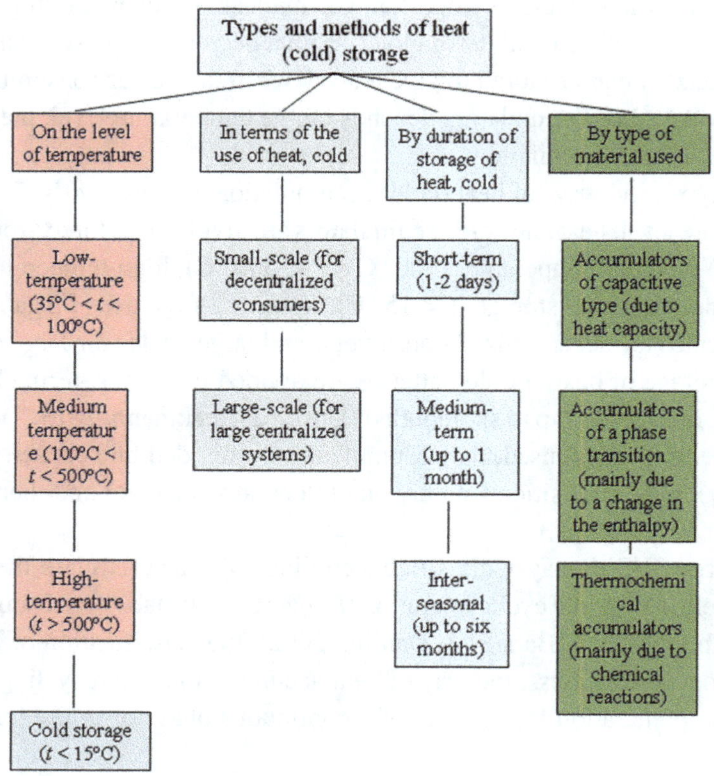

Figure 4. Classification of heat-storage materials

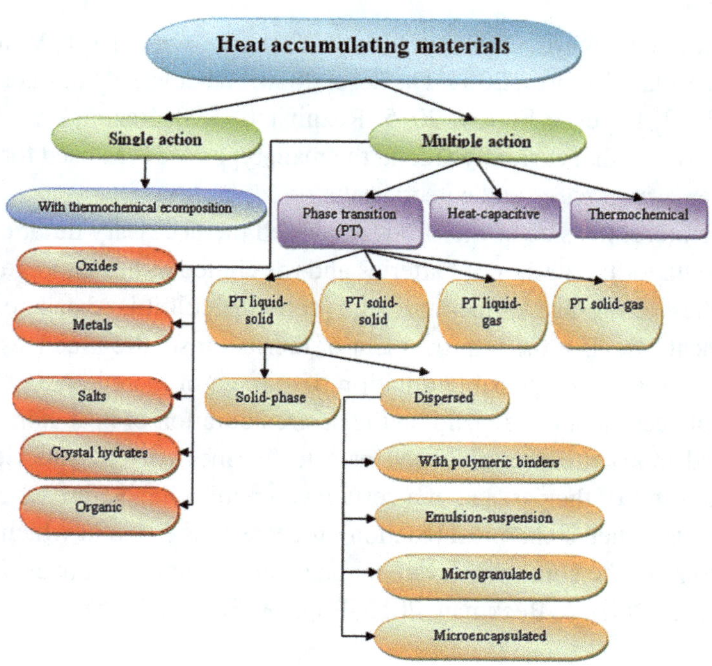

Table 1. Properties of the main classes of substances suitable for HSM

Classes	Thermal conductivity, W/(m·K)	Heat of fusion or decomposition, kJ/kg	Density kg/m³	Melting point or decomposition temperature, °C	Corrosion action
Organic matter or mixture	0,1 – 0,3	100 – 200	900 – 1000	160 – 200	Non-corrosive
Water-salt systems	0,4 – 0,6	100 – 200	1050 – 1300	90 – 120	Corrosion-active
Salts	0,4 – 0,6	150 – 800	1200 – 2000	150 – 900	Corrosion-active
Oxides	0,1 – 0,5	300 – 400	100 – 2000	800 – 2500	Non-corrosive
Metals and alloys	13 – 15	25 – 80	6000 – 10000	8 – 400	Low corrosion
The decomposing salts	0,3	2500	1200 – 2000	40 – 600	Non-corrosive
Degradable high-molecular compounds	0,1 – 0,3	2000 – 3000	900 – 1000	300	Non-corrosive

The success of the creation and introduction of thermochemical heat accumulators, which can be considered as seasonal, depend on the solution of thermophysical problems related to the selection of the necessary working bodies and the economic problems associated with their cost parameters.

PERSPECTIVE DIRECTIONS OF CREATION OF METHODS AND DEVICES FOR ACCUMULATION OF THERMAL ENERGY

The development of devices – heat accumulators (HA) for power plants (PP) on the basis of renewable energy sources (RES) should be made depending on their purpose. At the same time, the factors determining the choice of the method of energy storage should be taken into account.

The choice of the method of heat storage depends on the type of PP, its purpose and operating conditions, on the technological characteristics and cost parameters of the heat storage system (HSS).

The cost of HSS depends on:

- Capital costs for HA and HSM used to store energy, for equipment that provides the supply and removal of heat from the HA (including heat exchangers, pumps, pipelines, fittings, test equipment and automatic devices, as well as other devices serving and providing the work of the HSS);
- Values of energy losses (self-discharge);
- Temperature of the heat consumer;
- Connection schemes for HA (via a heat exchanger or directly, where HSM is also a heat carrier);
- The type of accumulation technology (due to the heat capacity, phase transitions or reversible chemical reactions);
- Operating mode of HA – in a constant load mode (in the basis of the load graph) or with variable load (in peak mode).

The total cost of HSS in solar power plants (SPP) for a number of projects is from 5 to 15 – 20% of the investment in SPP (Babaev, 2016a; Beckmann & Gilli, 1987; Duffy & Beckman, 2013).

An important characteristic of HSS is the specific energy capacity, which depends on the properties of HSM. With a high specific energy capacity, the capacity of HA decreases, and with a simple technology of its manufacturing, the construction of HSM is cheaper.

To date, a wide variety of HA types and designs is known, due to a wide range of tasks and areas of application (Grigorev, 2003; Da Rosa, 2010; Duffy & Beckman, 2013; Popel & Fortov, 2015; Sorensen, 2011). Many means and methods of heat accumulation lead to various technical and constructive solutions.

One can distinguish the following main types of HA (Babaev, 2016b; Reznitsky, 1998):

- Liquid HA;
- HA with a solid HSM;
- HA with liquid and solid HSM (gravel with oil, etc.);
- Steam HA;
- HA with electric heating element;
- Underground heat accumulators (UHA);
- HA with melting HSM;
- Thermochemical heat accumulators (TCHA).

In the United States, Japan, France numerous studies of HA have been carried out to justify the technological and economic conditions of their operation (Gigavat, n.d.). The experimental setups included a two-stage hot steam and cold tanks, mono- and two agent storage principles, and as the silicone oil used HSM, sand-gravel mix, together with the oil, oil molten salt ($NaNO_3 + NaNO_2 + KNO_3$), liquid Na, water under pressure with a molten salt, melts of salts KCl-LiCl and KF-LiF. The operating temperature was ranging from 210 °C to 827 °C (Gigavat, n.d.). The material obtained from the results of the experimental studies could be used in the development of HSS in a number of power plants based on RES.

At present, a wide range of substances has been studied, which provide an accumulation temperature from 0 to 1400 °C.

The use of phase transitions of hydrates of salts to store energy within the temperature range of 30 – 60 °C is complicated by a number of physicochemical processes – supercooling, nonequilibrium melting, low thermal conductivity, and deposition of matter on the heat exchange surface. Elimination of these shortcomings can be carried out by the introduction of chemical additives, the creation of a dynamic process due to the rotation of the cylinder, filled with salt hydrate.

At operating temperatures from 500 °C to 1600 °C, compounds and alloys of alkaline and alkaline-earth metals are generally used (Beckmann & Gili, 1987). A significant drawback of the use of metal compounds is considered to be a low coefficient of thermal conductivity, corrosion activity and a change in volume during melting.

Organic substances that are less corrosive, do not destroy the hulls, provide high energy intensity of HA, and have good technical and economic indices have proved to be good as melting phase-transfer heat-accumulating materials (PTHAM). The disadvantages of organic PTHAM include the fact that in the process of work, the heat of fusion decreases, and at a relatively low thermal conductivity of organic PTHAM, the creation and application of advanced heat exchange surfaces is required (Shishkin, 2005).

For the implementation of melting PTHAM, devices have been developed that exclude partially listed shortcomings (Reznitsky, 1998; Reznitsky, 1996, Shishkin, 2005). The main types of HA with PTHAM are: capsular; shell-and-tube; with scraping removal of PTHAM; with ultrasonic removal of PTHAM;

with direct contact and pumping PTHAM; with evaporation-convective heat transfer (Reznitsky, 1996, Shishkin, 2005).

Accommodation PTHAM capsules (Shishkin, 2005) provides a highly reliable structure, allows to create a developed heat-exchange surface, offset (using the capsules from flexible elastic material) changes in volume during phase transitions. However, due to the low thermal conductivity PTHAM need a numerous of small capsules, which leads to high manufacturing labor expenditure for HA production and to insufficiently efficient use of volume (for cylindrical capsules) low rigidity structure (for flat capsules). Because of this, it is advisable to use capsular HA in cases of heat dissipation of small heat fluxes from the heat exchange surface.

Especially promising looks to be HA, combined with a heat exchanger based on heat pipes. They are distinguished in view of construction by simplicity and include a reservoir (underground or aboveground) filled with granular mass. Uniformly, heat less pipes is distributed in the battery array. Some heat pipes (condensers) are placed in the collector with the heated medium, and the other part (evaporators) is put into the flue gas duct (Figure 5).

However, due to the wide range of the temperature potential of heat consumed in different industries, there is an additional need for the selection of appropriate PTHAM. Despite the wide variety of chemical compounds and mixtures possessing latent heat of phase transition, most of them are excluded for safety reasons in terms of toxicity, corrosion, and fire and explosion hazard, economy and due to inadequate physical, chemical and thermal properties.

Therefore, searches and creation of new compositions of salt mixtures are being conducted with the aim of their possible use as PTHAM in HSS.

Much attention has been paid in recent years to thermochemical heat accumulators.

The use of TCHA is based on the principle of using the thermal effect resulting from a reversible chemical reaction in a nonequilibrium state. Thermochemical cycles provide the principal opportunity to convert thermal, mechanical, light or electric energy into chemical potential.

Figure 5. Construction of a thermal battery with heat pipes: 1 – HA with a granular matrix; 2 – collector with heated medium; 3 – collector-flue with heating medium; 4, 5 – respectively, heat supply and heat removal heat pipes

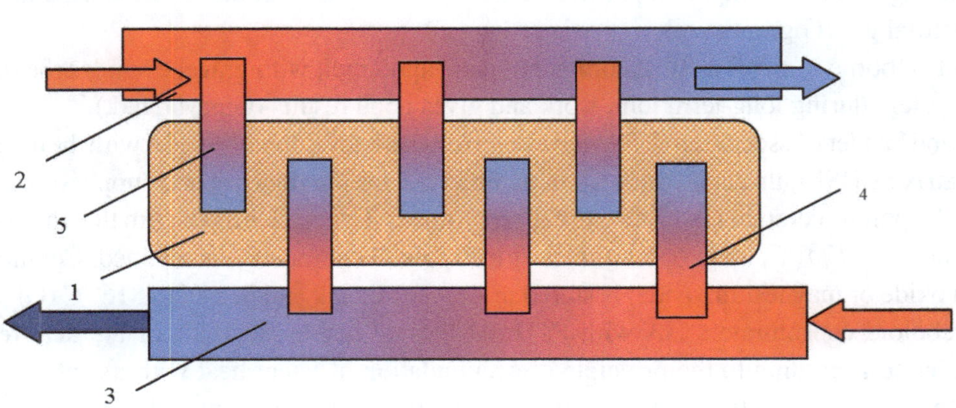

An important advantage of chemical methods of accumulating thermal energy in comparison with conventional ones is that the stored energy can be stored for a sufficiently long time without the use of thermal insulation; the problems of energy transport are alleviated for considerable distances without heat losses to the environment. This makes promising chemical methods of accumulating thermal energy.

The success of the creation and introduction of thermochemical heat accumulators, which can be considered as seasonal, depend on the solution of thermophysical problems related to the selection of the necessary working bodies and the economic problems associated with their cost parameters.

The most important direction in the field of thermal accumulation is the study and design of TCHA with granular heat-accumulating material (Reznitsky, 1996; Shishkin, 2005).

TCHA is proposed based on the reactions of hydrogenation of cyclohexane, on the thermal effect of the chemical reaction hydration – dehydration of hydrated salts of polyvalent metals (aluminum, beryllium, magnesium, iron, calcium and zinc), on the use of absorption properties of gases, reversible reactions for hydrogen and hydrate.

The search for new solutions based on reversible chemical reactions is continuing, which makes it possible to overcome a number of difficulties and shortcomings inherent in thermochemical HA (gas generation and storage, low heat of gas condensation, high cost of vessels for storage of non-condensible gases, etc.). However, the considered solutions testify to the prospects of such HA for their use not only in solar energy.

The analysis of the state of development of the HSS in the structure of the PP determined not only the scope of their application, but also the ways of their improvement. HSS have a relatively high efficiency (70% for indirect systems and more than 90% for direct ones, when HSS is both a heat carrier and there is no heat exchanger in the HSS circuit) (Gigavat, n.d.).

A promising direction of HSS improvement is the improvement of the design of heat-capacity HA, and solving one of the most difficult problems is the search for inexpensive PTHAM and TCHA.

The high efficiency of HSS is ensured by the use of substances such as sodium or molten salts, which simultaneously serve as PTHAM, as the heat transfer medium passing through the heat sink.

Successfully developed HA based on salts. Their advantages: inexpensive, have a high density of stored energy and operate at sufficiently high temperatures. Company Martin Marietta develops HSS based on nitrate melt. Hot salt (566 °C) is stored in a refractory tank lined with internal thermal insulation, and cold salt (288 °C) – in a tank of carbon steel, which is also protected by internal thermal insulation. The inner lining is made of liquid impermeable wafer membranes, similar to those used for storage of liquefied natural gas (Gigavat, n.d.).

It should be borne in mind that the combination of liquid melt with gravel excludes the degradation of the salt system during long-term joint work and gives good results (Gigavat, n.d.).

Boeing and Sanders Association (USA) have completed the development of HA with the use of a porous ceramic matrix as HSM, through which air is pumped out for the removal and supply of heat. HSM is retained in the porous ceramic matrix due to capillary forces. Experiments confirm that in such matrices at a temperature of 700 °C, up to 65% of the salt melt (alkali carbonates) is retained. Ceramic material (aluminum oxide or magnesium oxide) is stored in a pressure tank and heated to 816 °C (Gigavat, n.d.).

For paraboloid concentrators HA with PTHAM located inside the walls of the heat receiver are developed, based according to the principle of accumulation of latent heat with its subsequent use in engines with Rankine, Brighton, Stirling cycles that are mounted together with the HA on concentrators and are performed in the form of an integral receiver-battery design (Gigavat, n.d.).

Heat exchangers in HA with PTHAM are made of expensive alloys to protect them from high-temperature corrosion, therefore it is necessary to find new types of heat exchangers for PP based on renewable energy sources.

Under the scientific guidance of NASA in the United States, a high-temperature heat exchanger with direct contact between the HSM and the coolant is developed and investigated (Figure 6).

In it, refractory balls of complex oxides (40% SiO_2, 20% MgO, 35% CaO and 5% Al_2O_3) are fed to the solar heat receiver, where they are melted and then pumped to the HA tank When selecting a heat, the melt is fed into a high-pressure heat exchanger. In it, the melt is sprayed into the working gas stream under high pressure and gives it heat, and it hardens itself. The heated gas enters the turbine, and the solid melt in the form of pellets remains at the bottom of the heat exchanger. This system is economically justified, but some of its basic principles still need to be worked out, verified and confirm.

One of the promising areas for the development of new HSS is the development of a method or device for accumulating (conserving) the summer heat for the winter, and winter cold for the summer, since a large amount of energy is spent not only for heating, but also for air conditioning in the summer.

Figure 6. High-temperature heat exchanger with direct contact: 1 – heat-insulated wall; 2 – hot gas channel; 3 – supply of melt under pressure; 4 – hot gas outlet; 5 – partition with injectors; 6 – the throttle of pressure; 7 – exit of solid balls into an additional container; 8 – inlet of cold gas

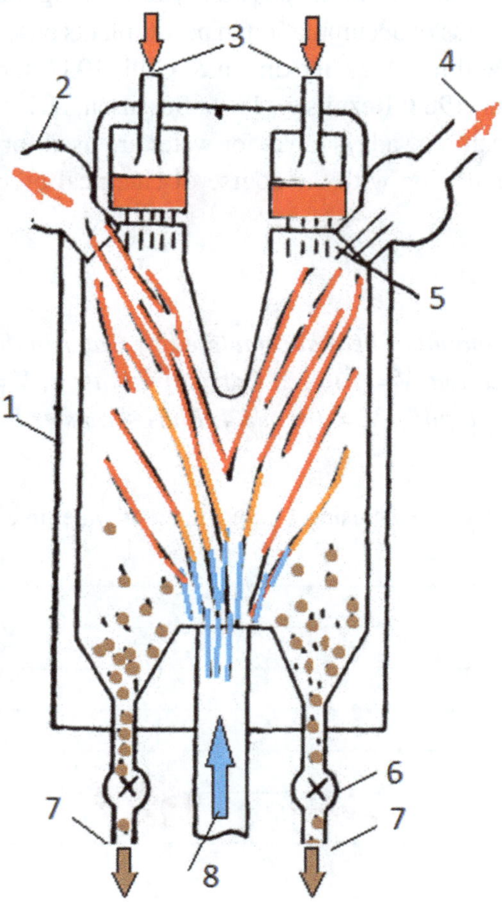

COMPARISON OF THE ENERGY CAPACITY OF HEAT ACCUMULATORS

When considering the different types of energy accumulators, many of their characteristics must be taken into account. In this case, the overall efficiency of the accumulation process is very important, and then the delivery of energy in one convenient form or another. At last, especially for large accumulators, the specific value has a unit cost, which includes both the cost of the active substance and the structural elements, auxiliary mechanisms, etc. Of course, the performance characteristics – temperature, pressure, magnetic and electric fields, safety, simplicity service and much more as well.

Figure 7 shows the comparative characteristics of the specific heat-storage capacity of 1 m^3 of water with its possible use in different types of HA. When accumulating in the form of mechanical energy, for example, in a pumped storage power plant with a rise of 100 m, the energy storage capacity is 1 MJ/m^3. When accumulating due to the heat capacity (with an increase in temperature by 1 °C), 4,2 MJ/m^3, due to the "solid-liquid" phase transition, 332,4 MJ/m^3 and due to the "solid-liquid" phase transition – 332,4 MJ/m^3, and due to the "liquid-gaseous" phase transition, the specific heat of accumulation is 2257 MJ/m^3.

As one can see, the best specific indicators have technologies that involve the involvement of smaller structural particles of the material into the process of accumulation. This indicates the relevance and prospects of developing technologies and devices for heat storage due to phase transitions of materials and chemical reactions (Babaev & Khalilullayev, 2005).

In recent years, studies have also been actively carried out to investigate the processes and thermal effects of reversible chemical reactions and the possibility of creating thermochemical HA.

Analysis of the applied methods of accumulation in power plants based on renewable energy sources (Babaev, 2016b; Babaev & Danilin, 2002; Beckmann & Gilli, 1987; Grigorev, 2003; Da Rosa, 2010; Duffy & Beckman, 2013; Popel, 1988; Reznitsky, 1996; Sorensen, 2011; Twidell & Weir, 1990; Shishkin, 2005) shows that in the main nowadays tanks for water are used for accumulation, and the search for efficient methods of accumulation with the course of chemical processes in multicomponent salt systems can give good results.

Figure 7. Comparison of the amount of heat accumulated by water in different types of HA (1. Hydro accumulating electric power station, V = 1 m^3; 2. Capacitive battery, V = 1 m^3, Δt = 1.0 °C; 3. Phase-transfer battery of heat (solid - liquid), V = 1,0 m^3; 4. Phase-transfer battery of heat (liquid-gaseous, at t = 100 °C), V = 1,0 m^3).

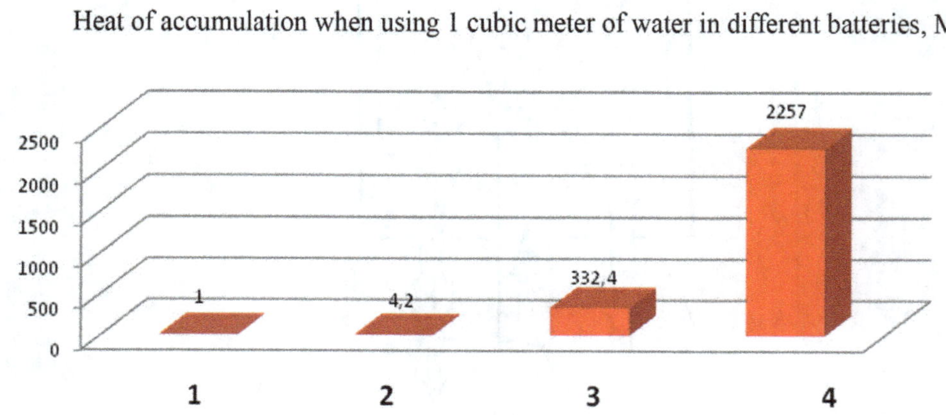

Thus, electrochemical (acid and alkaline) batteries are more effective for accumulating electric energy, and for the accumulation of thermal energy, the use of PTHAM and TCHA (Babaev, 2016a) is a more effective way of storing energy from renewable energy sources.

HEAT-RESISTANCE AND HEAT-RESISTANT THERMAL-CONTAINING MATERIALS

Storage at the expense of heat capacity is carried out by changing the temperature of the material without its phase or chemical transformation. The heat storage capacity or the amount of heat (kJ) that can be accumulated in a heat accumulator of a capacitive type is determined by the formula

$$\Delta H = \int_{T_1}^{T_2} C_p dT \tag{1}$$

where $[T_1, T_2]$ is the operating temperature range of the battery; C_p is the isobaric heat capacity of HSM.

There are semi-empirical laws, according to which the heat capacity can be estimated in the first approximation. For simple solids at room temperature, the heat capacity is in the range 25,9 – 26,8 J/(g-atom•K) (Dulong-Petit rule). This value increases almost linearly with increasing temperature and at the temperature of the first phase transition (melting or allotropic transformation) is approximately the same for all elements and is 29,3 – 30,3 J/(g-atom•K). To obtain the heat capacities of compounds, these values must be multiplied by the number of atoms in the molecule (Neumann and Kopp rule). Thus, assuming a linear increase in $C_p(T)$, it is possible to calculate with high accuracy the heat capacity of the compounds for a wide range of temperatures:

$$C_p(T) = \alpha + \beta T \tag{2}$$

where α and β are temperature-independent empirical constants that are found from the system

$$C_p(298^o K) = \alpha + \beta 298 \tag{3}$$

$$7,25 \cdot n = \alpha + \beta \cdot T_{ph.t.} \tag{4}$$

The values of C_p (298 °K) are known for most compounds, n is the number of atoms in the compound, $T_{ph.t}$ is the temperature of the first phase transition. Then the heat stored by the battery in the interval $[T_1, T_2]$ for a solid HSM is derived from the formula:

$$\Delta H \approx \int_{T_1}^{T_2} \alpha + \beta dT = \alpha \Delta T + \beta T_{av} \Delta T = Cp_{av} \cdot \Delta T \quad (5)$$

where $T_{av} = (T_1 + T_2)/2$, $Cp_{av} = \alpha + \beta T_{av}$.

It is seen that the greater the temperature difference and the average heat capacity, the more effective the heat accumulators.

The heat capacity in the liquid state often plays a more significant role in accumulation (Magomedov, 2000), while the experimental data on the heat capacity of liquid compounds are very limited. Therefore, at least an estimate of the heat capacity of the liquid phase is desirable. From a number of formulas given in the different papers, the heat capacity in the absence of experimental data is satisfactorily described by the following formulas:

$$C_p^{liq} = \frac{\Delta H_{boil}}{\dfrac{T_{boil}}{2 - \exp\left(0.237 \dfrac{\ln M}{M^{0.265}}\right)} - T_{boil}} + 16.76m - 90, \quad (6)$$

where M is the molecular weight, m is the number of atoms in the HSM.

$$C_p^{liq} = C_{p,\max}^{n} \left(\frac{1 - \dfrac{340}{T_{boil}} \cdot \exp(-0.0055 T_{boil}}{0.65} \right) \quad (7)$$

$C^n_{p,max} = R(3m - 5)$ – for linear molecules,

$C^n_{p,max} = R(3m - 6)$ – for nonlinear molecules,

where $C^n_{p,max}$ – the specific heat of the molecule in the ideal gas state, J/(mol·K); m is the number of atoms in the molecule of the compound.

With complete absence of experimental information on the thermophysical properties of the compound, the evaluation of its specific heat in the liquid state is carried out according to the Neumann-Kopp method:

$$C_p^{liq}\left(A_0 B_\varepsilon \ldots Z_z\right) = 0 C_p^{liq}\left(A\right) + \varepsilon C_p^{liq}(B) + \ldots + z C_p^{liq}\left(Z\right) \quad (8)$$

where $C^{liq}_p(A)$..., $C^{liq}_p(Z)$..., – reference data C^{liq}_p for atoms A, B, .., Z, or are considered equal to 30,33 J/(mol·K), which, of course, reduces the accuracy of the description of HA.

All three methods of calculating the heat capacity of the liquid phase provide quite satisfactory accuracy – 7.5 – 9%. For most substances, the heat capacity of the liquid in the interval between the melting and boiling points remains practically constant. The heat stored by the liquid due to its heat capacity is:

$$\Delta W^{liq} = C_p^{liq} \cdot \Delta T \tag{9}$$

Thus, in the absence of experimental data, a number of calculation formulas can be used that can accurately estimate the heat capacities of both the solid and liquid phases, which allows approximately describing the heat balance of HA for virtually any compound used as HSM.

The main characteristics of heat capacity HSM, which are mainly used at present, are given in Table 2.

For temperatures up to 100 °C, water is a very convenient heat-accumulating substance. It has a large specific heat $c = 4{,}2$ kJ/(kg·K), has many attractive performance characteristics, and, which is very important for large batteries, is very cheap.

The shortcomings of all capacitive HSM can be attributed to the relatively low heat-accumulating capacity and the inconstancy of the charging and discharging temperature of HA.

Thus, materials suitable for the role of HSM should have a high thermal capacity and be stable in the operating temperature range. In addition, they must have a high thermal conductivity or have a convenient way of adding heat to them and removing it from them.

From the above laws it is clear that the specific heat (per unit mass) will be the greater, the smaller the relative molecular mass of the substance. Therefore, it is advisable to use light solids or liquids as working bodies of thermal accumulators.

Table 2. The main characteristics of heat-storage materials (Sorensen, 2011)

Material	Temperature range, (°C)	Specific heat, kJ/(kg·°C)	Volumetric heat capacity, kJ/(m³·°C)	Thermal conductivity, W/(m·°C)
Solid				
Sodium chloride	< 800	0,92	2,0	9[1,2]
Cast iron	< 1500	0,46	3,6	70[2] – 34[3]
Stone, (granite)	< 1700	0,79	2,2	2,7[2]
Brick		0,84	1,4	0,6
Earth (dry)		0,79	1,0	1,0
Liquid				
Water	0 – 100	4,2	4,2	0,6
Oil (“thermo”)	from –50 to 330	2,4	1,9	0,1
Sodium	from 98 to 880	1,3	1,3	85[2] – 60[3]
Diethylene glycol	from –10 to 240	2,8	2,9	

[1]In the case of a granular material with air voids filled with air, less

[2]At 1000 °C

[3]At 700 °C

DEVICES FOR HEAT-ENERGY ACCUMULATION

The most well-known thermal storage devices are heat accumulators due to the heat capacity – tanks with water and with a solid HSM (Figure 8) (Babaev, 2016b; Babaev & Danilin, 2002; Beckmann & Gilli, 1987; Reznitsky, 1996; Twidell & Weir, 1990).

As a rule, a steel vertical tank is used that is 3 to 5 times larger than its diameter to ensure a thermal separation of water. To reduce the thermal losses of the tank, the insulation is made of glass wool. The inner surface of the tank, in contact with tap water, is protected from corrosion. In the tank, horizontal partitions can be provided (figure 8a, d), which divide the tank into sections with different water temperature levels in height. This increases the efficiency of heat accumulation and coolant circulation. The installation may comprise a float valve for supplying cold water (Figure 8c) and a pipe for its entry into the lower part of the tank, a heat exchanger in a two-circuit system for supplying heat (Figure 8b, d), an electric heater and a heat exchanger for transferring heat to the heating system (Figure 8d).

Large-scale heat storage systems successfully use reinforced concrete and steel tanks with a capacity of up to 100,000 m^3, in which hot water can store up to 8,000 GJ of heat at a temperature of 80 to 95 °C (Babaev, 2016b), and so-called "solar ponds" for example, in Miamisburg, Ohio, with a gradient of NaCl concentration from 0% at the top to 18,5% at a depth of 1,5 m (Sorensen, 2011) (Figure 8e). They

Figure 8. Tanks-batteries and types of HA capacitive type: a – a tank with cold water supply from below and internal partitions; b – a tank with heat input through the heat exchanger; c – a tank with a float valve and a pipe for supplying cold water to the bottom of the tank; d – sectionalized tank with electric heater; e – solar pond; f – a tank with a solid HSM; g – channel; 1 – heat-insulated housing; 2 – partition; 3 – float valve; 4 – standpipe; 5 – heat exchangers; 6 – electric heater; 7 – sectional partition; 8 – solid HSM (gravel, crushed stone, etc.); 9 – supports; 10 – HSM; 11 – water with increasing downward concentration; 12 – a layer of hot solution

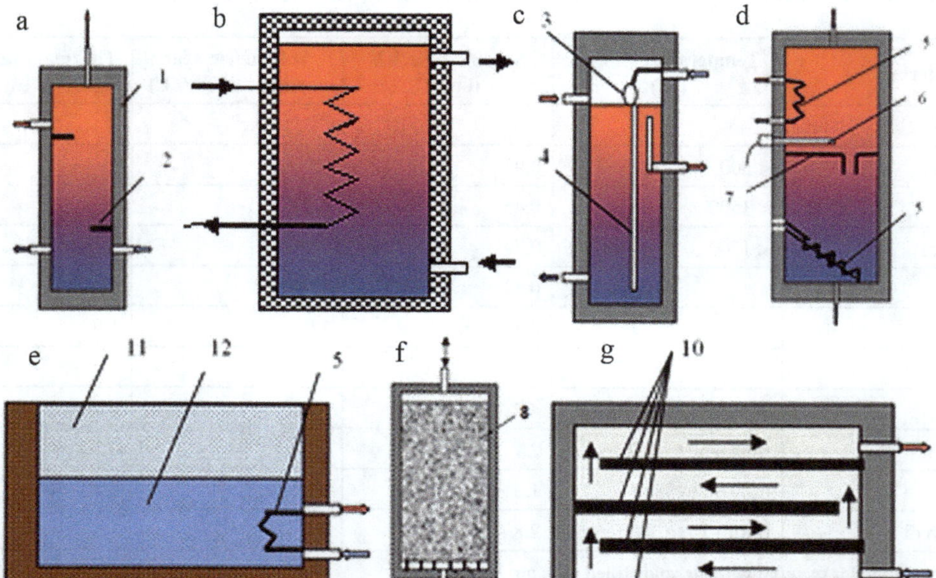

are fairly simple to operate, but require large investments. It is advisable to use them together with heat pumps, in this case their heat storage capacity can be doubled due to deeper (up to 5 ºC) cooling of water.

In the work of Popel (1988) a classification of solar ponds, the described processes occurring therein and given the temperature dependence of water solubility of some salts are considered. In the solar pond occurs simultaneously capture and accumulation of solar energy in a large volume of liquid.

It has been found that some natural saline lake water temperature at the bottom can reach 70 °C. This is due to the high concentration of salt. In the conventional pond absorbed solar energy mainly heats the surface layer, and this heat is lost rather quickly, especially at night and in cold rainy weather due to water evaporation and heat exchange with the ambient air. Solar energy, penetrating through the body of liquid in the solar pond is absorbed by dark-colored bottom and heats the adjacent layers of liquid, whereby its temperature can reach 90 - 100 °C, while the temperature of the surface layer may remain at 20 °C. Due to the high heat capacity of water in the solar pond for summer season accumulate a large amount of heat and the heat losses due to low temperature drop in the lower layer in the cold season is slow, so that the solar pond serves seasonal energy accumulator. Heat to the consumer is taken from the lower zone of the pond.

Usually the depth of such ponds is 1 – 3 m. On 1 m^2 of the pond area it is required 500 – 1000 kg of table salt (Popel & Fortov, 2015), it can be replaced with magnesium chloride.

A detailed description of the results of research and practical application of unstabilized solar ponds is devoted to works (Popel & Fortov, 2015; Popel, 1988). The study of heat supply systems based on solar ponds is devoted to (Popel, 1988).

Often, gravel, crushed stone, pheolite (iron ore), sea pebbles, etc. are used in HA with solid-grained HSM. (Figure 8f). These materials are mainly composed of silicon oxide, the specific heat of which slightly exceeds 1 kJ/kg·K. Such batteries can operate up to temperatures of several hundred degrees; Air, oil or inert gases may serve as the coolant. During charging, hot air is blown through the accumulating material layer, heating it, and during the discharge process, air or other gas is pumped in the opposite direction and takes this heat away. The advantage of such batteries is also their simplicity and cheapness (Grigorev, 2003; Sorensen, 2011; Shishkin, 2005).

Electroheating elements can be used in HA with liquid or solid HSM (Figure 8d). The channels HA are widely used in heat and power supply systems (Figure 8g) (Bekirov, 2011; Babaev, 2016b; Bekmann & Gilli, 1987; Duffy & Beckmann, 2013). Heat-accumulating material (chamotte, refractory brick, etc.) is heated during periods of minimal electricity consumption. Passing cold air through the HSM, it is possible to heat the premises. Accumulators of this type are produced abroad serially for individual and small-family houses (Duffy & Beckmann, 2013; Sorensen, 2011).

Special types of channel HA with solid HSM are thermal graphite batteries used as an energy source in autonomous power plants (Sorensen, 2011; Shishkin, 2005). The temperature of their heating can reach 3500 K, which provides good weight and size characteristics of the installation.

In passive systems of solar heat supply, the materials of enclosing structures and enclosed spaces of extensions, attics and cellars with HSM – air or gravel are used as HA.

Thermal accumulators with liquid and solid HSM are HA with a two-agent accumulation principle, where, for example, a sand and gravel mixture is used together with silicone oil (Gigavat, n.d.).

In steam HA, the use of water or other liquid materials at higher temperatures requires the use of pressure vessels, which significantly complicates and increases the cost of construction (Adadurov, 2005; Babaev, 2016b).

For the seasonal accumulation of heat, the use of underground reservoirs, soil, rock and other natural formations is promising.

The accumulation of energy in underground heat accumulators is developed on the basis of the following means:

- Deep wells with water injection;
- Deep wells with a bubbled layer of liquid;
- Heat-exchange solid backfill, water or a system of concentric pipes, air-blown, in an isolated underground cavity;
- Heat accumulation in aquifers.

Examples of such seasonal heat stores using the heat capacity of the soil (clay, pebble-sand mixture), which are systems of horizontal or vertical canals in the ground, made of metal or plastic pipes (Figure 9) are the systems operating in the Netherlands (Gronengen), Sweden (Sankley, Kullavik), Switzerland (Walruz), Austria (Kransbiten). The cost of energy in them is 0,03 – 0,08 USD/kW·h (Adadurov, 2005; Babaev, 2016b).

Figure 9. Underground HA. a, b – underground with vertical and horizontal channels; c – in the aquifer; d – the use of underground voids; 1 – thermal insulation; 2 – HSM; 3 – inductor; 4 – aquifer layer; 5 – waterproof layer; 6 – thermal insulation; 6 – brickwork

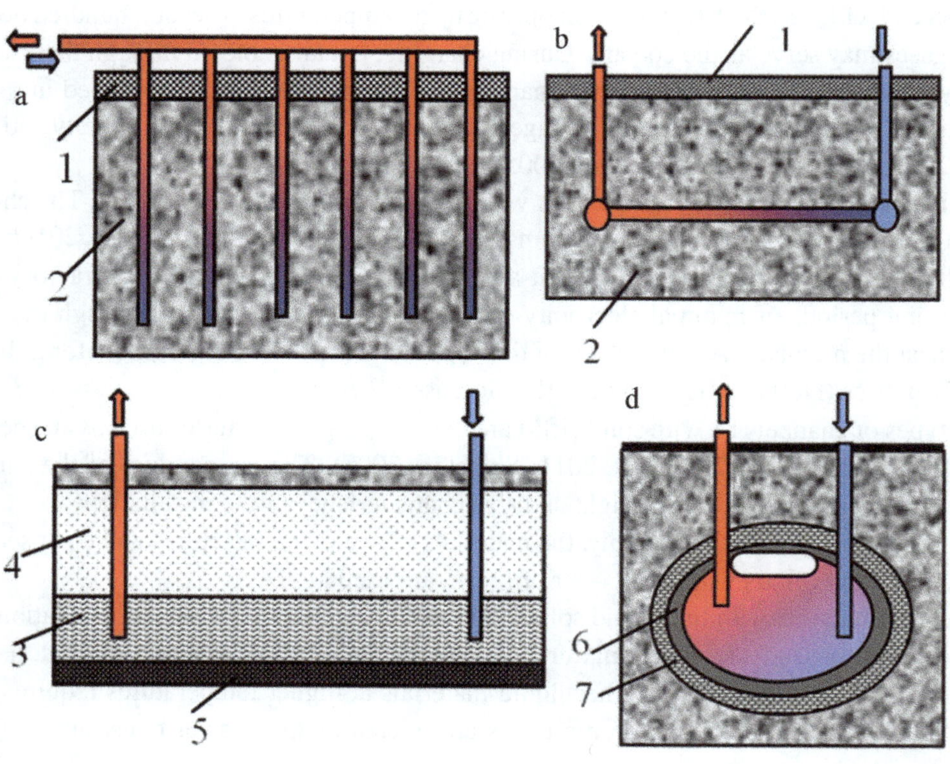

The seasonal UHA with a capacity of 100 thousand m^3, built at a depth of 30 m in Sweden, has an annular shape, its height is 30 m, the outer and inner diameters are respectively 75 and 35 m. The total surface of the collectors is 4,2 thousand m^2. In summer and autumn, UHA receives heated water up to 90 °C, and is taken from it with a temperature of 65 - 70 °C. The average annual efficiency was 30% (Gigavat, n.d.). The cost of heat from this UHA was equivalent to the cost of heat from the heat network. In this country, a UHA project with a capacity of 700 thousand m^3 is being developed (Gigavat, n.d.).

When using mine workings or cavities as a UHA, the questions of the effect of the coolant as a function of pressure and temperature on the contour zones, ecology, the conditions for the occurrence of fracture of the constituent arrays, and the loss of stability during thermal cycling are to be studied.

It is necessary to find new solutions for creating effective UHA. One such solution is the UHA, made in the form of an underground cavity lined with brickwork, into which a reservoir of rubber or plastic is placed (Figure 9d). Between them and the walls of the cavity a space is formed, filled with foam for thermal insulation (Gigavat, n.d.).

CONCLUSION

The inclusion of accumulating units in the power system improves the reliability and stability of the power supply to consumers, makes it possible to regulate the operation mode of power plants without strict dependence on the energy consumption mode by cutting the peak of the graph by means of discharging the battery and filling the graphs with charge of the battery.

More and more attention has been paid to thermochemical batteries, which ensure the storage of energy as a result of the passage of a certain reversible chemical reaction. The accumulation of heat in this case occurs when a chemical reaction proceeds to the side in which heat absorption take place, and the heat return to the consumer is due to the reverse reaction.

The success of the creation and introduction of thermochemical heat accumulators, which can be considered as seasonal, depend on the solution of thermophysical problems related to the selection of the necessary working bodies and the economic problems associated with their cost parameters.

The development of devices – heat accumulators for power plants on the basis of renewable energy sources should be made depending on their purpose. At the same time, the factors determining the choice of the method of energy storage should be taken into account.

When considering the different types of energy accumulators, many of their characteristics must be taken into account. In this case, the overall efficiency of the accumulation process is very important, and then the delivery of energy in one convenient form or another.

REFERENCES

Adadurov, E. A. (2005). Povyshenie ehffektivnosti ispol'zovaniya akkumulyatorov teploty s vozobnovlyaemymi istochnikami ehnergii [Increasing the efficiency of using heat accumulators with renewable energy sources]. Krasnodar: Dis. ... cand. tech. sciences. (in Russian)

Alekseev, V. V. (1991). Solnechnaya ehnergetika: perspektivy razvitiya [Solar energy: development prospects]. Knowledge. (in Russian)

Babaev, B. D. (2014). Principy teplovogo akkumulirovaniya i ispol'zuemye teploakkumuliruyushchie materialy [Principles of thermal accumulation and used heat storage materials] [in Russian]. *Thermophysics of High Temperatures*, *52*(5), 760–776.

Babaev, B. D. (2015). Resursy vozobnovlyaemyh istochnikov ehnergii Respubliki Dagestan: uchebno-spravochnoe posobie [Resources of renewable energy sources of the Republic of Dagestan: a manual]. Makhachkala: Publishing house "Rainbow". (in Russian)

Babaev, B. D. (2016a). Razrabotka i issledovanie ehnergosistem na osnove vozobnovlyaemyh istochnikov s fazoperekhodnym akkumulirovaniem tepla [Development and investigation of power systems based on renewable sources with phase-transfer heat accumulation]. Moscow: RSL. (in Russian)

Babaev, B. D. (2016b). Akkumulyatory tepla pri ispol'zovanii vozobnovlyaemyh istochnikov ehnergii. Perspektivnye napravleniya novyh razrabotok [Heat accumulators using renewable energy sources. Perspective directions of new developments]. *Energetik*, *3*, 19–22.

Babaev, B. D., & Danilin, V. N. (2002). EHnergoakkumuliruyushchie ustanovki. Uchebnoe posobie [Energy storage facilities. Tutorial]. Makhachkala: CPI of the DSU. (in Russian)

Babaev, B.D. & Khalilullayev, G.M. (2005). *Opisanie termohimicheskogo vzaimodejstviya v mnogokomponentnyh vzaimnyh sistemah v zavisimosti ot temperatury. Teplovoj ehffekt v zavisimosti ot temperatury* [Description of thermochemical interaction in multicomponent reciprocal systems as a function of temperature. Thermal effect in relation to temperature]. Patent of the Russian Federation N° 2005610201. (in Russian)

Beckmann, G., & Gilli, P. (1987). Teplovoe akkumulirovanie ehnergii [Thermal energy storage: Translation from English]. Mir. (in Russian)

Bekirov, E. A. (2011). *Avtonomnye istochniki ehlektropitaniya na baze solnechnyh batarej: uchebnoe posobie* [Autonomous power supplies based on solar panels: a tutorial]. Simferopol: Arial. (in Russian)

Da Rosa, A. (2010). Vozobnovlyaemye istochniki ehnergii. Fiziko-tekhnicheskie osnovy. Uchebnoe posobie [Renewable energy sources. Physical and technical basis. Tutorial]. Intellect. Ed. house of Moscow Power Engineering Institute. (in Russian)

Duffy, J., & Beckman, U. A. (2013). Osnovy solnechnoj teploehnergetiki [Fundamentals of solar thermal power]. Dolgoprudny: Intelligence. (in Russian)

Gigavat. (n.d.). *All about power plants.* Retrieved from http://www.gigavat.com/

Grigorev, V. A. (2003). Razrabotka akkumulyatorov teploty s zernistym teplonositelem i metody ih rascheta na osnove matematicheskogo modelirovaniya [Development of heat accumulators with a granular coolant and methods for calculating them on the basis of mathematical modeling]. RSL. (in Russian)

Magomedov, M. M. (2000). Kompleksnaya metodologiya razrabotki fazoperekhodnyh teploakkumuliruyushchih materialov na osnove mnogokomponentnyh system [Complex methodology of development of phase-transfer heat-accumulating materials on the basis of multicomponent systems]. Moscow: RSL. (in Russian)

Popel, O. S. (1988). Issledovanie sistem teplosnabzheniya na osnove solnechnyh prudov [Research of systems of a heat supply on the basis of solar ponds]. Moscow: IVTAN. (in Russian)

Popel, O. S., & Fortov, V. E. (2015). Vozobnovlyaemaya ehnergetika v sovremennom mire [Renewable energy in the modern world]. Publishing house of Moscow Power Engineering Institute. (in Russian)

Reznitsky, L. A. (1996). Teplovye akkumulyatory [Thermal accumulators]. Energoatomizdat. (in Russian)

Reznitsky, L. A. (1998). Teploakkumuliruyushchie veshchestva i processy [Heat accumulating substances and processes]. *Journal of Inorganic Chemistry*, *43*(8), 1288–1298.

Shishkin, N. D. (2005). Kombinirovanie i ehffektivnoe ispol'zovanie istochnikov teplovoj ehnergii v avtonomnyh teploehnergeticheskih kompleksah, vklyuchaya vozobnovlyaemye istochniki [Combination and effective use of thermal energy sources in autonomous heat and power complexes, including renewable sources]. Moscow: RSL. (in Russian)

Sorensen, B. (2011). *Preobrazovanie, peredacha i akkumulirovanie ehnergii: uchebno-spravochnoe posobie* [Transformation, transmission and accumulation of energy: educational-reference manual]. Dolgoprudny: Intelligence. (in Russian)

Twidell, D., & Weir, A. (1990). *Vozobnovlyaemye istochniki ehnergii* [Renewable energy sources]. Moscow: Energoatomizdat. (in Russian)

Vasilev, Yu. S., Elistratov, V. V., Mukhammadiev, M. M., & Pretro, G. A. (1995). Vozobnovlyaemye istochniki ehnergii i gidroakkumulirovanie [Renewable energy sources and pumped storage]. St. Petersburg: SPbSTU. (in Russian)

KEY TERMS AND DEFINITIONS

Heat Accumulator: Device for heat accumulation for the purpose of its further use.

Heat Storage Material: Material used for the accumulation of thermal energy and may be classified according to the material class, the way of accumulation and heat release as well as from the cyclic operation.

Heat Storage System: The system of devices intended for temporary storage of thermal energy by means of various heat-storage materials and including a number of auxiliary equipment including heat exchangers, pumps, pipelines, fittings, test equipment and automatic devices, as well as other devices serving and providing the work of the heat storage system.

Phase-Transfer Heat-Accumulating Material: The type of heat-accumulating material using latent heat of phase transitions of the heat-accumulating material.

Thermochemical Heat Accumulator: The type of heat accumulator based on the use of binding energy of reversible chemical reactions of chemical accumulation.

Underground Heat Accumulator: The thermal energy accumulator located under the ground surface with different types of thermal insulation and heat storage materials.

Chapter 6
Creating a Synergy:
Renewable Energy to Develop the Russian Arctic

Dmitry Alexandrovich Solovyev
Shirshov Institute of Oceanology, Russia

Maria Olegovna Morgunova
Russian Academy of Sciences, Russia & KTH Royal Institute of Technology, Russia

Alexander Alekseyevich Solovyev
Lomonosov Moscow State University, Russia

ABSTRACT

The chapter focuses on different aspects and challenges of power supply for remote energy consumers in the Russian Arctic. The authors discuss the potential use of renewable energy, some specified technological features and risks related to the broader deployment of decentralized renewable energy systems in the Arctic region. Even though there is limited experience of renewable energy systems installation in the Russian Arctic and greater technological challenges, the authors see it as a potential opportunity to contribute to innovative and sustainable development of the region. The authors underline the potential synergistic effect of broader deployment of renewable energy systems in the Russian Arctic. The key argument is that a sustainable and efficient energy system will open new development opportunities and stimulate future socioeconomic development of the region through the use of local and renewable energy resources and the implementation of new power generation modes and technologies.

INTRODUCTION

Arctic is a special geographical area with fragile nature, harsh climate and huge economic potential. It is attractive not only due to the presence of natural resources, which can be extracted, but the Arctic is becoming an international "playground" to explore innovative technology and business approaches. The latest examples of transforming Arctic challenges into opportunities are Arctic data centers (Kelion,

DOI: 10.4018/978-1-5225-9179-5.ch006

2017) to save energy on cooling, and floating LNG liquefaction plants (Shell, 2018) minimizing footprint during offshore natural gas production especially in the Arctic.

Even though Arctic is rich and prospective in many ways, the key issue is availability of efficient energy supply. Arctic is rich in energy resources of different kinds, but at the same time lacks sufficient and stable energy supply due to its remoteness, low population and harsh weather conditions. Energy supply insufficiency is a huge challenge for future economic development of the Arctic territories throughout the North Pole. At the same time, it stimulates business and science communities, as well as governments to find efficient solutions.

In this chapter the main focus is on the Russian Arctic, even though both challenges and opportunities connected to energy system development are valid for all the Arctic territories and also some sub-Arctic areas. The Russian Arctic is also one of the most extreme cases, since Russia possesses the vast Arctic territories, which are mostly uninhabited and have very limited communication and infrastructure. At the same time, the Russian Arctic is considered to be the richest in natural resources, especially oil and natural gas, and very attractive for international transit and transportation along the Northern Sea Route (Kovalenko et al., 2018). Nevertheless, the same challenges are present in the Arctic zones of other circumpolar countries (USA, Canada, Greenland, etc.).

Due to the growing interest towards the Russian Arctic, the issue of energy supply and energy infrastructure development in the region becomes more urgent. Vast energy resources in the Russian Arctic are imported from other parts of the country. Insufficient transport infrastructure and harsh climate complicate the logistics of energy resources supply. Transportation complexity, thus high transport costs of fuel affect the final cost of energy produced.

Owing to the latest technological advances in energy generation and distribution, there is a possibility to widely introduce renewable energy in the Arctic, using decentralized and combined power supply regimes (Gabderahmanova et al., 2016; Solovyev, 2017). The new technologies are able to ensure stable, sustainable and efficient energy supply, giving broad possibilities for the Arctic region economic development. The most promising in terms of reducing the cost of transporting fuel in the Arctic regions of Russia are local renewable energy sources, such as wind, solar, water and biomass (Morgunova and Soloviev, 2016). Besides, renewable energy systems are minimizing human impact on the environment.

An improvement within the energy sector can give a higher impact on reaching the strategic economic, social and environmental goals. By implementing new models of energy supply, Arctic communities can receive additional combined value from the sustainable development, new technologies and new business opportunities without harm to the unique and sensitive Arctic nature. Following this idea, the chapter is focused on experiences and examples from different circumpolar countries.

The chapter is structured as follows. First, there is an analysis of the renewable energy sources potential in the Russian Arctic and their current use. Second, it is followed by a discussion of the application of decentralized and combined modes of power generation based on renewable and fossil fuels in the Arctic. Third, there is a more detailed analysis of technical peculiarities of the application of renewable energy generation facilities in the Arctic and risks related to that. After that the chapter suggests some ideas on the possibility of creating a synergistic effect of energy system development in the Russian Arctic due to new opening opportunities. Finally, the chapter concludes on the perspectives of using renewable energy in the Russian Arctic.

RENEWABLE ENERGY RESOURCES POTENTIAL AND THEIR CURRENT USE IN THE RUSSIAN ARCTIC

Renewable energy potential in the Russian Arctic, as well as in the Arctic as a whole, is underestimated. This is tidal energy, solar, wind and waste of timber and woodworking industry. Based on statistical data from 2016 (e.g., Solovyev et al., 2017), there are not that many functioning generation facilities based on renewable energy, nor combined power plants (using diesel fuel) on the territory of the Russian Arctic. Renewable power plants are mostly small hydropower facilities (92%), as well as wind (about 5%) and solar power plants (approx. 1.5%) (Solovyev et al., 2017).

Even though Arctic from the climate perspective is a special region with polar nights and polar days, heavy ice conditions (further discussed in the chapter), there is a considerable potential for renewable energy. Besides, renewable energy technologies in general are becoming more efficient and cheaper (Gridasov et al., 2011), thus affecting electricity generation costs.

In international practice, wind and solar energy potential assessment is being done by the perennial actinometric surface and meteorological measurements (Gidrometeoizdat, 1997), reanalysis and mathematical modeling of the global databases (Morgunova and Solovyev, 2017). Detailed maps of the distribution of sun and wind resources make it possible to identify promising areas for the introduction of renewable energy (Morgunova and Solovyev, 2017; Gabderahmanova et al., 2016; Gabderahmanova and Shakun, 2016).

The highest solar energy potential in the Russian Arctic is in the Republic of Sakha (Yakutia). In the rest of the Arctic it is possible to use solar energy seasonally (Morgunova and Solovyev, 2017). In general the leaders in terms of the daily sum of total solar radiation and the total potential of solar energy coming to the inclined surface are the Republic of Sakha, the Krasnoyarsk Territory and the Yamalo-Nenets Autonomous District. In the Republic of Sakha solar energy technologies are being introduced at a very high rate and are already successfully operating. Besides, 64 solar power stations of a total capacity 50 MW are planned to be constructed there (Solovyev et al., 2017). Further development of renewable energy capacity also in areas with lower solar radiation is dependent on technical adaptation of solar installations to low temperature, icing, low sun declination angles, and wide ranges of azimuth.

Some areas with lower solar radiation have larger wind energy potential. The leaders in the wind flow density are the Krasnoyarsk Territory and Chukotka Autonomous District. Those are in greater degree Arctic Ocean coastal lines, which means wind potential can be utilized to power ports and navigation systems of the Northern Sea Route infrastructure.

The present state of renewable energy use in the Russian Arctic is shown in Figure 1 (Morgunova and Solovyev, 2017; GIS, 2018). The existing installations are mostly used to provide communication, remote surveillance, lightning and small equipment operation.

Some of the most interesting examples of power installations on renewable energy in the Russian Arctic are listed in Morgunova and Solovyev (2017):

- combined wind-diesel power (by OJSC "Peredvizhnaja Jenergetika") in the North and the Far East regions (ongoing development);
- 250 kW wind turbine in Labytnangi Yamalo-Nenets Autonomous District;
- 7 wind farms with a total capacity of up to 3.4 MW in Kamchatka (design stage);
- wind-diesel station in North Yakutia, Chokurdakh village (under development);

Figure 1. Renewable energy power installations in the Russian Arctic (Morgunova and Solovyev, 2017; GIS, 2018)

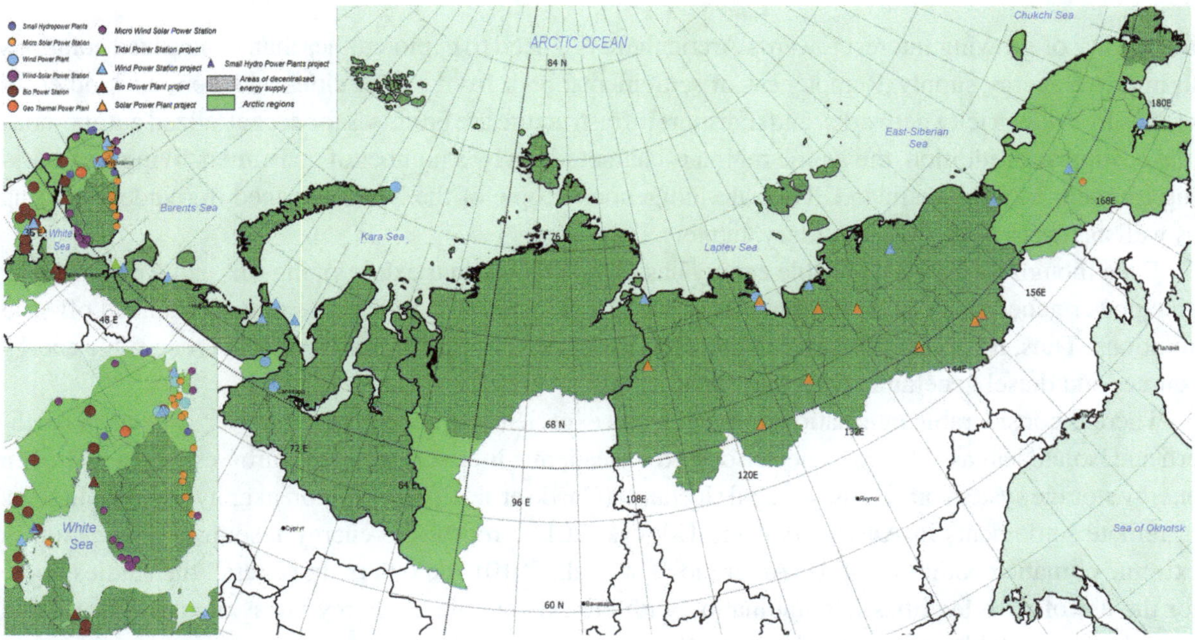

- 50 power plants (100-500 W) based on photovoltaic modules, nickel-cadmium batteries and small wind turbines on the banks of Okhotsk and the Bering Straits (Russian Scientific Centre "Kurchatov Institute" in frames of the radioisotope generators replacement programme);
- mobile photovoltaic systems up to 500 W from lead-acid batteries and low power diesel generators in the Republic of Sakha.

In total, cumulative installed power of renewable energy plants in the Russian Arctic is about 100 MW, where wind and solar energy is only around 6 MW. The other sources are micro hydropower (92 MW), tidal (1.7 MW) and geothermal energy (0.21 MW). (Gabderahmanova et al., 2016; Gabderahmanova and Tarasenko, 2016) Most of existing renewable power projects are pilot ones. Thus, the amounts of renewable energy use in the Russian Arctic are minor and does not really make any change to the established energy supply regime of expensive fuel deliveries from the mainland.

About 400 MW of installed power capacity of renewable energy is planned to be constructed in the coming years (374 MW wind energy, 7 MW solar energy, (Gabderahmanova et al., 2016; Gabderahmanova and Tarasenko, 2016), but still are considered to be very modest, even though this can give tangible results on reducing diesel fuel consumption, ecological footprint and electricity cost.

Currently, the use of local renewable energy sources in the Russian Arctic is non-systematic. A comprehensive approach to the development, as well as new technical solutions is required to realize the existing potential of small and medium scale power systems.

DECENTRALIZED POWER SUPPLY BASED ON LOCAL AND RENEWABLE ENERGY SOURCES

In a reality of growing interest towards Arctic there is a need to explore other than centralized solutions to improve energy supply. Remoteness of settlements, both from bigger cities and from each other, as well as from electricity networks, and transport infrastructure are premises for decentralized autonomous power supply. In addition, the Arctic region is characterized by a number of consumers living and working isolated from the mainland. They are indigenous people of the North engaged in reindeer herding as well as tourists, scientists, military, hunters, etc.

Even though we argue renewable energy has high potential in the developing Russian Arctic, renewable power generation can be unstable and insufficient, especially in the case of decentralized off-grid solutions. Thus, it requires a backup in the form of combined installations using local renewable energy sources and diesel generators (Morgunova and Solovjov, 2015).

There is considerable international research and experience in decentralized power systems in northern and isolated areas. The examples discussed in academic literature are wind turbines with a hydrogen energy storage system on Grimsay Island (Iceland) (Chade et al., 2015); windbreaker hybrid installations in remote settlements in Alaska (Fay and Udovyk, 2013); renewable energy resources application in extreme climatic conditions of the Antarctic (Tin et al., 2010) and others. In general, the studies argue for the use of such facilities in combination with the existing diesel energy infrastructure, since it can provide a more stable energy supply to remote consumers. Besides, those studies conclude on relatively short investment return (Chade et al., 2015), positive socioeconomic effect (Fay and Udovyk, 2013) and wide applicability of such installations all over the northern territories (Tin et al., 2010).

Decentralized power supply systems based on renewables are rare in the Arctic region, but has huge potential if can be adapted to the harsh climate and Arctic conditions (further discussed in the chapter). In this context, it is worth to mention the successful experience of using decentralized energy systems based on renewable energy sources for other remote energy consumers not only in the northern regions. They are, for example, solar and wind power generation with hydrogen energy storage systems on islands in Malaysia (Ashourian et al., 2013), and use of photovoltaic panels in combination with the hydrogen storage system in Brazil in the Amazon River area (Silva et al., 2010). Those examples, both in the northern territories and in the other remote areas, show the possibility of using combined power generation facilities.

In studies of Russian scientists and researchers, the issue of the renewable energy use in combined power plants for remote energy consumers is less discussed, even though it is considered as one of the possible solutions for Arctic energy infrastructure development. There are studies of Popel et al. (2015) to highlight on the use of renewable energy sources for the energy supply of the Russian Arctic, which describes the characteristics of energy consumers, the evaluation of renewable energy resources in the Russian Arctic on the basis of statistical data and modeling, the assessment of the potential of renewable energy supply and the project of a portable power plant for working in severe conditions of the Arctic.

In general, studies on decentralized renewable energy projects in the Arctic and ones with close to the Arctic climate are limited. This does not allow to study in depth the threats and opportunities of developing the Russian Arctic energy infrastructure, mainly due to the lack of widely available data on the technical and economic evaluation of such projects.

Currently, around 200'000 people living in the Arctic region in Russia are actually supplied by decentralized autonomous power plants of different scale (from 1% of the population in the Murmansk area up to 100% in the Republic of Sakha) (Gabderahmanova and Shakun, 2017). A wider use of decentralized energy systems can provide more industrial opportunities and serve as a basis for an integrated approach towards exploration and use of Arctic region energy potential (Bushuev et al., 2017). There are 21 ports along the Northern Sea Route, which can significantly benefit on using local energy sources, including renewables via utilizing decentralized power solutions. Besides, there are nature protected areas with growing supporting infrastructure, which demands energy (Morgunova and Solovyev, 2017). It can also accelerate the development of the Arctic infrastructure, including the Northern Sea Route.

Nevertheless, the feasibility of installing those decentralized power systems also depends on other factors than local energy resource potential. They are, as defined by Gabderahmanova and Shakun (2017): distribution of consumers; access to the grid; other existing power generation facilities nearby and their technical characteristics; land use regime including nature protected areas, military objects, etc.; future infrastructure development plans; transport network; and nature and climate conditions of the area. Thus, specific features of renewable energy installations play one of the crucial roles in their future deployment.

TECHNICAL FEATURES OF RENEWABLE ENERGY SYSTEMS IN THE ARCTIC

In general, the main qualities of power installations should be resistance to climate, high performance and rare maintenance need. The main qualities of renewable energy installations are reliability, robustness and cost-efficiency (Fay and Udovyk, 2013). Since power generation equipment and supporting installations are subject for adaptation to harsh climate conditions of the Arctic, it is necessary to underline the key challenges related to that. In case of solar power installations, those are, as it has been already mentioned, low temperature, low sun declination angles, as well as energy accumulation (suitable type of accumulator for low temperatures, etc.) and also icing.

For wind turbines, one of the main issues is icing. Special measures are required to adapt the technical equipment of the windmills for icing and increasing vibration, especially in the regions of high risk of icing, such as coastal lines (Figure 2). The specific measures to be taken with regard to the wind turbines are: hydrophobic coating on the blades of the wind wheel; replacement of pneumatic braking system by electromechanical; anti-corrosion coating of the stator and rotor of the generator; installation of the heating system of the gondola and wind speed and direction sensors; reservation of wind speed sensors, and others. Besides, other technological issues not related to icing are: logistics of delivery of the parts of the equipment, a need for a special type of foundation to prevent permafrost melting, as well as resistant construction materials are some of the other challenges renewable energy is facing in the Arctic region.

Some adaptation activities for wind power installations took place in Antarctica for polar stations and field camps (Tin et al., 2010). The adaptation measures include the replacement of gearboxes, shorter towers, control software modifications, and de-icing equipment. The results of implementation of vertical axis wind turbine were positive with only few replacements and high productivity, accounting for 5-12 year investment return (Tin et al., 2010). This experience is arguing for broader deployment of specially constructed commercial size wind turbines for the Polar region. Solar power, combined with wind turbines and diesel generators also worked well, even though icing and snow degrades their performance

Figure 2. Fighting the icing of wind turbine blades with a helicopter (Nilsen, 2015)

by covering the panels. In general, solar panels in the Polar region are only to supply electricity for some basic needs, and require to be combined with some other installations.

Needless to mention that technical adaptation can be costly. It is possible to economically justify the effectiveness of the renewable power installations adaptation by comparing the effect of reducing losses (S_{eff}) and the actual cost of adaptation measures (S_{ad}). With this purpose the following relations are used (Solovyev, 2017; Elistratov and Denisov, 2016):

$$S_{eff} = Sp_1 - Sp_2,\ Sp_i = P_{\Sigma i} * W * S_{ET},\ S_{ad} = S_{INST} + S_M + W_{EC} + S_{ET},$$

where S_{ET} is a tariff for electricity for a kWh; S_{Pi} is a cost of losses without application of adaptation measures (i = 1) and with their application (i = 2); $P_{\Sigma i}$ is a total losses without application of adaptation measures (i = 1) and with their application (i = 2),%; W is an annual output of electricity without losses from adverse environmental effects, kWh; W_{EC} is an annual electricity consumption for adaptation activities, kWh; S_{INST} is a cost of protective measures; S_M is a cost of a maintenance.

Using this assessment technique and taking into account the statistical information about the operation of a power plant and weather conditions at the location of the small power generation facility (for example, the number of days in a year with conditions conducive to the icing of the wind turbines), it is possible to conclude on the efficiency of adaptation activities of the particular renewable energy equipment. As the result, some of the adaptation measures might be ineffective in certain locations.

One of the examples of the efficiency of adaptation measures is active heating systems. For wind turbines it depends on the minimum number of days of effective use of systems and the number of days when icing takes place. For Amderma (a small port, around 570 inhabitants) those adaptation measures are inefficient, when for Anadyr (a port, 15'000 inhabitants) active adaptive systems are effective, since the minimum number of days for effective use of systems is much less than the number of days with icing (Solovyev, 2017). In the Russian Arctic in the locations with already existing wind generation systems the number of days in a year with an active icing of the wind turbine blades varies from 60 (Anadyr, Chukotka, 64° N) to 110 days (city Amderma, Nenets AR, 69° N).

An important element of the adaptation of various renewable energy installations to harsh climate conditions is intelligent protection systems, including special software, which can adjust the operation of an installation to icing, wind gusts and high air humidity. The main principles of the intelligent protection systems should be:

- to maximize real time energy production of the elements of renewable energy installations;
- to ensure fossil fuel economy at diesel power plants when covering the existing and peak loads;
- to provide remote monitoring of parameters and operating modes of hardware and software;
- to monitor the state of the equipment;
- to forecast the availability of the renewable energy resources;
- to schedule maintenance, make risk assessments and prevent emergencies.

One of the key tasks and also the main challenge for the intelligent protection systems is to chase the problem of the power generation stability, which arises due to the intermittent nature of renewable energy generation. The stability is then the ability to provide a continuous and stable flow of electricity.

The factor of stability of energy flow is also one of the risks to be considered at the design stage. Stability is not a risk factor for large centralized power systems, where a lot of automatic relay protection is used. However, it is crucial for local decentralized systems (for example, the power system of a small city or a settlement), where it is not possible to ensure a constant stable electricity flow during the coincidence of frequency and phase values. Thus, technical features, their adaptation and costs of adaptation measures influence the viability of renewable energy systems in the Arctic region, since this directly affects the quality of energy supply to end users.

RISKS OF USING RENEWABLE ENERGY SYSTEMS IN THE ARCTIC REGION

Naturally, the priority task is the development of renewable power plants using innovative technical solutions that ensure minimum environmental risks and reliable operation in different climatic conditions (Nefedova et al., 2016). It is worth underlining that renewable energy systems in the Russian Arctic are subject for specific types of risks due to the climate conditions and remoteness of the region: investment, resource and technical (or operational) risks. Moreover, it is necessary to highlight the risks associated with global climate change. The importance of stable cost-efficient energy supply, as well as compact energy storage devices, cannot be underestimated. Thus, in is important to pay attention to some potential mitigation measures for wider renewable energy deployment.

Investment risk, which is a risk of losing revenue due to economic inefficiency of an investment project (in our case – decentralized renewable energy systems), is considered to be relatively high. From this point of view, deployment of renewable energy systems in the Russian Arctic can be characterized by the limited number of potential investors and primarily a pilot character of those projects. Since the commercial value of those projects is limited, the investment risk is relatively high (if we omit technical and climate issues here). Nowadays, local communities and governments, as well as scientific institutions and probably international scientific organizations, are seen as main potential investors in the development of renewable energy potential in the region. It is hard to attract private investments on the beginning stage, when profitability and effectiveness of renewable energy systems in the Arctic are not fully investigated. In the beginning phase of deployment, the investment risk can be optimized primarily by constructing low-power combined power plants based on the existing energy infrastructure.

Resource associated risks are characterized by spatio-temporal changes in the parameters of renewable energy power plant generation, or, in other words, intermittent nature of renewable energy sources. This issue has been already broadly discussed previously in this chapter and in other before mentioned studies. The mitigation of resource risks is possible through the use of advanced back-up technologies and storage, as well as a comprehensive assessment of renewable resources potential with reference to the specific territories, where those installations are to be placed. When renewable energy systems are connected to the grid, the grid by itself has a stabilizing role. This type of operation, as a rule, does not require accumulation systems. However, if the total amount of energy from intermittent sources exceeds 15-20%, it affects the grid (Gabderakhmanova et al., 2015). Thus, mitigation of resource associated risks is to a certain extent dependent on the development of the accumulation technologies. Even though there are efficient batteries, which can serve the grid stabilization function (for example, lithiated iron phosphate batteries and vanadium redox batteries), they also need to be adapted to the low Arctic temperatures.

Despite the importance of the resource risks mitigation in the Arctic, where there is a high spatial and temporal variability in the resource characteristics of renewable energy sources, the emergence of technical risks is the greatest. Technical risks are equipment malfunction, errors arising in the selection and design of process equipment, and process failures. This means technical risks can arise during all the phases of renewable energy systems design, construction and use. For wind turbines most of the technical risks in the Arctic conditions are associated with operation phase. Those are productivity reduction due to blades and other equipment icing (Figure 3), increased vibration (ice imbalance), short circuits due to moisture, break of support structures, towers and blades.

Figure 3. Dependence of the rated power of wind turbines on wind speed under different conditions of icing of blades and equipment (Elistratov, 2017)

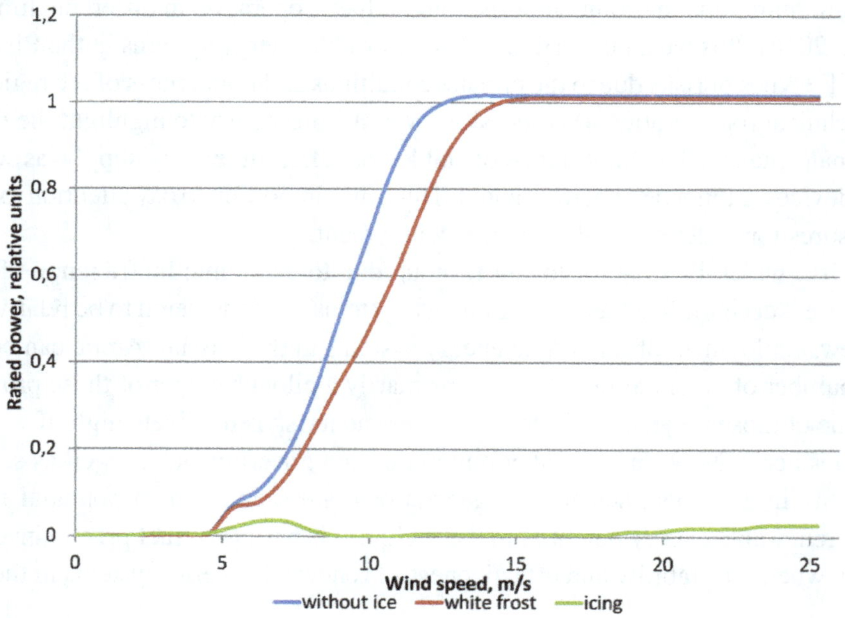

To avoid technical risks some security measures need to be implemented. The preventive measures are to use certified technical equipment, adaptive measures and timely maintenance. Special stabilization and protection measures are the block-modular principle of combining renewable energy systems with diesel generators, the availability of storage, radio electronic matching and control in containers with a high degree of protection from exogenous factors, and also in some cases an internal climate-control system.

In the case of surface icing, there is a sharp decrease in the nominal power of wind turbine equipment (Elistratov, 2017), up to its total failure (Figure 3). The disturbance in the operation of the wind turbines during icing is due to the fact that the surface ice crust reduces the aerodynamic characteristics of the blades. Ice also damages the balancing of the wind wheel, which leads to increased wear of bearings and the entire equipment of the wind turbine. The pieces of ice from the ends of the rotating blades can break and fly apart over considerable distances. Thus, gondola blades and wind turbines require additional special measures against icing and storm loads during the design and production phase.

Some emergent measures against icing can be quite expensive and questionably efficient. In some countries hot water is spayed by the helicopters to combat extreme icing and melt the ice crust on the surface of the blades (Figure 2). Comparing the costs of such measures, hot water spraying can cost up to the amounts of electricity received over several days of operation of the wind turbine. Therefore, preventive measures to mitigate technical risks are very important for the operation in the Arctic. Besides, parts of the wind generator must be made of special frost-resistant materials and equipped with an effective anti-icing system. All the technical fluids used in the generator should be nonfreezing.

Risks associated with global climate change can be classified as systemic thus critical, as they may affect operational principles of decentralized energy systems. Besides, climate changes on the territory of the Russian Arctic are considered as the most emerging. According to the climate report of Rosgidromet (2017), the average annual temperature on the territory of Russian Federation in 2016 continued to grow by more than 2.5 times faster than globally (by approximately 0.45 °C in 10 years). The change is especially noticeable in the Arctic region, where the growth rate has reached 0.8 °C in 10 years (Taymyr). Furthermore, due to warming, the annual amount of precipitation in the Arctic is continuing to grow and was 3.0% above the norm in 2016. Further increase of precipitation in the Arctic will lead to its water logging. This effect can already be observed in the central and eastern parts of the Arctic coast.

There is a reduction in both the total area of the Arctic ice and the ice cover of the Siberian seas (Kara, Laptev, East Siberian, and Chukchi seas). The ice cover of the Siberian seas has significantly decreased to 450’000 sq. km, when in the mid-20th century it was about 700’000-800’000 sq. km. Even more dramatic evidence comes from the reduction of perennial ice cover, which has decreased several times in volume, and signals about the long-term changes in the winter ice cover in general.

Another sign of the climate change is the melting of permafrost (Kokorin et al., 2008). More than 60% of the territory of Russia is a permafrost zone (the largest permafrost in the world under a single national jurisdiction). From the 1970s to the 1990s the clear tendency towards an increase in the temperature of the upper layers of permafrost was observed, which was primarily associated with the warming of the atmosphere. Although climate warming signs in the European part of Russia are weaker than in Siberia, the changes in the permafrost are significant. Over the past 20-30 years, the temperature of the upper permafrost layer in the European part of the Russian Arctic and the Subarctic has increased from +0.22 to + 1.56 °C. In parallel to that the number of underground thawed areas has increased. Observation data also indicate a progressive increase of 14-80% of those thawed areas in the permafrost over the past 10 years in some areas of the Russian Arctic (Kolyma Lowland, Eastern Chukotka, Bolshezemelskaya tundra).

Warming leads to an increased risk of dangerous cryogenic phenomena, such as solifluction, thermokarst and land surface subsidence (Burkett, 2011). Degradation of the permafrost represents a great danger for buildings, roads, and other engineering structures located in the Far North. Many structures are built on pile foundations, use long-term soil as bases and are designed for operation under specified temperature conditions. Over the past 30 years in Yakutsk, due to subsidence of frozen ground, more than 300 buildings have been seriously damaged. In recent years, the number of structures damaged by uneven subsidence of foundations has increased by 42% in Norilsk, by 61% in Yakutsk, by 90% in Amderma in comparison with the previous decade (Anisimov and Lavrov, 2004).

Studies show that the destruction of perennial frozen rocks does not go from south to north, but is much more complicated. Special investigations were carried out at the State Hydrological Institute (Anisimov and Reneva, 2006) and resulted in risk maps showing the geocryological hazard index. In many areas of the Arctic, and in particular in Chukotka, the risk for buildings and infrastructure is generally high. Even though the increase in summer temperatures is relatively small the risk of settling of the buildings is the highest (Figure 4).

In the Arctic region, where permafrost covers more than 90% of the surface, the depth of seasonal melting will mainly increase. Here, taliks can also arise and develop, mainly under large rivers and lakes, with permafrost from the surface and its preservation in deeper layers. Almost all of Chukotka (except for the northeastern part) has the highest values of the geocryological risk index, including the Bilibino nuclear power plant, power lines from the village Chersky on the Kolyma River to the village Peveka on the coast of the East Siberian Sea. This also threatens Yamal Peninsula with the massive development of infrastructure and oil and natural gas resource extraction activities. The increased salinity of water may significantly affect the Arctic coast of Yakutia and the Kola Peninsula.

During the last decade the destruction of the shores of the Arctic seas has been noted in addition to the permafrost melting. This happens due to an increase in summer temperatures and intensification of sea waves. In general, there is a possibility of two to three times increase in the frequency of strong winds, squalls, storms, etc.

Figure 4. Risk map for buildings and structures in the permafrost zone: 1 - weak, 2 - medium, 3 - high risk (Anisimov and Reneva, 2006)

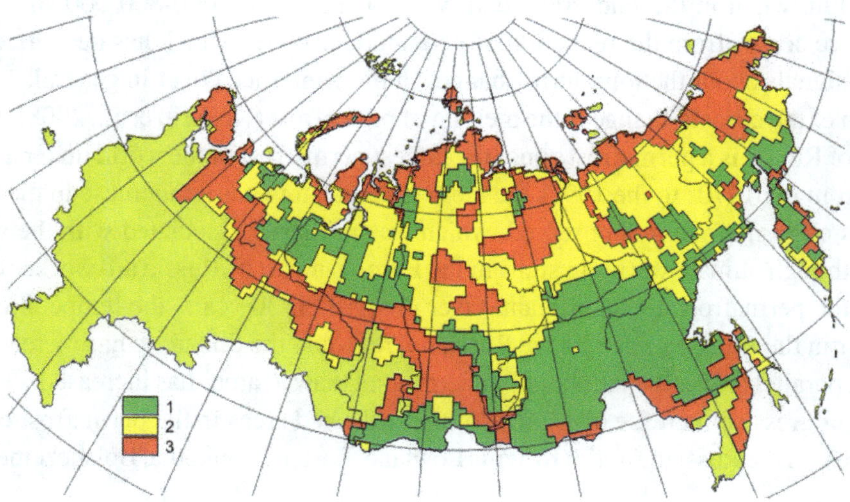

Despite the prevailing view that global warming and sea ice melting positively influence on the opportunities for navigating along the Arctic seas, it is too early to make final conclusions. It is considered that in the next 20-30 years with the continued warming trend the navigation along the Northern Sea Route will be more accessible. For example, the duration of navigation through the Vilkitsky Strait (the most difficult place on the Northern Sea Route) can increase by four times from just one month up to 120 days. However, it is important to take into account that warming will intensify the processes of glacier destruction, which certainly will dramatically increase the number of drifting ice floes and appearance of large icebergs. Even stronger negative effect on shipping can have greater mobility of the pack ice fields. Ice packs of tens of square kilometers can detach from a solid mass of ice covering at the North Pole and drift apart.

All these negative changes due to warming influence the landscape of future renewable energy systems installation and will certainly dramatically affect those few existing ones. The risk of such global changes for the existing Arctic infrastructure and prospects of energy infrastructure development cannot be underestimated. According to some sources, the adaptation cost of the impact of climate change on public infrastructure (roads, seaports, airports, transmission lines, etc.) can lead to significant extra investments. In Alaska those estimates in the development of infrastructure reach $ 13 billion (Larsen et al., 2008). More rapid warming will require less adaptation measures (thus less investment) than currently existing slow climate change. At the same time, renewable energy systems are more environmentally friendly, and will reduce hydrocarbon footprint, thus combatting the global warming.

To conclude, global climate change will undoubtedly influence all sectors of the economy in the Arctic region, including the energy sector, and increase the overall development costs. It will be necessary to allocate additional investments for adapting the existing infrastructure to these changes.

Creating a Synergy: New Development Opportunities in the Russian Artic

Arctic region is not only rich in natural resources, as it is mainly seen nowadays (Morgunova, 2015). Arctic is a place opening new horizons for radical innovations and new business models. The energy industry is experiencing considerable transformation under the Arctic conditions and far greater transformation than ever before. Nevertheless, the energy industry can serve as a potential driver to other industries and region development by introducing new technological advances, thus creating synergy.

Synergy, as a notion, means elements combined in a certain way give greater total effect than just a sum of those elements. New development opportunities in the energy sector in the Arctic can potentially result in the synergistic effect for the economy, environment and people through the use of local resources, implementation of new power generation models and technologies, and, finally, through new business opportunities based on the energy infrastructure development, such as potential savings or surplus of energy. This combination can result in an innovative development model for the future of the Arctic communities internationally.

The energy sector is one of the dominating sectors in the Russian economy, and is extremely important for the country. However, energy transition in Russia, which accounts both for innovation diffusion and deeper renewable energy penetration, is very slow due to the established socio-technical regime. For example, renewable energy in Russia (not considering hydropower) accounts for 0.2 million tons of oil equivalent, which is far less than 1% of total primary energy consumption (BP, 2017). Furthermore, the existing energy system architecture in the Russian mainland is currently inflexible, since it is heavily centralized and dependent on huge electricity generation facilities.

On the contrary, the Russian Arctic, as well as most of the Arctic territories internationally, does not have an established energy architecture. Arctic gives all the premises for an innovative start through the challenging environment, underdeveloped infrastructure and growing socioeconomic activity. The transition towards alternative and unconventional resources, decentralized energy systems, etc., which become possible due to technological advances, opens new horizons. From the technological perspective, the Arctic may be seen as a certain kind of playground, where threats can be transformed into opportunities. Under pressure of external factors, such as harsh climate and remoteness, it is possible to design a modern energy system of the future and boost both demand and deployment of renewable energy. Besides, adaptation measures for wider renewable energy systems deployment in the Arctic, broadly discussed in this chapter, requires the intensification of technological studies and research, which will also positively influence Russian economy and science.

Renewable energy technologies can be a powerful tool to stimulate socioeconomic and infrastructure development, technological modernization, academic exchange, etc. Energy sector investment is actually seen as a key driver of the current GDP growth (IRENA, 2016). Some studies are arguing for a considerable impact of renewable energy deployment on GDP, where by doubling the renewable share the global GDP can increase between 0.6% and 1.1% in 2030 (IRENA, 2016). The economic effect is seen as considerable, since renewable energy technologies offer some economic advantages, such as, for example, technology intensive jobs. Deployment of renewable energy, where energy availability can be limited, can incentivize further technological advances in energy efficiency. At the same time, integration of existing energy networks and combined decentralized renewable energy systems, as the most suitable solution in pursuing stable energy supply, can create new business models.

The synergistic effect is also visible in the connection between Arctic warming and the broader introduction of renewable energy systems. The cleaner is energy used, the less are greenhouse gas emissions, thus slower the global warming. This is particularly important for the Russian Arctic due to the climate change risks and rapid melting of ice and permafrost.

And finally, wider deployment of renewable energy is going to stimulate industrial and socioeconomic development of the Arctic region, resulting in cheaper electricity tariffs for households and production facilities, including natural resources and manufacturing industries, as well as ports and other consumers.

It is worth underlining that major investments in the establishment of the renewable energy systems in the Arctic should be done by the government. The governance instruments will define the pace of those complex changes. In later stages with higher investment attractiveness, the investors from the private sector will come.

The synergy from the broader use of renewable energy in the Russian Arctic can be significant. To maximize the synergistic effect, there is a need in greater governmental support, an understanding of technological features and risks of renewable energy systems, as well as knowledge of renewable energy potential of the Arctic region. It is important to frame the Arctic region as the innovative place, taking into account threats and opportunities of the renewable energy systems and energy sector as whole, as well as inter-linkages between other sectors.

CONCLUSION

Arctic in general and the Russian Arctic in particular have high renewable energy potential, which can be widely utilized for the industrial and household needs. Still, the use of these resources is non-systematic and limited, partly due to the specific climate conditions and other corresponding challenges.

Owing to the latest technological advances in the energy industry, there is a possibility to widely introduce renewable energy in the Arctic, using decentralized and combined power supply regimes. Adapted technical solutions are required for the effective use of renewable energy facilities to ensure high system performance and resistance to ice conditions and low temperatures. In general, the main qualities of power installations should be resistance to climate, high performance and rare maintenance.

A comprehensive approach to the development, as well as new technical solutions are required to realize the existing potential of small and medium scale power systems in the Arctic. Identifying and prioritizing risk mitigation, understanding their acceptable levels, maximizing opportunities are those necessary requirements in the broader deployment of renewable energy systems in the Russian Arctic. Some of the risks discussed, e.g. investment, resource, technical risks and the risks associated with global climate change, are seen as the ones with high priority.

An improvement within the energy sector can give a high impact on reaching the strategic economic, social and environmental goals in the Arctic region both in Russia and internationally. By implementing new models of energy supply, Arctic communities can receive additional combined value from sustainable development, new technologies and new business opportunities without harm to the unique and sensitive Arctic nature.

New development opportunities in the energy sector in the Arctic can potentially result in the synergistic effect for the economy, environment and people through the use of local resources, implementation of new power generation models and technologies, and, finally, through new business opportunities based on the potential savings or surplus of energy.

ACKNOWLEDGMENT

The study was supported by the state assignment (theme No AAAA-A16-116032810088-8; theme No AAAA A16-116051810068-1; theme No 0149-2018-0001) and the RFBR grant (project No. 18-05-60252).

REFERENCES

Anisimov, O., & Reneva, S. (2006). Permafrost and changing climate: the Russian perspective. *AMBIO: A Journal of the Human Environment*. Retrieved from http://www.bioone.org/doi/abs/10.1579/0044-7447(2006)35%5B169:PACCTR%5D2.0.CO%3B2

Anisimov, O.A., & Lavrov, S.A. (2004). Global warming and melting of permafrost: assessment of risks for production facilities of the fuel and energy sector. *Technologies of Fuel and Energy Complex*, (3), 78-83.

Ashourian, M. H., Cherati, S. M., Mohd Zin, A. A., Niknam, N., Mokhtar, A. S., & Anwari, M. (2013). Optimal green energy management for island resorts in Malaysia. *Renewable Energy*, *51*, 36–45. doi:10.1016/j.renene.2012.08.056

BP. (2017). *Statistical Review of World Energy*. Retrieved from www.bp.com

Burkett, V. (2011). Global climate change implications for coastal and offshore oil and gas development. *Energy Policy*, *39*(12), 7719–7725. doi:10.1016/j.enpol.2011.09.016

Bushuev, V. V., Zaichenko, V. M., Morgunova, M. O., Soloviev, D. A., & Chernyavsky, A. A. (2017). The potential of renewable energy in the Arctic: new challenges. In *Proceedings of the V International Conference "Renewable Energy: Problems and Prospects" and the X School of Young Scientists "Actual Problems of Development of Renewable Energy Resources" named after Corr. RAS EE Shpilrayna* (vol. 1, pp. 94-99). Makhachkala: FGBUN Institute of Geothermal Problems, Dagestan Scientific Center, Russian Academy of Sciences.

Chade, D., Miklis, T., & Dvorak, D. (2015). Feasibility study of wind-to-hydrogen system for Arctic remote locations - Grimsey island case study. *Renewable Energy*, *76*, 204–211. doi:10.1016/j.renene.2014.11.023

Elistratov, V. V. (2017). *Problems and solutions to create effective energy supply systems for consumers in the Arctic with the use of innovative renewable energy technologies. In Arctic-2017 (Arctic: offshore projects and sustainable development of regions) II International Conference* (pp. 1–41). Moscow: System Consulting. Retrieved from https://energy.s-kon.ru/arktika-2017/

Elistratov, V. V., & Denisov, R. S. (2016). Estimation of energy losses in the selection of wind turbine equipment for severe climatic conditions. In *Polar Mechanics Materials of the Third International Scientific Conference* (pp. 896-907). Vladivostok: Far East. Fed. un-t.

Fay, G., & Udovyk, N. (2013). Factors influencing success of wind-diesel hybrid systems in remote Alaska communities: Results of an informal survey. *Renewable Energy*, *57*, 554–557. doi:10.1016/j.renene.2013.02.021

Gabderakhmanova, T. S., Director, L. B., Popel, O. S., & Tarasenko, A. B. (2015). Comparative analysis of electrochemical energy storage devices. *Alternative Energy and Ecology*, *23*, 184–195.

Gabderakhmanova, T. S., Kiseleva, S. V., Popel, O. S., & Tarasenko, A. B. (2016). Some aspects of the development of renewable energy in the Arctic zone of the Russian Federation. *Alternative Energy and Ecology*, (19-20), 41-53. Retrieved from http://elibrary.ru/item.asp?id=26935418

Gabderakhmanova, T. S., & Shakun, V. P. (2016). Geoinformation mapping of resources and renewable energy facilities in the Arctic zone of the Russian Federation. In S. V. Kiseleva (Ed.), *Renewable Energy Sources: Proceedings of the All-Russian Scientific Conference with International Participation and the Xth Scientific Youth School: A Compilation* (pp. 348–353). Moscow: University Book. Retrieved from https://lomonosov-msu.ru/eng/event/3654/

Gabderakhmanova, T. S., & Shakun, V. P. (2017). Criteria for the prospects of the territories of the Arctic zone of the Russian Federation for the introduction of solar and wind generation facilities. In *Proceedings of the V International Conference "Renewable Energy: Problems and Prospects" and the X School of Young Scientists "Actual Problems of Development of Renewable Energy Resources" named after Corr. Of the Russian Academy of Sciences EE Shpilrayna* (vol. 1, pp. 114-117). Makhachkala: IP Ovchinnikov (ALEF). Retrieved from http://www.ipgdncran.ru/confayer

Gabderakhmanova, T. S., & Tarasenko, A. B. (2016). On the possibility of using solar energy in the charging infrastructure of electric transport. In *Construction - the formation of the environment of life: a collection of materials of the XIX International Interuniversity Scientific and Practical Conference of students, undergraduates, graduate students and young scientists* (pp. 1025-1027). Moscow: NIU MSSU. Retrieved from http://elibrary.ru/item.asp?id=26275764

Gidrometeoizdat. (1997). *Scientific and Applied Guide to Russia's Climate (Arctic Region): Solar Radiation*. St. Petersburg: Author.

GIS. (2018). *Renewable Energy Sources in Russia*. Retrieving January 15, 2018, from http://gisre.ru/

Gridasov, M. V., Kiseleva, S. V., Nefedova, L. V., Popel, O. S., & Frid, S. E. (2011). Development of the Geoinformation System "Renewable Sources of Russia": Statement of the Problem and Choice of Solution Methods. *Thermal Engineering*, *58*(11), 924–931. doi:10.1134/S0040601511110073

IRENA. (2016). *Renewable energy benefits: measuring the economics*. Abu Dhabi.

Kelion, L. (2017). Record-sized data centre planned inside Arctic Circle. *BBC News*. Retrieved July 13, 2018, from https://www.bbc.com/news/technology-40922048

Kokorin, A. O., Karelin, D. V., & Stetsenko, A. V. (2008). *Impact of climate change on the Russian Arctic: analysis and ways to solve the problem*. WWF Russia. Retrieved from http://narfu.ru/upload/medialibrary/2af/vozdeistvie_na_arcticu.pdf

Kovalenko, A. S., Morgunova, M. O., & Gribkovskaia, V. (2018). Infrastructural Synergy of the Northern Sea Route in the international context. *Jenergeticheskaja Politika*, *4*, 57–67.

Larsen, P., Goldsmith, S., Smith, O., & Wilson, M. (2008). Estimating future costs for Alaska public infrastructure at risk from climate change. *Environmental Change, 18*(3), 442–457. Retrieved from http://www.sciencedirect.com/science/article/pii/S0959378008000216

Morgunova, M. O. (2015). *Arctic Offshore Hydrocarbon Resource Development: Past, Present and Vision of the Future*. KTH Royal Institute of Technology.

Morgunova, M. O., & Soloviev, D. A. (2016). Energy supply of the Russian Arctic: hydrocarbons or renewable energy sources? *Jenergeticheskaja Politika*, (5), 44-51. Retrieved from http://www.energystrategy.ru/editions/EP.htm

Morgunova, M. O., & Solovjov, D. A. (2015). Decentralized power supply based on local and renewable energy sources: a case of Russian Arctic. In *International Geographical Union Regional Conference: Geography, Culture and Society for Our Future Earth*. Moscow, Russia: IGU.

Morgunova, M. O., & Solovyev, D. A. (2017). Challenges to overcome: Energy supply for remote consumers in the Russian Arctic. *Journal of Physics: Conference Series*, *891*, 012157. doi:10.1088/1742-6596/891/1/012157

Nefedova, L.V., Solovyev, A.A., Shilova, L.A., & Solovyev, D.A. (2016). Risk factors during construction of power plants using renewable energy sources. *Vestnik MGSU*, (12), 79–90. doi:10.22227/1997-0935.2016.12.79-90

Nilsen, J. (2015). *Her spyler helikopteret bort et tykt lag med is - Tu.no*. Retrieved June 27, 2017, from https://www.tu.no/artikler/her-spyler-helikopteret-bort-et-tykt-lag-med-is/222975

Popel, O. S., Kisileva, S. V., Morgunova, M. O., Gabderakhmanova, T. S., & Tarasenko, A. B. (2015). The use of renewable energy sources for energy supply to consumers in the Arctic zone of the Russian Federation. *Arctic: Ecology and Economics, 1*(17), 64–69.

Rosgidromet. (2017). *Report on the peculiarities of climate on the territory of the Russian Federation for 2016*. Retrieved from http://www.meteorf.ru/press/news/13595/

Shell. (2018). *Floating LNG*. Retrieved July 13, 2018, from https://www.shell.com/energy-and-innovation/natural-gas/floating-lng.html

Silva, S. B., de Oliveira, M. A. G., & Severino, M. M. (2010). Economic evaluation and optimization of a photovoltaic-fuel cell-batteries hybrid system for use in the Brazilian Amazon. *Energy Policy, 38*(11), 6713–6723. doi:10.1016/j.enpol.2010.06.041

Solovyev, D.A. (2017). Small Energy in the Arctic: Adaptation Problems and Risks. *Jenergija: Jekonomika, Tehnika, Jekologija*, (11), 14–21.

Solovyev, D.A., Morgunova, M.O., & Gabderakhmanova, T.S. (2017). Adaptation of the energy infrastructure in the Arctic to climate change using renewable energy sources. *Jenergeticheskaja Politika*, (4), 72–80.

Tin, T., Sovacool, B. K., Blake, D., Magill, P., El Naggar, S., Lidstrom, S., ... Berte, J. (2010). Energy efficiency and renewable energy under extreme conditions: Case studies from Antarctica. *Renewable Energy, 35*(8), 1715–1723. doi:10.1016/j.renene.2009.10.020

KEY TERMS AND DEFINITIONS

Arctic: Is a territory north of the Arctic Circle (66°32'N).

Decentralized Energy Generation: Energy generation facilities are placed close to a consumer. This term also may include that energy generation is "off-grid," which means it is stand-alone and not connected to the grid, and "autonomous," as it is not dependent on other generation facilities or grid and as a rule are based on an energy mix (combined power supply).

Established Energy Supply Regime: In transition theory, this means existing and broadly functioning principles of energy generation and supply, including technologies, existing infrastructure and institutional setting.

Intermittent Energy Source: Is an energy source that is not continuously available or not providing a constant flow of electricity generated, such as wind and solar energy.

Local and Renewable Energy Sources: Include energy resources available locally (including oil and natural gas) and renewable energy sources as tidal energy, solar, wind, and waste of timber and woodworking industry.

Power Generation Stability: Means a constant and sufficient flow of electricity coming from an energy generation facility, satisfying customer needs and covering consumption peaks.

Renewable Energy Potential: Is the amount of energy from renewable energy sources available in a given area (not specifying technical or economic potential).

Socio-Technical Regime: Is a dominance of certain social and institutional routines, as well as technical solutions, which are linked together into a one functioning piece.

Synergy: Means elements combined in a certain way give greater total effect than just a sum of those elements.

Chapter 7
Renewable Energy Sources Development in Rural Areas of African Countries

Liudmila V. Nefedova
Lomonosov Moscow State University, Russia

Alexander Alexsvitch Solovyev
Lomonosov Moscow State University, Russia

Olena Popova
University of Cote d'Azur, France

ABSTRACT

The prospects of increasing access to electricity for the population of rural areas of Africa are considered. The main international funds and organizations aimed at sustainable energy development in Africa are described. An analysis of the state and possible options for using renewable energy sources for this purpose in decentralized energy supply through the creation of mini-grids or stand-alone systems is given. The risks by developing renewable energy sources in rural areas and modern mechanisms for financing in solar energy are presented.

INTRODUCTION

In the 21st century, the use of renewable energy sources (RES) is stimulated in economically developed countries by the desire for environmentally oriented energy development and the need to ensure energy security of states. In developing countries, the development of renewable energy resources is the determining lever for overcoming poverty and stimulating the development of the economies of these states. The carbon-free path of energy development in Africa is an important contribution to solving such problems of global warming, greenhouse gas emissions, forest destruction. The problem of access to electricity is one of the priority conditions for the economic and social development of African countries. In African countries with high potential resources of renewable energy, development of renewable

DOI: 10.4018/978-1-5225-9179-5.ch007

energy resources is an important factor of social and economic development of these countries, increase of energy and economic independence, as well as a lever to overcome the spatial heterogeneity of the population's access to electricity, improve ecological conditions and fight against deforestation. The problem of low people's electricity access in rural Africa is one of the most acute problems of our time. About 600 million rural people do not have access to grid electricity, of which 98% are concentrated in the Sub-Saharan region (IEA, 2014). Access to electricity is extremely low: as of 2016 - on average in Sub-Saharan Africa (SSA) – 43%. In 30 countries, the indicator is between 11 to 39%, and in 10 countries - less than 10% (WB DATA, 2016). Rural population spend about US$14 billion each year on lighting and mobile phone charging with kerosene, candles, batteries or other fossil fuel-powered technologies. That is very expensive, hazardous, polluting and harmful to health (SEAR, 2017).

In rural areas of SSA countries, the level of electrification does not exceed 10%, while in general in developing countries this figure has increased in recent years to 50% (IRENA, 2017a). The most important problems of sub-Saharan Africa are the shortage of electricity generation and the weak development of the energy infrastructure, which deprives rural residents of access to electricity. Africa has the world's lowest per capita energy consumption: with 16% of the world's population (1.18 billion people out of 7.35 billion) it consumes about 3.3% of global primary energy. With current trends, it will take Africa until 2080 to achieve full access to electricity. Energy from biomass accounts for more than 30 per cent of the energy consumed in Africa and more than 80 per cent in many sub-Saharan African countries. Indoor pollution from biomass cooking — a task usually carried out by women — will soon kill more people than malaria and HIV/AIDS combined. Nearly 60 per cent of refrigerators used in health clinics in Africa have unreliable electricity, compromising the safe storage of vaccines and medicines; half of vaccines are ruined due to lack of refrigeration (UNEP, 2017). The poorest African households spend 20 times more per unit of energy than wealthy households when connected to the grid Power supply even in cities is characterized by frequent outages and interruptions, which requires the installation of duplicating diesel generators. Electric transmission lines are in a dilapidated state and are characterized by losses 2-3 times higher than the standards. Economic development and access to energy are connected very closely. Improving electricity supply and distribution boosts economic growth, creates jobs, and expands the reach of educational and health services (AFREA, 2014). Foreign companies of economically developed countries are interested in the technology transfer and technical participation in the construction of such energy projects, created primarily with considerable investment support from international financial and environmental funds. At the household and community scale, distributed renewable energy (DRE) technologies include small-scale solar PV and stand-alone lighting systems; wind, biodiesel generators, and micro- and pico-hydro stations for electricity generation; and solar and biomass heating and cooling units and cooking devices. Many of these technologies provide productive or mechanical energy for commercial purposes as well. For the purposes of this section, renewable energy-based micro- and mini-grids also qualify as DRE technologies.

A number of international environmental and financial organizations, individual developed countries of the world, carry out programs to support the use of renewable energy sources in Africa, mainly in the poorest region of the continent - Sub-Saharan Africa (*Tab. 1. Some Programmes furthering energy access in Africa by RES development*). Of great importance is Global Initiative of former UN Secretary-General Ban Ki-moon - Sustainable Energy for All (SE for ALL) - with three objectives for 2030: achieving universal access to electricity and clean cooking solutions; doubling the share of the world's energy supplied by renewable sources; and doubling the rate of improvement in energy efficiency.

Table 1. Some Programmes furthering energy access in Africa by RES development

Programmes	Description
Lighting Africa	An IFC and World Bank programme to accelerate the development of sustainable markets for affordable, modern off-grid lighting solutions for low-income households and micro-enterprises across Africa
Energy Sector Management Assistance Program (ESMAP).	A global knowledge and technical assistance program administered by the World Bank. It provides analytical and advisory services to low- and middle-income countries to increase their know-how and institutional capacity to achieve environmentally sustainable energy solutions for poverty reduction and economic growth.
Scaling Up Renewable Energy in Low Income Countries (SREP)	A Strategic Climate Fund programme that was established to expand renewable energy markets and scale up renewable energy deployment in the world's poorest countries. To date, USD 264 million has been approved for 23 projects and programmes. Additional USD 1.9 billion in co-financing is expected from other sources.
African Rural Energy Enterprise development (AREED)	The Program for Africa under UNEP with funding of $ 250 million, operates in Senegal, Mali, Ghana, Tanzania and Zambia and is aimed at supporting small and medium-sized enterprises in energy supply based on RES.
Sustainable Energy Fund for Africa (SEFA)	A fund administered by the AfDB and anchored by a Danish government commitment of USD 57 million to support small and medium-scale clean energy and energy efficiency projects in Africa through grants for technical assistance and capacity building, investment capital and guidance.
GreenClimateFund (GCF)	A fund established by 194 countries party to the UN Framework Convention on Climate Change in 2010 that aims to mobilize funding at scale to invest in low emission and climate-resilient development in developing countries. The fund is to mobilize USD 100 billion per year by 2020.

Source: (REN 21 GSR, 2018)

Without energy, it is challenging, if not impossible, to promote economic growth, overcome poverty, expand employment, and support human development. Sustainable energy is the seventh goal of the 17 UN Sustainable Development Goals (SDGs), with a call to "ensure access to affordable, reliable, sustainable and modern energy for all." Its five targets indicate areas where policies can be designed—such as boosting the share of renewable energy in the global energy mix and doubling the global rate of improvement in energy efficiency (UNECA, 2017).

Targets for Sustainable Development Goal 7 by 2030

- Ensure universal access to affordable, reliable and modern energy services;
- Increase substantially the share of renewable energy in the global energy mix;
- Double the global rate of improvement in energy efficiency;
- Enhance international cooperation to facilitate access to clean energy research and technology, including renewable energy, energy efficiency and advanced and cleaner fossil-fuel technology, and promote investment in energy infrastructure and clean energy technology;
- Expand infrastructure and upgrade technology for supplying modern and sustainable energy services for all in developing countries, in particular least developed countries, Small Island Developing States, and land-locked developing countries, in accordance with their respective programs of support.

DEVELOPMENT OF SOLAR AND WIND POWER RESOURCES

The countries of Africa are characterized by a very high potential for renewable energy resources with marked differentiation depending on the geographic location. The UNEP Solar and Wind Energy Resource Assessment (SWERA) project funded by the Global Environment Facility (GEF) in 2001-2007 is focused on preparation of reliable information on renewable energy resources by country and region for the purpose of developing policies for their development and justification of investments based on satellite materials (mainly NASA), taking into account the comparison with the available data of a fairly rare network of meteorological observations. For the countries of Africa, open databases and maps of solar resources and wind energy were compiled in the countries where the distribution of mean annual values of direct solar radiation (kWh / m^2 / day) and average annual specific wind power (V/m^2) at height 50 m (SWERA). The project was launched in 13 pilot countries in Africa, and then in 2006. expanded to a global level. At the moment the GIS "SWERA RREX" has been created, allowing to obtain in a free access the necessary databases, maps and calculation methods for solar and wind power resources of a number of developing countries in Africa, South America and southeast Asia.

The influx of solar radiation is high and amounts to about 2200-2600 kWh/m^2 per year, with the exception of the equatorial Africa region and the countries of the Gulf of Guinea. According to expert estimates in the territory of the SSA countries by 2011, remote rural areas are allocated, where the use of photovoltaic cells is more economically efficient than diesel generators with imported fuel (EC JRC, 2011). Last three years when the development of thin-film technology has led to a rapid decline in the price of photovoltaic panels, such areas have become much more extensive and for almost the whole SSA, it has become more economic effective to install photovoltaic panels on the roofs of residential buildings and office buildings and the creation of local power grids than the use of diesel generators, the construction of new large energy facilities and the creation of networks of power lines to remote areas. The high proportion of the rural population in the total population makes this problem especially acute. The largest share of the rural population in East Africa (76%), while in South and North Africa it is less than half (41% and 49%, respectively). In Africa, rapid growth of the rural population (1.6% per year) persists, but the urban population on the continent is growing even faster (3.4% per year) due to the urbanization process.

Due to the higher cost and significant primary costs by the use of renewable energy, government support is required, which can be expressed in preferential taxation, higher tariffs, mandatory sales volumes, and other legally imposed measures (Ferroukhi et al., 2016). All economically developed countries in the development of renewable energy used some or other support measures. A number of countries in Africa are already using government support measures and prepared state plans for the development of renewable energy sources. The plans include both the creation of large power facilities with their connection to the energy networks, and the placement of photovoltaic installations to provide electricity to remote distributed energy consumers in rural areas, solar collectors for water heating or for the use of biofuels in transport. According to World Bank experts, in order to achieve universal access to electricity in Sub-Saharan countries by 2030, in accordance with the UN "Sustainable Energy for All" Program, about 40% of all investments under this program should be directed to the creation of mini-networks in rural areas with the installation of photovoltaic panels and Wind Power Generators (with the necessary wind power potential) (SEAR, 2017). Foreign companies are actively involved in this process. For example, the British company in the field of solar energy "Solarcentury" installed Solar Plant 1 MW - one of the largest in East Africa - on a tea farm in Western Kenya. Of great importance are government initiatives: the Government of Tanzania has announced a plan to equip a million Tanzanian homes with access to reliable solar electricity by the end of 2017. It is expected that this project will provide 10% of energy supply to the population of the country through solar energy and will create 15 thousand jobs in industry related to solar energy. Currently in Tanzania 86% of home lighting depends on kerosene and candles. Since 2013 the United States "Energy Africa" Program is underway to create in the region power stations with a total capacity of 30 GW, with planned financing of $ 7 billion from public and private investments. "Gigawatt Global" is a partner in this program: a total capacity 8.5 MW Solar Plants in Rwanda has already been commissioned, two international grants for solar projects in Burundi with a total capacity of 7.5 MW have been signed, which after their implementation will increase electricity generation in Burundi by 15% and provide electricity to about 60 thousand farms in the district.

Of great importance is the creation of cooperatives for the acquisition and equipping of rural settlements with photovoltaic systems. In Africa, there are a number of international associations of entrepreneurs aimed at selling, installing and servicing renewable energy off grid facilities of various sizes. Here are some of them:

- Alliance for Rural Electrification (ARE) - an international business association that represents the decentralized energy sector and works towards the integration of renewables into rural electrification markets in developing and emerging countries. It has more than 90 members along the whole value chain of off-grid technologies.
- Global Off-Grid Lighting Association (GOGLA) - an independent, not-for-profit industry association that represents over 125 members as the voice of the off grid solar energy industry and promotes the solutions they offer. GOGLA was founded in 2012, borne out of the IFC/World Bank's Lighting Global programme.
- African Renewable Energy Alliance (AREA) - a global multi-stakeholder platform to exchange information and consult about policies, technologies and financial mechanisms for the accelerated uptake of renewable energy in Africa.
- AKON Lighting Africa - an initiative launched in February 2014 that seeks to provide a concrete response at the grassroots level to Africa's energy crisis and to lay the foundations for future development. It aims to develop an innovative solar powered solution that will provide African villages with access to a clean and affordable source of electricity
- Africa Mini-grid Developers Association (AMDA) - an association representing efforts by the mini-grid development industry to alleviate the problem of energy access with sustainable and environmentally friendly renewable energy mini-grids. AMDA is active in Kenya and Tanzania through advocacy, promotion and co-ordination.

In rural and remote regions improved access to energy can generally be acquired in three ways (*Table 2. Technological tracks for expanding energy services*)**:**

1. grid-based centralized electrification, where the grid is extended beyond urban and peri-urban areas.
2. community-level or renewable energy-based mini- or micro-grid systems;
3. household-level use of isolated devices and stand-alone systems to generate power lighting and heat for space and water heating, cooking and productive uses.

Table 2. Technological tracks for expanding energy services

Systems	Centralized	Micro-grids and Mini-grids	Stand-alone systems
Scale	National, regional, and even international	Community	Household
Geographic radius	> 50 square kilometers	1 to 49 square kilometers	< 1 square kilometer
Number of customers	Thousands to millions	Dozen to hundreds	Usually a dozen or less
Installed capacity	> 10 MW	20 kW to 10 MW	< 20kW
Technologies involved	Large-scale and centralized	Medium-scale and small-scale	Very small-scale
Investment required	Billions of dollars	Millions of dollars to hundreds	Thousands of dollars

Source: (SEAR, 2017)

ARACTERISTICS GRID ELECTRIFICATION OFF-GRID ELECTRIFICATION

At utility scale, they provide electricity to meet the diverse needs of grid-connected urban and rural customers. IRENA estimates that by 2035, the cost of electricity generation from a solar PV mini-grid will be as low as $0.20/kWh (IRENA, 2017b). In Sub-Saharan Africa, South, and East Asia, mini grids are rapidly emerging as a viable option for providing energy services, thanks to both technological and institutional innovations and cost reductions. Grid-connected clean energy technologies can range from a few kilowatts of roof top solar photovoltaics (PV) systems connected to the low voltage distribution network, to 10 to 1,000s of megawatts of large centralized utility scale power plants. Examples include hydropower, solar parks, wind farms, geothermal power plants, or biomass-fueled plants connected to medium and high voltage substations. These utility scale renewable energy plants are in place in Ethiopia, Kenya, Rwanda, and elsewhere, providing better quality service to existing customers, as well as widening the reach of the grid to those previously without access. In mini-grid configurations, clean energy technologies can meet the needs of communities of owner than waiting for grid extension. A clean energy technology mini-grid can be a single power source—such a small hydropower plant, or a hybrid system with renewable energy sources with batteries or a diesel generator. When communities are small or dispersed and electricity demand is limited, stand-alone systems—such as solar home systems (SHS)—can be more cost effective especially when coupled with new business and financial models. Increasingly, small PV systems, known as pico-solar systems (ranging from a few watts to tens of watts of solar PV) provide high value lighting and mobile phone services (SEAR, 2017). Solar lighting systems (solar lanterns) have seen the greatest development in recent years—and they are now considerably cheaper than conventional kerosene-based lighting systems (depending on existing subsidies). Solar lanterns, often priced as low as $10, provided lighting to 28.5 million people across the African continent by end-2014: in Kenya, these lanterns provided lighting to about 12 percent of the population, and in Tanzania's Lake region, about 50 percent of the population (IRENA, 2017b).

The Sustainable Energy Fund for Africa (SEFA) has approved in 2015 a US $ 1.0 ml grant to launch the first phase of a Green Mini-Grid Market Development Program (GMG MDP) which will be implemented by the SE for ALL Africa Hub. Green Mini-Grids are village or district level electrical distribution networks. GMGs serve the needs of communities too distant to be economically connected to the grid in the near to medium term. The International Energy Agency (IEA) has estimated that 40% of investment required to achieve universal electricity access by 2030 can be economically achieved through mini-grids. Universal power access in Africa will require as many as 350,000 mini grids (IEA, 2017).

During the 5-th SE for ALL Forum in Lisbon in May 2018, the African Development Bank announced the launch of the redesigned Green Mini-grid (GMG) Help Desk of the Bank's. GMG Market Development Programme funded by the Sustainable Energy for Africa (SEFA). The new Help Desk offers a wider range of knowledge products and provides technical assistance services to public sector stakeholders for the design of GMG policy and regulations complementing the support to mini-grid developers. The Help Desk has been providing technical assistance to some 60 developers in 30 African countries to date. Solar micro-grids are playing an important role in ensuring that communities living in remote areas away from main power have access to electricity in East Africa in a region where nearly 80 percent of the population have no access to grid power. But the biggest and most unique feature of the growing concept is the ability of companies offering the service to leverage and use mobile phone technology to run their businesses and save on costs. Consumers use mobile money — M-Pesa in Kenya, for example — to purchase power units as well as monitor usage and get an alert when the amount of power purchased is going low. That option means that no staff is deployed to read power meters and no bills are due at the end of the month — users purchase power units in a pre-pay arrangement that uses crowd computing.

Systems of all the above energy services types can be both fossil fuels and renewable energy sources. When choosing the option of electrification in rural areas and designing the construction of power lines or the creation of mini-networks, it is necessary to conduct cost estimates of all options, including diesel generators and RES systems. Three rural electrification options are analyzed showing the cost optimal conditions for a sustainable energy development applying renewable energy sources in Africa. A spatial electricity cost model has been designed to point out whether diesel generators, photovoltaic systems or extension of the grid are the least-cost option in off-grid areas. The resulting mapping application offers support to decide in which regions the communities could be electrified either within the grid or in an isolated mini-grid. Donor programs and National Rural Electrification Agencies (or equivalent governmental departments) could use this type of delineation for their program boundaries (Szabo, 2011). Multi-dimensional cost analysis of social and environmental impacts using GIS highlight that PV/hybrid mini-grids offer a unique opportunity to create a standardized framework for quantifying costs of PV/hybrid mini-grids in SSA, that can support decision-making processes for designing the variable business models (Moner-Girona et al., 2018). (Maudi et al., 2018) determined a least cost electricity solution for Sub-Saharan Africa. The power system discussed in this study is hourly resolved and based on 100% Renewable Energy technologies. Sub-Saharan Africa was subdivided into 16 sub-regions. Four different scenarios were considered involving the setup of a high voltage direct current (HVDC) transmission grid. An integrated scenario that considers water desalination and industrial gas production was also analyzed. This study reveals that RE is sufficient to cover 866.4 TWh estimated electricity demand for 2030 and additional electricity needed to fulfill 319 million m^3 of water desalination and 268 TWh_{LHV} of synthetic natural gas demand. Existing hydro dams can be used as virtual batteries for solar PV and wind electricity storage, diminishing the role of other storage technologies. The results for total levelised cost of electricity (LCOE) decreases from 57.8 €/MWh for a highly decentralized to 54.7 €/MWh for a more centralized grid scenario. For the integrated scenario, including water desalination and synthetic natural gas demand, the levelised cost of gas and the levelised cost of water are 113.7 €/MWh_{LHV} and 1.39 €/m^3, respectively. A reduction of 6% in total cost and 19% in electricity generation was realized as a result of integrating desalination and power-to-gas sectors into the system. (Mentis et al., 2017) presented least-cost electrification strategies on a country-by-country basis for Sub-Saharan Africa by a geographic information systems approach coupled with open access. The electrification options include grid extension, mini-grid and stand-alone systems for rural, peri-urban, and urban contexts across the economy. At low levels of electricity demand there is a strong penetration of standalone technologies. However, higher electricity demand levels move the favourable electrification option from stand-alone systems to mini grid and to grid extensions.

RISK ANALYSIS

National rural electrification strategies in Africa countries need to be backed by dedicated and stable policy and regulatory frameworks. Traditional, centralized electricity sector frameworks have to be adapted to support the deployment of off-grid solutions (stand-alone and mini-grids). Policies, including incentive structures, need to be designed for long-term, sustainable market development. The participation of the private sector will be key to complement electrification efforts by governments and development agencies, and to substantially scale-up deployment. To enable this, a robust policy and regulatory environment are needed to reduce investment risks, improve viability and increase the overall attractiveness of the sector. Stability in policy must be highlighted to be particularly important. The example of changes in taxation policies was given wherein unscheduled changes can dramatically alter the economics resulting in a slow-down in uptake. Adequate standards and quality control measures were highlighted to be important to avoid proliferation of low-quality systems and market spoilage. The viability and sustainability of renewable energy-based mini-grids largely depend on well-designed tariff regulations. One approach to setting the tariffs for private sector mini-grids is to impose a national uniform tariff (which is often too low for sustainable operation) and provide sufficient viability gap funding. Other approach is to allow mini-grid tariffs high enough to cover costs but still well beneath current customer spending on conventional energy in off-grid areas, as also recommended by the Southern Africa Regulators Association. The private sector prefers the minimal touch points with governments to improve efficiency and reduce transaction costs. By bringing greater regulatory certainty, clearly defining institutional roles and responsibilities, investment risks for mini-grids can be reduced (IRENA, 2017c).

Energy supply using small hydropower plants is effective in equatorial Africa, in other areas it is complicated by high resource risks due to the dry season, a sharp decrease in runoff during this period, and sometimes the drying out of the majority of rivers that have rainwater. High temperatures cause a rapid decline in the level of seasonal reservoirs due to evaporation. Resource risks in the use of solar and wind power resources for the countries of the African continent are low due to the low level of their temporary variability and reliable estimates of the energy potential of these resources, conducted under the UNEP SWERA program, funded by the Global Environment Facility in 2001-2007 on the basis of the NASA data. At present, the geoinformation system "SWERA RREX" has been created, allowing to obtain for free the necessary databases, maps and calculation methods for solar and wind power resources for a number of developing countries in Africa, South America and southeast Asia.

The types of risks of developing renewable energy sources (Nefedova, 2015) are characterized for rural areas of Africa by economic, social and political risks. From economic claims it is necessary to allocate investment and currency and financial risks. The risks of investment risks are especially high for the development of geothermal resources, which is associated with the high cost of primary exploration drilling with the probability of negative results. Such projects require large investments with the involvement of international financial structures. In connection with the installation of predominantly imported equipment at RES facilities in Africa, as well as with the general volatility of the economies of these countries, the currency, financial and exchange rate risks are of great importance.

Due to the low level of education of the rural population in Africa, the role of accounting and social risks is increasing. Informing the population about the benefits of renewable energy, about the characteristics of the planned facilities is a prerequisite for creating a social climate for supporting the construction and uninterrupted operation of facilities for RES, preventing damage to energy facilities. Reducing the level of operational risks with high reliability of mainly imported equipment requires the organization of training of specialists for work on large RES facilities and mini-networks, as well as a network of points and instructors for the use of autonomous photovoltaic installations. It should be noted, that by developing renewable energy sources in the region, it is impossible to ignore the political risks caused by the possible instability of the political situation in the countries of the African continent, which hinders long-term plans and investments, in violation of international contracts for political reasons (IRENA, 2016).

A significant achievement in the information plan to ensure increased access of the rural population to electricity, which also allows to reduce risks due to updated data is "The Atlas of Africa Energy Resources". The Atlas was developed by the African Development Bank (AfDB) and the Infrastructure Consortium for Africa (ICA) in cooperation with the United Nations Environment Program (UNEP). It is designed for use by the governments of the African states, investors in the energy sector of various types and the private sector. In the atlas maps, satellite images and data for 54 countries in Africa are given. Estimates and calculations are presented on the possibilities of providing the population of countries with affordable reliable energy supply (UNEP 2017, Atlas).

TRENDS IN RES DEVELOPMENT IN RURAL AREA OF SUB-SAHARAN COUNTIES

The perspectives for the development of off-grid energy supply using RES, do not exclude the creation of large power plants. High wind energy resources (average annual wind speeds of more than 6-8 m/s) are confined to open spaces of savannas and semi-deserts, coastal and mountainous areas. The leader of wind energy in Sub-Saharan Africa is Ethiopia: three large wind farms have already been commissioned - Adama I (52 MW in 2013), Ashegoda (120 MW in 2013) and Adama II (153 MW-2015). Six more wind farms with a total capacity of about 950 MW are planned [Scaling Up REP 2012]. In Kenya the construction of the Lake Turcan Windfarm 300 MW will be completed in 2018. There are also projects for three more wind farms (250 MW) at different stages of development. Geothermal resources are concentrated in the zone of the East-African Valley with the length of 6000 km, in the territories of Eritrea, Djibouti, Ethiopia, Kenya, Uganda, Rwanda, Tanzania, Zambia. The geothermal potential of the zone of the East African Rift Valley is estimated, 20000 MW. The greatest success in geothermal energy was achieved by Kenya, having occupied by 2015 the eighth place in the world according to the total installed capacity of Geothermal Power Plants (576 MW). Since 1999 the pilot Geothermal Power Plant Aluto Langano, with a capacity of 7.3 MW, operates in Ethiopia. In 2020 it is planned to commission six more Geothermal Power Plants (450 MW).

Due to the higher cost and significant primary spending when using RES, state support is required. The support may be expressed in preferential taxation, higher tariffs, mandatory sales volumes and other legally imposed measures. All economically developed countries by the development of renewable energy used some of these or other support measures. In Africa, government support measures have also been applied and government plans for the development of renewable energy sources have been prepared. A number of countries (Benin, Djibouti, Kenya, Ghana, Guinea Bissau, Lesotho, Mozambique, Nigeria, Rwanda, Sudan, Tanzania and Ethiopia) have already included in their plans the development of renewable energy sources by 2030, as the creation of large energy facilities with their connection to the energy networks, and the placement of photovoltaic installations to provide electricity to remote distributed energy consumers in rural areas and solar collectors for water heating. Biomass supplies a substantial share (roughly a third) of Africa's industrial-based heat. South Africa's solar thermal market has grown relatively quickly, although it dropped in 2015 due to a delay in government tenders linked to an improved solar hot water programme. In 2015 Lesotho, Mozambique and Zimbabwe formulated new policies to support solar hot water. South Africa and Tunisia were among the top 20 countries worldwide for installations of solar collectors during 2017, ranking 14th and 18th, respectively. The use of biogas for cooking continued to grow in sub-Saharan Africa, namely in Ethiopia, Kenya and Tanzania. Countries in the Great Rift Valley, where there are significant geothermal resources (as in Kenya), have begun to tap direct geothermal heat for use in greenhouses, for example (as well as for electricity). Clean cook stoves, many of which use biogas as a source, are used increasingly in Africa, notably in Ethiopia, Kenya and, to a lesser extent, in Nigeria and Rwanda (REN 21 GSR, 2018).

In sub-Saharan Africa, which has a predominantly high potential for solar resources (except countries in the equatorial belt), a number of researchers have estimated the comparison of the cost of electrification of settlements in rural areas of Africa in the absence of power line infrastructure and taking into account the environmental consequences of using diesel generators and the cost of diesel fuel, dependence of the feasibility of using photovoltaic systems from the distance to power lines and transport routes. The most cost-effective is the creation of local mini-systems with PV installations in several localities or farms. The programs for the contracting of photovoltaic installations are carried out mainly in rural areas with international financial support and technical equipment by European firms. The power of the units is predominantly 10-50 kW, to supply isolated consumers (schools, health centers, greenhouses). In the region, African Carbon Asset Development (ACAD) is a public-private partnership organized by the United Nations Environment Program and its Rice center. ACAD, founded in 2009 with Standard Bank, serves as a platform for investment promotion, provides initial financing and consulting for "green" entrepreneurs on issues of obtaining carbon credits and investing in low-carbon initiatives.

Among the renewable sources of energy, currently the most developed for Sub-Saharan Africa are hydro resources, which account for about 12% of their global volume, although about 7% of the potential is currently exploited. The hydropower potential of the SSA countries is concentrated mainly in the basins of the Congo, Zambezi, Nile, Niger, Limpopo, Orange, Senegal. The hydropower plants account for more than 90% of energy balances of the following sub-Saharan African countries: DRC, Ethiopia, Mozambique, Namibia and Zambia. More than 70% of electricity is generated by hydroelectric power plants (HPP) in these five countries, as well as in Angola, Cameroon, Sudan, Togo and Zimbabwe. The largest hydroelectric power stations in the region are: Cahora Bassa Hydro Power Plant in Mozambique on the river. Zambezi (3.6 GW), HPP Inga I in the DRC on the river Congo (1.42 GW), Meroe hydroelectric power station on the river Nile in Sudan (1.25 GW), Kariba Hydroelectric Station on the Zambezi River 1.32 GW of Zimbabwe and Zambia). In Ethiopia, a hydroelectric power station on the Omo River was commissioned: Gilgel Gibe I, II and III (184 MW, 420 MW, 1870 MW), the largest hydroelectric power station in Africa, the Giant Power of the Ethiopian Renaissance Dam, with a capacity of 5.25 GW is being built on the Blue Nile. The construction of large hydropower hubs in Sub-Saharan Africa, on the one hand, seems to be a prerequisite for ensuring the vital needs of the economy and the population, and, on the other, can cause irreparable harm to the environment. Therefore, where feasible and cost-effective, it is more rational, in accordance with the principles of sustainable development, to build and install small HPPs that are close to the rural consumer and reduce costs and losses in the transmission of electricity.

FINANCING MECHANISMS

Expanding the use of off-grid installations for renewable energy requires the development of various adapted financing models, taking into account the capabilities of the end user. The last few years have been characterized by positive market trends and increased investment in renewable energy development in Africa. Innovative business models of financing continue to develop and improve. Use of mobile payments and scratch cards, "Powerhive" business model (literally "energy hive"), providers of integrated services with a set of products from solar facilities - lamps, radios and mobile phones on solar energy to TV sets are becoming more frequent. Distributed generation for renewable energy was supported through various measures: by holding auctions, setting goals for electrification, and initiatives to organize environmentally friendly cooking using renewable energy. Private investors substantially increased their investment in renewable energy projects in 2016. The number of large banks involved in financing renewable energy has increased, and loans also increased. International investment companies have made a major contribution to the financing of renewable energy and energy efficiency. New investment instruments - "green" bonds, crowdfunding (the formation of funds by collecting money from individuals and legal entities) and yield Cos (financial companies), were distributed in 2016 (REN21 GSR 2018). Yield Cos are typically solar and wind companies that are focused on returning cash flows to investors. The main financial structures and structures that create funds began to move actively towards developing countries' markets, as companies (especially solar panel manufacturers) and investors enter these markets for high profits, despite the high risks. Electrification programs and initiatives address the problem of relatively high initial costs for stand-alone systems through end-user financing schemes, taking into account their income and demand. In Tanzania, this has led to the expansion of microfinance services to households to purchase solar home systems. Preliminary payments are usually 15% of the system cost. The total cost, as a rule, was provided with 36-month loans, similar to the monthly expenses for kerosene. In other East African countries, the "pay-as-you-go" (or PAYGO) model is becoming more common, when households follow a payment scheme for various products and services by agreement of the form of perpetual lease or the possible ownership of the system after a certain period of time. The allocation of the cost of solar installations for a long time of payments allows consumers to access electricity off-grid.

Creating a local pool of capital through country-led initiative and making local finance available for energy access programmes could be a way to facilitate sector development and scale-up. Unlocking low-cost, local financing will also help address the risk of foreign currency. Guarantee schemes are important instruments for backing up financing and addressing key investment risks, including political, currency and off-take risk. Financial support programmes need to be designed with long-term market development, sustainability and scalability in mind. Subsidies should be integrated within a wider financing support systems rather than a pure subsidy system, which might jeopardise natural scalability. Bundling of mini-grid projects can be an effective way to reduce financing costs and project-specific risks. Bundling could help access a wider pool of financing that is otherwise inaccessible for low value transactions. Practical ways should be developed to link rural consumers to asset-based financing. These could be tied with a specific target for banks (with low interest rates and long-term financing) and creating a risk guarantee facility for financing institutions (IRENA, 2017c).

To finance facilities for RES, international banks (WB, EBRD, AfDB) open credit lines to local banks that want to participate in the program and who meet the standard requirements for partner banks. The mechanisms of these financial structures encourage investments in sustainable energy projects through a combination of funding and technical support. The size of the loan varies depending on the size of the credit line provided to the bank, and the prospects of the project. Some projects may also be considered by experts in terms of the potential for carbon financing under the mechanisms of the Kyoto Protocol (Joint Implementation, Clean Development Mechanism or within the EU Emissions Trading Scheme).

CONCLUSION

The countries of Africa, including those located to the south of the Sahara, have a very high potential for renewable energy resources. The main factor for its development is electricity deficit and extremely low access of the population to electricity, development of these resources is necessary for sustainable social and economic development of the region. According to the scenario of accelerated progress of UNDP, the largest forecasted growth of the Human Development Index in the world by 2050 will occur in sub-Saharan Africa. The demand for electricity in industrial countries in developing countries is growing on a very large scale. The construction of new large coal-fired power plants in the required volumes will lead to serious environmental and climatic consequences. The development and strengthening of the concept of sustainable development contributes to the increase of renewable energy sources role in the global energy sector. A number of countries in the region have introduced government measures to support the development of electricity for RES, and have drawn up plans for their use, and have carried out rural electrification programs. The process of developing renewable energy sources in the region is possible only with the support of international environmental and financial structures, which are already active in this direction. A number of countries in the region have introduced government measures to support the development of electricity for RES and have drawn up plans for their use. Many countries in Africa have already begun the process of developing RES, receiving support from international structures in this process. It is planned to provide total access to electricity by the year 2030 at the initiative of the UN in the region, using RES at the same time. Two countries in the region have significant results on renewable energy development: Kenya is the leader of geothermal energy in the African continent and Ethiopia is the leader of Sub-Saharan Africa in wind energy. The volume of the forthcoming works is colossal. In the preparation of renewable energy projects, African governments organize tenders for work and equipment supplies. Socio-economic conditions in Africa, the state of the study of RES resources lead to the fact that the highest risks in their development in this region are investment, monetary, social, and political risks.

At present, the importance of autonomous electrification solutions, including home solar power plants and cooperative mini-grids in Africa, working on renewable energy sources, mainly on the basis of photovoltaic installations, is increasingly recognized. Off-grid solar energy products provide a safer, cleaner, cheaper and reliable alternative to conventional solutions. With international financial support, including carbon funds, major projects are developed for the construction of solar power plants, geothermal PPs and wind farms in Africa, which provides sustainable development in the region without increasing greenhouse gas emissions. A number of effective business financing models have been developed, the acquisition of stand-alone installations and the creation of mini-networks for RES. New investment instruments - green bonds, crowdfunding (the formation of funds by collecting money from individuals and legal entities) and yield Cos (financial companies), the "down payment" model (or PAYGO) - have developed in Africa in the field of renewable energy.

REFERENCES

AFREA. (2014). Energizing Africa: Achievements and Lessons from the Africa Renewable Energy and Access Program (AFREA). The International Bank for Reconstruction and Development/WB Group.

BNEF (Bloomberg New Energy Finance). (2016). *Off Grid Solar Market Trends Report 2016*. Retrieved from https://about.bnef.com/white-papers/off-grid-solar-market-trends-report-2016

EC JRC. (2011). *Renewable energies in Africa – current knowledge. Scientific and technical Report. The European Commission Joint Research Center*. Luxembourg: Publications Office of the European Union.

Ferroukhi, R. (2016). *Renewable Energy Benefits: Measuring. The Economics*. Abu Dhabi, UAE: IRENA.

IEA. (2014). *Africa Energy Outlook. A focus on the energy prospects in sub-Saharan Africa*. Paris, France: International Energy Agency.

IEA. (2017). *World Energy Outlook 2017. From Poverty to Prosperity. Special Report*. Paris, France: International Energy Agency.

IRENA. (2016). *Unlocking Renewable Energy Investment: The Role of Risk Mitigation and Structured Finance*. Abu Dhabi: International Renewable Energy Agency.

IRENA. (2017a). *REthinking Energy 2017: Accelerating the global energy transformation*. Abu Dhabi: International Renewable Energy Agency.

IRENA. (2017b). *Electricity storage and renewables: costs and markets to 2030*. Abu Dhabi: International Renewable Energy Agency.

IRENA. (2017c). *Accelerating Off-grid Renewable Energy: Key Findings and Recommendations from IOREC (2016)*. Abu Dhabi: International Renewable Energy Agency.

Maudi, B. (2018). A cost optimal resolution for Sub-Saharan Africa powered by 100% renewables in 2030. *Renewable & Sustainable Energy Reviews*, *92*, 440–457. doi:10.1016/j.rser.2018.04.110

Mentis, D., Howells, M., Rogner, H., Korkovelos, A., Arderne, C., Zepeda, E., ... Scholtz, E. (2017). Lighting the World: The first application of an open source, spatial electrification tool (OnSSET) on Sub-Saharan Africa. *Environmental Research Letters*, *12*(8), 085003. doi:10.1088/1748-9326/aa7b29

Moner-Girona, M., Solano-Peralta, M., Lazopoulou, M., Ackom, E. K., Vallve, X., & Szabó, S. (2018). Electrification of Sub-Saharan Africa through PV/hybrid mini-grids: Reducing the gap between current business models and on-site experience. *Renewable & Sustainable Energy Reviews*, *91*, 1148–1161. doi:10.1016/j.rser.2018.04.018

Nefedova, L.V., & Soloviev, A.A. (2015). Analysis of risks of development of renewable energy sources in Russia. *Problems of Risk Analysis, 12*(6), 56-63.

REN21 GSR. (2018). *Renewables 2018. Global status report. REN21Secretariat.* Paris: UNEP.

Scaling - Up Renewable Energy Program. (2012). *Ethiopia Investment Plan (Draft Final).* Federal Democratic Republic of Ethiopia, Ministry of Water and Energy.

SEAR. (2017). *State of Electricity Access Report 2017.* International Bank for Reconstruction and Development / The World Bank. Retrieved from http://www.worldbank.org/en/topic/energy/publication/sear

SWERA - Solar and Wind Energy Resource Assessment. (n.d.). Retrieved from https://openei.org/doe-opendata/dataset/swera

Szabó, S. (2011). Energy solutions in rural Africa: mapping electrification costs of distributed solar and diesel generation versus grid extension. Environ. Res. Lett., 6.

UNECA. (2017). *Africa Sustainable Development Report: Tracking Progress on Agenda 2063 and the Sustainable Development Goals. African Union, Economic Commission for Africa.* Addis Ababa, Ethiopia: African Development Bank and United Nations Development Programme.

UNEP. (2017). *Atlas of Africa Energy Resources.* Nairobi, Kenya: United Nations Environment Programme.

World Bank Data. (2016). *Access to electricity, rural.* Retrieved from https://data.worldbank.org/indicator/EG.ELC.ACCS.RU.ZS

Chapter 8
Designing Solar Power Plant Due to Consumer Load Schedule, Solar Energy Potential, and Electricity Prime Cost

Yuliia Daus
https://orcid.org/0000-0001-9120-7637
Don State Agrarian University, Russia

Igor Yudaev
Don State Agrarian University, Russia

Valeriy Kharchenko
Federal Scientific Agroengineering Centre VIM, Russia

Vera Dyachenko
Zaporizhzhia National Technical University, Ukraine

Shavkat Klychev
Academy of Science, Uzbekistan

ABSTRACT

The object of research in the chapter is the solar power plant as the source of additional economically expedient power supply of the electrical energy consumer. The purpose of this research is to analyze the options for the layout of solar power plant, taking into account the solar energy potential of the district, the design features of the proposed location, the load curve of the consumer, and the cost of the generated electrical energy. The chapter presents the results of calculation and selection of the parameters of solar power plant elements on the roof of the consumer's production building. The chapter presents the results of research of the dependence of the cost of the electricity generated by the solar power plant on the number of installed panels, which in order to increase the realized solar energy potential of the district also allows adding photoelectric modules and accumulating devices in the layout of the operating station at tariff growth. The chapter presents the results of researching these areas, that are conducted by the authors and which are completely original.

DOI: 10.4018/978-1-5225-9179-5.ch008

INTRODUCTION

The amount of energy that falls on the Earth's surface from the Sun within a week exceeds the energy of all the world's reserves of such hydrocarbon fuels as oil, gas, coal, and uranium. The transformation of the power coming from the Sun will almost completely satisfy humanity's need for energy for many years. At the same time, solar energy is "clean" and does not adversely affect the environment.

It should also be noted that silicon, used as raw material for the manufacture of photovoltaic cells, is one of the most frequently encountered elements in the bowels of the Earth. Despite this, experts consider solar energy to be not entirely environmentally friendly, since the production of pure silicon for photovoltaic batteries is chemically high-tech and energy-intensive production (Elistratov & Aronova, 2009). In addition, ensuring the efficient operation of power system containing solar power plants is possible:

- in the presence of significant traditional reserve capacity, to work in overcast and nighttime periods of the day;
- after the modernization of electrical networks or power supply systems.

Despite the above facts, solar energy continues to develop intensively both in the Russian Federation and around the world. If the production of photovoltaic modules and other components of the solar power plant becomes cheaper, in the next decade photoenergy will constitute a serious competition for oil and gas power engineering.

At the moment, two types of solar power plants are used in the world: they are installations that convert solar energy into electrical energy using direct conversion method - photoelectric setups, and thermodynamic installations, in which solar energy is first converted into thermal energy and then to mechanical and electrical one.

Solar power plants of any capacity as the source of electrical energy can be used in industry, agriculture, household and municipal sector, building spheres, autonomous systems of video surveillance, lighting, in the space industry, etc.

The South of Russia is one of the most promising regions for the use of local renewable resources in the Russian Federation, as only 2% of Russian oil reserves, 7% of gas and 3.5% of coal reserves are concentrated on its territory, the energy deficit here is covered by the use of predominantly steam coal from Siberian Federal District and natural gas from the Urals Federal District. In addition, the South of Russia is a region of developed production of a variety of agricultural products, often grown at remote livestock sites, camps in the steppe zone, as well as extended plantations, crops and plantations located at a significant distance from populated areas and centralized power supply. At the moment, the district already has a successful experience of economically expedient power supply to remote agricultural facilities of small capacity by means of renewable energy sources.

Now there is a large number of tables and maps of the solar energy potential, both for the territory of Russia, and for its individual subjects

There are also being developed a number of maps to allocate areas that are favorable for the building solar power plants on the territory of the studied region. The analysis takes into account a number of natural, ecological, technical, economic, social factors, as well as maps of land resources; transport networks; specially protected areas; map of the population's living; maps of energy consumption dynamics.

Such maps allow to analyze the solar energy potential throughout the region, highlighting the most favorable locations to place of solar power plants, maximizing the projected source of electric power generation to the consumer, thus reducing the costs of its transmission and conversion through electrical poer networks. However, in order to calculate and select the parameters of solar power plant elements, it is necessary to know the hourly sums of all the components of incoming solar radiation during the year, which will also allow to predict the electricity generation schedule and its coordination with the load curve of the consumer. That is why it is necessary to conduct additional experimental and mathematical studies.

The primary issue of designing generating facilities is the pre-project study of the solar energy resource in the places of their intended location. (Adomavicius et al., 2013; Amerhanov et al., 2008; Bezrukih, 2008)

Examples of calculating the gross, technical and economic potential of solar energy for particular regions and for the country as a whole, which are presented in a number of works, allow to identify only promising and favorable areas for the location of solar power plants, because they are not a source of information sufficient to determine their composition, power and to predict their operating mode.

As it is known from (Aronova & Grilihes, 2006; Atwater & Ball, 1978), the methods of calculating solar radiation intensity that is coming at arbitrary oriented surface at any time are based on the average long-term observations and are based on the following principle: the data for a horizontal receiving surface for a point with certain geographical coordinates are recalculated at an arbitrarily oriented surface according to empirical formulas proposed in the works of various authors, and taking into account the variety of influencing factors.

Data for the design and selection of parameters for solar power units can also be obtained by analytical methods. Methods for calculating gross solar energy resources at specific geographical point on horizontal surface are divided into methods with a limited amount of information and with full information (Butuzov, 2009; Vissarionov et al., 2008).

In the absence of data from long-term meteorological measurements, empirical formulas and expressions for determining the density of the incoming solar radiation flux in terms of the number of hours of sunshine are used. In the modern literature, there are seven main approaches of different authors (Rihab Mahjoub Essefi, Mansour Souissi, & Hsan Hadj Abdallah, 2014; Sen, 2008; Hosseini et al., 2017; Shyaml & Rajeev, 2011; Winter, 1991; Yang, Huang & Tamai, 2001) on the calculation of this value. All these methods require an estimation of their accuracy in comparison with data from the meteorological stations that are close to the calculated geographical point. In addition, some of them allow us to calculate only the direct component of solar radiation, while others give an overestimate of the total solar radiation, which indicates that they cannot be used directly for solar engineering calculations. The algorithms themselves, which have the greatest accuracy, require the introduction of a variety of correction factors.

In the methods, in the presence of complete information for determining the sums of coming solar radiation at certain geographical point, the results of actinometric measurements for the selected averaging period are used as the initial data, the results of which plot the arrival of solar radiation, integrating the obtained data over time and in space, obtaining the values of the gross solar energy potential of the studied area of future placement of the solar generating electrical installation. Using calculation methods, if full information is available, it is necessary to determine the probability of variability for each indicator.

Thus, in addition to the main task of assessing the solar energy potential for the location of power equipment, a number of problems arise related to the limited information on the initial data, the lack of a complete necessary volume of information for solar engineering calculations, the need to take into account the climatic features of the region and territories, and the carrying out a large amount of calculations.

There is also a wide variety of climate reference books, atlases and electronic databases. But the most attractive one among all existing sources of meteorological measurements (Daus, Kharchenko &

Yudaev, 2016 a; Daus, Kharchenko & Yudaev, 2016 b) to use in southern Russia is the database created by National Aeronautics and Space Administration NASA. It contains an array of data obtained as a result of satellite measurements with verification of them with terrestrial actinometric data from weather stations. The calculated valuesof solar radiation characteristics are interpolated to the grid of 1°x1°, which allows to expand the range of variation and identify the dependences of the hourly sums of direct, diffused and reflected solar radiation, not only on latitude, but also on longitude.

BACKGROUND

In the available reports based enlarged indicators of the energy sector functioning, so-called "utility-scale" solar power plants – that are plants connected to the Unified Energy System of the country and generating electricity in industrial scale are often represented (Kharchenko, Nikitin & Tikhonov, 2010; Popel' et al., 2010). However, many researchers argue that today the most promising direction is the use of low-power plants, both network and autonomous, used for the needs of the housing and communal sector, small businesses, social and health facilities, recreational zones, remote from centralized power supply systems of facilities and farms producing agricultural products. In the above categories there are also objects of distributed generation. The introduction of such small objects of "solar" generation makes it possible to reduce the load on the power system, and also to reduce losses in its elements at transporting electricity to the consumer (Budzko & Zul', 1990; Belenov et al., 2016; Daus, 2016; Daus & Kharchenko, 2018).

The use of low-power plants is the most widely covered in the special technical literature, which describes two aspects of their operation: the first ones are solar power plants whose photovoltaic panels are located at given tilt angle to the horizon on the flat roof of the building; second ones are integrated in the construction of the building photovoltaic panels.

However, the use of such power supply systems is limited by the structural features of the roofs of buildings. According to the first approach to the formation of the structure and parameters of a solar power plant, it is necessary to take into account the mutual shading of the series of photoelectric panels and structural elements (Belenov et al., 2016; Butuzov, 2009). This leads to the limitation of the power of the installation and to limiting the use of the technical solar energy potential of the geographical point of its placement. Integration of photovoltaic panels into the construction of the building provokes decrease in the amount of generated electric energy caused by deviation from their optimal spatial orientation (Kharchenko, Nikitin, & Tikhonov, 2010; Popel' et al., 2010), which leads to the reduction in the coefficient of the installed capacity of electrical equipment, that is, the low efficiency of converting solar energy into electrical energy.

Apart from choosing method for assembling photovoltaic panels of solar power plant, one of the important tasks of their design is to converge the graphs of generation and consumption of electrical energy. The graph of electric energy generation is completely conditioned by the laws of the arrival of solar radiation on the receiving surface of the panels. This is, first of all, schedule that has one maximum – at noon. But the graphs of loads of electric consumers most often have two or three peak consumption, occurring in the morning and evening hours, when the intensity of solar radiation is low. In the literature, there are two ways of compensating these differences in the schedules of generation and consumption of electrical energy: the first one is the use of storage equipment to accumulate excess energy during periods of high insolation and utilize in the period of low solar activity or in the absence of it; the second way is the unification of consumer groups in a micro-network. The use of the first approach is associated with high costs for the acquisition of storage equipment, which are commensurate with the cost of the PV panels themselves, and consequently with the growth in the prime cost of generating electricity (Vissarionov et al., 2008; Winter, 1991).

Connecting several consumers and linking them to a single microgrid facilitates the realization of a greater technical solar potential of the region, however, it requires remote computer monitoring of all areas, and, consequently, a decrease in the reliability of the system as a whole, in view of excessive reliance on computational capacities and communication channels.

The introduction of generation systems based on renewable energy sources is associated with the costs of their installation and commissioning, which requires the analysis of each option layout in terms of economic feasibility. In the practice of power supply, the choice of cost-effective projects is based on the implementation of preliminary technical and economic calculations, that is the determination of the following indicators: capital (initial) investments and current (annual) expenditures.

For the express assessment of the economic feasibility of the introduction of solar power plants, it is convenient to use such a concept as the cost price of the generated electric energy. If we compare it with the current electricity tariff, then this decision will convincingly indicate the profitability or unprofitableness of the solar power plant project.

MAIN FOCUS OF THE CHAPTER

The object of this study is the solar power plant as a source of additional economically feasible power supply of the electrical energy consumer.

The purpose of the study is to determine the layout and general parameters of the power plant equipment, for additional power supply of the consumer, taking into account the solar radiation intensity during the year, the structural features of the roof of the building, the load schedule features to reduce the costs of payments for electricity.

Methodology of the Research

The proposed approach to the definition of the parameters of solar power plant that makes it possible to provide economically feasible generation of electric energy presupposes preliminary solution of the issue of searching for the number of photovoltaic panels taking into account their spatial orientation, the solar energy potential of the area, and the size of the field on which the photo panels are presumed to be placed. The criterion here is the minimum value of the prime cost of the generated electric energy (Daus, Kharchenko & Yudaev, 2016 c; Daus, Kharchenko & Yudaev, 2018):

$$\sum_{i=1}^{L}\left(\frac{K_i}{T_{ei}}\right) / \sum_{n=1}^{365}\sum_{t=t_1}^{t_2}\left[Rsum_{\beta\gamma}(t,n)\cdot N\cdot S_{ph}\cdot \eta_{ph}\right] \to \min \tag{1}$$

where $\mathrm{Rsum}_{,,}\left(\mathrm{t,n}\right)$ is hour sums of total solar radiation for the geographical point with coordinates (φ,°N, λ,°E) for the moment t during the day n, кWh/m^2; $N, S_{\Phi\Pi}, \eta_{\Phi\Pi}$ are number (pcs.), surface area

(m^2) and coefficient of efficiency (r.u.) of photovoltaic module; t_1 и t_2 are time of the start and finish of daytime; K, L is number of elements of i-type (photovoltaic panels, inverters, charge controllers, rechargeable batteries); K_i, T_{ei} are cost ($) and service life (year) of the i-th element of the solar photoelectric power plant.

Limitations in the process of solving this optimization problem are the following:

1. The value of the tilt angle of the receiving surface with respect to the horizon β lies in the range from 0° to 90°.
2. The value of the orientation angle of the receiving surface along the sides of the world γ lies in the range from -90° to 90°.
3. The dimensions of the placement field are determined by its length and width – a x b, m.

For computer calculations, there are used the hourly sums of total solar radiation $Rsum_{\beta\gamma}(t,n)$ obtained with the help of the program for estimating the solar energy potential on the example of concrete decree (46.8°N 40.6°E), which is based on express method for estimating the solar energy potential at the specific geographical point in the southern Russia (Daus & Yudaev, 2016; Daus, Yudaev & Stepanchuk, 2018). Here, the geographic coordinates and spatial orientation of the receiving area are entered as initial data . The search for the optimal tilt angle of the receiving surface relatively to the horizon β at the given geographical point of the Southern Federal District was also carried out with the help of the designed computer program (Daus & Yudaev, 2016; Daus, Yudaev & Stepanchuk, 2018). The estimated operating time of solar batteries was taken from 8^{00} to 17^{00}. The required parameters as a result of the calculation are the number of photoelectric modules, as well as the list and parameters of equipment in the solar power plant (photoelectric modules, inverters, charge controllers, batteries).

EVALUATION OF SOLAR ENERGY POTENTIAL IN THE PARTICULAAR GEOGRAPHICAL POINT

To determine the operating characteristics of solar power plant, and also to model its operating modes, it is necessary to determine the hourly sums of the total solar radiation on a receiving surface of a photovoltaic panel oriented in a predetermined way. The calculation begins from the determination of this value on the horizontal surface. The total energy flux of solar radiation that is incident on a horizontal surface is determined by the flux of direct radiation scattered from the sky. And here it should be noted that it is important to take into account both components, as the peculiarity of incoming solar radiation, for example, in the southern part of Russia is a high proportion of the diffused component in the radiation balance, which ranges from 25% in summer to 60% in the winter months. The characters of the diurnal course of direct and diffused radiation are similar in this case and depend, first of all, on the altitude of the Sun above the horizon increasing from zero during the hours of sunrise to a maximum at around noon. It should be noted that cloudiness only affects the symmetry of the course.

According to Bird's method (Atwater, & Ball, 1978), the calculation of the intensity of direct and diffused solar radiation is conducted on the basis of the following system of equations:

$$\begin{cases} R^{cs}_{drH\,tn} = E_c \cdot \cos\theta_{tn} \cdot \tau_{dr\,tn}, if \cos\theta_{tn} > 0, \\ R^{cs}_{dfH\,tn} = E_c \cdot \cos\theta_{tn} \cdot \tau_{df\,tn} \\ R^{cs}_{drH\,tn} = 0, if \cos\theta_{tn} < 0, \end{cases} \tag{2}$$

where $E_c = 1,367\ kW / m^2$ is normal flux density of solar radiation in cosmos; θ_{tn} is angle of sunlight incidence on the receiving surface that is horizontal to the surface of the Earth (hereinafter - horizontal receiving surface or area), which depends on the latitude of the terrain φ, time *t* and day ordinal number n; $\tau_{dr\,tn}$ is coefficient that takes into account the refraction and absorption of direct solar radiation in the atmosphere by ozone, a mixture of gases, water vapor, and also Rayleigh and aerosol scattering of solar radiation).

The presented method allows to determine the intensity of direct, diffused and reflected solar radiation at any point, knowing only its latitude under the clear cloudless sky. However, equations (1) do not take into account real atmospheric conditions, which have a significant effect on the characteristics of solar radiation.

In order to extend the scope of the proposed method application for estimating the solar energy potential at given point in remote areas, according to (Bezrukih, 2008), the hourly values of total solar radiation for day n at time t under real cloud conditions are made with respect to an arbitrarily angled receiving surface with angle β and azimuth γ:

$$R_{sum\,tn} = R_{dr\beta\gamma\,tn} + R_{df\beta\gamma\,tn} + R_{rf\beta\gamma\,tn} = R_{drH\,tn} \cdot \frac{\cos u_{tn}}{\cos\theta_{tn}} + R_{dfH\,tn} \cdot \left(\frac{1+\cos\beta}{2}\right) + (R_{drH\,tn} + R_{dfH\,tn}) \cdot r_7(n) \cdot \left(\frac{1-\cos\beta}{2}\right) \tag{3}$$

where $R_{dr\beta\gamma\,tn}, R_{df\beta\gamma\,tn}, R_{rf\beta\gamma\,tn}$ are hourly sums of direct, diffused and reflected intensity of solar radiation for day n at time t at an arbitrary reception surface at angle β with an azimuth γ, kWh / m², $r_7(n)$ is coefficient of reflection of the earth's surface (albedo), $\cos u(t, n)$ is angle of incidence of direct solar radiation at an arbitrary surface at angle β>0 with azimuth γ.that is defined according to (Vissarionov et al.,2008) by formula:

$$\cos u_{tn} = (\sin\phi(\sin\delta\cos\beta + \cos\delta\cos\gamma\cos\omega\sin\beta) + \cos\phi(\cos\delta\cos\omega\cos\beta - \sin\delta\cos\gamma\sin\beta) + \cos\delta\sin\gamma\sin\beta\sin\omega) \tag{4}$$

The simplest way to place solar power plant in are flat roofs, as their shape repeats the surface of horizontally located heliopoly fields on the Earth's surface, in relation to which the design requirements are now extensively developed. There are calculated the monthly and annual sums of direct and diffused solar radiation coming to the horizontal receiving surface for the point with geographical coordinates of concrete decree (46.8°N 40.6°E). The annual solar potential of this settlement is 1 250.2 kWh/m^2. The magnitude of the components of the total solar radiation on the horizontal surface by the month is presented at Figures 1 –2.

Figures 1 and 2 show that the fraction of diffused radiation is large, and from the middle of autumn to the middle of spring it even exceeds direct radiation and ranges from 51.0% in autumn and spring to 63.55% in winter. In the rest of the time, the diffused radiation is commensurate with the direct one and is between 43.5% and 46.9%. That is, for every region it is very important to take into account all the components of solar radiation when calculating solar energy potential, as this will reduce the capacity of the selected equipment and reduce capital costs.

During the year, the monthly amount of incoming solar radiation varies from 31.15 kWh/m^2 in December to 180.3 kWh/m^2 in July, while in the period of the lowest solar activity from November to February it is 31.1 - 37.6 kWh/m^2, and in the period of high activity - 159.7-180.3 kWh/m^2.

Figure 1. Monthly sums of direct and diffused components of total solar radiation on the horizontal receiving surface

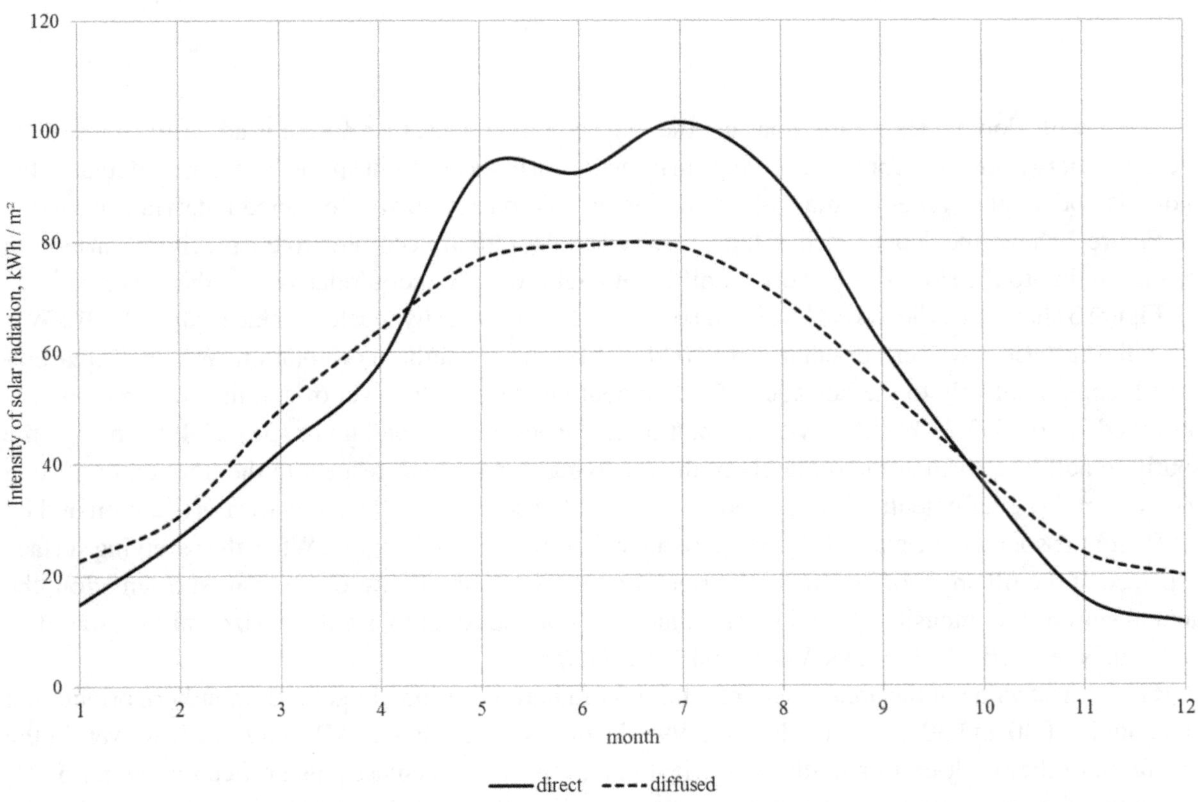

Figure 2. Monthly sums of total solar radiation on the horizontal receiving surface

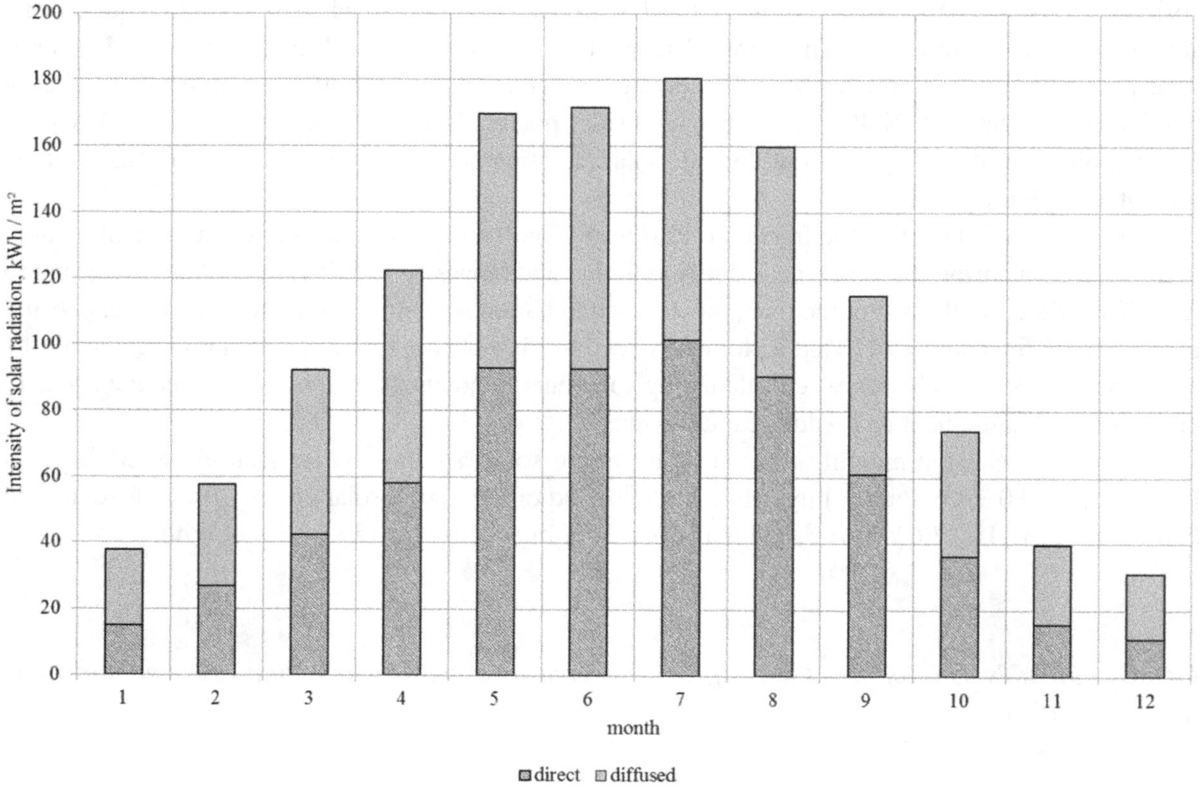

Roofs with sloping design solutions are the most common ones, therefore it is advisable to estimate the solar energy potential for the receiving surface arbitrarily oriented in space. There are calculated the monthly and annual values of total solar radiation on an arbitrary surface-oriented receiving surface.

Figure 3 shows the dependence of the intensity of solar radiation coming to the receiving surface on its orientation to the sides of the world at different angles of inclination relatively to the horizon.

Figure 3 shows that the lowest level of insolation falls on vertically located surface from 792.01 kWh/m^2 in the eastern or western orientation to 1020.25 kWh/m^2 at southern orientation. With a decrease of the tilt angle relatively to the horizon to 50°, gradual increase in the level of coming solar radiation is observed up to 1093.29 kWh/m^2 with orientation to the east or west and up to 1381.72 kWh/m^2 - to the south. In addition, even at the tilt angle of the receiving surface with respect to the horizon of 50°, an increase in the solar potential is observed by 4.2% for the south-east and south-west orientation and by 10.5% for the southern orientation in comparison with its horizontal location. When the receiving surface is placed at the tilt angle of 70° of south orientation or 60° of south-east or south-west orientation, the annual sum of the intensity of total solar radiation is commensurate with the horizontal location of it, and equals respectively 1254.4 kWh/m^2 and 1252.8 kWh/m^2 .

The highest value of the annual sum of total solar radiation was obtained for the southern orientation at an angle of 40 and 30, respectively, 1405.99 kW • h / m^2 and 1404.41 kW • h / m^2. However, in the transition to the south-east or south-west orientation, this value decreases insignificantly by 4.4-5.1%, and with orientation strictly to the east or west - significantly by 15.6 - 18.6%, since the reception the surface falls on the direct component mainly only in the morning or evening hours of the day.

Figure 3. Dependence of the intensity of solar radiation coming on the receiving surface on its orientation to the sides of the world at different angles of inclination relatively to the horizon

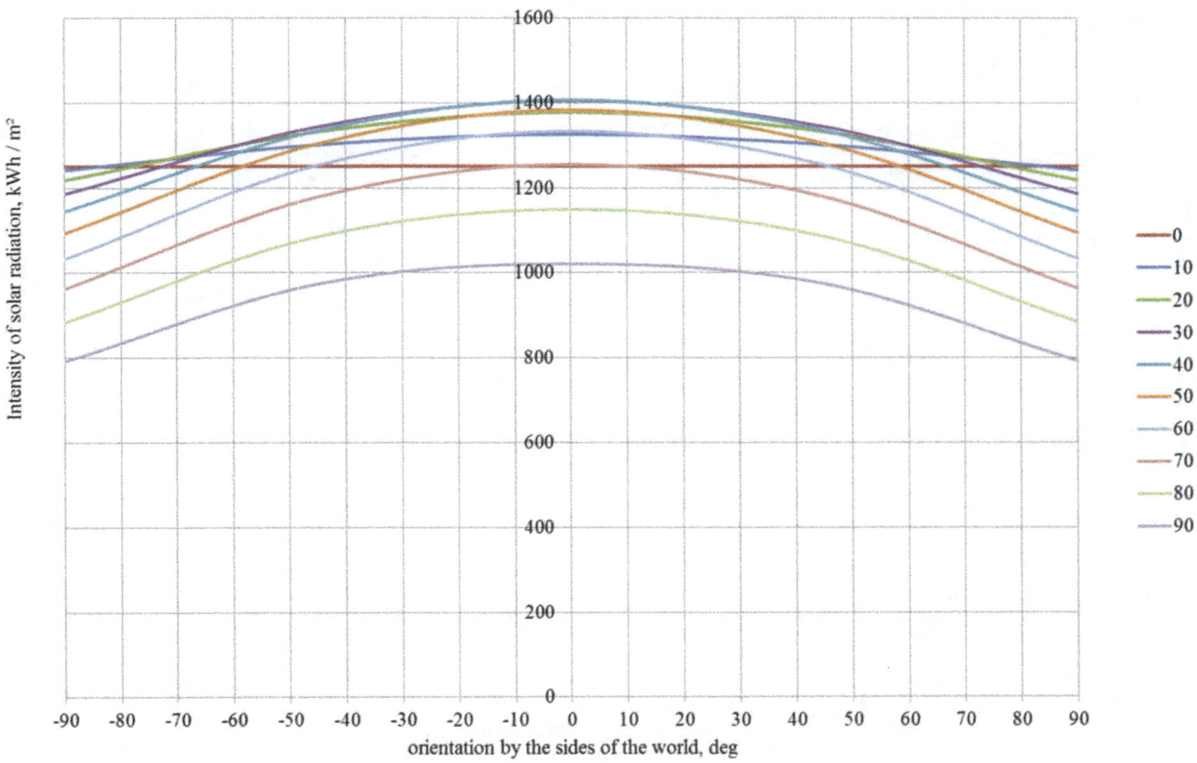

Figure 4 shows the dependence of the intensity of solar radiation coming on the receiving surface on its inclination angle with respect to the horizon with its different orientations to the sides of the world.

The data at Figure 4 allow with the fixed orientation along the sides of the world, to identify the optimal tilt angles of the receiving surface with respect to the horizon. So it is evident that when orienting to the east or west it is advisable to place the receiving surface horizontally. If its vertical position is used this is the option of placing PV modules on the walls of the building, then the incoming flow will decrease by 36.6%.

South-east or south-west orientation allows to utilize less amount of solar energy. The maximum value of 1342.44 kWh/m² corresponds to the angle of inclination relatively to the horizon of 30°. In this case, transiting to larger or smaller tilt angle with the step of 10° this value falls slightly - by less than 1%. Using the extreme values of the tilt angle relatively to the horizon for the horizontal and vertical placing of the receiving area the gross solar energy potential drops by 6.8% and 27.5%, respectively.

The best orientation for the studied region is the southern orientation. Maximum solar energy flow to the surface of the receiving area for a year is obtained at the tilt angle of 30° – 40° relatively to the horizon and equals from 1404.13 to 1405.99 kWh/m². Then with the increase in the tilt angle to 90 ° the annual amount drops to 1020.246 kWh/m², which is the largest value among all orientations by the sides of the world. That is, when placing photovoltaic panels on the walls of the building, it is advisable to use walls of southern, south-eastern or south-western orientation. When the angle of inclination is reduced in comparison with the optimal ones to the horizontal position, the potential of the utilized solar radiation decreases by 10.7%.

Figure 4. Dependence of the intensity of solar radiation coming on the receiving surface on its inclination angle with respect to the horizon with its different orientations to the sides of the world

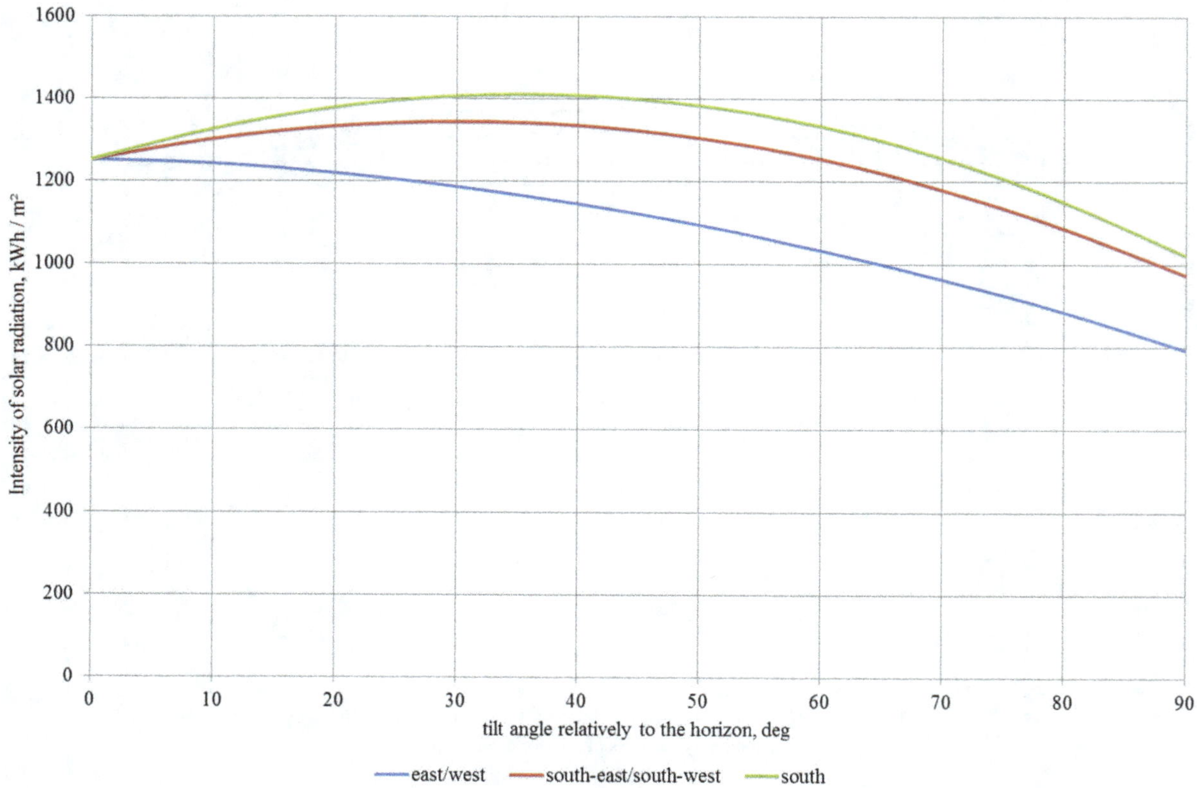

There were defined the optimal angles of the inclination of the receiving surface with respect to the horizon for different orientations along the sides of the world, which ensure the maximum flux of incoming solar radiation during the studied period (Table 1). There were also studied photoelectric generating systems without tilt angle adjustment or with seasonal tilt angle adjustment.

The calculation results showed that for eastern or western orientation, it is not advisable to adjust the angle of inclination relatively to the horizon, and the maximum flow can be obtained only on the horizontal surface.

Table 1. Optimum angles of inclination of the receiving surface at its various spatial orientations

Tilt adjustment period relatively to the horizon	Orientation by the sides of the world		
	east/west	south-east/ south-west	south
year (without adjustment)	0°	30°	36°
winter	0°	67°	75°
spring	0°	23°	28°
summer	0°	13°	16°
autumn	0°	45°	52°

Table 2. Monthly sums of total solar radiation at optimal angles of inclination of the receiving surface of different spatially orientation

Month	east/west	south-east/south-west	south
January	37.62	51.20	60.98
February	57.60	73.50	85.06
March	92.12	103.64	112.00
April	122.07	127.09	130.55
May	169.58	168.29	167.32
June	171.61	166.09	162.33
July	180.34	176.73	174.12
August	159.78	164.53	167.54
September	114.82	127.44	136.25
October	74.02	90.04	101.71
November	39.54	51.29	59.90
December	31.15	42.60	50.82
Year	1250.24	1342.44	1408.56

Table 2 presents the monthly values of total solar radiation for unregulated tilt angle with respect to the horizon.

The greatest effect from the use of the optimal unregulated tilt angle of the receiving surface relatively to the horizon is observed with the southern orientation, it is 1408.56 kWh/m^2. At the same time, the monthly amount of incoming solar radiation during the period of high activity of the Sun from May to September is from 162.33 kWh/m^2 to 174.12 kWh/m^2. In the same period, for slightly different orientations, with an optimal angle of inclination relatively to the horizon, a little bit higher values solar radiation are observed: they are from 168.29 kWh/m^2 to 176.73 kWh/m^2 for the south-east and south-west orientation, from 169,58 kWh/m^2 to 180.34 kWh/m^2 for the east and west orientation. However, during the rest of the year, for example, in winter months, the use of receiving surfaces located at tilt angle of 36° to the horizon of the southern orientation makes it possible to utilize by 13.6% - 16.1% more than for the reciving surface located at tilt angle of 30° to the horizon of south-east and south-west orientation and by 32.3% - 38.7% more than for the surface that is located horizontally.

Figure 5 shows the monthly sums of all the components of solar radiation for the optimal angle of inclination with respect to the horizon of the receiving surface of the southern orientation.

Figure 5 shows that the proportion of the direct component of solar radiation coming on the receiving surface located at the tilt angle of 36° to the horizon of the southern orientation is 53% -64%, and the reflected component is negligible and equals 3% – 6%. That is, this spatial arrangement allows to utilize most of the incident direct sunlight flux.

Figure 6 shows the monthly sums of total solar radiation for the optimal angle of inclination relatively to the horizon of the receiving surface of different orientations

Figure 6 allows to analyze the effect of adjusting the angle of inclination relatively to the horizon by season during the year.

Figure 5. Monthly sums of all components of solar radiation for the optimal angle of inclination relatively to the horizon of the receiving surface of southern orientation

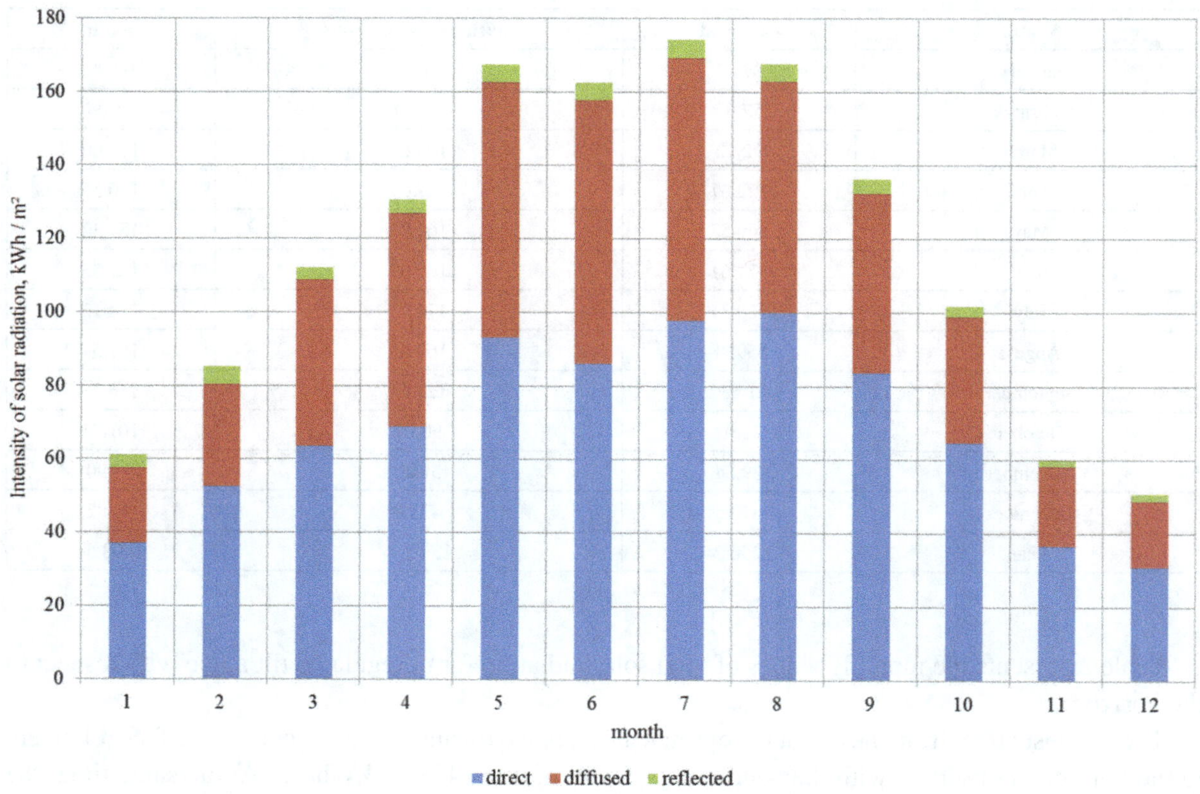

From March to November, with the orientation to the south, the use of seasonal regulation makes it possible to increase the flux of utilized solar radiation only by 0.5% to 6%. Only in winter the flow increases by 13.4% - 21.7%. As a result, for a year the potential of the utilized solar radiation is increased by 4.5% in comparison to an unregulated tilt angle to 1472.53 kWh/m^2. That is, it is impractical for the southern orientation to adjust the angle of inclination relatively to the horizon every season - the effect is negligible.

For the south-east and south-west orientation, during the period from March to September, with seasonal regulation, the flux of solar radiation to the receiving surface remains practically unchanged (0.1% to 3.5%). However, in the winter months, the amount of incoming solar radiation increases significantly by 10.1% in February and by 16.9% in January, in the autumn months it increases by 39.7% in September and by 50.4% in November. As a result, for a year the value of the annual solar potential, which can be used seasonally by adjusting the angle of inclination relatively to the horizon, increases by 11.2% to 1492.03 kWh/m^2.

Figure 7 shows the annual sums of total solar radiation for the optimal angle of inclination relatively to the horizon of the receiving surface of different spatial orientations.

Figure 7 illustrates the possibility of expanding the potential of recycled solar radiation by controlling the spatial orientation of the receiving surface. The minimum value of this value was obtained on the horizontal surface. With the southern orientation of the receiving area, the annual amount of coming solar

Figure 6. Monthly sums of total solar radiation for the optimal angle of inclination relatively to the horizon receiving surface of different spatial orientation

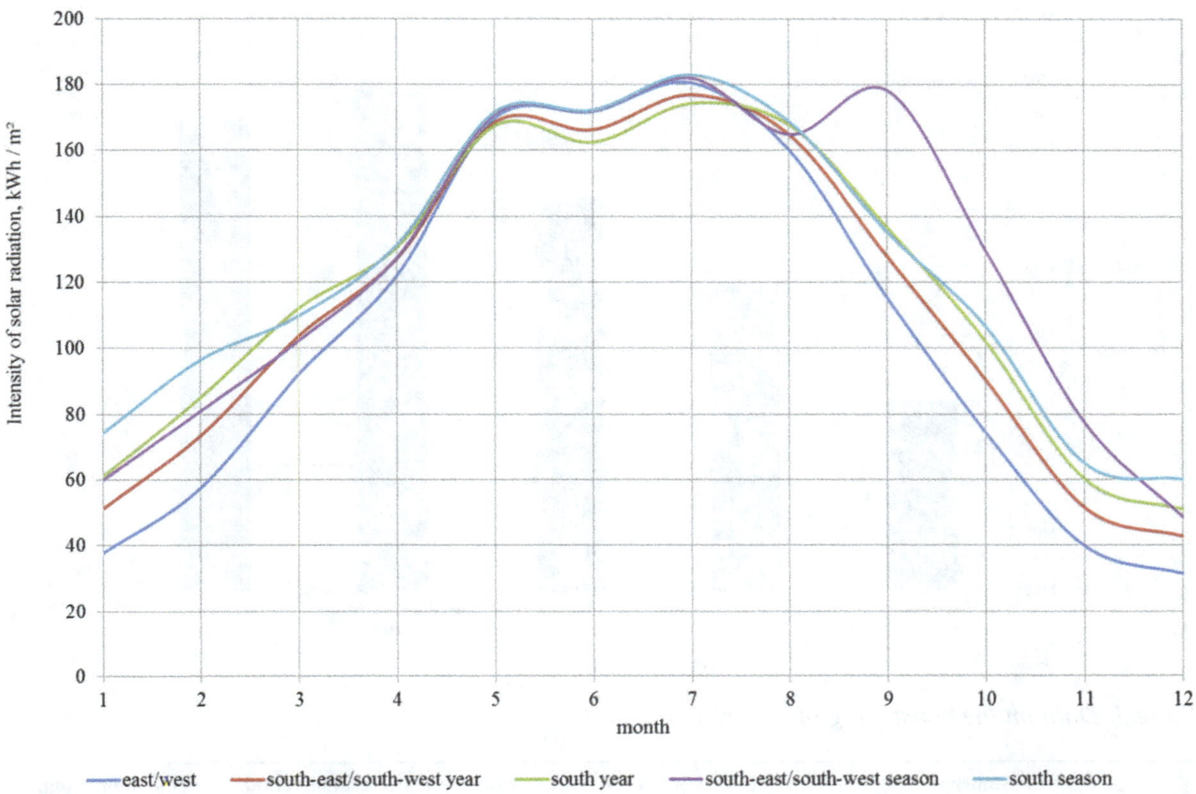

radiation can be increased to 1408.6 kWh/m^2 for the non-adjustable tilt angle of inclination relatively to the horizon and 1472.5 kWh/m^2 for seasonal regulation. With the south-east and south-west orientation of the receiving area, the annual amount of solar radiation coming on the receiving surface can be increased to the maximum among the analyzed spatial orientation options up to 1492.2 kWh/m^2 with its seasonal regulation and 1342.4 kWh/m^2 for unregulated angle of inclination relatively to the horizon.

JUSTIFICATION OF PARAMETERS OF THE SOLAR PHOTOELECTRIC POWER PLANT FOR REDUCING THE COSTS FOR THE LIGHTING SYSTEM OF THE PROCESSING ENTERPRISE

The object of the study is the system of intra-shop lightening of two branches of the mill for baking in concrete decree (46.8°N 40.6°E). Data on the characteristics of the consumer are presented in Table 3.

The annual volume of consumed electricity is 10275.75 kWh/year. The tariff for payment for consumed electric energy is b = 8.18 rubles/kWh (0.12 \$/kWh).

The choice of solar power plant parameters includes determining of the nominal power of the solar modules, their number, the capacity of the batteries, the capacities of the inverter and the charge-discharge controller.

Figure 7. The annual sums of total solar radiation for the optimal angle of inclination relatively to the horizon of the receiving surface of different spatial orientations

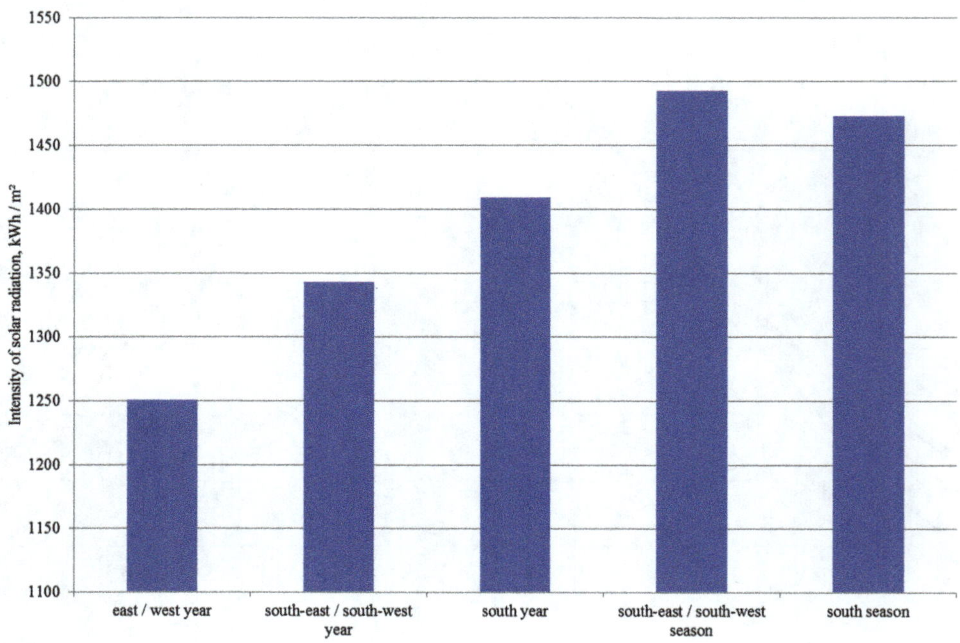

Table 3. Data on the consuming of electricity

Consumer	I, A	U, V	cosφ	Operation teriod	Operation regime
Branch Nº 1	3.1	230	0.75	day and night	constant
Branch Nº 2	3.7	230	0.75		

The choice of the number of photovoltaic panels is based on the consumed power and the geometric dimensions and spatial orientation of the location of their future placement.

The choice of the inverter power is carried out at the peak load power in the presence of consumers with starting currents and at rated power in the remaining cases.

The choice of charging batteries is determined by the selection of its operational parameters, such as capacity, energy density, self-discharge, temperature and atmospheric conditions, type. The capacity of the battery is determined by the amount of charge that is measured when the energy is supplied to consumers from the full charged state to the minimum permissible value of the output voltage and the amount of excess output of electrical energy.

The choice of the technical characteristics of the charge controllers is determined by the input voltage and current values.

There was conducted choice of parameters of the solar photo-electric power plant on thr flat part of a roof of production building

There were defined the parameters of the components of the solar power plant when it is placed on the flat part of the roof of the baking mill having small tilt angle, this part of the roof is 13 m x 39 m, and it is oriented narrowly to the south and it is 0.5 m high.

To organize the power supply of the intra-shop lighting system of the mill compartments, it is necessary to calculate the number of photovoltaic modules that can be placed on the roof, taking into account mutual shading from the rows of panels and elements of the roof structure. The calculation was carried out using photovoltaic monocrystal batteries of the Solarland brand SLP180-24 Silver Poly Solar Panel, the characteristics of which are indicated in Table 4.

The number of photovoltaic panels, which can be located on the roof of the building, is limited by its design features. First of all, there must be noted the fact that the roof is oriented with the narrow part to the south, that is, the rows of batteries will be placed parallel to this line. Basing on the dimensions of the Solarland SLP180-24, vertically in one row can be accommodated no more than 13 pcs.

The distance between the rows and from the roof edge to the first row, at which the surface of the solar module would not be obscured by the formula according to Figure 8:

$$\begin{bmatrix} kw = (\cos\beta + \dfrac{\sin\beta}{tg(90-\theta)}) \cdot w \\ w_{sh} = 0 \end{bmatrix} \quad (5)$$

where θ is angle of sunlight incidence on the receiving surface, located horizontally to the surface of the Earth; w is height of photovoltaic module, m.

The minimum distance between the lower points of the modules in adjacent rows for the time t = 9 hours on December 22 (the time of the rising of the sun above the horizon for the shortest day of the year) with the optimum tilt angle of the receiving platform will be 21.36 m. And the shortest distance from the roof edge to the first row at this time moment is 20.83 m. Basing on the geometric dimensions of the roof, it is possible to place 2 rows of 13 PV solar panels SLP180-24 parallel to its narrow side.

There were considered three options for the layout of solar power plant photoelectric panels: 1 option is maximum use of available space and placement of modules, taking into account the conditions

Table 4. The characteristics of SLP180-24 Silver Poly Solar Panel

Power	180 W
Open circuit voltage	43,2 V
Voltage at work on load	34,4 V
Rated voltage	24 V
Current at work on load	5.23 A
Size	1482 x 992 x 40 mm
Temperature of operation and storing	-40..+85 °C
Protection class	IP 65
Solar elements	monocrystal
Weight	20 kg

Figure 8. Geometric model for determining the length of the shadow from the photoelectric module

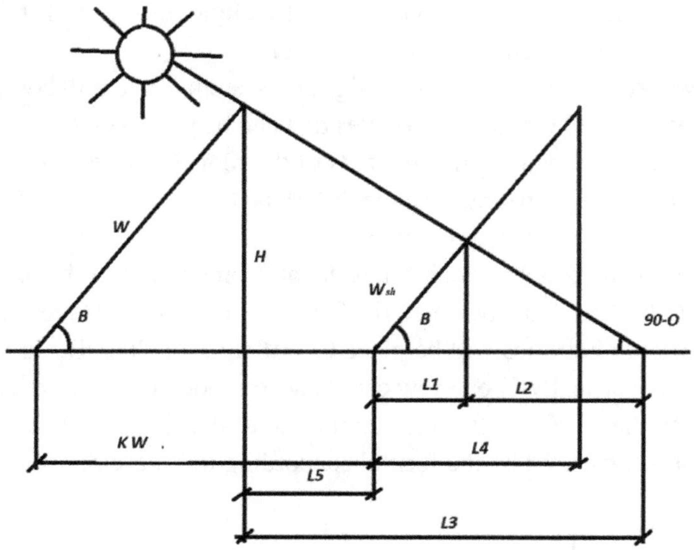

of shading (2 rows of 13 modules); option 2 is generation mode without the use of electrical energy storage; 3 option is 1 row of 13 modules with the system of electrical power accumulation. The results of calculation and selection of the parameters of solar power plant elements for the selected options are presented in Table 5.

Figure 9 presents the daily graph of the electrical energy generation for the first option of the photovoltaic modules' layout of the solar power plant on the example of January 1, April 1, July 1 and October 1.

The solar power plant with the layout of 26 photovoltaic modules arranged in 2 rows at the tilt angle of 38 ° to the horizon and oriented to the south, produces the greatest output during the period of solar activity maximum at 14^{00} in summer 3.01 kWh, in spring and in autumn 2.57 – 2.83 kWh, and in winter 1.02 kWh. At the same time, the consumption of electrical energy is fully provided from the power plant in the summer in the period from 10^{00} to 18^{00}, at noon the output exceeds the load by 2 times. In the

Table 5. List of selected equipment for various options of the layout of photovoltaic panels on the solar power plant

Layout	Nph, pcs.	Type of photovoltaic panels	Ninv, pcs.	Type of inverter
1 option	26	Solarland 180 Silver Poly	1	Sunny Boy 5000TL-21 ST
2 option	11	Solarland 180 Silver Poly	1	MAP·HYBRID·24 В 2 кВт
3 option	13	Solarland 180 Silver Poly	1	Sunny Boy 2500TL -21 ST
Layout	Nbat, pcs.	Type of charging batteries	Ncon, pcs.	Type of charge controller
1 option	23	HZY12 -230	26	CM2024 12/24B
2 option	-	-	0	CM2024 12/24B
3 option	2	6-GP-150	13	CM2024 12/24B

Figure 9. Daily schedule of generation and consumption of electrical energy for 1 option of its layout by photoelectric modules

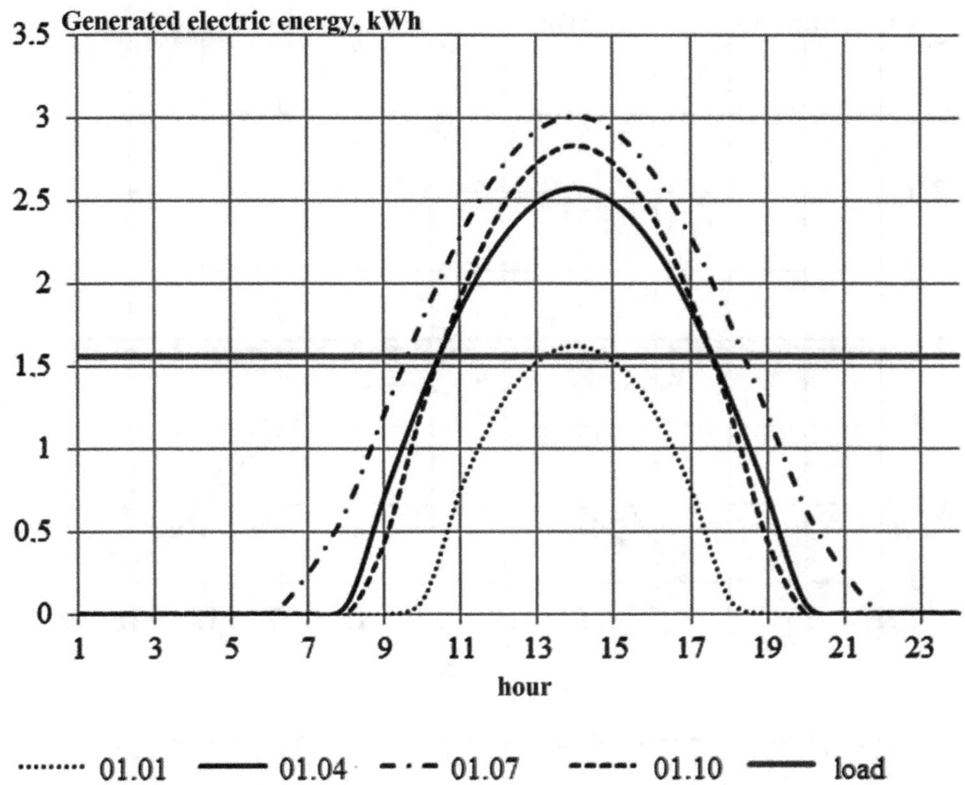

autumn-spring period, the lighting system receives power only from the solar power plant from 11^{00} to 17^{00}, with over-generation ranging from 59% to 81%. In winter, only 2 hours per day, the power supply from the modules is sufficient to supply the load, that is from 13^{00} to 15^{00} (98% – 100%). That is why, this approach requires almost all year except for the winter months application of additional storage equipment or connecting additional consumers to utilize the excess of generated electrical energy.

Figure 10 presents the daily graph of the electrical energy generation for various variants of the layout of photovoltaic modules of the solar power plant on the example of July 1.

The minimum mode of electrical energy generation is provided according to the second variant of the arrangement. It does not require the use of rechargeable batteries. Using the same layout with 13 photovoltaic modules already involves the accumulation of excess electricity produced in the summer months.

Figure 11 shows the annual amount of generated electrical energy for various variants of the layout of photovoltaic modules at the solar power plant when they are oriented at the optimum tilt angle to the horizon (38 °) using the angle of inclination of the roof structures (10 °).

It can be concluded from Figure 11 that the use of optimal tilt angles on the flat part of the roof allows more fully realize the available solar energy potential of the district (by 8%). Using less photovoltaic modules (option 2), despite the fact that it provides less generation of electricity per year, reduces the number of additional equipment that is needed to be used, and does not change the configuration of the existing power supply system of the consumer.

Figure 10. Daily schedule of generation and consumption of electrical energy for various options of its layout by photoelectric modules on the example of July 1

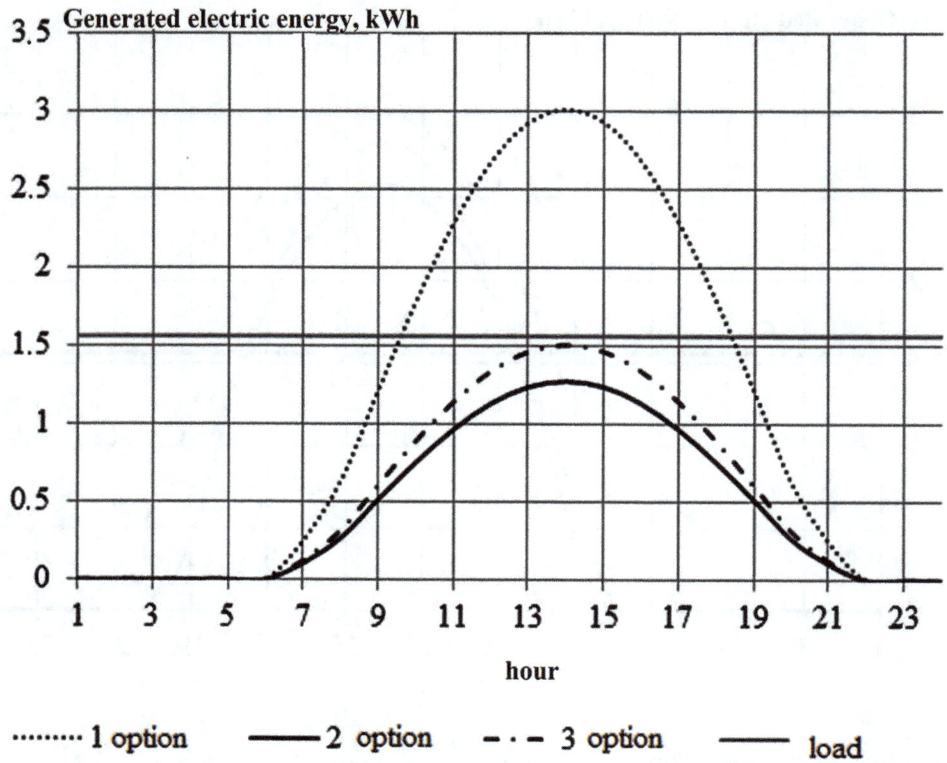

Figure 11. Daily schedule of generation and consumption of electrical energy for various options of its layout by photoelectric modules on the example of July 1

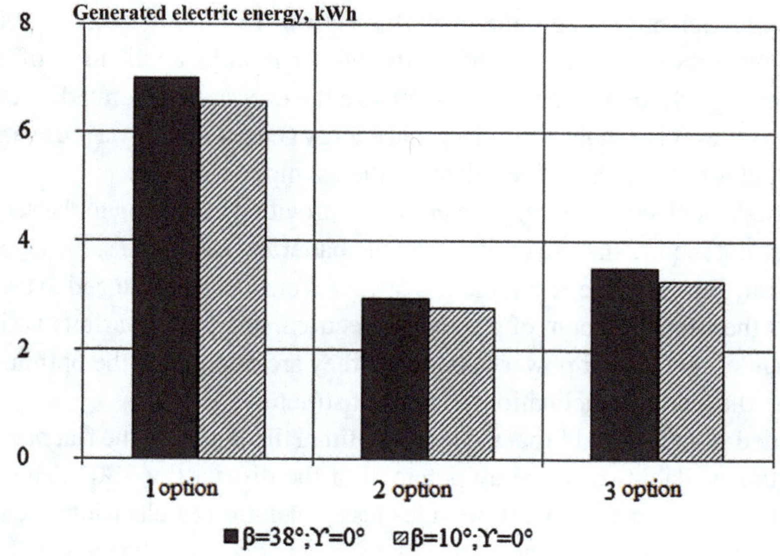

TECHNO-ECONOMIC CALCULATION OF THE VARIANTS OF SOLAR PHOTOVOLTAIC POWER PLANT LAYOUT

The construction of new generation systems based on the conversion of solar energy into electricity is associated with the costs on their acquisition, installation and commissioning, and therefore there is required the analysis of solutions from the point of view of economic feasibility. In power supply practice, the selection of economically viable projects is carried out on the basis of the implementation of technical and economic calculations, namely the determination of such indicators as capital investments and current costs.

Under conditions of the planned economy for the generation of electric power, the following recommendation exists: the payback period should not exceed eight years which is determined by the normative energy efficiency factor of capital investments. This value was obtained on the basis of analysis of calculations, which revealed that the increase in the payback period of more than 7 to 8 years does not significantly reduce the operating costs per unit of additional capital investment. Thus, the main criterion was the cost-effectiveness of investments, and not to reduce the cost of paying for electricity.

As for the prime cost of generated electricity at the solar photovoltaic power plant, its value should be compared and be lower than the current tariff for electricity consumed from the electric grid, that is, it would be more expedient for the enterprise to produce than buy it. Therefore, this value can be recommended for a preliminary assessment of the economic feasibility of using solar-to-electricity conversion systems, as it clearly shows the profitability or unprofitability of the project.

Thus, for the technical and economic evaluation of the solar power plant project as the source of additional generation of electric power in the power supply system of the enterprise, it is necessary to determine the following technical and economic indicators: investment and current costs of operating the power plant, the total amount of generated electric energy. To calculate the initial costs for the construction of the solar power plant, prices for electrical equipment, the cost of its installation and building structures for each of its elements are necessary. To determine the cost of operating the system there are needed the norms of deductions for depreciation, maintenance and repair of electrical equipment.

The results of the feasibility study will allow to choose such a layout and parameters of the solar power plant that will provide economically feasible generation of electric power, which ensures maximum realization of the technical potential of the solar energy of the area where it is planned to be located.

Calculation of technical and economic indicators is carried out according to the following procedure.

Total costs, \$:

$$Ktot = kins \cdot \left(k{>} \cdot \kappa ph + \kappa inv + \kappa bat + \kappa con\right).$$

where $\kappa ph, \kappa inv, \kappa bat, \kappa con$ .are capital costs for the purchase of photovoltaic modules, inverters, batteries and charge controllers respectively, \$; ko is coefficient, taking into account the cost of supports for fixing photovoltaic modules; kins is a coefficient that takes into account the cost of installing a solar power plant.

Current operating costs of the system, \$:

$$Otot = krep \cdot \left(\frac{ko \cdot \kappa ph}{Tph} + \frac{\kappa inv}{Tinv} + \frac{\kappa bat}{Tbat} + \frac{\kappa con}{Tcon} \right) \quad (6)$$

where $Tph, Tinv, Tbat, Tcon$ .are service life of photovoltaic modules, inverters, batteries and charge controllers respectively, years; krep is coefficient that takes into account the costs of current repair of equipment;

The prime cost price of electric energy produced by the solar power plant, $ / kWh:

$$c = \frac{Otot}{Wyear} \quad (7)$$

where Wyear .is annual amount of generated electric energy, kWh/year.

Specific capital investments in the solar power station, $/kW:

$$\mathrm{ktot} = \frac{\mathrm{Ktot}}{\mathrm{Nph} \cdot \ \mathrm{nom\ ph}} \quad (8)$$

where $Pnom\,ph$ .is nominal power of used photovoltaic panels, kW; Nph is number of used photovoltaic modules, pcs.

Payback period of capital investments, years:

$$Tpb = \frac{Ktot}{Wyear \cdot \beta} \quad (9)$$

where β is current tariff for consumed electric energy, $ / kWh.

The results of calculation of technical and economic parameters of the solar power plants on the flat part of the roof of the production building are presented below. There are calculated the technical and economic indicators for the options for on-roof solar power plant with photovoltaic panels on the flat part of the roof of the production building.

Table 6 presents the total capital costs and individual elements of the solar power plant according to the options for its layout by photoelectric modules. Figure 12 shows the proportion of the initial costs of acquired the components of the power plant.

Table 6. Capital investments in solar cells of the solar power stations

layout	Kph, $	Kinv, $	Kbat, $	Kcon, $	Ktot, $
1 option	9867	1380	12949	650	27330.6
2 option	41745	685	0	0	534545
3 option	49335	1221	598	325	778525

The layout of the solar power plant by photoelectric modules according to option 1, despite the fact that it allows to realize more fully the technical solar energy potential, requires the greatest capital investment during the construction phase. At the same time, the cost of storage equipment reaches 47% of the cost of the entire power plant. Using the minimum generation mode –that is, when the solar power plant provides the power supply to the consumer partially without overgeneration of electrical energy –the largest part of the initial investment is required by the modules (78%). With an insignificant increase in the number of installed solar energy converters (modules) to electric energy, the initial costs of building the solar power plant are increased by more than 1.5 times due to the purchase of accumulator batteries, the share of investments in which amounts to 8% of the total capital expenditure. That is, the application of storage batteries into the solar power plant, despite the fact that it provides a higher level of utilization of the available potential of solar energy, leads to the significant increase in the cost of the studied project.

Table 7 presents the total and specific capital costs and the annual volume of electricity generation from the solar power plant according to the options for its layout by photoelectric modules.

Specific capital investments in option 3 increased by 26.9% comparing to option 2, although the installed capacity of the station increased only by 18.9%. The maximum possible increase in the capacity of the solar power station to 4.68 kW, defined basing on shading conditions of the flat part of the roof, requires unit costs of 2.1 times greater than for option 2.

Figure 12. The share of the initial costs of acquired components of the solar power plant

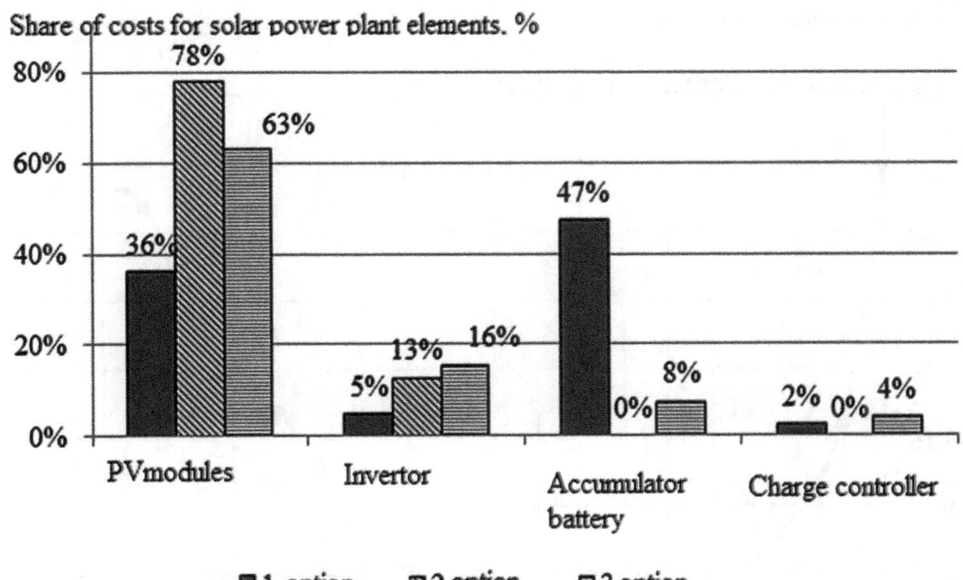

Table 7. Specific and total capital costs for options of the solar power plant

Layout	Ktot, $	P, kW	ktot . $/kW
1 option	27330.6	4.68	5839.9
2 option	5345.45	1.98	2699.7
3 option	7785.25	2.34	3327.0

Table 8 presents the current costs for repairs and maintenance and the payback period of the solar power plant, the cost of generated electricity by options for its layout by PV modules. Figure 13 shows the dependence of the cost of generated electrical energy on the number of photovoltaic modules used in the solar power plant.

The results of the calculation, presented in Table 8, make it possible to distinguish those variants of the layout of the solar power plant by photoelectric modules that allow economically feasible generation of electric energy and a partial refusal of consuming it from the electric power grid. The lowest value of the prime cost price of the generated electric energy was obtained for the second variant – 0.06 $ / kWh,

Table 8. The current costs for repairs and maintenance and the payback period of the solar power plant, the cost of generated electricity by options for its layout by PV modules

layout	Otot, $	c, $/kWh	Tpb, year
1 option	3506.228	0.34	22.2
2 option	259.028	0.06	10.3
3 option	518.694	0.1	12.7

Figure 13. The dependence of the cost of generated electrical energy on the number of photovoltaic modules used in the solar power plant

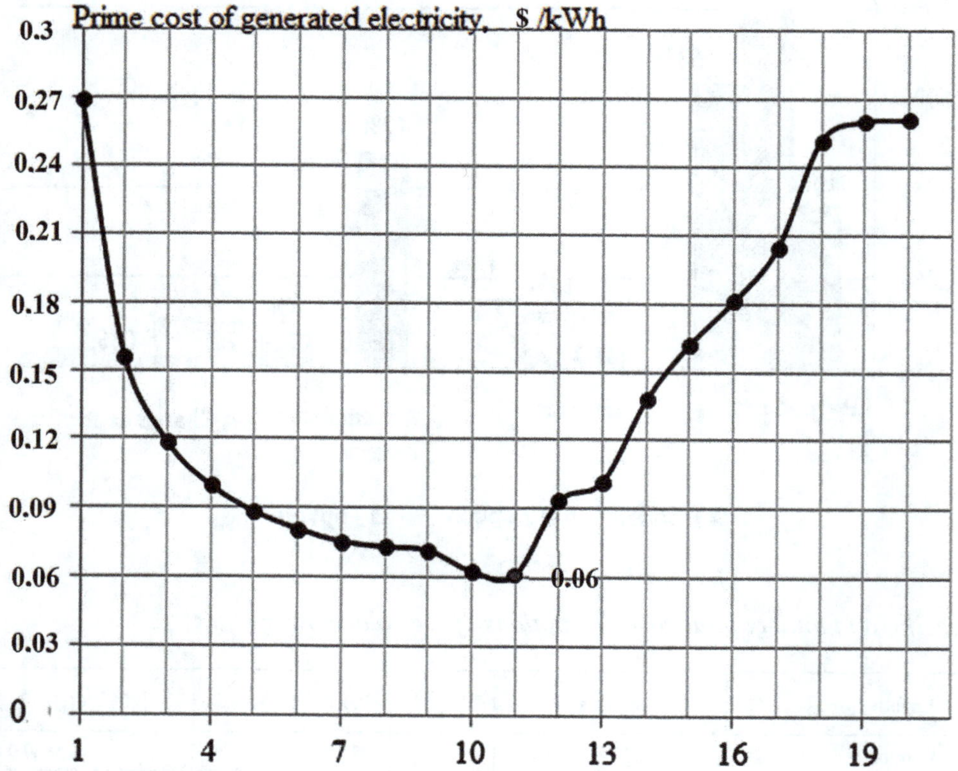

this value is also minimal for the solar power plant with any number of photovoltaic modules located on the flat part of the roof of the production building. However, the third option can also be selected for additional power supply of the lighting system. Although it is characterized by a higher prime cost value (0.1$ /kWh), this value is lower than the current tariff for electricity (0.12 $/kWh). That is, during periods of solar high activity, it makes sense to supply the consumer by the solar power plant.

The calculated dependence of the prime cost of the electric power generated by the solar power plant on the number of installed panels also makes it possible to distinguish a number of variants of the solar power plant layout by modules that can be introduced into the power supply system of consumers. This is all projects with the number of panels up to 13 pcs, since they provide prime cost price lower than the current tariff for the consumer. With the growth of tariffs, it is possible to increase the realized solar potential of the district by adding photoelectric modules and accumulating devices to the existing power plant.

SOLUTIONS AND RECOMMENDATIONS

The proposed research is recommended to be used as the element of the systems of automated design of photovoltaic power plants and microgrids.

FUTURE RESEARCH DIRECTIONS

Further increase in the efficiency of the solar power plant is possible through the harmonization of consumption and power generation schedules, as well as in the framework of the comprehensive modernization of the consumer's power supply system, and the choice of the location of the solar power plant in terms of reducing losses in the grid elements.

CONCLUSION

Assessment of the solar energy potential in concrete decree (46.8°N 40.6°E) identified the region as the territory that is promising for the use of facilities based on the conversion of solar energy into electricity.

The annual potential of solar energy for concrete decree (46.8°N 40.6°E) on the horizontal receiving surface is 1250.2 kWh/m^2. At the same time, the fraction of diffused radiation is large, and from the middle of the autumn to the middle of the spring it exceeds the direct solar energy. During the year, the monthly amount of incoming solar radiation varies from 31.15 kWh/m^2 in December to 180.3 kWh/m^2 in July.

The analysis of the possibility of expanding the potential of utilized solar radiation by controlling the spatial orientation of the receiving surface showed that the level of insolation coming to the vertically located surface is 792.01 kWh/m^2 with the eastern or western orientation and is equal to 1020.25 kWh/m^2 with south orientation.

The highest value of the annual sum of total solar radiation was obtained for the southern orientation at the tilt angle of 40° and 30°, and equals respectively 1405.99 kWh/m^2 and 1404.41 kWh/m^2. However, transiting to the south-east or south-west orientation, this value decreases insignificantly by 4.4–5.1%, and with orientation strictly to the east or west – significantly by 15.6 - 18.6%.

In addition, with orientations to the east or west, it is advisable to locate the receiving surface horizontally. If there is used its vertical position – and there is the option of placing PV modules on the walls of the building, then the incoming solar flux will decrease by 36.6%.

South-east or south-west orientation allows to utilize less amount of solar energy. The maximum value corresponds to the tilt angle of 30° relatively to the horizon and equals 1342.44 kWh/m^2.

The best orientation for the considered region is the southern orientation. The maximum flow on the surface of the receiving area for a year is obtained at the tilt angle of 30° – 40° relatively to the horizon and equals 1404.13 to 1405.99 kWh/m^2.Calculation and analysis of optimal tilt angles of receiving surfaces relatively to the horizon with different orientations along the sides of the world that provide the maximum flux of incoming solar radiation without angle adjustment or with seasonal angle adjustment revealed that at the eastern or western orientation it is not advisable to adjust the inclination angle with respect to the horizon and the maximum flow can only be obtained on the horizontal surface. The greatest effect from the use of the optimal unregulated tilt angle of the receiving surface relatively to the horizon is observed with the southern orientation and equals 1408.56 kWh/m^2.

Seasonal tilt regulation comparing to the unregulated tilt angle increases the annual flow by 4.5% to 1 472.53 kWh/m^2. For the south-east and south-west orientation, the annual value of the solar energy potential, which can be utilized, increases seasonally by adjusting the angle of inclination relatively to the horizon by 11.2% to 1492.03 kWh/m^2.

That is, despite the high gross potential of the studied region, its value can be both increased and reduced significantly due to structural limitations of the buildings on which it is proposed to place the designed solar power plant

From the point of view of the maximum realization of the available potential of the solar energy of the studied area, the use of the flat part of the roof of the building is preferable, since it allows to arrange photoelectric modules at the optimal tilt angle to the horizon, which provides the greatest annual flux of total solar radiation to the receiving surface. However, the number of installed converters is limited by the features of the roof structures, namely the shading conditions from both the roof elements and the mutual shadowing of the PV series from each other, which significantly reduces the level of realized technical potential of solar energy.

The use of the inclined part of the roof of the building for installing generating elements of the solar power plant allows to place photovoltaic modules without the use of complex orientation and attachment systems. However, the amount of utilized solar energy flow is reduced in view of the use of the tilt angles of the receiving surfaces relatively to the horizon corresponding to the slope angles of the roof, which often do not correspond to their optimum values, that provides the greatest annual flux of total solar radiation to the receiving surface.

In order to realize greater amount of incoming solar radiation, larger number of photovoltaic modules can be used in the layout of the solar power plant, but this leads to the regeneration of electric energy and results in necessity of accumulating it with the help of rechargeable batteries to provide electricity in the evening, or to connect additional consumers of electrical energy, which often requires the reconstruction of the existing electrical network or complete change of configuration of the power supply systems for consumers.

Therefore, each considered variant of the solar power plant arrangement by photoelectric modules requires additional technical and economic evaluation.

Calculation and analysis of options for the layout of the solar power plant by PV panels on the flat or inclined part of the roof made it possible to find the most economically attractive option that provides minimum generation mode on the flat roof of the building (option 2). It is characterized by minimum values of such technical and economic parameters as: the prime cost of generated electric energy is 0.06 \$/kWh; payback period is 10.3 years. However, the results of the conducted studies make it possible to identify the number of attractive options of solar power plant projects that are also attractive for implementation – these are all options for assembling photovoltaic panels in quantities of up to 13 pcs. both on the flat and inclined part of the roof of the production building. The revealed dependences of the prime cost of generated electric energy on the number of photovoltaic panels also make it possible in the future to increase the capacity of the solar power plant gradually with the growth of the tariff for consumed electric energy. That is, with the growth of tariffs, it is possible to increase the realized potential of the district's solar energy by adding photoelectric modules and accumulating devices to the already functioning solar power plant.

REFERENCES

Adomavicius, V., Kharchenko, V., Valickas, J., & Gusarov, V. (2013). RES-based microgrids for environmentally friendly energy supply in agriculture. *Proceedings of 5th International Conference TAE 2013, Trends in Agricultural Engineering*, 51-55.

Amerhanov, R. A., Butuzov, V. A., Tomarov, G. V., & Butuzov, V. V. (2008). Jenergeticheskij potencial solnechnoj radiacii i jekonomicheskaja celesoobraznost' primenenija gelioustanovok v Krasnodarskom krae i Jakutii [Energy potential of solar radiation and economic feasibility of solar installations in the Krasnodar Territory and Yakutia]. *Works of the Kuban State Agrarian University*, *1*, 26–32.

Aronova, E. S., & Grilihes, V. A. (2006). Metodika rascheta real'noj plotnosti solnechnogo izluchenija pri proektirovanii fotojelektricheskih ustanovok [Method for calculating the real density of solar radiation in the design of photovoltaic installations]. *SPbSU Scientific and Technical Statements*, *6*(1), 62–66.

Atwater, M., & Ball, J. T. (1978). A numerical solar radiation model based on standard meteorogical observation. *Solar Energy*, *21*(3), 163–170. doi:10.1016/0038-092X(78)90018-X

Belenov, A. T. Kharchenko, V.V., Rakitov, S.A., Daus, Y.V., Yudaev, I.V. (2016). The Experience of Operation of the Solar Power Plant on the Roof of the Administrative Building in the Town of Kamyshin, Volgograd Oblast. *Applied Solar Energy*, *52*(2), 105–108.

Benkaciali, S., Benkaciali, S., & Gairaa, K. (2012). Comparative study of two models to estimate solar radiation on an inclined surface. *Revue des Energies Renouvelables*, *15*(2), 219–228.

Bezrukih, P. P. (2008). Renewable energy sources and reliability of power supply. *Energy Policy*, *3*, 3–8.

Budzko, I. A., & Zul', N. M. (1990). *Jelektrosnabzhenie sel'skogo hozjajstva* [Electricity supply of agriculture]. Moscow: Agropromizdat.

Butuzov, V.V. (2009). Raschetnye znachenija intensivnosti solnechnoj radiacii dlja proektirovanija gelioustanovok [Calculated values of the solar radiation intensity for the design of solar power plants]. *Alternative Energy and Ecology*, *11*, 75–80.

Daus, Ju. V. (2016). Opredelenie optimal'nogo ugla naklona panelej solnechnyh jenergoustanovok v uslovijah Rostovskoj oblasti [Determination of the optimal inclination angle of solar energy panels under the conditions of the Rostov region]. *Proceedings of the Higher Educational Institutions. The North Caucasus region. Series: Engineering, 2*(190), 67-71.

Daus, J. V., Kharchenko, V.V., & Yudaev, I.V. (2018). Solar Radiation Intensity Data as Basis for Predicting Functioning Modes of Solar Power Plants. In Handbook of Research on Renewable Energy and Electric Resources for Sustainable Rural Development. Hershey, PA: IGI Global.

Daus, Yu. V., & Kharchenko, V. V. (2018). Evaluating the applicability of data on total solar-radiation intensity derived from various sources of actinometric information. *Applied Solar Energy, 54*(1), 71–76. doi:10.3103/S0003701X1801005X

Daus, Y. V., Kharchenko, V. V., & Yudaev, I. V. (2016a). Evaluation of Solar Radiation Intensity for the Territory of the Southern Federal District of Russia when Designing Microgrids Based on Renewable Energy Sources. *Applied Solar, 52*(2), 124–129.

Daus, Y. V., Kharchenko, V. V., & Yudaev, I. V. (2016b). *Ocenka potenciala solnechnoj jenergii v zadannoj tochke Juzhnogo federal'nogo okruga* [Estimation of solar energy potential at a given point in the Southern Federal District]. FGBOU VO "Donskoj GAU".

Daus, Y. V., Kharchenko, V. V., & Yudaev, I. V. (2016c). *Poisk optimal'nogo ugla naklona prijomnoj ploshhadki otnositel'no gorizonta v zadannoj tochke Juzhnogo federal'nogo okruga* [Search for the optimal inclination angle of the receiving area relative to the horizon at a given point in the Southern Federal District]. FGBOU VO "Donskoj GAU".

Daus, Yu. V., & Yudaev, I. V. (2016). Designing of Software for Determining Optimal Tilt Angle of Photovoltaic Panels. In *Proceedings of International Conference on Education, Management and Applied Social Science (EMASS2016)*. Beijing: DEStech Publications, Inc.

Daus, Yu. V., Yudaev, I. V., & Stepanchuk, G. V. (2018). Reducing the costs of paying for consumed electric energy by utilizing solar energy. *Applied Solar Energy, 54*(2), 137–141. doi:10.3103/S0003701X18020056

Elistratov, V. V., & Aronova, E. S. (2009). Solar photo energy technologies for electric power consumers. *Applied Solar Energy, 45*(3), 143–147. doi:10.3103/S0003701X09030025

Essefi, R. M., Souissi, M., & Abdallah, H. H. (2014). Maximum Power Point Tracking Control Using Neural Networks for Stand–Alone Photovoltaic Systems. *International Journal of Modern Nonlinear Theory and Application, 3*(03), 53–65. doi:10.4236/ijmnta.2014.33008

Grishko, V. I., & Misjura, L. I. (Eds.). (2007). *Spravochnik po resursam vozobnovljaemyh istochnikov jenergii Rossii i mestnym vidam topliva. Pokazateli po territorijam: spravochnoe posobie* [A guide to renewable energy resources in Russia and local fuels. Indicators for territories: reference book]. Moscow: IAC Jenergija.

Guemard, C. (2003). Direct solar transmittance and irradiance prediction with broadband models. Part I: Detailed theoretical performance assessment. *Solar Energy, 74*(5), 355–379. doi:10.1016/S0038-092X(03)00195-6

Gupta, S., & Chandrashekar, M. (1995). A unified approach to modeling photovoltaic powered systems. *Solar Energy*, *55*(4), 267–285. doi:10.1016/0038-092X(95)00033-N

Hosseini, S. M., Aslani, A., Hafeznia, H., & Naaranoja, M. (2017). Analysis of Energy System in Sweden Based on Time series Forecasting and Regression Analysis. *International Journal of Energy Optimization and Engineering*, *6*(3), 97–114. doi:10.4018/IJEOE.2017070105

Jarymbash, D.S., & Daus, Ju.V. (2014). Osobennosti identifikacii intensivnosti solnechnoj radiacii pri proektirovanii solnechnyh jelektrostancij [Features of identifying the solar radiation intensity in the designing solar power plants]. *Power and Electrical Engineering, 1,* 74-78.

Kharchenko, V. V., Nikitin, B. A., & Tikhonov, P. V. (2010). Estimation and forecasting of PV cells and modules parameters on the basis of the analysis of interaction of a sunlight with a solar cell material. *Proceedings of 4th International Conference TAE 2010, Trends in Agricultural Engeneering*, 307-310.

Liu, B. Y. H., & Jordan, R. C. (1961). Daily insolation on surfaces tilted towards the equator. *ASHRAE Journal*, *3*(10), 53–63.

NASA Atmospheric Science Data Center. (n.d.). Retrieved from https://eosweb.larc.nasa.gov/sse

Novakovskij, B. A., Prasolova, A. I., Kiseleva, S. V., & Rafikova, Yu. Yu. (2011). Geoinformacionnye sistemy po vozobnovljaemoj jenergetike [Geoinformation systems for renewable energy]. In *Proceedings of the International Conference INTERKARTO / INTERGIS. Belokuriha: Denpasar.* Retrieved from Official website "Quantum", at: http://npp-kvant.ru

Popel', O. S., Frid, S. Y., Kolomiets, Y. G., Kiseleva, S. V., & Terekhova, E. N. (2010). Atlas resursov solnechnoj jenergii na territorii Rossii [Atlas of solar energy resources in Russia]. Moscow: OIVT RAN.

Sen, Z. (2008). *Solar Energy Fundamentals and Modeling Techniques, Atmosphere, Environment, Climate Change and Renewable Energy* (1st ed.). Springer.

Shyam, S. C., & Rajeev, K. A. (2011). Estimation of Hourly Solar Radiation on Horizontal and Inclined Surfaces in Western Himalayas. *Smart Grid and Renewable Energy*, *2*(01), 45–52. doi:10.4236gre.2011.21006

Vissarionov, V.I. Burmistrov, A.A., Deryugina, G.V., Kuznetsova, V.A., Kunakin, D.N., Malinin, N.K., & Pugachev, R.V. (2008). *Solnechnaja jenergetika: Uchebnoe posobie dlja vuzov* [Solar energy: A textbook for high schools]. Moscow: Izdatel'skij dom MJeI.

Volkova, N. G. (Ed.). (1989). Matematika i SAPR. Vychislitel'nye metody. Geometricheskie metody [Mathematics and CAD. Computational methods. Geometric methods]. Moscow: Mir.

White, M., & Ahad, A. (2016). Solar PV Residential Grid-Tie System for Optimizing Installation and Performance. *International Journal of Energy Optimization and Engineering*, *5*(1), 48–67. doi:10.4018/IJEOE.2016010104

Winter, C.-J. (1991). Solar Power Plants. Berlin: Heidelberg. doi:10.1007/978-3-642-61245-9

Yang, K., Huang, G. W., & Tamai, N. (2001). A hybrid model for estimating global solar radiation. *Solar Energy*, *70*(1), 13–22. doi:10.1016/S0038-092X(00)00121-3

ADDITIONAL READING

Darhmaoui, H., & Lahjouji, D. (2013). Latitude Based Model for Tilt Angle Optimization for Solar Collectors in the Mediterranean Region. *Energy Procedia*, *42*, 426–435. doi:10.1016/j.egypro.2013.11.043

Głuchy, D., Kurz, D., & Trzmiel, G. (2013) Studying the Impact of Orientation and Roof Pitch on the Operation of Photovoltaic Roof Tiles. *Przegląd elektrotechniczny, 6*, 281 – 283.

Jo, J. H., Loomis, D. G., & Aldeman, M. R. (2013). Optimum penetration of utility–scale grid–connected solar photovoltaic systems in Illinois. *Renewable Energy*, *60*, 20–26. doi:10.1016/j.renene.2013.04.008

Markvart, T., & Castaner, L. (2003). *Practical handbook of photovoltaics: fundamentals and applications*. Elsevier.

Petrakopoulou, F., Robinson, A., & Loizidou, M. (2016) Simulation and evaluation of a hybrid concentrating–solar and wind power plant for energy autonomy on islands. *Renewable Energy, 96, A*, 863–871.

KEY TERMS AND DEFINITIONS

Actinometrical Data: Results of long-term meteorological observations at weather stations, processed and systematized by specialized organizations in the form of climate reference books and databases.

Geographical Coordinates: Angular values, latitude and longitude, which determine the position of objects on the earth's surface and on the map.

Optimum Inclination Angle of the Receiving Surface: Inclination angle of the receiving surface relative to the horizon, which allows obtaining the maximum solar radiation flux on its surface for a given period of time.

Receiving Surface: The surface of the photoelectric device/photovoltaic part of the device, which receives solar radiation.

Solar Radiation Intensity: The density of solar radiation (energy illumination), coming per unit area of the photoelectric module.

Standard Conditions for Testing the Solar Cell: Test conditions, regulated by the density of the solar energy flux of 1000 W/m^2 and the temperature of photovoltaic solar cells of 25 °C.

Verification of Data: Verification of theoretical results by comparing them with experimental data.

Chapter 9
The Exploitation of Wind Systems in Rural Electrical Network

Volodymyr Kozyrskyi
National University of Life and Environmental Sciences of Ukraine, Ukraine

Andrii Petrenko
National University of Life and Environmental Sciences of Ukraine, Ukraine

Mykola Trehub
National University of Life and Environmental Sciences of Ukraine, Ukraine

Yangibay Charyev
Turkmen State Institute of Transport and Communications, Turkmenistan

ABSTRACT

The activity of farm enterprises is directly connected with the efficient use of resources such as electrical energy and water. Taking such conditions as factors of economic expediency and ecological safety into account, it is more reasonable to use wind stations in order to provide consumers with energy and water. The use of conventional wind and electrical stations is an easy and reliable solution. However, annual wind velocity on most settled territories does not exceed 6 m/s or even 4 m/s. It makes the efficient use of wind electrical stations more complicated. One of the solutions can probably be the use of wind and electrical stations on the basis of slow speed non-transmission arc-shaped-stator inductor-type generators with an integrated radial and ring-shaped rotor. Another efficient solution to provide areas with water and electrical energy is to use a combined wind station with a crank-and-rod mechanism and the rod of the driving mechanism of the back-and-forth motion of the core of a magnetic and electrical linear generator and the piston of a plunger pump.

DOI: 10.4018/978-1-5225-9179-5.ch009

INTRODUCTION

Electrical and mechanical complexes and systems of electromechanical transformation of wind energy make the foundation of autonomous wind and energy systems (AWES). They usually refer to rural wind energetics. Nowadays their power is not as strong as the power of large-sized power supply network of wind and electric stations (WES). However, market demand for AWES can be higher if to enhance their technical and economic indicators and genericity as well as legislative support of the state, energetic needs and financial capacity of producers of agricultural products of various forms of management. In different countries of the world, legislature supports the development of small wind energetics. Here, a general focus of genericity (Amirat, J., & Benbouzid, M.E.H., & Bensaker, B., & Wamkeue, R. (2007)) should become the expansion of technical capacities of electromechanical WES both for autonomous energy consumers and for the work in electro energetic system where they can realise the pros of dispersed generating. Main advantages of energy sources of dispersed generating in comparison, for instance, with modern wind parks of large-sized WES, are the possibilities to exploit not only territorially limited powerful continental wind flows but local winds as well almost on all areas which have fewer problems of affecting eco system and energy transfer to peripheral electric networks which does not cause significant power fluctuations of energy flows. Systems (Viswanadh, M.M.G., Sekhar, A.S.R. (2012)) of dispersed generating are programmed to use autonomous WES of smaller sizes (Mueller, M.A. (2005)) which can cause fewer human-induced changes of natural landscapes than, for instance, those ones which are caused by constructing wind parks of large-sized million watt WES on mountain plateaus or sea shelves.

BACKGROUND

Considering economic feasibility and ecological safety, especially for the area remote from the centralized electrical networks and the systems of water supply, installations for providing consumers with electric energy and water must use as much as possible existing renewables.

Wind pump, wind electric pump and wind electrical units belong to such means of power supply and water supply. Wind pump installations are the simple and reliable decision but have low coefficient of loading of the wind turbine. For use of wind electric pump installations characteristic there is the general low efficiency. For electric pumps generally, it doesn't exceed 60%. Besides it is necessary to have a reserve of rated power if additional electroconsumers, except the electric pump are connected to wind electroinstallation.

One of the drawbacks of conventional non-transmission designs of a wind energy station is placing a continuous generating turbine with a large diameter on the horizontal axis, behind which (Sergeev, V.D., & Kuleshov, E.V., & Proskurenko, S.S, & Chernyshova, A.S. (2000); RMA "Південне". (1999)) an axial whirl is produced, and reverse air flows occur, thus, reducing significantly the kinetic wind energy caught with blades. Therefore, a conventional construction of a wind energy station has unacceptable limitations without a mechanical intensifier with a direct installation of a continuous housing of an electric generator on a horizontal axis. These are more unacceptable (Lebsir, A., Benbouzid, M.E.N., Bentonnsi, A., Salah, I.B. (2016)) for large-sized wind turbines.

MAIN FOCUS OF THE CHAPTER

The Methodology of an Efficient Improvement of Non-Transmission Wind and Electric Complexes

The results of an analytical review of literature and a theoretical study of processes of electro mechanical transformation of wind energy give grounds to state that it is necessary to develop methods of building non-transmission low-speed electromechanical complexes (Chen, J., Chen, J., Gong C. (2014)) with high frequency of a mechanical change of a magnetic flow in a working air gap between the rotor and the stator. It will enhance energy efficiency of exploiting local wind and energy potential. It can be done by increasing the speed of polar elements of the rotor and increasing their number and at the same time by keeping amplitude values of a magnetic flow.

Technical problems having been studied, we figured out that the most efficient non-transmission wind energy station can be an integral construction with a cyclic rotor of a generator, made coaxially on a horizontal axis of a wind turbine. Researchers have already offered such a similar technical idea of using a cyclic wind turbine. However, this construction does not use a ring as the rotor of an electric generator. Scientists gave profound evidence of pros of its total capacity, but they did not describe the methods of improving geometrical parameters of a coaxial circular rotor which will meet the requirements of both aero mechanical and electromagnetic systems. That is why, a number of new efficient variants of a wind energy station (Trehub, M.I. (2004); Trehub, M.I., Kozyrskyi, V.V., Hrebenikov, V.V., Pryimak, M.V. (2013)) were offered, containing a circular rotor, set coaxially on horizontal axis of a wind turbine and arch stator on moving horizontal spar. This aero-electro-mechanical system of a wind energy station constructed integrally which combines functional elements of a wind turbine and the rotor of an electric generator of a large diameter in one construction is presented in *Figure 1.*

However, the process of setting a ring-shaped rotor even of a flowing shape on the part of a wind engine which is affected by wind impacts on its aero dynamic parameters. That is why we have found out a prior succession of main conditions to minimalize energy loss in the whole aero-electric-mechanical system of transforming wind energy.

The Analysis of how a coaxial ring-shaped rotor of an electric generator impacts on the process of receiving wind energy by an aero mechanical system of a wind energy station. We used an analytical method to study conditions of how a vortex flow is created behind the blades of a horizontal-axial wind turbine with an open coaxial ring rotor of an electric generator of a non-transmission wind station which is presented schematically in *Figure* 2.

As the velocity of wind flows which affects the ring rotating cylinder is relatively low (Mach number is almost never over 0.1), then the air density can be considered to be unchanged. However, a mathematical model of

a complex relative movement of a wind flow through a usual plane of a rotating ring can be explained with the help of complicated dependences. So, we can utilise a simplistically modified version of the principle of super positions of individual simpler movements. The circulation of the speed of a wind flow around a motionless profile of a ring-shaped cylinder can be explained by means of the following equation without taking into consideration surface effects:

$$G_C = V_2 \ell_2 - V_1 \ell_1 = 0, \tag{1}$$

Figure 1. A general construction scheme of a non-transmission wind energy station with an electric generator with the ring-shaped rotor and the arc-shaped stator: 1 – a blade, 2– a ring-shaped rotor, 3– an arc-shaped stator, 4 – a swinging spar, 5 – a spider of the swinging spar, 6 – a bearing pier, 7 – pedestal, R – a radius of a circle of the rotation of blades ends, r – a radius of a circular rotor.

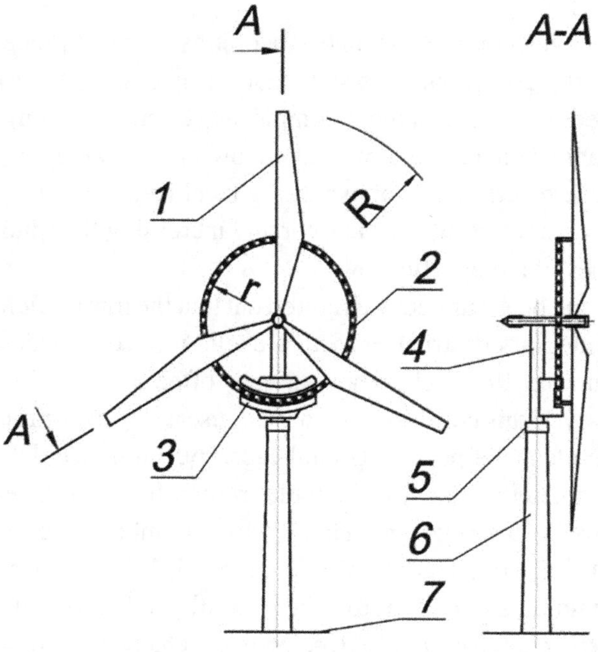

Figure 2. A scheme of the possible motion of a wind flow through the plane of blades rotation with a ring-shaped rotor of an electric generator: 1 – a profile of a coaxial ring-shaped cylinder of the rotor of an electric generator, 2 – blades of a wind turbine.

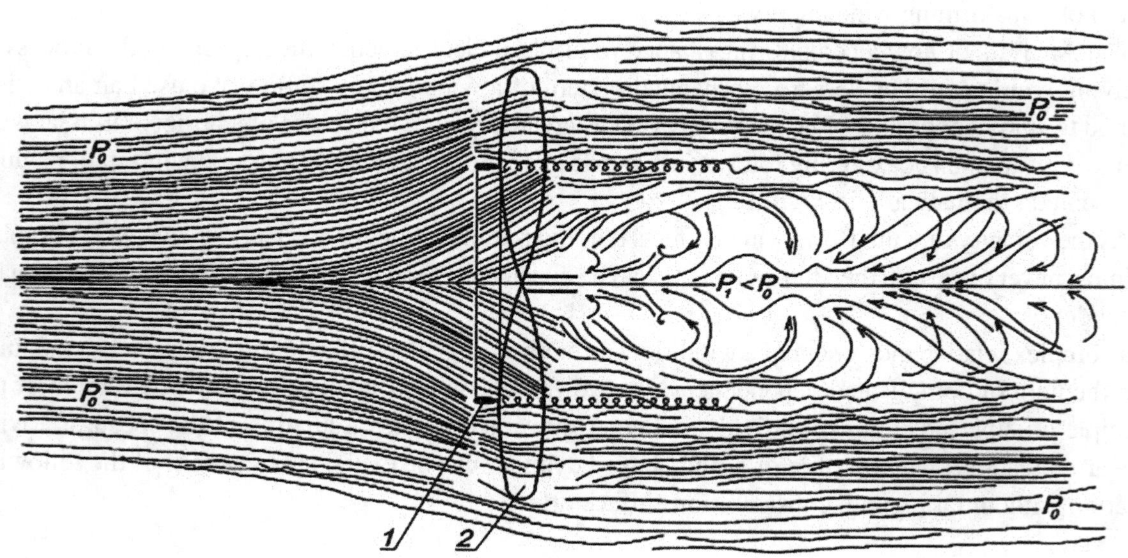

G_C – the circulation of the movement on the surface of a ring-shaped cylinder which is the product of the path length of streamline and the speed of an air flow and which has the same unit of measurement as kinematic air viscosity, m²/s, V_1, V_2 –speeds of inner and outer profile surface flows of a ring-shaped cylinder, m/s, ℓ_1, ℓ_2 – the length of profile flows (circuits) of the inner and outer surface, m.

If to take $V_1 = V_2$ into account, then the equation (1) explains a perfect laminar air flow on the absolutely symmetrical surface. However, in the case of the inner and outer surface of a ring-shaped cylinder it is impossible to achieve the effect of a symmetrical surface due to different value of radii of curvature in a usual to the direction of wind plane. Thus, the most expectable air flow behind an elementary profile of the ring is a chaotic swirling flow which, within the relatively low wind speed range, can be explained by means of:

$$\partial\Omega_w / \partial t = \xi\Delta\Omega_w, \qquad (2)$$

Ω_w – angular velocity of a circular movement of an individual vortex as regard to the direction of the air initial motion; rad/s; t – time; ξ – air kinematic viscosity, m² /s; $\Delta\Omega_w$ – the Laplace operator for the function of a spatial angular velocity of the swirl relative to the axis.

The Laplace operator for the angular velocity of a swirling movement of air masses according to Cartesian coordinate system can be explained by means of:

$$\Delta\Omega_w = \partial^2\Omega_w / \partial X^2 + \partial^2\Omega_w / \partial Y^2 + \partial^2\Omega_w / \partial Z^2 \qquad (3)$$

X, Y, Z, – coordinates of air movement in three-dimensional plane.

The shape and the size of the surface of inner and outer profiles of a ring-shaped cylinder have a big influence on the process of how wind flow moves through the cylinder. Researchers present theoretical explanations of the principles of how Board and Brix-Kort nozzles work and of specific profiles of air intake, gradually narrowed channels (confusers) which are often used to enhance flow capacity.

It occurs and changes when the profile of a ring-shaped cylinder rotates with an angular velocity Ω_c. Then the direction of a wind flow can be found out by angle β, which can also be found with the help of:

$$\beta = arcctg\lambda_c, \qquad (4)$$

λ_c – the velocity of a ring-shaped cylinder of the rotor.

$$\lambda_c = \Omega_c r_c / V_0, \qquad (5)$$

V_0 – the velocity of a wind flow before the plane of blades rotation, m/s; r_c – a radius of a ring-shaped cylinder of the rotor, m; Ω_c – an angular velocity of the rotation of a ring-shaped rotor, rad/s; $\Omega_c = 2\pi n/60$, n – the frequency of rotation of a ring-shaped rotor, m⁻¹.

In a perfect incompressible air medium with density ρ, the movement of an axial flow can be examined along concentric circular trajectories with a circular velocity V_{in}, which depends on r_a, taking only centripetal acceleration into consideration, which equals V_{in}^2/r_a. We can get the dependence of the projection of the movement in a radial plane when we use the Euler equation:

$$\rho\alpha_r = -\rho V_{in}^{\ 2} / r_a = -\ \partial p / \partial r_a \tag{6}$$

α_r – the diametrical size of the vortex; r_a – an alternative radius of rotation of elementary layers of air masses of the vortex; p –variable pressure inside the vortex flow.

If we take into account that the pressure in the surrounding air plane has the value p_0 at a faraway distance from the centre of the vortex then the difference of pressures at different distances from the centre of the vortex can be explained by means of:

$$p_0 - p = \int_{\infty}^{r} \rho V_a^{\ 2} / r_a \partial r_a \tag{7}$$

This expression can show that the pressure decreases monotonically when it approaches the centre of the vortex from a distant plane. We can explain the change of the pressure by means of the given below dependence if $r_a \geq \alpha_r$ and the density of the environment do not change:

$$p = p_0 - \rho\Omega_w^2 \alpha_r^2 / 2r_a^2 \tag{8}$$

The pressure can be determined depending on its geometrical location inside the vortex using the given below formula if to use a similar to a perfect gas medium formula $r_a \leq \alpha_r$:

$$p = p_0 - \rho\Omega_w^2 \alpha_r^2 + 0,5\rho\Omega_w^2 r_a^2 \tag{9}$$

The minimal pressure in the centre of the vortex can be explained by the following equation if to take $r_a = 0$ into consideration:

$$p_{\min} = p_0 - \rho\Omega_a^2 \alpha_r^2 \tag{10}$$

While analysing this equation, we can learn that the value of the minimal pressure in the centre of the vortex depends on the size of the vortex and the angular velocity of air masses inside the vortex. Here, we should take into consideration the square of these interdependent parameters and constant air density.

Due to the fact, that the vector of the vortex is solenoidal, $div\Omega_w = 0$ and $divV_{in} = 0$, the opposite flows moves in the closed loop that is, first, it approaches the plane of blades rotation through the axial part of the vortex, then, it is thrown away from the centre by centrifugal means and after it mixes with new air masses, the part of it can move to the peripheral vortex and the other part can again enter the reverse flow of the axial vortex. This results in decreasing the flow capacity.

The design of summarised intersystem requirements for geometrical parameters of a ring-shaped rotor and an arc-shaped stator of a wind energy station. We have done an analysis of a possible impact of a ring-shaped rotor on general aerodynamic parameters of a wind turbine. It helps us to outline general requirements for sizes and profile of the constructive and electromagnetic functional part of an aeromechanical system.

Thus, we have formulated the following requirements in some priority sequence for the design of a ring-shaped electrical generator:

- A ring-shaped cylinder of the rotor of a generator with an arc-shaped stator should be elongated towards the wind streamlined profile with a minimum aerodynamic cross section which is limited only my mechanical strength and radial polar sizes;
- A mechanical strength of a model with a ring-shaped rotor should be provided by building a monolithic structure of electromagnetic functional and body workpieces of a ring-shaped rotor which is set on the horizontal axis and joined with blades with axial forces and the contactless force of magnetic field between the rotor and the stator.

All generators are limited in weight and size indicators, the largest numbers of induction and current density, the number of windings. In addition, this generator is also limited in the minimum rotation frequency. That is why, these indicators are regulated. Thus, there is only one thing of an integral constructive design which can change, that is radius r_c (*Figure* 1) of an axial ring-shaped rotor. Hence, a target function of receiving the strongest electrical driving force can become as follows: $r_c \rightarrow$ MAX, which has a definite option of setting rotor parts on the ends of blades that is, $r_c \rightarrow R$. It is technically possible only in case of the smallest sizes of WT and demands confirming its limiting conditions and rules of building models of different sizes.

Due to the fact that setting a ring-shaped rotor even if it is streamlined on the part of a wind turbine which is affected by the wind affects aerodynamic parameters, all variants of geometrical configurations of an electromechanical system were assessed according to target functions of limitations for different systems, which can be presented by two blocks of requirements of different vectors (*Figure* 3).

The analysis of principal technical possibilities of different variants enables us to synthesize the configuration of an optimised electromagnetic system of a wind and electrical generator with a ring-shaped rotor.

Figure 3. The chart of mutual dependency of different requirements for an aeromechanical and an electromechanical system

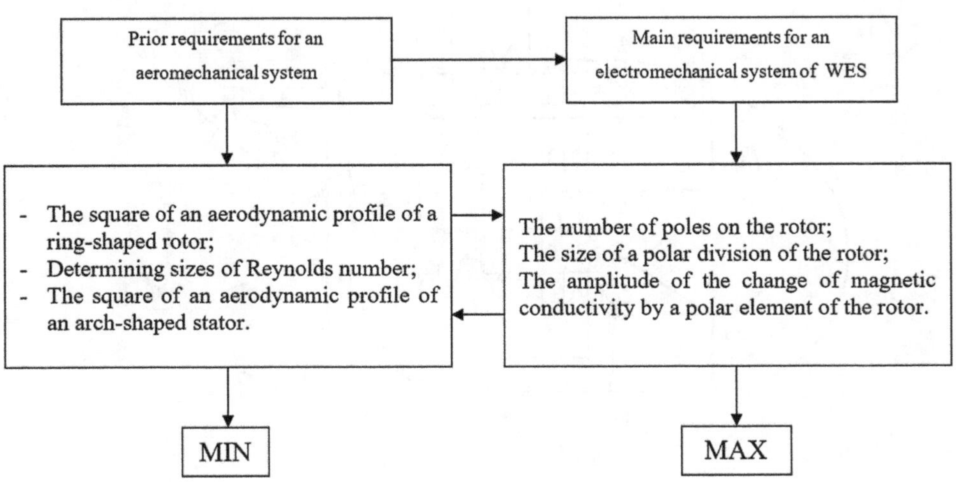

Each module is practically an individual independent contactless electrical device (Jebaseeli, E.A. E., Susitra, D. (2010)) which can work in different modes. If we apply the principle of mutual substitution we can install a magnetic core with excitation winding on the installation site of permanent magnets. It will convert a module of a magneto electrical type into electromagnetic. There are different variants of designing an electromagnetic system of an induction type with permanent magnets on an arc-shaped stator. We chose a radial and ring model with a two-packet magnetic core on a sector of the rotor which is presented in *Figure* 4.

Methodological principles of the number limitation for factors were to minimize the number of variables of geometrical parameters. Only after that a system can be considered to be applicable. We determined the sizes of the rotor elements in a cylindrical system of coordinates in the following way: *b* – the arc length specified by radius r and the central angle of the pole division; *ℓ* – the axial length; *h* – the difference between radii of the inner to the centre and outer from the centre surfaces of a ring-shaped rotor. Here, a wind flow moves in an axial direction towards the rotor axis and a magnetic flux moves in a radial direction towards the rotor axis.

Figure 4. The design of a radial and ring-shaped configuration of the system of a two-packet magneto electrical module of an arc-shaped stator of an inductor-type generator: 1 – the right, on the side of a north pole of the magnet, packet of the magnetic core of a sector of the stator; 2 – the left packet of the magnetic core of a sector of the stator; 3 – a uniform area of the magnetic core between packets; 4 – magnetically non-conductive part of the rotor; 5 – ferromagnetic element of the rotor; 6 and 7 – anchor windings of the right and the left packet of magnetic cores; Φ_1 and Φ_2 – the path of magnetic excitation fluxes in packets; δ – a working airy gap; r – a radius of a ring-shaped rotor; b, ℓ, h – overall sizes of ferromagnetic elements of the rotor; τ – a polar division of the rotor, τ = b; V_a – an approximate wind direction, e_1, e_2 – an electric driving force in the windings of the corresponding packets.

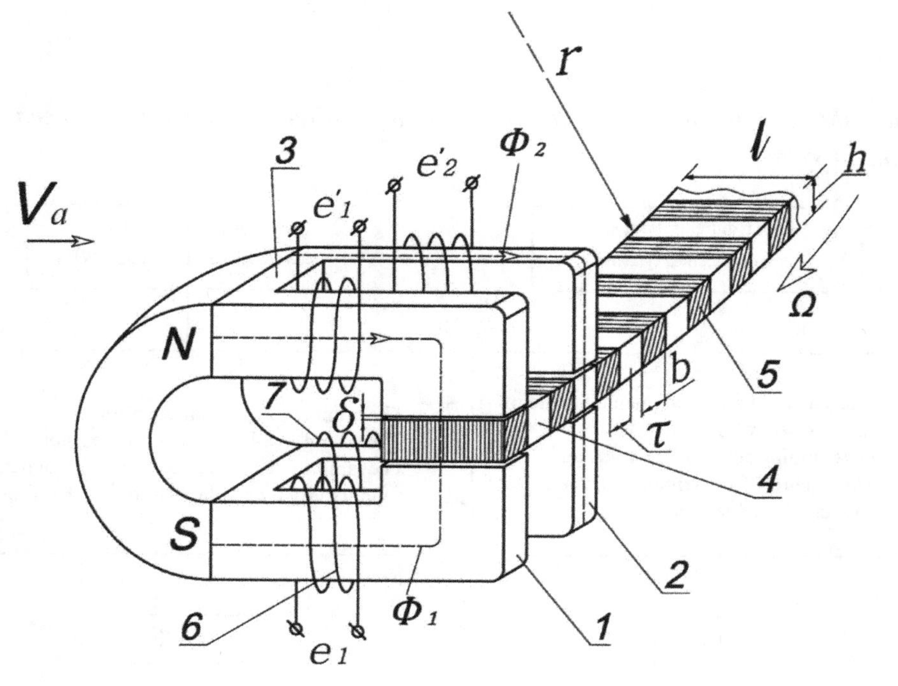

An alternative geometrical parameter of a ring-shaped rotor is a radial size h (*Figure* 4). Sizes b and ℓ are given fixed values. The value b is a part of solid ring housing so it does not influence on parameters of aerodynamic resistance. We evaluated the impact on aerodynamic parameters by means of the force of aerodynamic resistance of a ring profile:

$$F_a = C_X S_M \rho V^2 / 2, \tag{11}$$

C_X – the coefficient of the force of air resistance (reference values, constant), S_M – the areas of radial crossing the housing of a ring-shaped rotor which is determined by the square of a ring with an inner diameter which is $2r$ and outer one which is $2(r + h)$:

$$S_M = \pi \cdot h \cdot (2r + h), \tag{12}$$

r – a radius of an inner to the centre surface of a ring-shaped rotor.

Accordingly, the force of aerodynamic resistance can be presented by means of the following equation:

$$F_a = C_X \pi h(2r + h)\rho V^2 / 2 = C_X \pi \rho V^2 (2rh + h^2) / 2 \tag{13}$$

These show that the value of aerodynamic resistance of a ring-shaped rotor and the area of a wind engine which is shaded by the wind have, if to take $r >> h$, into consideration a hardly expressed parabolic dependence on the radial height h and acquire the smallest values when they are at the origin of coordinates.

Reynolds number can be explained with the help of the given below equation if the wind has a specific speed and both temperature and air viscosity do not change:

$$\mathrm{Re} = Vh / \eta_a, \tag{14}$$

V – the wind speed, η_a – kinematic viscosity of the air.

A limitation parameter is the one known critical value of Reynolds number which characterises the flow transition from laminar to turbulent, but which must not exceed 10^4. The reverse transition from the turbulent flow to the laminar one occurs when $\mathrm{Re} = 2{,}3 \cdot 10^3$.

The methods of improving geometrical parameters of ferromagnetic polar elements of a ring-shaped rotor of a wind and electrical station which has no gear. The radial size of a ring-shaped rotor in an electromagnetic system of an induction type generator influences on the value of electrical driving force due to the change of magnetic conductivity which is inversely proportional to magnetic resistance because of a mechanical movement of a ferromagnetic element through an air gap of a magnetic circle. This principle is basic for the known methods of calculating induction generators according to Faraday law.

$$E = -\, d\psi\left(\Lambda\right) / dt, \tag{15}$$

$\psi(\Lambda)$ – a variable flow linkage of an electromagnetic system of an induction type generator which is caused by a periodic change of magnetic conductivity by means of mechanical movement of ferromagnetic elements of a ring-shaped rotor.

As $\psi = Iw^2\Lambda$, then, taking I= const and w= const which occurs in a generator of an induction type into consideration, we get the following equation:

$$E = -Iw^2 d\Lambda \,/\, dt, \tag{16}$$

The equation of a rotation system in which the change of magnetic conductivity occurs with some rotation angle of the rotor θ is as follows:

$$E = -\; Iw^2 d\Lambda \,/\, d\theta \tag{17}$$

Λ – the magnetic conductivity of a circle is an inverse value of the total magnetic resistance of the circle can be found with the help of known values (Benbouzid, M.E.N., & Beltran, B., & Mangel, H., & Mamoune, A. (2012)) of magnetic values of an electromagnetic system:

$$\Lambda = 1 \,/\, R_M, \tag{18}$$

$$\Lambda = \Phi \,/\, F, \tag{19}$$

$$\Lambda = \psi \,/\, I \cdot w^2 \tag{20}$$

$$\Lambda = L \,/\, w^2, \tag{21}$$

R_M – the total resistance of a magnetic circle, R_M = F/Φ, Φ – magnetic flux of a circle, F – magnetomotive force, L – inductivity.

These equations present the influence of the total magnetic conductivity on all main indicators of an electromagnetic system. But in case of an inductor-type generator it is also necessary to analyse physical and geometrical values which can be calculated totally and on specific parts of the circle by means of this formula:

$$\Lambda = \Sigma \mu_i S_i \,/\, \ell_i, \tag{22}$$

μ_i – magnetic permeability of the material i of a part of a magnetic circle, S_i – the square of a usual crossing site i of a part of a magnetic circle, ℓ_i – the length i of a part of a magnetic circle on the line of the magnetic flux.

When we accept the condition of non-change motionless part of a magnetic circle of the stator of a contactless induction-type generator with a ring-shaped rotor, we can only study the changes of magnetic conductivity of a moving part of a magnetic circle. To achieve this goal, it is necessary to analyse the design of the part of magnetic circle of a four-modular mono-packet arc-shaped-stator wind and electric generator with a ring-shaped rotor (*Figure* 5).

This schematic figure of four mono-packet modules with magnetic moments in the state of balance, taking $r \gg \tau$ into consideration, makes it possible to use laws which approximate square forms to consider the configuration of a usual part of a magnetic circle. According to known methods of presenting flat-parallel magnetic fields (Sergeev, V.D., & Kuleshov, E.V. (2001)), this version belongs to usual typical forms. The nearest to these forms value is reference value of the principle of the flux distribution through the magnetic circuit with rectangular cross-section.

The law of the graphical construction of force lines of flat-parallel magnetic field, using Richter's method is to present the surface layer of parts of magnetic circuit which are separated by an air gap or non-magnetic area made from the material which is not magnetically conductive, and to make parallel divergent lines which are perpendicular, starting with the shortest distane. We have built a graphical figure of force lines of flat-parallel magnetic field for two end positions of areas of a polar division of a ring-shaped rotor (*Figure* 6).

This drawing shows that when there occurs full overlapping of air by a ferromagnetic element of the rotor between polar ends of magnetic circuit of the stator, magnetic conductivity will be the largest. The value of this magnetic conductivity can be calculated by means of its linear dependence on the square of overlapping and distances of air gaps if they are uniform ($b = \tau$). We can make a conclusion (*Figure* 6) that provided the sizes of a polar distribution of the rotor do not change; the increase of size h will not cause the decrease of magnetic conductivity as it is expressed by a theoretical equation. It will have a less influence on the absolute values.

The impact of radial size h of a ring-shaped rotor on the parameters of an electromagnetic system for better values of *emf* can be found out by means of the following target function:

Figure 5. The design of the configuration of the part of a magnetic circle of an arc-shaped wind and electrical generator with a ring-shaped rotor

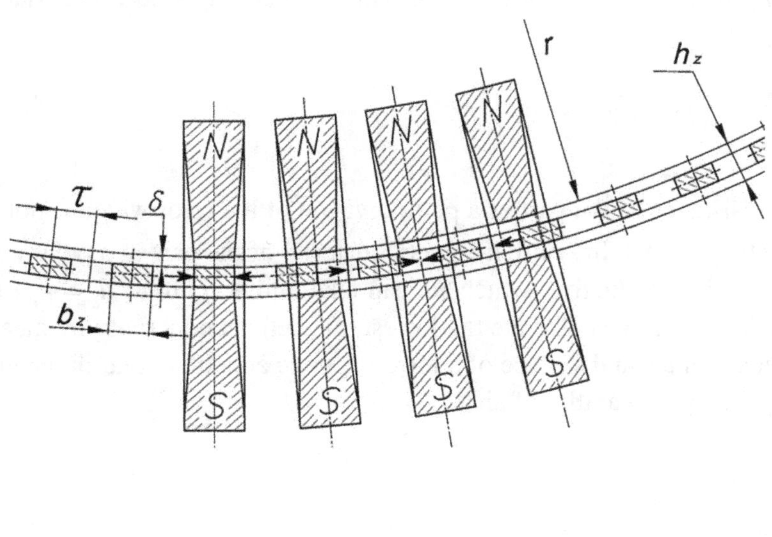

Figure 6. The line drawing presentation of force lines of magnetic field without loading for end positions of ferromagnetic elements of the rotor within one polar division

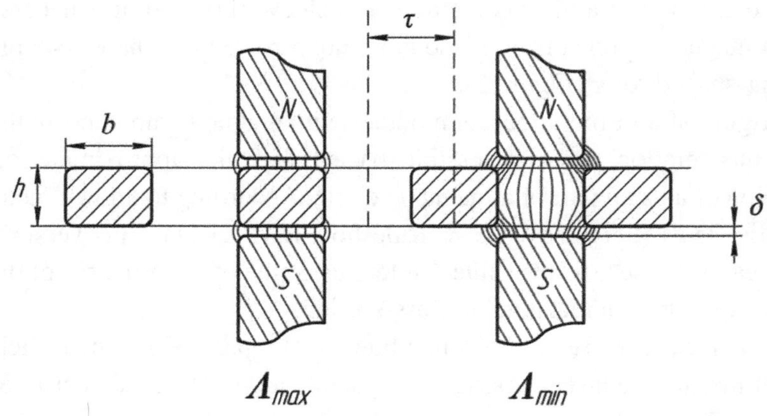

$$\Delta\Lambda(h) = (\Lambda_{max} - \Lambda_{min}) \rightarrow MAX, \tag{23}$$

Λ_{max} – the largest number of magnetic conductivity of the circle which will in the position of full overlapping by a ferromagnetic element of the rotor the gap of the magnetic core of the stator and can be calculated using the following formula:

$$\Lambda_{max} = (R_\delta - R_f)^{-1} = (2\delta / \mu_0 b\ell + h / \mu_f b\ell)^{-1} = b\ell(2\delta / \mu_0 + h / \mu_f)^{-1}, \tag{24}$$

R_δ, R_f – magnetic resistances of an air gap and a ferromagnetic element of the rotor correspondingly; μ_0, μ_f – magnetic permeability of an air gap and a ferromagnetic element of the rotor correspondingly.

As $\mu_0 << \mu_f$, then Λ_{max} depends mainly on the sizes of a working air gap δ, and a radial height h of a ferromagnetic element does not influence on this parameter.

The lowest value of magnetic conductivity Λ_{max} can be calculated by the following formula if this gap is fully overlapped with the part of the rotor which is not magnetically conductive, the height of which is h:

$$\Lambda_{max} = (R_\delta - R_d)^{-1} = (2\delta / \mu_0 b\ell + h / \mu_d b\ell)^{-1}, \tag{25}$$

R_d – magnetic resistance of the part of a polar division of the rotor which is not magnetically conductive, μ_d – magnetic permeability of diamagnetic material of the part of a polar division of the rotor which is not magnetically conductive, which we can take into account as $\mu_d = \mu_0$ to acquire more accurate calculations. Then we can calculate the lowest magnetic conductivity by means of a simple dependence on the square of a usual surface of a ferromagnetic area of a polar division of the rotor τ and the constant of magnetic permeability of air:

$$\Lambda_{\min} = b\ell\mu_0 / (h + 2\delta), \tag{26}$$

When we analyse the equations of the largest and the lowest conductivity we can see that at the origin of coordinates when $h=0$, they have the same largest values. That is why it is enough to analyse only the last dependence (26). When there is no magnetic loading of a magnetic circuit, the dependence of the lowest magnetic conductivity is similar to the change of the conductivity of magnetic circuit by the change of the size of an air gap between edge surfaces of packets which have the sizes between 2δ and $(h+2\delta)$.

Figure 7 presents a theoretical dependence of magnetic conductivity on the radial size h of the part of a ring-shaped rotor which is not magnetic if the fixed value of the working air gap is $\delta = 0{,}001\text{m}$ for different values of the square of a usual crossing of a polar division.

The graph (*Figure* 7) shows that the change of the conductivity of magnetic circuit by means of a mechanical circle of a ring-shaped rotor of an induction-type generator depends in direct proportion on the square of a usual crossing of magnetic and non-magnetic areas of a polar division and in inverse proportion on its radial size. The joint chart (*Figure* 8) shows acquired dependences of three functions (13), (14), and (26) on the same parameter.

The values of magnetic conductivity on the graph do not depend on the velocity of the wind and are calculated in absolute units. It enables us to evaluate this geometrical size using the requirements for mechanical rigidity of the construction without which it is technically impossible to keep a working air gap. These coordinates can be found when we find the root of two systems of equations:

Figure 7. The influence of radial sizes of a ring-shaped rotor and the square of cross section of a ferromagnetic element on the value of magnetic conductivity of a circuit

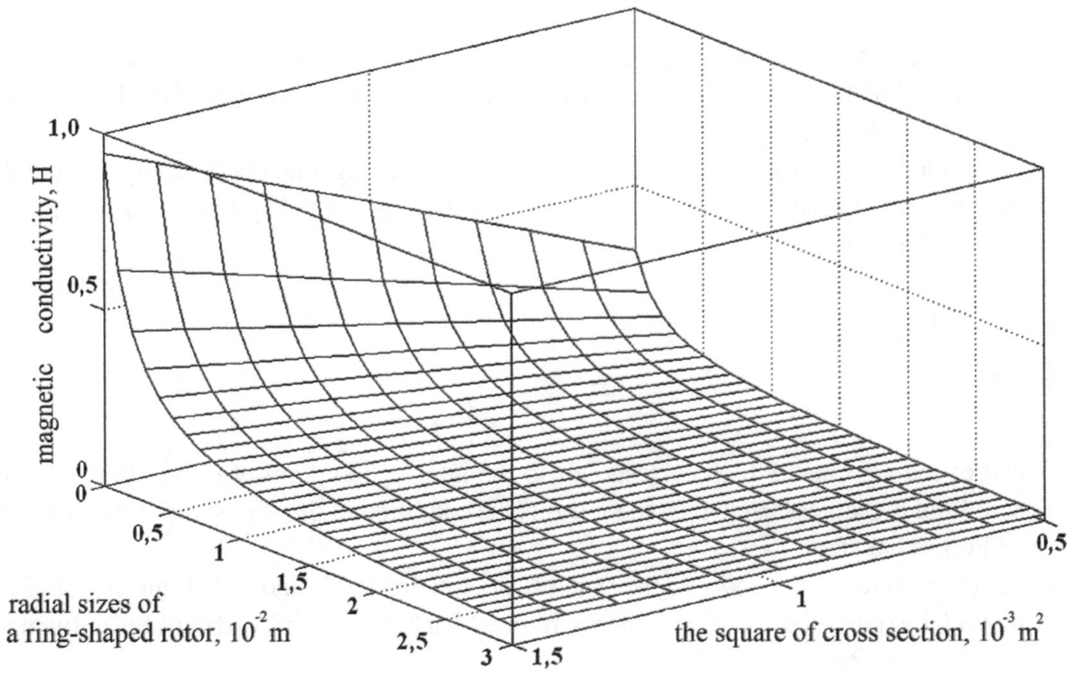

Figure 8. The dependence of magnetic conductivity Λ and the force of aerodynamic resistance F_a and Reynolds number Re on the size h of non-magnetic area of the ring-shaped rotor which is between poles when the velocity of the wind is 8 m/s and fixed sizes b, ℓ and δ which are 0.02, 0.05 and 0,001 m.

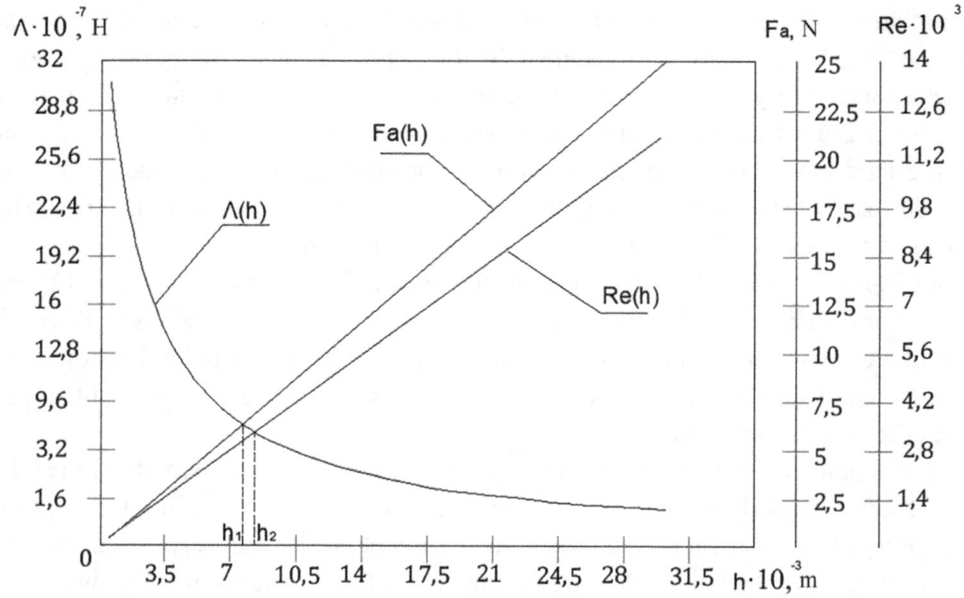

$$\begin{cases} \Lambda_{\min} = b\ell\mu_0 / (h + 2\delta) \\ \mathrm{Re} = Vh / \xi_a \end{cases} \tag{27}$$

This value of the system of equations gives us the smallest radial size which is limited by the condition of the start of the turbulent motion of circular flows of air and permissible values of the lowest magnetic conductivity which is h_1 on the graph.

The second intersection point of the function of the smallest magnetic conductivity and the dependence of the force of the head resistance will be a real positive root of the system of equation when the speeds of the wind are the same:

$$\begin{cases} \Lambda_{\min} = b\ell\mu_0 / (h + 2\delta) \\ F_a = C_X \pi \rho V^2 (2rh + h^2) / 2 \end{cases} \tag{28}$$

The value of the argument which corresponds to the intersection point of curves in this system of equations will limit the largest size which more meets the condition of the lowest magnetic conductivity than the smallest radial size.

Thus, numeric values of the roots of the systems of equations (27) and (28) limit the multitude of optimal values of the radial size h of a ring-shaped rotor which helps to use it to determine optimal values of this multitude using Pareto method.

The development of the algorithm of projecting an integral non-transmission electromechanical converter with a ring-shaped rotor. All geometrical parameters are interdependent and inter-subordinated in this integral construction, starting from the size of a wind turbine and up to the calculation capacity.

The formula shows that the diameter of a ring-shaped rotor depends on the diameter of wind turbines according to the condition of proportional compliance of the values of their high speeds:

$$D_c = D_b \lambda_c / \lambda_b, \tag{29}$$

D_b, λ_b – the diameter of the circle and high speed (the number of modules) of ends of blades WT, λ_c – a high speed of wind turbines on the area with the diameter of a ring-shaped rotor $D_c = 2r$.

$$\lambda_c = 2\pi nr / V = \lambda_b r / R \tag{30}$$

n – the frequency of the rotation of wind turbines, s^{-1}

As the cylinder of a ring-shaped rotor is set on the intra-ring blades, so their number and parameters of aerodynamic quality also depend on high speed λ_c.

So, the first step to calculate an arc-shaped stator wind and electrical generator is to determine the diameter of a wind turbine using the size of main blades D_b = *2R*, which is determined by the given capacity provided by the permissible calculated wind velocity and main aerodynamic parameters of an aeromechanical system. All other geometrical parameters can be found in a subordinated priority succession according to the set ratios.

We offer a general methodology of determining optimal geometrical parameters of a wind and electrical generator with a ring-shaped rotor and an arc-shaped stator using the results of the theoretical analysis on the basis of obtained interdependences. The succession of steps of the algorithm of their calculation is as follows:

- Basic geometrical sizes of an aeromechanical system are calculated using permissible values of the working wind speed;
- The best diameter of a ring-shaped rotor is determined using the given values of the high speed and the sizes of blades of a wind turbine;
- The best geometrical sizes of the cross of a ring-shaped rotor are determined due to compromise requirements for an aeromechanical and electromagnetic systems;
- The calculation of inter-subordinate sizes of a polar division of the rotor and corresponding distances between the sections of the stator;
- The optimization of the shape and the size of a magnetic core and windings of a section of the stator using the known square of crossing sector of magnetic core and distances between sections;
- The calculation of the number of sections of an arc-shaped stator according to the calculation capacity.

Figure 9 gives a general scheme of the calculation of an integral electromechanical converter with a ring-shaped rotor.

As there exists a definite dependence of the sizes of a wind turbine (Choi, D.W., Byun, S.I., Cho, Y.H. (2014)) on it initial capacity and some calculated wind speed so we can evaluate specific parameters of linear load not along the axial length of the rotor but along the length of the arc of a ring-shaped rotor overlapped by an arc-shaped stator. For instance, taking some axial size of a ferromagnetic element of the rotor as ℓ=0,1m it is necessary to know specific parameters of power kW/m in order to calculate the length of the arc of the stator.

Figure 9. The algorithm of projecting an integral non-transmission electromechanical converter with a ring-shaped rotor

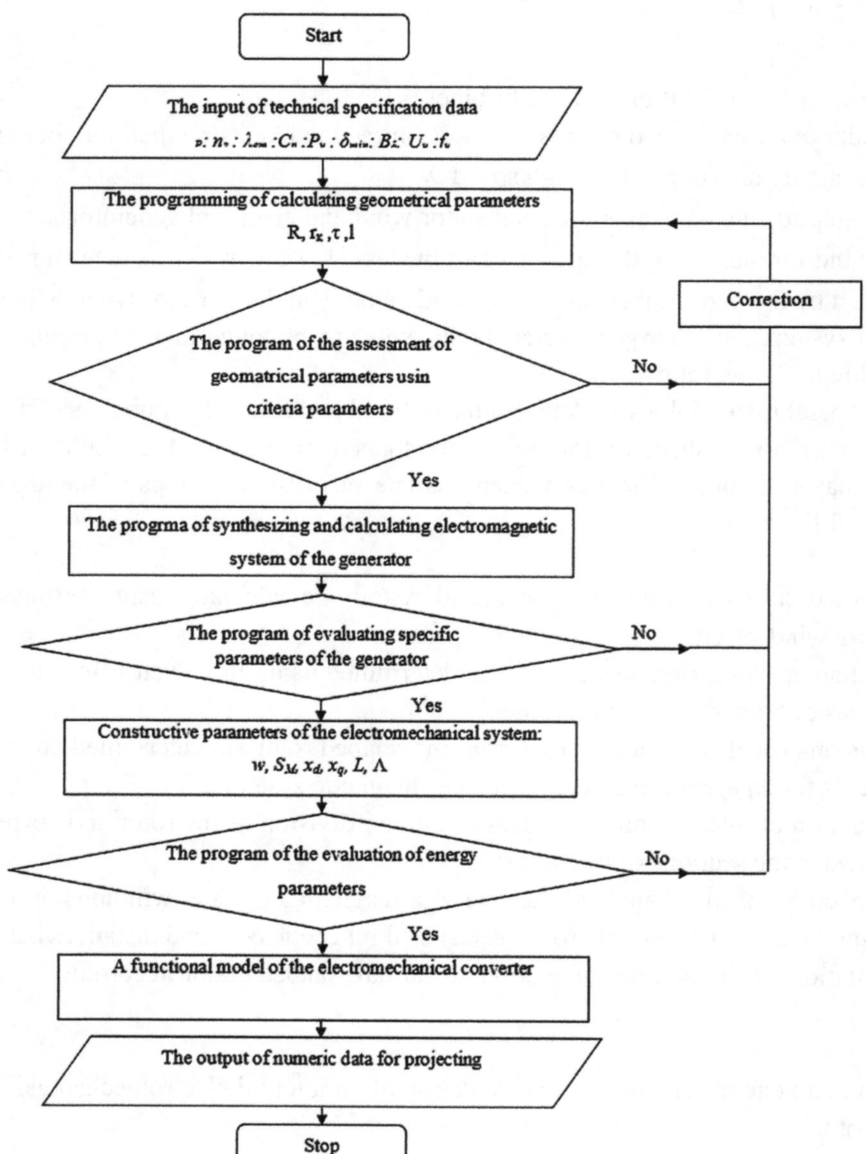

Combined Wind and Energy Stations With a Magneto Electric Linear Generator

One of the foci of the enhancement of energy efficiency of wind stations is the use of wind and energy systems with a horizontal axis of the rotation with a magneto electric linear generator of back-and-forth motion between a crank-and-rod mechanism and a massive water pump. It will enlarge the functionality and enhance the efficiency of the use of the mechanical energy of the wind engine. That is why a combined wind station will not only pump water but also generate electric energy to charge storage batteries. This energy can charge an electric machine when it is necessary to start the engine in calm, windless weather and when the wind gets stronger to load the wind engine by starting the electric generator. It will increase the coefficient of the use of the fixed capacity of a wind station. By developing a structural and functional design of a combined electrical and water-supplying wind station (CEWWS) (Kozyrskyi, V.V., Kaplun, V.V., Petrenko, A.V. (2007)) with a magneto electrical linear generator (MELG) with a reciprocal motion (Hrebenikov, V.V., Petrenko, A.V. (2009)), we have improved the design of a wind and pumping station and increased the number of its functions.

Figure 10 shows a basic combined electric water-supplying wind station with a magneto electric linear generator of back-and-forth motion and pumping equipment.

This basic functional and kinematic schematic figure shows (*Figure* 10) the structure of CEWWS as follows: the wind engine 1 moves the shaft of which is joined with the crankshaft 2 of the crank-and-rod mechanism 3. The crankshaft 2, set up with its root necks in trunnions of holder bearings 4 of the turbine 5 of a wind station, is hinged with the rod 6 which is fixed in the leading cylinder of the tubular prop 7 by means of crosshead 8, which is rigidly joined by means of the stock 10 and a linking coupler with the inductor 12 of the magneto-electric linear generator 13 through the support-and-rotary mechanism. The generator 13 is located in the tubular prop of the wind station. The inductor 12 of the linear electric generator 13 is rigidly connects the stock 16 with the forcer 17 of the water pump 18 by means of connecting-and-disconnecting couplers 14, 15.

The systems which supply with electric power consist of: MELG 13, rectifier 19, autonomous power invertor with the system of regulation 20, storage batteries 21 and energy consumers 22; storage tank for water 23; 24 – voltage division for electric consumers of permanent voltage; 25 is the unit of controlling couplers11, 14, which connects an inductor of MELG with the core of the crank-and-rod mechanism and the stock of the displacement pump; 27 – sensors of the top and bottom level of water which are located in storage tank for water 23.

Figure 11 gives the algorithm of the control of the developed system presented in *Figure* 10. It describes the work of the system in the following way: 1 – before the start of the work RB1 and RB2 are discharged *70%* of their minimum capacity; 2 – before the start of the work RB1 and RB2 are charged *100%* of their minimum capacity; 3 – before the start of the work RB1 is discharged *70%* of its nominal capacity, and RB2 is charged *100%* of its nominal capacity; 4 – before the start of the work RB2 is discharged *70%* of its nominal capacity, and RB1 is charged *100%* of its nominal capacity; 5 – before the start of the work the storage tank for water is full; 6 – before the start of the work the storage tank for water is empty.

After we have analysed peculiarities of functioning CEWWS, we have developed the following principles of building such an electric and technical complex:

Figure 10. A basic functional-and-kinematic design of a combined electrical and water supplying wind station.

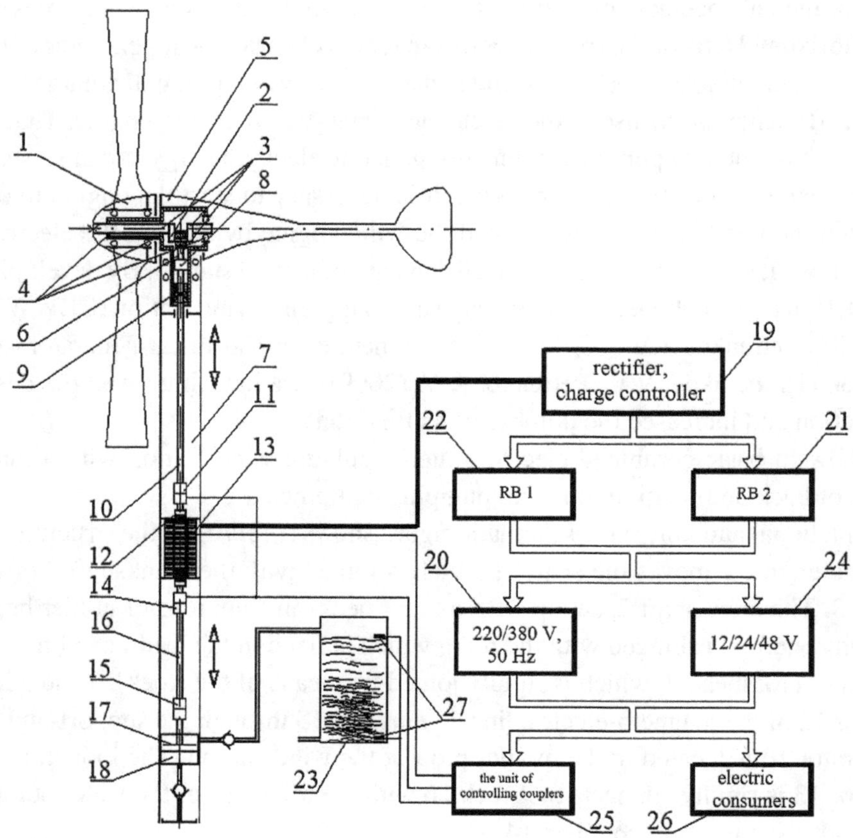

1. The distribution of mechanical energy of a wind engine is done due to the consensus of modes of electricity and water consumption and is divided into two energy flows which are simultaneously used to start the pump and magneto electrical linear generator;
2. The priority of the use of the mechanical energy of wind engines belongs to the pump with the fixed capacity which fills the storage tank with water. Magneto electrical linear generator is used to load the wind engine when wind speeds are high by generating electric energy to store it in a storage battery;
3. The start of the pump in calm, windless weather is done by a magneto electrical linear machine which can consume electrical energy of the storage battery and work in the mode of an electrical energy; here, magnetic connection is absent between the wind engine and the linear machine;

We can use this basic functional and kinematic design of CEWWS (*Figure* 10), to build and analyse a structural and functional scheme which will present the balance of the power of primary energy of an air flow in a combined electric and water supply wind station (*Figure* 12).

The capacity of primary energy of an air flow for round calculation crossing of this flow is:

Figure 11. The algorithm of autonomous functioning of an electrical and technical system of a combined electrical and water supplying wind turbine

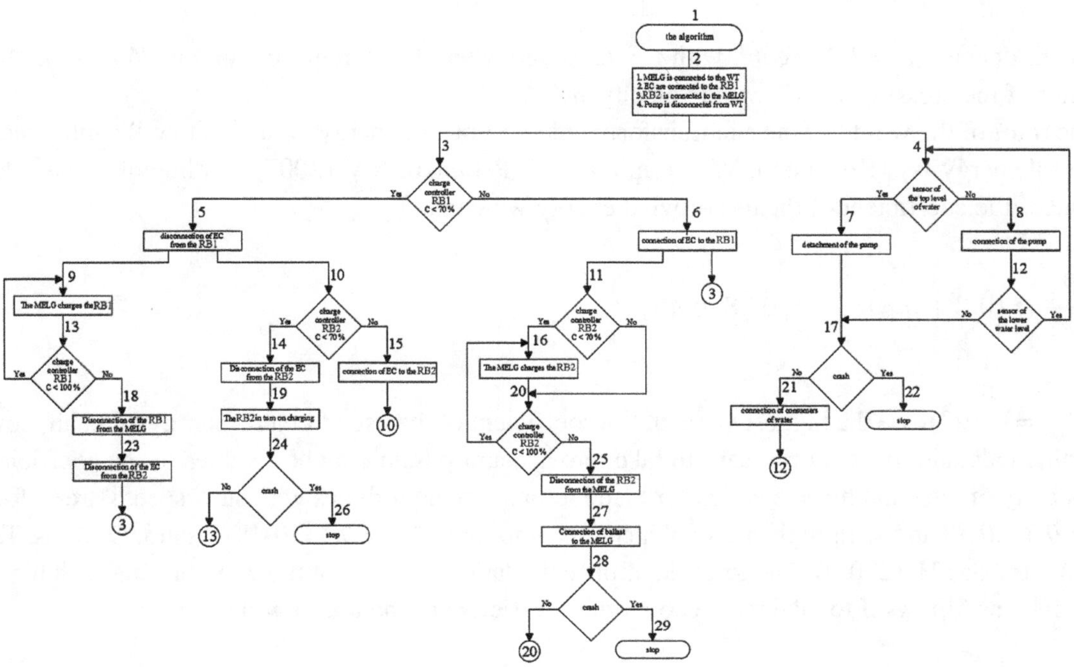

Figure 12. A structural and functional scheme of the balance of capacity of CEWWS

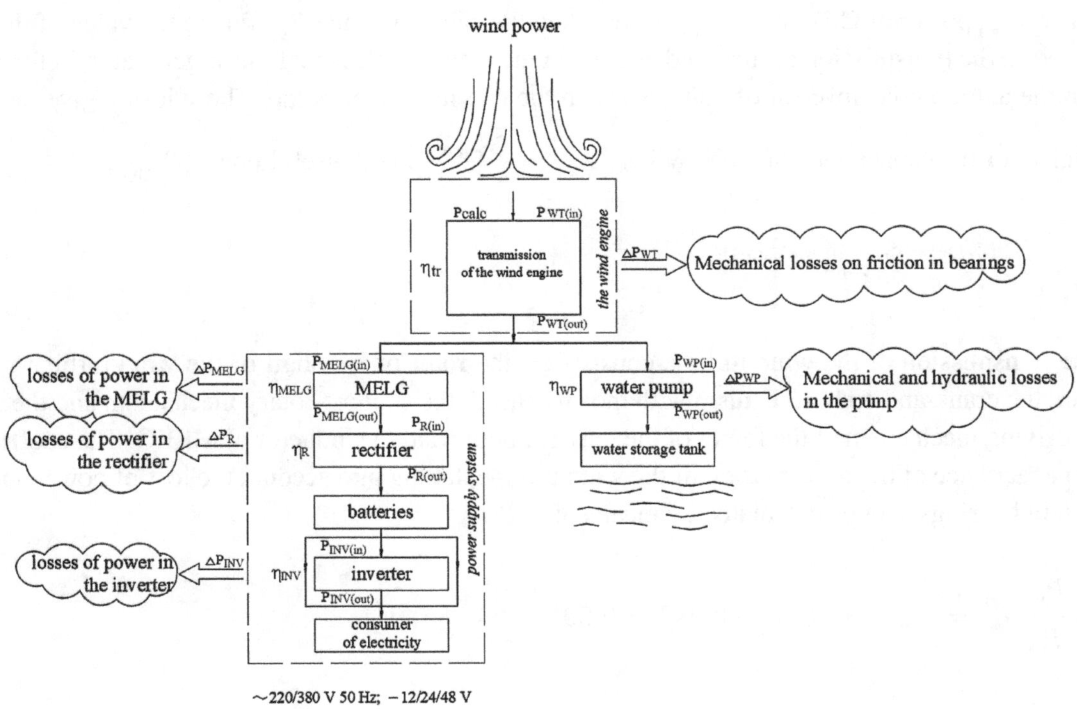

$$P_W = \frac{\pi}{8} \rho D^2 V^3 10^{-3} kW, \tag{31}$$

ρ – air density (ρ = $1{,}23$ kg/m^3, with t = 15^0C and when the mercury column is *760 mm.*); D – the diameter of the crossing, m; V – wind velocity, m/s.

The rotor of the wind turbine can transform only part of full energy of an air flow Wa into mechanical useful energy W_1. (Kozyrskyi, V.V., Kaplun, V.V., Petrenko, A.V. (2007); Podhurenko, V.S. (2007)) introduces the coefficient of the use of wind energy which is

$$\xi = \frac{W_1}{W_a} = \left(0,3 \div 0,8\right) \xi_{\max} \approx 0,18..0,48\,,$$

with ξ_{max} = $0{,}593$ to be the highest value of the coefficient of the use of kinetic energy of an air flow. To do further calculations it is necessary to take into account possible highest values of the coefficient of the capacity of the wind turbine which for blade rotors of wind turbines with slow speeds are as follows ξ_{slow} = $0{,}33..0{,}38$ and with high speeds which are as follows ξ_{fast} = $0{,}45..0{,}48$ (Yahno, O.M., & Taurit, T.H., & Hrabar I.H. (2002)). The formula of the calculation capacity of the wind turbine with blade rotors will be as follows if to take into account the coefficient of the use of wind energy:

$$P_{calc} = \frac{\pi}{8} \rho D^2 V^3 \xi 10^{-3} kW. \tag{32}$$

The useful power of CEWWS P_{CEWWS} is less than the calculated one P_{calc} due to the value of the loss of power: in the transmission of the wind engine, in the magneto electrical linear generator, in the rectifier, in the autonomous invertor of voltage and in the massive water pump. These losses give the total coefficient of the capacitance of CEWWS $\eta_{CEWWS} = \frac{P_{CEWWS}}{calc}$, with useful power P_{CEWWS}:

$$P_{CEWWS} = \frac{\pi}{8} \rho D^2 V^3 \xi \eta_{CEWWS} 10^{-3} kW. \tag{33}$$

The transmission of the wind turbine consists of: the rotor of the shaft of the wind turbine with a coupler, the crank-and-rod mechanism with the crosshead, the support-rotary mechanism and the stock of the driving mechanism of the forcer of the massive pump and the inductor of MELG. The coefficient of the capacitance of the transmission of the wind turbine, taking into account the loss of power and the friction in bearings, can be calculated by means of:

$$\eta_{tr} = \frac{P_{WT(out)}}{P_{WT(in)}} = \eta_{roll} \eta_{slid} = 0,99^n \cdot 0,98^m = 0,99^4 \cdot 0,98^1 = 0{,}941,$$

$P_{WT(out)}$ – the useful capacity of the wind turbine, kW; $\eta_{roll} = 0{,}99..0{,}995$ – the energy conversion efficiency of rolling bearings with the liquid lubricant; $\eta_{slid} = 0.98..0.985$ – energy conversion efficiency of sliding bearings with liquid lubricant; n – the number of rolling bearings; m – the number of sliding bearings.

The energy conversion efficient should be reduced by *0,015..0,01* if to use the semiliquid lubricant.

So, the mechanical energy conversion efficiency of the transmission of the wind engine of CEWWS is $\eta_{tr} = 0{,}941$.

Accordingly, the useful capacitance of the wind turbine is:

$$P_{WT(out)} = \frac{\pi}{8}\rho D^2 V^3 \xi \eta_{tr} 10^{-3}\, kW. \tag{34}$$

The useful capacitance $P_{WT(out)}$ of the wind turbine, as it was mentioned above is distributed into two simultaneous energy flows which are used by the forcer $P_{WP(in)}$ of the water pump and the inductor $P_{MELG(in)}$ of MELG. So, the formula $P_{WT(out)} = P_{MELG(in)} + P_{WP(in)}$ is reasonable.

The balance of the capacity of the piston pump is:

$$P_{WP(in)} = P_{WP(out)} + \Delta P_h + \Delta P_m,\ kW, \tag{35}$$

$\Delta P_h = (0{,}1 \div 0{,}13)P_{WP(in)}$, kW – the losses of the hydraulic origin; $\Delta P_m = (0{,}04 \div 0{,}06)P_{WP(in)}$, kW – the losses of the mechanical origin.

The formula of the balance of the capacity gives the value of the useful capacitance of the pump which is as follows:

$$P_{WP(out)} = P_{WP(in)} - 0{,}13\, P_{WP(in)} - 0{,}06\, P_{WP(in)} = 0{,}81\, P_{WP(in)},\ kW. \tag{36}$$

Hence, the approximate value of the energy conversion efficiency of the piston pump equals:

$$\eta_{WP} = \frac{P_{WP(out)}}{P_{WP(in)}} \approx 0{,}81.$$

Correspondingly, the power $P_{WP(in)}$ of the pump needed to lift the water equals:

$$P_{WP(in)} = \frac{Qp_{WP}}{\eta_{WP}} \cdot 10^{-3}, kW \tag{37}$$

Q – the actual pump flow, m^3/s; p_{WP} – the pressure of the pump, Pa.

We will calculate the power of the pump with the actual delivery $Q = 2\cdot 10^{-3}\ m^3/s$ and the pressure $p_{WP} = 0.29\cdot 10^6 Pa$, which equals the fully supply of water for drip watering and partly for the irritation. In this case, the full capacity of the pump is:

$$P_{WP(in)} = \frac{2 \cdot 10^{-3} \cdot 0,29 \cdot 10^{6}}{0,81} \cdot 10^{-3} = 0,716\,kW.$$

Taking into consideration a parallel character of the mechanical energy transfer to the stock of the pump piston and the inductor of MELG it is necessary to calculate the energy conversion efficiency of simultaneous (parallel) transfer using the following equation:

$$\eta_p = \frac{P_{MELG(out)} + P_{WP(out)}}{P_{WT(out)}} = \frac{P_{MELG(out)} + P_{WP(out)}}{P_{MELG(in)} + P_{WT(in)}}. \tag{38}$$

This formula confirms that low quality of some elements when they are joined parallel has a less impact on the energy conversion efficiency than when they are joined continuously, thus enhancing the efficiency of the whole station.

We can calculate energy conversion efficiency of MELG taking into account all losses of its capacity as the ratio between the useful electrical capacity of MELG $P_{MELG(out)}$ and mechanical capacity of the driving mechanism of the inductor of MELG $P_{MELG(in)}$.

$$\eta_{MELG} = \frac{P_{MELG(out)}}{P_{MELG(in)}} = \frac{P_{MELG(out)}}{P_{MELG(out)} + \Delta P_{MELG}}, \tag{39}$$

$P_{MELG(out)}$ – the useful electrical power of MELG, *kW*; $P_{MELG(in)} = P_{WT(out)} - P_{WP(in)} = P_{MELG(out)} + \Delta P_{MELG}$ – the mechanical power of the driving mechanism of the inductor of MELG, *kW*; ΔP_{MELG} – total losses of power in the MELG, *kW*.

The total losses of the power in MELG equal:

$$\Delta P_{MELG} = \Delta P_{EL} + \Delta P_{ML} + \Delta P_{PL} + \Delta P_{AL,}\ kW, \tag{40}$$

ΔP_{EL} – electrical losses of power in windings of the stator of MELG, *kW*; ΔP_{ML} – magnetic losses of power in the magnetic core of MELG, *kW*; ΔP_{PL} – the mechanical power loss on the friction of the inductor core of MELG, *kW*; ΔP_{AL} – additional losses, *kW*.

Theories of electrical machines state that power losses in electrical machines equal $\Delta P_{EL} = (0.4..0,5)\ \Delta P$, and the values of power loss for MELG which have magnetic excitation are the lowest that is $\Delta P_{EL} = 0,4\Delta P_{MELG}$. Then, the product of the rest of losses will equal:

$$\Delta P_{MELG} - \Delta P_{EL} = \Delta P_{ML} + \Delta P_{PL} + \Delta P_{AL} = \Delta P_{MELG} - 0,4\Delta P_{MELG} = 0,6\Delta P_{MELG}.$$

Taking into account all calculated data, we can calculate that mechanical power of the driving mechanism of the inductor of .$P_{MELG(in)}$ after transforming the mechanical energy into the electrical one equals the product of the electromagnetic internal capacity of MELG and non-electrical power losses, or the product of useful capacitation and the power of all losses:

$$P_{MELG(in)} = P_{EM} + 0,6\Delta P_{MELG} = P_{MELG(out)} + \Delta P_{MELG} kW,$$

whereas electromagnetic power equals:

$$P_{EM} = P_{MELG(out)} + \Delta P_{MELG} - 0,6\Delta P_{MELG} = P_{MELG(out)} + 0.4\Delta P_{MELG} \; kW.$$

Taking into account the ratio of the voltage drop $k_V \approx 0,9$ in windings of the generator, we receive the value of the useful capacitation of MELG $P_{MELG(out)}$ and the relative value of the total power loss

ΔP_{MELG}: $P_{MELG(out)} = 0.9P_{EM} = 0.9(P_{MELG(out)} + 0.4\Delta P_{MELG}) = 0.9P_{MELG(out)} + 0.36\Delta P_{MELG}$, $0.36\Delta P_{MELG} =$.,$1P_{MELG(out)}$,

$$\Delta_{MELG} \approx \frac{0,1}{0,36} \approx 0,278 P_{MELG(out)}.$$

In this case energy conversion efficiency of MELG will equal:

$$\eta_{MELG} = \frac{P_{MELG(out)}}{P_{MELG(in)}} = \frac{P_{MELG(out)}}{P_{MELG(out)} + \Delta P_{MELG}} = \frac{P_3}{P_3 + 0,278_{\;3}} = \frac{1}{1,278} = 0,782.$$

Thus, we can use the known ratio of power loss in electrical machines and known coefficient of voltage drop in windings of the stator of MELG to receive the anticipatory calculation of energy conversion efficiency of MELG. When we make more thorough calculations, we will receive more accurate values of the energy conversion efficient.

The useful power $P_{R(out)}$ of the rectifier is less than the one of the generated power $P_{MELG(out)} = P_{R(in)}$ of MELG by one unit of power loss in the rectifier:

$$P_{R(out)} = P_{R(in)}\eta_R. \tag{41}$$

Storage batteries charge power (RB group) should equal output power $P_{R(out)}$ of the rectifier and storage batteries discharge power should equal input power $P_{INV(in)}$ of the autonomous voltage invertor.

The autonomous voltage invertor is used in the system to invert permanent current into alternating to supply energy consumers of the alternating current of standard voltage ~ *220 V* with frequency *50 Hz*. The value of the energy conversion efficiency of the invertor η_{INV} of modern commercial machines is nearly $\eta_{INV} \approx 0.9$. The electrical power loss is *10%*. Hence, the useful capacitation of the autonomous voltage invertor is as follows:

$$P_{INV(out)} = P_{INV(in)}\eta_{INV}. \tag{42}$$

If the value of the power of electricity consumers is P_{EC} = *1,0 kW*, which is supposed to provide electrical power of controllers, electromagnetic valves of an automated irrigation system and possible additional electrical equipment, we can calculate the power of structural elements of an electromechanical system and the value of total energy conversion efficiency applying the mentioned above method.

The structural flow chart in *Figure* 13 presents the results of the calculation when power values of both the massive water pump and electricity consumers are known:

1. The useful power of the massive water pump with the valid flow

$Q = 2 \cdot 10^{-3}\ m^3/s$ and pressure $p_{WP} = 0{,}29 \cdot 10^6 Pa$:

$$P_{WP(out)} = Qp_{WP}10^{-3} = 2 \cdot 10^{-3} \cdot 0,29 \cdot 10^{6} \cdot 10^{-3} = 0,58\,kW.$$

2. The total power of the massive water pump is

$$P_{WP(in)} = \frac{Qp_{WP}}{\eta_{WP}}10^{-3} = \frac{2 \cdot 10^{-3} \cdot 0,29 \cdot 10^{6}}{0,81}10^{-3} = 0,716\,kW.$$

3. The total power of the autonomous voltage invertor is

$$P_{INV(in)} = \frac{P_{EC}}{\eta_{INV}} = \frac{1,0}{0,9} = 1,11\,kW.$$

4. The total power of the voltage rectifier is

$$P_{R(in)} = \frac{P_{INV(in)}}{\eta_{R}} = \frac{1,11}{0,85} = 1,307\,kW.$$

5. The total power of MELG is

$$P_{MELG(in)} = \frac{P_{MELG(out)}}{\eta_{MELG}} = \frac{1,307}{0,782} = 1,67\,kW.$$

6. The useful power of the wind engine is:

$$P_{WT(out)} = P_{MELG(in)} + P_{WP(in)} = 1,67 + 0,716 = 2,386\,kW.$$

7. The total power of the wind engine is:

$$P_{WT(in)} = \frac{P_{WT(out)}}{\eta_{tr}} = \frac{2,386}{0,941} = 2,54\,kW.$$

8. The value of the total energy conversion efficiency of CEWWS is

$$\eta_{CEWWS} = \eta_{tr} \frac{P_{MELG(in)}\eta_{MELG}\eta_R\eta_{INV} + P_{WP(in)}\eta_{WP}}{P_{MELG(in)} + P_{WP(in)}}, \quad (43)$$

$$\eta_{CEWWS} = 0,941 \cdot \frac{1,67 \cdot 0,782 \cdot 0,85 \cdot 0,9 + 0,716 \cdot 0,81}{1,67 + 0,716} = 0,623\,.$$

9. The useful power of CEWWS is

$$P_{CEWWS} = P_{EC} + P_{WP(out)} = P_{calc}\eta_{CEWWS} = 1,0 + 0,58 = 2,54 \cdot 0,623 = 1,58\,kW. \ (44).$$

The value of the useful power P_{CEWWS}, which is supplied to the electricity consumer and the value of the total energy conversion efficiency of CEWWS η_{CEWWS} help us to calculate the diameter of the rotor of the wind engine (for example, for the high speed rotor with wind velocity which is *8 m/s*):

$$D = \sqrt{\frac{P_{CEWWS}}{\frac{\pi}{8}\rho V^3 \xi \eta_{CEWWS} 10^{-3}}} = \sqrt{\frac{1,58}{0,393 \cdot 1,23 \cdot 8^3 \cdot 0,33 \cdot 0,623 \cdot 10^{-3}}} \approx 5,6\,m.$$

Figure 13. The structural flow chart of the balance of powers of CEWWS for electricity consumers, the value of the power being 1.0 kW, with the valid water flow $2{\cdot}10^{-3}$ m³/s, and the pressure $0.29{\cdot}10^6$Pa

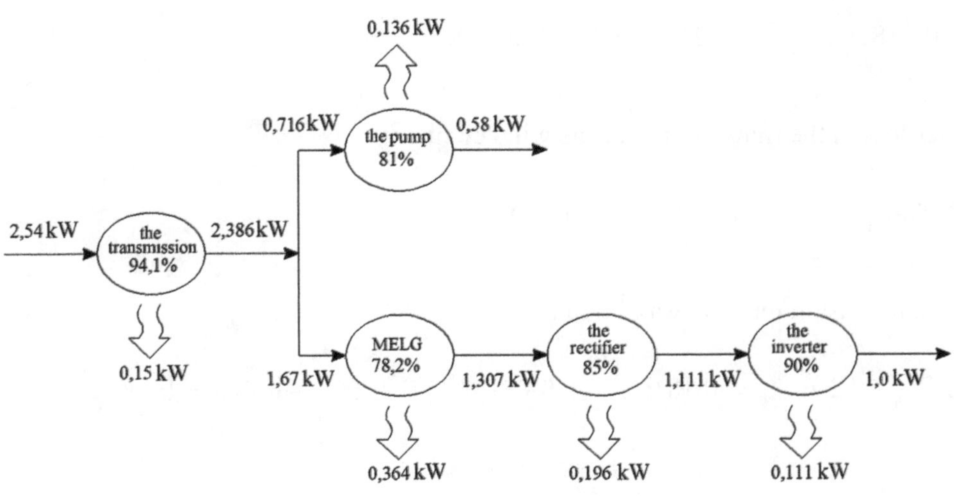

We can check the accuracy of the diameter of CEWWS when we know the value of the calculation power of the wind station P_{calc}, which is the power of the wind flow which is consumed by the rotor:

$$D=\sqrt{\frac{P_{calc}}{\frac{\pi}{8}\rho V^{3}\xi 10^{-3}}}=\sqrt{\frac{2,54}{0,393\cdot 1,23\cdot 8^{3}\cdot 0,33\cdot 10^{-3}}}\approx 5,6\,m.$$

It is necessary to take into account that the calculated diameter of the rotor of the wind engine D is an approximate solution because this formula includes the determined velocity of the wind which does not meet real natural and climate conditions. In real, wind velocity is a random value. The calculation which will take into account the wind velocity as a random value will be discussed in the next chapter of this dissertation.

To summarise the mentioned above methodical material, we will analyse the balance equation of the power of a combined electrical water supplying wind station:

$$P_{calc(in)}=P_{WP(out)}+P_{EC}+\Delta P_{CEWWS}\,kW, \tag{45}$$

P_{EC} – the power of electricity consumers, kW; $\Delta P_{CEWWS}=\Delta P_{WT}+\Delta P_{MELG}+\Delta P_{R}++\Delta P_{INV}+\Delta P_{WP}$ – the total value of power loss in CEWWS, kW.

The power loss in the autonomous voltage invertor is

$$\Delta P_{INV}=0,1P_{INV(in)}=0,1\cdot 1,11=0,111\,kW.$$

The power loss in the rectifier is

$$\Delta P_{R}=0,15P_{R(in)}=0,15\cdot 1,307=0,196\,kW.$$

The power loss in magneto electrical linear generator is

$$\Delta P_{MELG}=0,218P_{MELG(in)}=0,218\cdot 1,67=0,364\,kW.$$

The power loss in the transmission of the wind engine is

$$\Delta P_{WT}=0,059P_{WT(in)}=0,059\cdot 2,54=0,15\,kW.$$

The loss in the piston massive water pump is

$$\Delta_{\ WP}=\Delta P_{WPG}+\Delta P_{WPM}=0,13P_{WP(in)}+0,06P_{WP(in)}=0,093+0,043=0,136\,kW.$$

Thus, the total value of power loss in CEWWS is

$\Delta P_{CEWWS} = 0.15 + 0.364 + 0.196 + 0.111 + 0.136 = 0.957$ kW.

When we substitute the values of P_{calc}, $P_{WP(out)}$, P_{EC} and ΔP_{CEWWS} into the equation (45), we receive

2.54 kW ≈ 0.58 + 1.0 + 0.957 kW;

2.54 kW ≈ 2.54 kW.

Thus, the balance equation confirms the accuracy of the received results and the efficiency of the developed methodology.

We took into account the highest values of the energy conversion efficiency of the used elements in systems when we compared the energy conversion efficiency as the value of energy efficiency of structural elements, two electrical and technical systems, a conventional and the mentioned above methodology. Moreover, the first version of the system that is a wind electrical station with the electrical pump presents three values of the energy conversion efficiency of electrical pump equipment caused by different kinds of electrical pump. The values of the energy conversion efficiency of elements of wind stations which are being compared are taken from nominal data, literature sources and the results of our own calculations are presented in Table 1.

When we substitute the values from Table 1, according to *Figure* 14 in the corresponding equations, we receive the values of the energy conversion efficiency for a wind and electrical station which can work with electrical pumps:

$$\eta_{WS} = \eta_{tr} \cdot \eta_G \cdot \eta_R \cdot \eta_{INV} \cdot \eta_{EP}, \tag{46}$$

η_{tr} - energy conversion efficiency of the transmission of the wind engine; η_G - the energy conversion efficiency of an electrical generator; η_R - the energy conversion efficiency of the voltage rectifier; η_{INV} - the energy conversion efficiency of the autonomous voltage invertor; η_{EP} - the energy conversion efficiency of the electrical pump.

Table 1. The values of the energy conversion efficiency of elements of an electrical and technical system of wind stations

N°	Elements of the System	An Electrical Pump Wind Station, %	CEWWS, %
1	The transmission of the wind engine	96.1	94.1
2	The electrical generator	90	78.2
3	The voltage rectifier	85	85
4	The autonomous voltage invertor	90	90
5	The electrical pump: 1) the vibrational electrical pump 2) the centrifugal electrical pump 3) the centrifugal submerged electrical pump The piston-type, plunger massive pump	 22.6 41 51	81

Figure 14. The structural flow chart of a wind electrical station with the electrical pump.

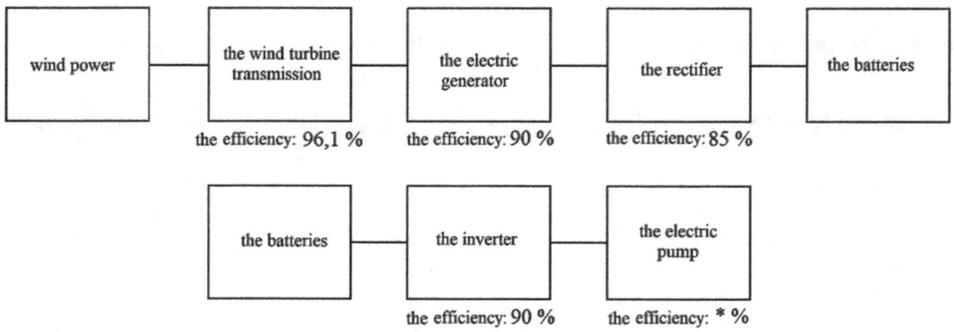

After we do simple mathematical actions in equation 43, we get the formula of the calculation of the energy conversion efficiency of a combined electrical and water supplying wind station:

$$\eta_{CEWWS} \approx \frac{\eta_{tr}\eta_{MELG}\eta_{R}\eta_{INV}\eta_{WP}\left(P_{EC} + Qp_{WP}\right)}{P_{EC}\eta_{WP} + Qp_{WP}\eta_{MELG}\eta_{R}\eta_{INV}}, \tag{47}$$

P_{EC} - the power of electricity consumers, *kW*; *Q* – the valid flow of the pump, m^3/s; p_{WP} – the pressure of the pump, *Pa*; η_{MELG} - the energy conversion efficiency of the magneto electrical linear generator; η_{WP} - the energy conversion efficiency of the piston-type (plunger) plump.

According to the explained above and *Figure* 15, we get the calculated value of the energy conversion efficiency of a combined electrical, water supplying wind station which equals $\eta_{CEWWS} \approx 62,3\%$ with water flow which equals $2 \cdot 10^{-3} m^3 / s$, the pressure which equals $0.29 \cdot 10^6$ *Pa*, and the generated electrical power which equals *1 kW*.

Table 2 presents the results of the calculation of the energy conversion efficiency in both versions and the efficiency of parts of this system from the source to the consumer.

Table 2 and flow charts (*Figures* 14, 15) present the value $\eta_{WS} \approx$ *(15%, 27,1%, 33,7%)* of a wind and electrical station with three options of using different kinds of electrical pump and $\eta_{CEWWS} \approx 62,3\%$ of CEWWS which uses massive water pumps. These values assess the efficiency of each version of wind stations.

Thus, the value of the energy conversion efficiency of CEWWS $\eta_{CEWWS} \approx 62,3\%$ exceeds the similar value of a conventional wind and electrical station with the power $\eta_{WS} \approx 15\%$ which uses vibrational electrical pumps by *47,3%* and the value of the energy conversion efficiency of a wind and electrical station with centrifugal submerged electrical pumps $\eta_{WS} \approx 33,7\%$ by *28,6%*.

CONCLUSION

1. The results of the analysis of technical errors show that the promising type of a non-transmission wind and energy system can be an integral construction with a ring-shaped rotor of the generator

Figure 15. The structural flow chart of a combined electrical and water supplying wind station

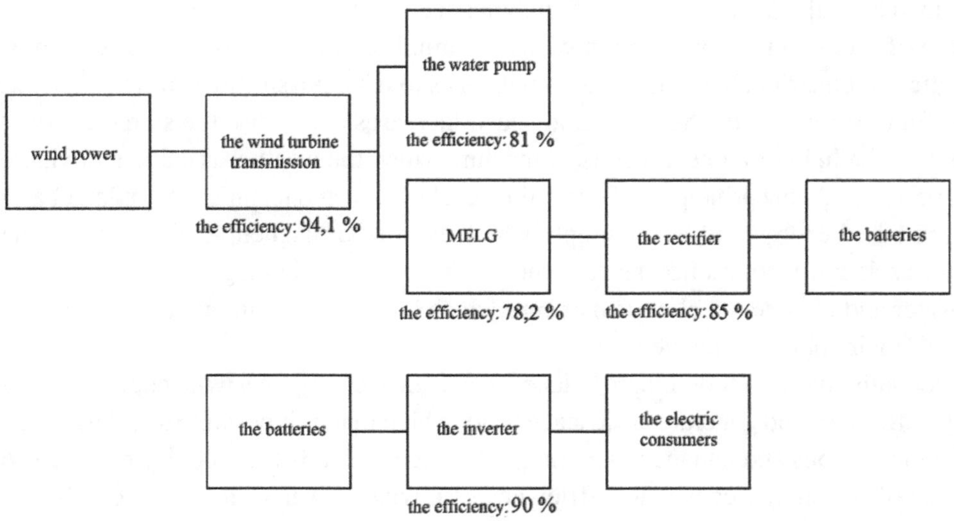

Table 2. The energy conversion efficiency of electrical and technical systems of wind stations

Nº	Electrical and Technical System	ECE, %	
		The Calculated Value	The Efficiency of CEWWS
1	The wind and electrical station with the electrical pump: 1. the vibrational electrical pump 2. the centrifugal electrical pump 3. the centrifugal submerged electrical pump	15.0 27.1 33.7	62.3 – 15.0 ≈ 47.3 62.3 – 27.1 ≈ 35.1 62.3 – 33.7 ≈ 28.6
2	CEWWS with the piston-type (plunger) pump	62.3	-

which is placed coaxially on a horizontal axis of the wind turbine which is actually the only possible contactless non-transmission generators which can achieve the needed values without any transferring mechanical link and kinematic complications.

2. We have synthesized a configuration of an electrical and according to summarized requirements for an aeromechanical system with a ring-shaped rotor in the form of an empty cylinder which is placed coaxially to a horizontal axis of the wind turbine with a radial shortest direct focus of the magnetic flow through its polar elements between working surfaces of the stator toroidal magnetic core.
3. The best diameter of the ring-shaped rotor can be calculated due to aerodynamic requirements for a wind turbine. The size of a polar division and an axial length and a radial thickness of a ferromagnetic element of the rotor can be found out using the ratio which provides the necessary indicators of a balanced change of magnetic conductivity and power.
4. We have explained a general methodology of the algorithm of the succession of actions to project a wind and energy station with a non-transmission contactless arc-shaped-stator inductor-type electrical and mechanical convertor with a coaxial ring-shaped rotor. All sizes starting from radial

sizes of blades with permissible wind velocity and finishing with the values of specific power and linear load are calculated by means of analytical equations.

5. The use of a combined construction of wind pumping system of back-and-forth motion with a magnetic and electrical linear generator enables us to offer possibilities of its effective controlling the useful electrical load when the wind speed increases and to use the stored energy to start the pump with the help of a linear electrical machine when the weather is calm and windless.
6. We have outlined new principles of designing an electrical and technical complex of a wind system with combined energy and water supply which exploits a magnetic and electrical linear generator with back-and-forth motion such as the use of mechanical energy of a wind engine to supply with water and to generate electrical energy which enlarges the functionality of such systems and enhances their energy efficiency.
7. We used mathematical modelling to balance mechanical energy of a wind engine to simultaneously supply with water and generate electrical energy which makes it possible to enlarge the functionality of wind stations and enhance their energy efficiency that is the calculated value is *62.3%*. It is the value of the energy conversion efficiency of a wind power system with combined electricity supply and water pumping with the water pumping mode which equals $2 \cdot 10^{-3} m^3 / s$ and the pressure which equals $0.29 \cdot 10^6$ *Pa*, and the producing electrical power which equals *1 kW*. These values are higher by *28.6%* in comparison with a conventional wind energy-supplying and water-pumping system.

ACKNOWLEDGMENT

The authors are deeply grateful to the editors of the book for the Prof. Valeriy Kharchenko and Pandian Vasant and the reviewers for valuable comments that were taken into account in the preparation and finalization of the chapter.

REFERENCES

Amirat, J., Benbouzid, M. E. H., Bensaker, B., & Wamkeue, R. (2007). Generators for wind energy conversion systems: State of the art and coming attractions. *Journal of Electrical Systems*, *3*(1), 26–38.

Benbouzid, M. E. N., Beltran, B., Mangel, H., & Mamoune, A. (2012). A high-order sliding mode observer for sensorless control of DFIG – based wind turbine. IEEE IECON, 7670 – 7674.

Chen, J., Chen, J., & Gong, C. (2014). On optimizing the aerodynamic load actingon the turbine shaft of PMSG – bared direct-drive wind energy conversion system. *IEEE Transactions on Industrial Electronics*, *61*(8), 4022–4031. doi:10.1109/TIE.2013.2284148

Choi, D. W., Byun, S. I., & Cho, Y. H. (2014). A study on the maximum power control method of switched reluctance generator for wind turbine. IEEE Trans. Magnetics, 50(1).

Hrebenikov, V.V., Petrenko, A.V. (2009). *Mahnitoelektrychnyi liniinyi henerator* [A Magnetic electrical linear generator]. Patent of Ukraine, IPC (2009) H02K 35/00, N° 41883; publ. 10.06.2009. Bul. N° 11.

Jebaseeli, E. A. E., & Susitra, D. (2010). *Performance Analisis of various configurations of switched reluctance machine for wind energy applications.* IEEE RSTSCC.

Kozyrskyi, V.V., Kaplun, V.V., Petrenko, A.V. (2007). *Vitroustanovka* [A Wind system]. Patent of Ukraine, IPC (2006) F03D 11/00, N° 23461; publ. 25.05.2007. Bul. N° 7.

Lebsir, A., Benbouzid, M. E. N., Bentonnsi, A., & Salah, I. B. (2016). Switched reluctance generator models for wind turbine systems. *ELECTROMOTION*, *23*, 3–9.

Mueller, M. A. (2005). *Design and performance of a 20 kW, 100 rpm, switched reluctance generator for a direct drive wind energy converter.* IEEE IEMDC. doi:10.1109/IEMDC.2005.195701

Podhurenko, V. S. (2007). *Netradytsiini dzherela enerhii: navch. posib* [Alternative energy sources: textbook]. NUC, Mykolaiv.

RMA "Південне". (1999). Vitroelektrychnyi ahrehat ВЕУ-500 [A wind power station ВЕУ-500]. Prospect. Dnipropetrovsk.

Sergeev, V.D., & Kuleshov, E.V. (2001). Optimizatsiya magnitoelektricheskih sinhronnyih vetrogeneratorov [The improvement of magnetic and electrical synchronic wind generators]. *Trudyi dalnevostochnogo gosudarstvennogo tehnicheskogo universiteta*, (129), 243-246.

Sergeev, V. D., Kuleshov, E. V., Proskurenko, S. S., & Chernyshova, A. S. (2000). Metodika rascheta magnitoelektricheskogo sinhronnogo vetrogeneratora [The methodology of calculating a magnetic and electrical synchronic wind generator]. *Trudyi dalnevostochnogo politehnicheskogo instituta im. V.V. Kuybyisheva,* (127), 181-185. Retrieved from https://elibrary.ru/item.asp?id=13042317

Trehub, M.I. (2004). *Lopat vitrodvyhuna* [The blade of a wind engine]. Patent of Ukraine, IPC F03D 1/06, F04D 29/30, N° 68628; publ. 16.08.2004. Bul. N° 8.

Trehub, M.I., Kozyrskyi, V.V., Hrebenikov, V.V., Pryimak, M.V. (2013). *Aksialnyi mahnitoelektrychnyi henerator* [An axial magnetic and electrical generator]. Patent of Ukraine, IPC H02K21/12, H02K21/24, N° 79084; publ. 10.04.2013. Bul. N° 7.

Viswanadh, M.M.G., Sekhar, A.S.R. (2012). Minimization of power loss and improvement of voltage profile by optimal placement of wind generator in distribution network. *International Journal of Engineering Research and Applications, 2*(6), 987-993.

Yahno, O.M., & Taurit, T.H., & Hrabar I.H. (2002). Vetroenergetika: konstruirovanie i raschet VEU: ucheb. posob [Wind energy: the design and calculation of WES: textbook]. *ZhSTU, Zhytomyr,* 255.

KEY TERMS AND DEFINITIONS

Aero Mechanical System: The system of the rotary aerodynamic elements connected to shaft which transforms kinetic wind power to energy of rotation of a shaft.

Arc-Shaped Stator: Not rotation part of the electrical generator.

Autonomous Wind and Energy Systems (AWES): An electrical and technical system which transforms kinetic energy into electrical and supplies a separated consumer with electricity and which is not a part of a total energy system.

Combined Electrical and Water-Supplying Wind Station (CEWWS): A system which transforms kinetic wind energy into mechanical back-and-forth motion one to start a magnetic and electrical linear generator and a plunger (piston-type) pump to supply water and electricity.

Electromechanical Transformation: The transformation of mechanical energy of a wind engine into electrical energy by an electrical generator.

Magneto-Electrical Linear Generator (MELG): An electrical generator of back-and-forth-motion with excitation caused by permanent magnets.

Non-Transmission Wind and Energy System: A system which transforms kinetic wind energy into a different type of energy without a mechanical link between an aeromechanical and an electromagnetic system.

Ring-Shaped Rotor: The rotation part of the electrical generator.

Wind and Electric Stations (WES): A station which transforms kinetic energy into different energy.

Chapter 10
Factors Affecting the Power-Efficient Operation of Wind-Diesel Complexes

Dmitry Chernov
Moscow Power Engineering Institute, Russia

Galina Deryugina
Moscow Power Engineering Institute, Russia

Nikita Karpov
Moscow Power Engineering Institute, Russia

Mikhail Tyagunov
Moscow Power Engineering Institute, Russia

Haiyang He
Moscow Power Engineering Institute, Russia

ABSTRACT

The purpose of this chapter is to compare the design and operational parameters of energy efficiency (power generation, capacity utilization factor) of a wind power plant (WPP) in order to improve the accuracy of the design parameters estimation at the preliminary design stage. Recommendations have been given for accounting different factors influencing on energy efficiency and economic indicators of a WPP. In this chapter, the research is carried out on the example of existing wind-diesel complexes (WDC) in the settlement of Ust-Kamchatsk (Kamchatka Krai), Novikovo (Sakhalin Oblast), and projected WDC on Popova Island (Vladivostok city district).

DOI: 10.4018/978-1-5225-9179-5.ch010

INTRODUCTION

One of the significant factors in the development of distributed energy systems with renewable energy sources (RES) is to determine the mutual influence of various parameters of generating, transmitting and consuming power units on the local power system mode (or load node) (Afonin et al, 2012; Deryugina et al, 2012; Kovalenko et al, 2014).

At present, there is a rather large amount of calculated data on the efficiency of the use of power plants based on RES as a part of the generating sources of local power systems (Scott et al, 1984; Stiller et al, 1983; Katiraei et al, 2007). However, the design and operation of wind-diesel complexes (WDC) for the conditions of Russia have been poorly studied and almost not investigated, since only in recent years several WDC have been built, for which there is little experience of their operation (Gevorgian et al, 1999; IRENA, 2017; Gsanger et al, 2017). Technical solutions in the design and operation of the WDC currently do not have a unified methodology for assessing the technical and economic efficiency of their use. At the preliminary stage of design, various assumptions and simplifications are adopted in modeling the wind resource. Design errors can be detected only during the operation stage. There are many different influencing factors, the neglect of which leads to significant discrepancies in the design and operational parameters of the WDC energy efficiency.

Influencing factors can be attributed to 3 groups:

- Inaccuracies in assessing the wind potential of the area - this is influenced by the choice of the information source and the methodology for modeling wind speed at the location of wind power plant at the hub height;
- Assumptions and neglection of various factors in estimation of design energy output;
- Not taking into account of limitations imposed by the operator of the local power system.

SOURCES OF WEATHER INFORMATION FOR THE SITE OF WPP "UST-KAMCHATSK"

The data of the met mast (wind-measuring complex, WMC) installed on the site, previously chosen for the location of the wind power plant, was used as the main information source. The WMC data are compared with the archive long-term data of the nearest ground-based meteostation (MS) "Ust-Kamchatsk" from the "Reliable Prognosis" website (n.d.).

Basic Information About WMC

The wind regime of the terrain can change at short distances. Such orographic elements such as hills, forests, mountains, steep embankments, as well as sea and lakes influence the local wind regime. Therefore, in accordance with international design standards for wind farms, it's necessary to perform a wind monitoring cycle in the area of the proposed construction for:

- Obtaining accurate and reliable wind characteristics and a correct estimate of the wind conditions of the terrain;
- Feasibility of the choice of the wind turbine model;

- Determining the optimal layout of the wind power plant;
- An accurate assessment of the energy output of the wind farm.

Wind measurements are carried out in accordance with international standards for measuring wind speed for estimating wind potential and WPP designing (IEC 61400-1, IEC 61400-12, IEC 61400-13, IEC 61400-14, etc.).

Wind monitoring cycle is performed using a set of meteorological instruments installed at certain altitudes at WMC. The WMC site is located near the settlement of Ust-Kamchatsk in the Ust-Kamchatsky District of the Kamchatka Territory. It is located between the main part of the settlement and the micro-district "Pogodny", near the substation "Pogodnaya", which provides power supply to the settlement. The location of the WMC was identified as the best way to meet the conditions: ensuring WMC power supply (the necessary condition of installed heated sensors) and maximum openness of the site (no obstacles).

WMC is located in an open area. In the south-west direction of the WMC is the substation «Pogodnaya», but since this direction is not the main direction of the wind (figure 1), the effect of shading from the substation on the measurement results is insignificant. In other directions, there are no obstacles. Over the observation period, north-north-east and south wind directions predominate (Figure 1).

The nearest neighborhoods of the WMC site are shrub vegetation ranging in height from 2.5 to 3.5 m. The influence of the surrounding vegetation on the wind shear profile was taken into account when analyzing the results of wind measurements. The height of the WMC base relative to the sea level is 1 m. WMC "Ust-Kamchatsk" (tubular type, height 34.5 m) was installed on June 23, 2010. Wind measurements are carried out to the present. The mast is equipped with anemometers at heights of 34, 21, 11 meters and wind vanes at heights of 31 and 18 meters. At heights of 34 and 21 m, 2 anemometers are installed to ensure reliable data measurement. Anemometers NRG # 40C and IF3 (heated), wind vane # 200P and IF3 (heated) are installed. Based on wind monitoring data, the wind speed averaged over 3 years at a height of 34 m was 6.2 m/s.

Specialized Database (SDB) "Reliable Prognosis"

The specialized "Reliable Prognosis" database presents climate information on the weather in 152 countries, of which 46 countries are placed on separate sites. In general, this source presents a weather forecast for the next 6 days and information about the actual weather. Forecasts are compiled using the UKMET MetOffice model (UK) and are provided on the site under the appropriate license. For the selected settlement from the archive of data it is possible to obtain information for the period of time (from January 1, 2005 to the present) on all days or only for the specified month or date.

The nearest MS to the settlement of Ust-Kamchatsk is MS «Ust-Kamchatsk» (56°14.22' N, 162°41.52' E). The MS is located at the airport of the settlement of Ust-Kamchatsk and is located at an altitude of ~18 m above sea level. The distance to WMC is about 8 km. Time series data from the SDB "Reliable Prognosis" have a measurement height of 10 m and are available for a period of 13 years (2005-2017).

As a primary information source on the site of wind power plant (WPP) as a part of WDC in Ust-Kamchatsk settlement can be used measurements of met mast. Also near the settlement of Ust-Kamchatsk there is a meteorological station (MS), which can be used to obtain long-term series of observations for the period from 2005 to the present at an altitude of 10 m.

When designing the WPP, as a rule, the design energy output of WPP is estimated for the long-term average year. It is analyzed whether the met mast data refer to the conditions of the long-term average

Figure 1. Distribution of wind speed in directions according to WMC data for the period from July 2010 to June 2013 (measuring height 34 m)

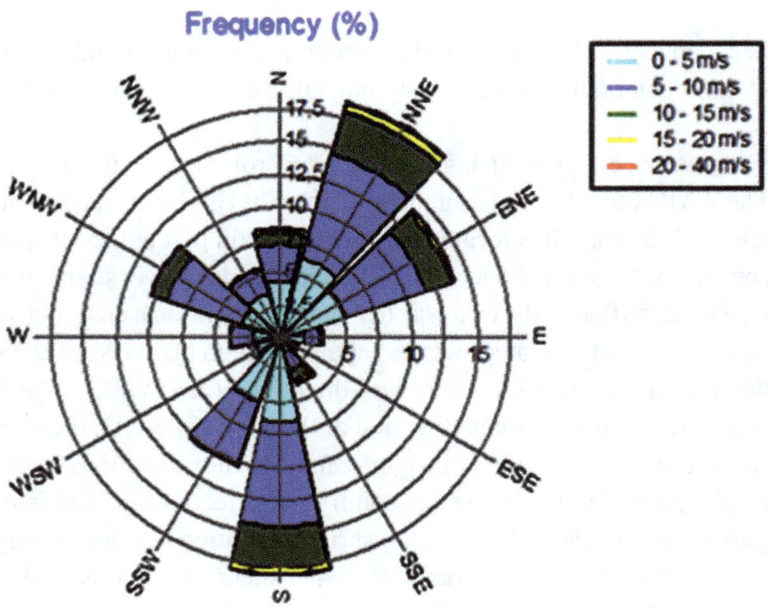

year. The year of observations is chosen under conditions close to the long-term average one. The choice of the characteristic year is made among the years with the completeness of the series not less than 95% (MEASNET et al, 2016), according to the criterion of the minimum relative deviation of the average annual wind speed from the long-term annual average.

The Table 1 presents the average annual wind speeds at an altitude of 10 m, the completeness of the annual series and the relative deviations in the values of the average annual wind speed from the long-term annual average one at the site of the MS «Ust-Kamchatsk».

The long-term variation of the average annual wind speed at the MS «Ust-Kamchatsk» site is shown in the Figure 2.

Figure 2 shows that the average annual wind speeds change insignificantly in the range of ±10%. Relative deviation of the average annual wind speed for 2016 (the year of observation at the met mast) from the long-term average annual is 5%.

The data on the met mast and MS sites were compared at a measurement height of 10 m for the same period of operation of the WPP, when the operation of the WPP was not limited by the local network operator - from 04.02.2016 to 31.10.2016, assuming that the data at 10m and 11m are the same.

Monthly average values of wind speed at a height of 10 m at the sites of met mast and MS are shown in the Table 2.

The analysis of the results revealed that the relative deviations of the average monthly wind speed values for MS and met mast ΔV_{0i} do not exceed 10%, which is comparable to the accuracy of anemometers of MS and met mast. One of the reasons is the different accuracy of the measuring instruments on the MS (± 0.5 m/s) and met mast (± 0.1 m/s), as well as the influence of various local factors, for example, slightly different terrain conditions. Therefore, MS data can be used on the met mast site for a

Table 1. Analysis of the 13-year series of observations at the site of the MS «Ust-Kamchatsk»

Year	Δt, hrs	Completeness of Data Series, %	V_k	δ_v,%	σ_r, %
2005	3	85,45	-	-	-
2006	3	96,97	4.36	-6.32	3.2
2007	3	99,17	4.11	-0.16	3.2
2008	3	97,12	4.08	0.46	3.0
2009	3	98,58	4.02	1.89	1.2
2010	3	98,82	4.05	1.33	2.7
2011	3	99,07	3.85	6.00	4.1
2012	3	98,86	3.99	2.61	3.0
2013	3	99,75	4.32	-5.33	3.7
2014	3	99,93	4.14	-1.05	3.6
2015	3	99,79	3.84	6.40	3.1
2016	3	99,90	3.90	4.89	3.5
2017	3	99,86	4.54	-10.73	9.6
Long-term annual average	3		4.10	-	-

Figure 2. The long-term variation of the average annual wind speed at the MS «Ust-Kamchatsk» site

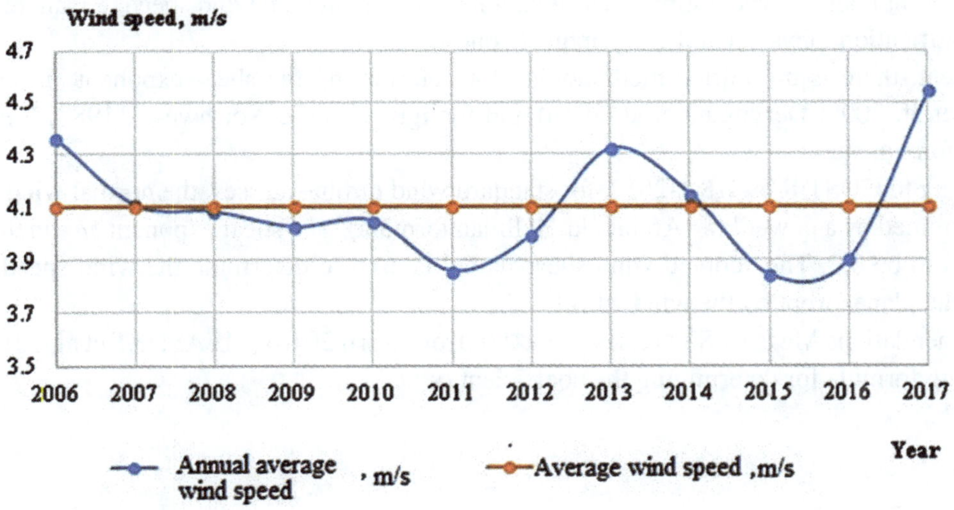

Table 2. Comparison of monthly averages of wind speed for MS and met mast

Month	February	March	April	May	June	July	August	September	October
Met mast, V_{0i}, m/s	3,88	3,56	4,55	3,65	4,26	4,46	3,55	3,96	3,45
MS, V_{0i}, m/s	3,57	3,56	5,11	3,94	4,54	4,29	3,44	4,08	3,12
ΔV_{0i}, %	-8,7	0,2	11,0	7,2	6,2	-4,0	-3,4	2,8	-10,5

long period. The analysis also revealed that the year of observation at the met mast site in its intensity is close to the long-term average year on the MS site, which allows using its data to estimate the design energy output of WPPs for the average long-term conditions.

INFLUENCE OF THE CHOICE OF THE WIND SHEAR PROFILE MODEL

The change in wind speed in height depends on many factors and is usually modeled by a power-law or logarithmic formula. For the near-surface and boundary layers of the atmosphere, these issues were considered in many domestic (Nikolaev et al, 2008; Deryugina et al, 2016) and foreign ("Pacific Northwest", 1987; Wegley et al, 1980) studies of scientists. They show that wind shear profiles in the atmosphere layer up to a height of several hundred meters are more accurately modeled by a power-law formula than a logarithmic formula. Therefore, in this paper models of the wind shear profile (WSP) are considered in terms of the power-law dependence represented by the empirical dependence:

$$\frac{V(Z_{hub})}{V(Z)} = \left(\frac{Z_{hub}}{Z}\right)^{m} \tag{1}$$

where $V(Z_{hub})$ and $V(Z)$ are the wind speed respectively at hub height Z_{hub} and Z above ground level at the point under consideration; – shear exponent, depending on wind speed, terrain relief, atmospheric stratification, physical and geographical conditions, etc.

At present, there is no unified methodology for determining the shear exponent m in domestic (Nikolaev et al, 2008; Deryugina et al, 2016) and foreign ("Pacific Northwest", 1987; Wegley et al, 1980) and foreign studies.

According to GOST R 54418.1-2012, for standard wind turbine classes, the normal wind shear profile is determined by a power law (Archibald-Hellman formula). The shear exponent m (in the standard α) is taken to be 0,2. The adopted wind shear model is used to determine the wind speed along the normal in the plane swept by the wind wheel.

Recommendations Measnet Site Assessment 2.0 (from April 2016) (MEASNET et al, 2016) suggest the following formula for determining the coefficient α:

$$\alpha = \frac{\ln\left(\frac{V_1}{V_2}\right)}{\ln\left(\frac{Z_1}{Z_2}\right)} \tag{2}$$

where: V_1 and V_2 – the wind speed measured at a height of Z_1 and Z_2, respectively.

According to paragraph 8.1. of European Wind Atlas (Troen et al, 1989), at high wind speeds the wind profile over flat and reasonably homogeneous terrain is well modeled using the logarithmic law:

$$U\left(Z\right)=\frac{U_{*}}{k}\cdot\ln\frac{Z}{Z_{0}} \tag{3}$$

where:

- $U\left(Z\right)$ is the wind speed at height;
- Z above ground level;
- Z_0 is the surface roughness length;
- k is the von Karman constant, taken here as 0,4 and;
- U_* is the so-called friction velocity related to the surface stress;
- τ through the definition:

$$\left|\tau\right|=\rho\cdot U_{*}^{2} \tag{4}$$

Thus, it can be concluded that now there is no unified methodology reflected in state standards that allows recalculating wind speeds for minor intervals from one height to another.

Modern Methods for Modeling Wind Speed in Height

Modern methods for modeling wind speed in height are presented in the Table 3.

When designing a wind power plant as part of a local energy complex, it's necessary to take into account the time factor, i.e. dependence of the parameter *m* on the effect of mesoscale phenomena on the development of heat and moisture exchange processes, turbulent friction occurring in the boundary layer of the atmosphere. Such studies require a time series of synchronous observations at various levels, which can only be obtained from a met mast. The known models of the WSP developed earlier are valid for the characteristics of a multi-year regime in the form of an average wind speed, for which the time factor does not play a big role. However, currently remain open issues related to the mathematical description of average wind speeds at various altitudes of the surface layer of the atmosphere (Ryholv et al, 2011).

The most reliable data on wind shear profiles can be obtained using WMC. The advantage of WMC data over aerological is the possibility of synchronous observations at different heights of the placement of the same type of meteorological devices.

A comparative analysis of the applicability of various models of the wind shear profile in accordance with the power law at the site of the WPP in the settlement of Ust-Kamchatsk is carried out. Verification of the models is carried out by comparing the actual energy output of the WPP in the settlement of Ust-Kamchatsk with the energy output determined by the modeled wind speed series. The calculation of the WPP energy output is carried out over a period of time from 04.02.2016 to 31.10.2016, when there were no restrictions on the operation of the WPP by the local power system operator in Ust-Kamchatsk.

In the paper, 5 models of the wind shear profile are considered, of which 3 models were obtained from met mast data in Ust-Kamchatsk, taking into account and without taking into account the time factor (Alihodgina et al, 2017) and 2 models, obtained from the AS data of the Kamchatka Peninsula, without taking into account the time factor (Nikolaev et al, 2008). Below is a brief description of these models:

1. Model of the wind shear profile based on the average monthly wind speed at the minimum and maximum height of measurements for the placement of WMC «Ust-Kamchatsk» (Model N°1).

Model N°1 was developed from the calculated values of the monthly averages: m_l (l is the number of the month) by the formula (1) and the monthly average wind speed V_l at the minimum WMC measurement height. An approximating dependence $m_l\left(V_l\right)$ was obtained with the correlation coefficient ($R = 1$) in the form:

$$m_l = 0,679 \cdot V_l^{-0,44} \tag{5}$$

2. Model of the wind shear profile based on the average values of the wind speed for the period of wind measurements on the sites of 6 WMCs of the Kamchatka Peninsula (Model N°2).

The model N°2 was based on data of 6 WMC: Ust-Kamchatsk, Ust-Khayryuzovo, Palana, Tilichiki, Manila, Ossora. The values of the average wind speed from the data of unheated anemometers at the minimum and maximum height of the WMC and the m_0 exponents calculated from them are shown in the Table 4.

Based on the results of the calculation, an approximating dependence $m_0(V_0^{h_1})$ was obtained with the correlation coefficient 0,7 (Model N°2):

$$m_0\left(V_0^{h_1}\right) = 0,142 \cdot \left(V_0^{h_1}\right)^{0,339} \tag{6}$$

Comparison in the form of δ_V and δ_m values of the exponent m_0^M and the values of the average wind speed at the maximum height of the WMC measurements $V_0^{h_2,M}$, calculated from Model N°2, with the calculated from the WMC data is presented in Table 4.

3. Model of the wind shear profile for 10-minute wind speed values in WMC «Ust-Kamchatsk» conditions (Model N°3), taking into account the time factor.

As Model N°3, the dependence of the exponent m on the turbulence intensity I and the averaged 10-minute wind speed V is proposed as:

$$m\left(I,V\right) = c\left(I\right)V^{d(I)}, \tag{7}$$

where the coefficients $c\left(I\right)$ and $d\left(I\right)$ are determined by the formulas (10) - (11) developed by the authors (Alihodgina et al, 2017) depending on the intensity of turbulence and the physico-geographical conditions of the WMC:

$$c\left(I\right) = 17,98 \cdot I^6 - 55,07 \cdot I^5 + 67,21 \cdot I^4 - 41,49 \cdot I^3 + 13,45 \cdot I^2 - 2,097 \cdot I + 0,549 \tag{8}$$

Table 3. Basic approaches and methods of modeling and approximation of wind shear profiles V(h) in the surface 100-200-meter layer of the atmosphere (where h - hub height, h_0 - height above ground level)

Nº	Name of the Methodology, Developer	Used Meteorological Data	Used Aerological Data	Approximation of the Distribution f(V)	Approximation of the Wind Shear Profile
1	Universal model	Nearby MS (2200 MS of RF)	Not used	Any known, appropriate to speed and height	$V(h) = V_0 \cdot (h / h_0)^m$ $m = 1/7$ for land $m = 1/10$ for sea
2	WASP, RISO, Denmark	Nearby MS + Z_0 model (332 MS of RF)	Not used	Weibull function of meteorological data with the rise to a height	$V(h) = V_o \cdot \ln(h / h_o)$
3	NASA, NREL, USA	Model with a resolution of 1x1 km	Not used	Rayleigh function of meteorological data	$V(h) = V_0 \cdot (h / h_0)^m$
4	TSAGI, 1960	Nearby MS-analogues (≈500 MS)	Not used	Functions of Kolodin and Pomortsev	$V(h) = V_0 \cdot (h / h_0)^m$
5	GGO, 1989	Nearby MS (≈1100 MS of RF)	Not used	G-functions from the nearest MS	$V(h) = V_0 (h / h_0)^m$
6	FLUGER 3.0, TSAGI, 1994	Nearby MS of class 7b & 6b (≈1000 MS of RF)	V at the level of 100 m of nearby MS	Tabulated regional functions of Grintsevich	$V(h) = V_0 (h / h_0)^m$ $m = \ln(V_{100} / V_0) / \ln(H_{100} / H_0)$
7	JSC «VIEN», 2002	Nearby MS (≈1100 MS of RF)	Not used	The shear of empirical repetitions	$V(h) = V_0 (h / h_0)^m$
8	FLUGER 2000, SIC «ATMOGRAPH»	Nearby MS of class 7b (≈1000 MS of RF)	V at the level of 100 m of region's AS	Tabulated regional functions of Grintsevich	$V(h) = V_0 (h / h_0)^m$
9	FLUGER XXI, SIC «ATMOGRAPH»	Average statistics for the area with clearing (≈1100)	V at the level of 100, 200, 300 and 600 m of all AS in region	Tabulated regional functions of Grintsevich	With the modeling $V_{land}(h) = V_A \cdot \ln(h / h_c)$ $V_{sea}(h) = V_M \cdot \ln(h / h_M)$
10	NRU "MPEI", Department of HRES	Nearby MS from the internet site "Reliable Prognosis" (≈1341 MS of RF) taking into account landscape features	The long-term average monthly V at the level of 10 m and 100 m of the nearby AS	Weibull function of meteorological data with the shear to a height according to AMS data	$V(h) = V_0 (h / h_0)^m$, where m is determined by the empirical dependence of the nearby AS or AS group of the region

Table 4. Model N°2: comparison of model and actual indicators

Name of WMC	$V_0^{h_1,a}$, m/s	$V_0^{h_2,a}$, m/s	m_0	m_0^M	$V_0^{h_2,M}$, m/s	δ_V, %	δ_m, %
Manila	4,5	5,0	0,123	0,236	5,5	-10,1	-91,7
Ossora	3, 5	4,4	0,203	0,217	4,5	-1,6	-7,0
Palana	2,8	3,6	0,223	0,201	3,5	2,4	9,7
Tilichiki	3,9	4,5	0,127	0,223	5,0	-11,5	-75,9
Ust-Kamchatsk	4,1	6,0	0,337	0,229	5,3	11,5	32,1
Ust-Khayryuzovo	4,5	5,7	0,209	0,236	5,9	-3,0	-12,7

$$d\left(I\right) = -164,5 \cdot I^6 + 504,2 \cdot I^5 - 601,6 \cdot I^4 + 350,4 \cdot I^3 - 101,3 \cdot I^2 + 12,98 \cdot I - 0,795 \qquad (9)$$

where I is the intensity of turbulence, is determined from the empirical dependence:

$$I = aV^b, \qquad (10)$$

where $a = 0,18$ and $b = -0,07$ are the approximation coefficients for the physico-geographical conditions of the WMC "Ust-Kamchatsk".

4. Wind shear profile according to aerological station (AS) (Model N°4)

The wind shear profile on the Ust-Kamchatsk WPP was obtained from the data of the AS-analogue "Klyuchi". Selection of an AS-analogue of the WMC site in question was performed in accordance with the methodology developed at the Department of HRES of NRU «MPEI», from the specialized database (SDB) «Wind shear profile» (Vaskov et al, 2012).. The approximate dependence between m_l and monthly mean of wind velocity V_l on the height 10m have been identified as:

$$m_0\left(V_0^{h_1}\right) = 0,76 \cdot \left(V_0^{h_1}\right)^{-0,95}. \qquad (11)$$

5. Dependence of the wind shear profile for the conditions of the Far East of Russia (Model N° 5)

In the SDB «Wind Shear Profile» data are presented on the sites of 40 AS of the Far Eastern Federal District of Russia (FEFD). As a result of statistical processing of these data, an empirical dependence of the mean annual Hellman coefficient m_0 on the average annual wind speed V_0 with a correlation coefficient $R = 0,8794$ was obtained:

$$m_0 = 0,6827 \cdot V_0^{-0,914} \qquad (12)$$

which confirms a good correlation between the parameters considered.

Estimation of Theoretical Energy Output at the Hub Height of the WTG as Part of the WPP «Ust-Kamchatsk»

Since the initial information on wind conditions by the MS data is given at an height of 10 m, and by WMC data at heights of 11, 21 and 34 m, it is necessary to shear the wind speed to a height either by the models of the wind shear profile described above, either by using specialized software WindPRO by WMC data (using a matrix of coefficients). In the above paragraphs it was shown that WMC data are typical for the average long-term conditions and were accepted as calculated.

Theoretical energy output for the period from 04.02.2016 to 31.10.2016 was estimated by WMC data on the power curve of the Komai wind turbine using a matrix of wind speed shear coefficients from an altitude of 11 m to a height of 41.5 m (a total of 144 coefficients depending on the time of day and month in a year).

Similarly, an estimation was made of the energy output by the power curve of the Komai wind turbine according to the models of the wind shear profile described above (see Table 5).

Also for this period of time was estimated the actual energy output of 3 wind turbines Komai.

Analysis of the data also showed that for the wind shearing to the hub height for various presented source data, a model is required for the averaging interval corresponding to the initial data. The elements of the coefficient matrix were obtained for a time interval $\Delta t = 10$ min, while the energy output by model N°3 is understated by 3,5% (interval $\Delta t = 10$ min), and by average and annual average values to $\pm$ 25% (model N°1 and N°2, respectively).

The topographic conditions of the AS differ from the topographic conditions of the WMC, so the energy output is reduced to 30% (similarly for the FEFD model, as a generalized model for all AS FEFD).

Influence of the Choices Calculation Year on the Design Energy Output

As a rule, the design output of WPP is estimated for a series of meteorological measurements of WMC, which in the long-term profile may not always correspond to the long-term average year. If there are significant variations in wind speed in the long-term section, significant errors may occur in estimation of the design energy output. The influence of the choice of the calculation year on the value of the design WPP energy output was estimated by the example of the WDC "Ust-Kamchatsk". Calculation of energy output was performed based on the power curve of the WTG Komai KWT300 for the period of observations at the WMC, at those times when the operator of the local power system did not impose limitations on the output of the WPP (04.02.2016 - 31.10.2016).

Comparison of wind speed data for WMC and MS at a height of 11 m and height of 41.5 m for the observation period from 04.02.2016 to 31.10.2016 is given in Table 6. It can be seen that the discrepan-

Table 5. Electrical generation from a single WTG by Komai according to various WSP models

	The Coefficient Matrix	Model N°1	Model N°2	Model N°3	Model N°4	Model N°5
WPP energy output, MWh	498,3	597,75	383,61	480,85	358,03	344,88
Relative deviation, %	-	-19,96	23,01	3,50	28,15	30,79

cies in the values of the annual average wind speed at the WMC and MS sites at a height of 41.5 m do not exceed 5% (which is comparable to the accuracy of measuring instruments whose accuracy at WMC is 0.1 m/s and at MS is ± 0.5 m/s), and the difference in the energy output by a single WTG Komai KWT300 is 11%. Therefore, to analyze the effect of the choice calculation year on the energy output value, homogeneous data obtained at the MS site should be used. Table 7 presents data on the annual average wind speed at the MS site at heights of 11 and 41.5 m for the same period of time (04.02.2016 - 31.10.2016) for 2013, 2015, 2016, 2017. For all these years, an estimate of the WTG Komai KWT300 energy output using power curve was also evaluated and compared with the energy output by the characteristic year (see Table 7).

Based on the calculations, the empirical dependence of the energy output of WTG E_{WTG} was constructed as a function of the annual average wind speed V_0 at the hub height of the WTG and its approximating dependence in the form:

$$E_{WTG} = 29,202 \cdot \left(V_0\right)^{1,605}, \quad (13)$$

where E_{WTG} is energy output of a single WTG, V_0 is the annual average wind speed at the hub height of the WTG.

Table 6. Comparison of wind speeds and energy output by WMC and MS data

Parameter Name	WMC Data	MS Data	Relative Deviation $\Delta_{WMC\text{-}MS}$, %
Wind speed at a height of 11m	3,92	3,98	-
Wind speed at a height of 41,5m	5,91	6,19	5
Energy output at a height of 41,5 m, MWh	498,3	557,5	11

Table 7. Comparison of the values of energy output by years of observation data

Parameter Name	WMC Data	MS Data			
		2016	2017	2013	2015
Wind speed at a height of 11m	3,92	3,98	4,63	4,15	3,74
Wind speed at a height of 41,5m	5,91	6,19	7,25	6,50	5,84
Energy output at a height of 41,5 m, MWh	498,3	557,5	695,5	590,6	485,6
Relative deviation of energy output from 2016 $\Delta Э^{MS}_{2016\text{-}i}$, %	-	-	24,7	5,9	-12,9
The relative deviation of wind speed from 2016 at a height of 11 m by the MS data $\Delta V^{MS}_{2016\text{-}i}$, % (11m)	-	-	14,0	4,1	-6,4
The relative deviation of wind speed from 2016 at a height of 41,5 m by the MS data $\Delta V^{MS}_{2016\text{-}i}$, % (41,5m)	-	-	14,7	4,83	-6,0

INFLUENCE OF THE AVERAGE AIR DENSITY ON THE POWER CURVE OF WTG

The density of air depends on the altitude of the area above sea level and the ambient air temperature, which must be taken into account when designing the WPP. The air density can be determined from the actual WMC (air temperature) and long-term data of the nearest MS, or calculated according to the ideal gas law for dry air:

$$\rho\left(t^{\circ},h\right)=3,4837\cdot\frac{P}{t^{\circ}}, \tag{14}$$

where t° is ambient temperature [$^{\circ}K$], P is atmosphere pressure [kPa], depending on the altitude above sea level, which is approximated to the height of 5000 m with a high degree of accuracy by the equation:

$$P=101,29-0,011837\cdot\nabla+4,793\cdot10^{-7}\cdot\nabla^{2} \tag{15}$$

where ∇ [m] is elevation above sea level (Deryugina et al, 2012).

In accordance with IEC 61400-12, if the average air density on the site differs by more than 0.05 kg/m^3 from 1.225 kg/m^3, it is necessary to correct the representation of the power curve.

On the site of the WPP "Ust-Kamchatsk" according to formulas (14-15), the mean air density for the observation period at the WMC was determined, which was 1.278 kg/m^3, which is more than 0.05 kg/m^3 from the standard value. Also, according to the long-term data of the nearest MS, the average air density was determined, which was 1.292 kg/m^3.

For the WTG Komai KWT300 with pitch-system, the power curve correction is performed only for wind speeds in the zone from the starting speed (V_{min}) to the nominal speed ($V_{@}$) according to the formula:

$$V\left(\nabla_{1}\right)=V\left(\nabla_{0}\right)\cdot\left(\frac{\rho_{0}\left(\nabla_{0}\right)}{\rho_{1}\left(\nabla_{1}\right)}\right)^{\frac{1}{3}}, \tag{16}$$

where $V\left(\nabla_{1}\right)$ is corrected wind speed on power curve, $V\left(\nabla_{0}\right)$ is wind speed by power curve.

According to the passport and adjusted power curves of WTG Komai KWT300 for the operated WPP "Ust-Kamchatsk" (see figure 3), energy output was determined according to the WMC data for the observation period from 04.02.2016 to 31.10.2016 (see Table 8).

Based on the calculations, the empirical dependence of the energy output of the WTG E_{WTG} was constructed as a function of the air density and its approximating dependence in the form:

$$E_{WTG}=327,78\cdot\rho_{0}+96,789, \tag{17}$$

where E_{WTG} is the energy output of WTG, ρ_{0} is the value of the air density.

Since the deviation of the actual average air density from the standard value can reach up to ± 15% (depending on the average temperature and altitude above the sea level), an incorrect consideration of this factor can lead to an error in the determination of the WPP energy output to ± 12% (Deryugina et al, 2017).

FACTORS AFFECTING THE SHARE OF DISPLACEMENT OF THE DPP ENERGY OUTPUT FROM THE LOAD CURVE OF THE LOCAL CONSUMER

An analysis of the effect of the limitations on the DPP regime in the composition of the WDC was carried out at 2 sites: the planned WDC on the Popov Island, and the operated WDC in Ust-Kamchatsk. WDC of Ust-Kamchatsk consists of 3 WTG Komai KWT300 with an installed capacity of 900 kW, working in parallel with the DPP, consisting of 10 DG with rated capacity of 800 kW. According to the questionnaires, the specific consumption of diesel fuel at the DPP is 375 g/kWh, the maximum load demand is 5400 kW, the minimum is 1130 kW, respectively. The share of capacity of the existing WPP in Ust-Kamchatsk is about 16% of the maximum load of the consumer.

The DPP on Popov island consists of 3 DG total installed capacity of 1700 kW (2 DG for 800 kW, 1 DG - 100 kW), the maximum load demand is 780 kW, a minimum of 170 kW.

Figure 3. Passport and adjusted for air density power curves of WTG

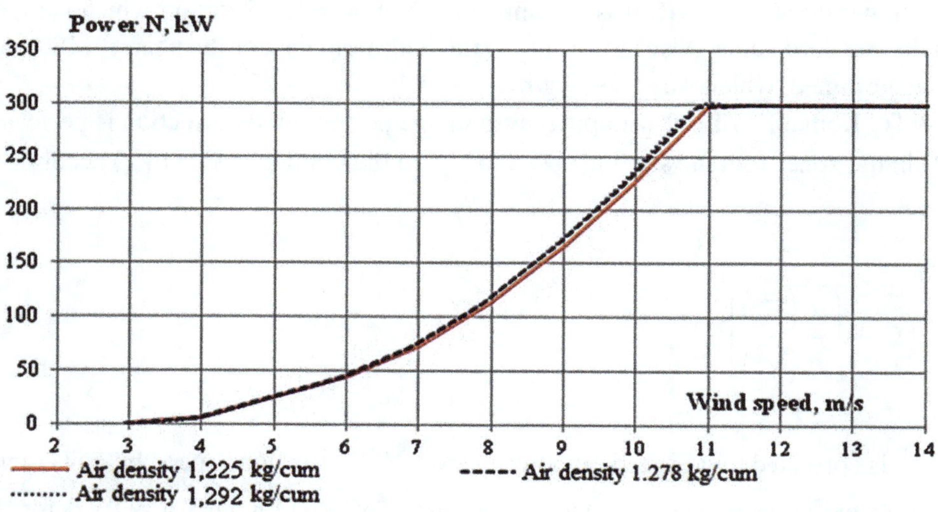

Table 8. The energy output of WTG according to WMC, MS and depending on the density of air

Air Density	Standard Atmospheric Conditions (1.225 kg/m³)	In Accordance With the Ideal Gas Law for Dry Air (1.278 kg/m³)	According to the MS Long-Term Data (1.292 kg/m³)
Energy output, MWh	498,3	515,8	520,2
Relative deviation from standard conditions, %	-	3,4	4,2

The impact of the actual fuel consumption on the energy efficiency of the WPP in the composition of the WDC was analyzed at the planned WDC on Popov island and the operated WDC in Ust-Kamchatsk; also at WDC «Ust-Kamchatsk», the effect of the minimum-load limitation of the DG on displacing the DPP from the load schedule, and, correspondingly, the economy of diesel fuel, was analyzed. The analyzes were performed in accordance with the equation of power balance of the local power system.

Description of the Local Grid Power Balance

The basis for calculating the power balance of the power system is the equality of the generated and consumed power, taking into account losses and own needs at any time t:

$$P_{auton.cons.}\left(t\right) \leq N_{WPP}\left(t\right) + N_{DPP}\left(t\right) - \Delta N_{aux}\left(t\right) - \Delta N_{loss}\left(t\right) \tag{18}$$

where *t* - the time interval, $P_{auton.cons.}\left(t\right)$ – total need of an autonomous consumer in electric power in the *t*-th hour, $N_{WPP}\left(t\right)$, $N_{DPP}\left(t\right)$ – the issuance of electric power for an autonomous consumer from the WPP and the DPP, respectively, $\Delta N_{loss}\left(t\right)$ – power loss value, $\Delta N_{aux}\left(t\right) = \Delta N_{aux\,WPP}\left(t\right) + \Delta N_{aux\,DPP}\left(t\right)$ – the power needed for auxiliaries (for the WTG and DPP losses for auxiliaries are equaled 8% of $N_{WPP}\left(t\right)$ and 10% of $N_{DPP}\left(t\right)$, respectively).

In this paper it is assumed that the time of switching on the DG for the load or its change (increase or decrease) is not taken into account, since it's less than 15-20 s, and the calculated time interval is 1 hour. The load range of the DG is taken from $N_{DG}^{\min} = \left[20\%\right] N_{DG}^{instal}$ to $N_{DG}^{\max} = \left[100\%\right] N_{DG}^{instal}$ (limitations on the permissible minimum and maximum load during long-term operation of the DG, recommended by the manufacturer).

When planning the qualitative and quantitative composition of WPP for feeding an autonomous consumer, it's necessary to take into account the limitations on the mode of operation of the main, reliable energy source - DPP, in order to save fuel and resource of the energy source.

On the basis of the foregoing, when drawing up the power balance of the autonomous consumer on equation (18), a limitation on the current power of the DPP $N_{DPP}\left(t\right)$:

$$N_{DG}^{\min} \leq N_{DPP}\left(t\right) \leq n_{DG} \cdot N_{DG}^{\max} \tag{19}$$

In addition to the limitation (19), it's necessary to take into account the main criteria for optimality of these energy calculations, i.e.:

- Minimum number of on/off DG per day (not recommended more than 2);
- Minimum consumption of diesel fuel for the calculation period;
- Minimum free energy by the WTG and DG for the calculation period $E_{free}\left(T\right) \rightarrow \min$.

Power Balance Calculation Algorithm

At the first stage, it's necessary to quantify the hourly average current power of the WTG $N_{WTG}(t)$. Quantification of the average current power of the WTG $N_{WTG}(t)$ is carried out according to the Komai passport characteristic $N_{WTG}(V)$ for known hourly wind speeds at the hub height of the WTG $V_{hh}(t)$.

Next, quantity the number of operating WTG $n_{WTG}^{oper}(t)$ at time t:

$$n_{WTG}^{oper}(t) = n_{WTG}\text{, if } \begin{cases} n_{WTG}^{P}, if\, n_{WTG}^{P}(t) \le n_{WTG}; \\ n_{WTG}^{P}(t) > 0; \\ 0, if\, N_{WTG}(t) = 0, \end{cases} \tag{20}$$

where $n_{WTG}^{P}(t)$ is the number of WTG required to cover the load curve in the *t*-th hour, excluding the share of the output of the DPP and operating with $N_{WTG}(t)$:

$$n_{WTG}^{P}(t) = \frac{P_{auton.cons.}(t) - N_{DPP}^{share}(t)}{N_{WTG}(t)} \tag{21}$$

Then the power of the WPP in the *t*-th time interval:

$$N_{WTG}(t) = \gamma \cdot n_{WTG}^{oper}(t) \cdot N_{WTG}(t), \tag{22}$$

where $\gamma = 0,92$, p.u. – technical power losses of WPP; $n_{WTG}^{oper}(t)$ – number of operating WTGs as a part of WPP.

The balance of power is verified by the formula (23):

$$dP_{auton.cons.} = P_{auton.cons.}(t) - N_{WTG}(t) - \left(N_{DPP}^{share}(t) - \Delta N_{DPP}^{share}(t)\right), \tag{23}$$

where $\Delta N_{DPP}^{share}(t) = 0,1\Delta N_{DPP}^{share}(t)$ – technical power losses and for the DPP auxiliaries.

Depending on the power ratio $P_{auton.cons.}(t)$, $N_{WTG}(t)$ and $N_{DPP}^{share}(t)$, there is a power deficit in the system or its excess, i.e.:

1. $dP_{auton.cons.} = 0$ – the power balance of the autonomous consumer is fulfilled, go to (t + 1) hour (go to step 1);
2. $dP_{auton.cons.} > 0$ – power shortage in the energy complex;
3. $dP_{auton.cons.} < 0$ – free power in the energy complex.

The annual energy consumption of the autonomous consumer $E_{auton.cons.}\left(T\right)$, as well as the annual actual energy output at the DPP and WPP ($E_{DPP}\left(T\right)$ $E_{WPP}\left(T\right)$), as well as the free energy $E_{free}\left(T\right)$ based on the following formula:

$$E\left(T\right)=\sum_{1}^{8760}N\left(t\right) \tag{24}$$

The annual volume of diesel fuel consumption at the DPP $B_{DPP}\left(T\right)$, as well as the amount of diesel fuel savings per year ΔB_{save}^{year} are quantified by the formulas (25) - (26):

$$B_{DPP}\left(T\right)=b_{spec}\cdot E_{DPP}\left(T\right), \tag{25}$$

$$\Delta B_{save}^{year}\left(T\right)=b_{spec}\cdot\left(E_{auton.cons.}\left(T\right)-E_{DPP}\left(T\right)\right). \tag{26}$$

where b_{spec} is specific fuel consumption;

Research of the Mutual Influence of the Consumption and Production Schedules of the WPP

An important indicator necessary for assessing the efficiency of the WDC is the amount of penetration of the WPP energy generation into the DPP energy generation. To calculate the efficiency and feasibility of construction of a WDC, a mathematical model of the WDC has been developed, with the help of which calculations have been made for WPP energy output penetration into DPP energy output according to the projected WDC on Popov Island.

Currently, the electricity supply to the Popov Island is from a DPP, which includes 3 diesel generators (DG) with a total capacity of 1.7 MW. Estimation of penetration effect of guaranteeing DPP is carried out for 6 variants of WTG (see Table 9) for 10- minutes-averaged data series of met mast from May 2009 to May 2010.

WPP installed capacity ranges (number of WTG) for different types of WTG considered in the calculation of penetration levels are presented in Table 10.

Figures 4 and 5 show the WPP energy penetration values and the free energy of the variants of the WTG in the entire range of power variation.

The analysis of the results showed that with an increase in the number of installed WTG (n_{WTG}), the share of penetration increases and for various of WTG is approaching from 47% to 50% of the maximum consumption, and free energy increases in direct proportion to the number of WTG in the WPP. Thus, the researchers can talk about the presence of some optimal number of installed WTG in which the effect of increasing the penetration is compensated by the negative effect of free energy growth.

Table 9. Information on the planned WTG at the site on the Popov Island

Parameter	KWT33-300	Vergnet GEV MPC 30-275	Vergnet GEV MPR 30-275	Vestas V27	GHRepower FD21-100	Vestas V29
Rated capacity, kW	300	275	275	225	100	225
Rotor diameter, m	33	30	30	27	21,5	29
Hub height, m	41,5	55	32	30	37	31,5

Table 10. Variants of WTG brands in the layout of the wind-diesel complex on the site of the Popov Island

WTG Brand	Number of Turbines
Komai KWT33-300, 300кВт	2-200
Vergnet GEV MP-R 30-275, 275 кВт	2 -300
Vergnet GEV MP-C30-275, 275 кВт	2 - 350
Vestas V27, 225кВт	2 - 200
GHREpower FD21-100, 100 кВт	2 -700

Influence of the Load Demand and Wind Resource Changes on the Amount of Fuel Economy of the Wind Power Plant

It is common knowledge that the electrical energy output from a WPP depends significantly on the intensity of the wind flow, but with excessive production of WPP and inadequate consumption, the energy output of WPP has to be limited.

Let us consider the effect of two factors on the amount of penetration of the energy output of the WPP into the DPP energy output in the load demand:

Figure 4. WPP energy penetration, % of annual consumption

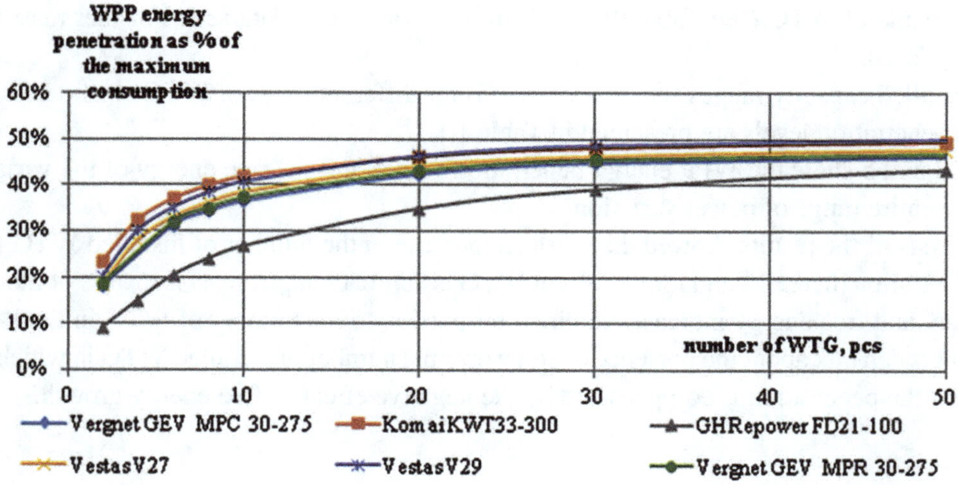

Figure 5. Free energy of WPP, % of annual consumption

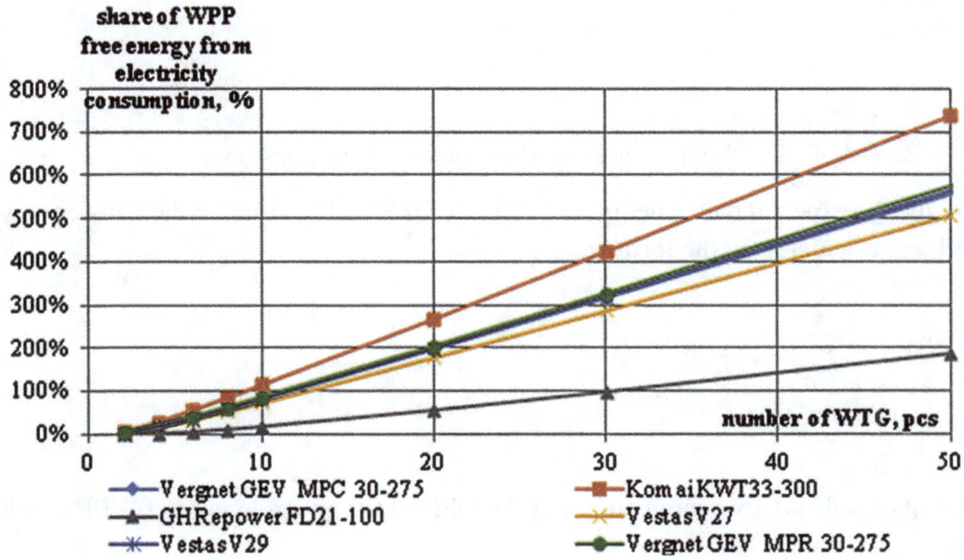

- Coincidence of the consumption schedule with the schedule of wind resource supply;
- Intensity of wind resources.

Let us first consider the influence of only the first factor, comparing the daily schedules of consumption and energy output of the WPP. As a criterion for the coincidence of the daily consumption schedule with the daily schedule of the wind resource, the researchers introduce the root-mean-square deviation for *j* day of the hourly values of the power consumption from the power of the WPP – r_j, calculated by the formula:

$$r_j = \sqrt{\frac{\sum_{i=1}^{24}\left(K_{N_{WPP_i}} - K_{N_{consum_i}}\right)^2}{24}}\left[p.u.\right], \tag{27}$$

where *i* (*i* =1, 2, ..., 24) is the number of the hour on the *j*-th day (*j* = 1, 2, ..., 365), $K_{N_{WPP_i}}, K_{N_{consum_i}}$ - the average hourly values of the power of the consumer and the power of the WPP relative to the average daily power values and calculated by the formulas for each *i*-hour of the *j*-day of the year:

$$K_{N_{WPP_i}} = \frac{N_{WPP_i}}{\bar{N}_{WPP}}\left[p.u.\right], \tag{28}$$

where $\bar{N}_{WPP} = \frac{\sum_{i=1}^{24} N_{WPP_i}}{24}\left[kW\right]$ - average daily power of WPP and

$$K_{N_{consum_i}} = \frac{N_{consum_i}}{\bar{N}_{consum}}, \tag{29}$$

where $\bar{N}_{consum} = \frac{\sum_{i=1}^{24} N_{consum_i}}{24} [kW]$ - average daily power of the consumer.

Accordingly, the expelled daily energy output of the DPP will be equal to the daily energy output of the wind turbine, calculated by the formula:

$$E^{DPP}_{expelled\,j} = E_{WPPj} = \sum_{i=1}^{24} N_{WPP_i} \Delta t_i [kWh], \tag{30}$$

where Δt_i =1 hour.

The share of expelled daily energy output of the DPP by the energy output of WPP is calculated by the formula:

$$E_{expelled_j} = \frac{E_{WPPj}}{E_{consumj}} \cdot 100\% [\%]. \tag{31}$$

Calculations were carried out according to data of WDC in Ust-Kamchatsk (Kamchatka Krai) and Novikovo (Sakhalin Region). The initial information was the actual time series of hourly-average values of electric energy consumption and the energy output of WPP: for the WDC in the settlement of Ust-Kamchatsk in 2017, obtained from the data of the ballast loading control system of the WDC and the collection of information from MicroGrid Systems; for the WDC of the village of Novikovo for 2017-2018, obtained from the data of the automated control system of the WDC of the village of Novikovo.

According to the calculations, empirical tabulated characteristics of the dependence of r_j on $E_{expelled_j}$ (see Figure 6 and 7) and approximating their dependences with a good correlation dependence were constructed:

- For WDC in the settlement of Ust-Kamchatsk: $r = 0,18 \cdot E_{expelled}^{-0,453}$ with $R = 0,7$;
- For WDC of the village of Novikovo: $r = 0,35 \cdot E_{expelled}^{-0,47}$ with $R = 0,89$.

The analysis of the dependences of r_j on $E_{expelled_j}$ revealed that the share of replacement of the energy output of the DPP by the WPP energy output depends on the coincidence of the consumption schedule with the schedule of the wind resource supply. Comparing the approximating dependences of r_j on $E_{expelled_j}$ for two objects (see Figure 8), the researchers can conclude that a significant discrepancy between these dependences is observed with a small share of energy output replacement of the DPP by WPP energy output (up to 5%) and depends primarily on the different character discrepancy in the consumption and production of WPP. With an increase in the share of DPP energy output replacement, the approximating dependences approach each other, i.e. the share of replacement of the DPP is primarily affected by the intensity of the wind flow.

Let us analyze the influence of the second factor - the intensity of the wind resource, which affects the daily energy output of WPP and, accordingly, the daily capacity utilization factor (*CUF*) of the WPP, which for the composition of the WPP under consideration will vary only from the intensity of the wind flow.

Figure 6. The empirical and approximating dependence of r_j on $E_{expelled_j}$ according to the data of WDC in the settlement of Ust-Kamchatsk, the Kamchatka Krai

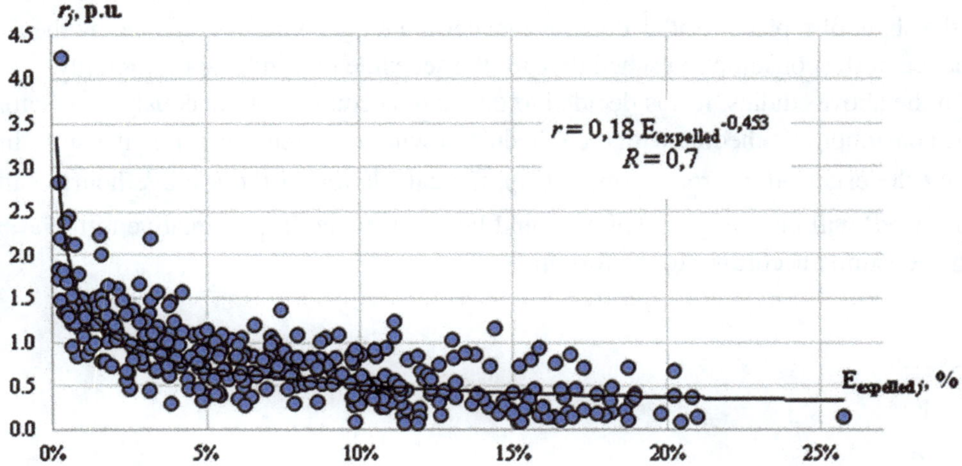

Figure 7. The empirical and approximating dependence of r_j on $E_{expelled_j}$ according to the data of WDC of the village of Novikovo, Sakhalin Region

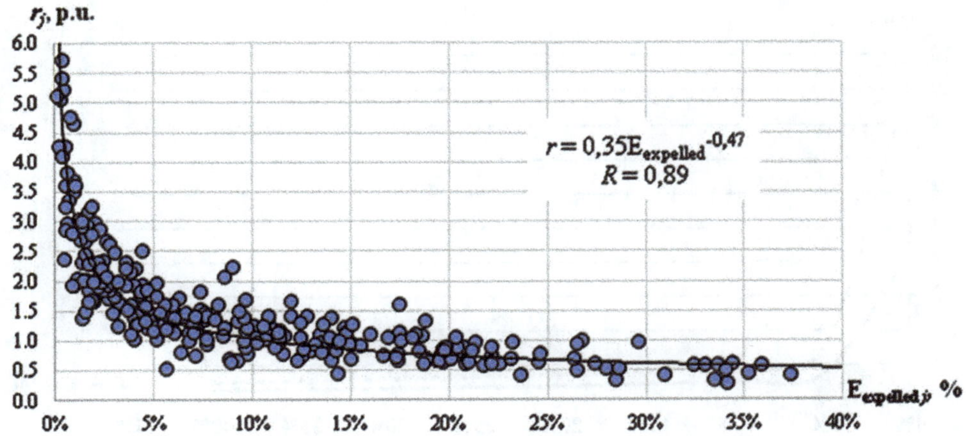

Preliminarily calculate CUF_j for each day of the year according to the formula:

$$CUF_j = \frac{E_{WPPj}}{N_{WPP}^{installed} 24}, \tag{32}$$

where $N_{WPP}^{installed}$ – installed capacity of WPP, E_{WPPj} – daily energy output of WPP.

Calculations were carried out according to the data of the WDC of the village of Novikovo (Sakhalin region) for 2017. The results of the calculations are shown in figure 9.

It can be seen that the share of replacement of the energy output of the DPP by the WPP energy output increases with increasing intensity of wind resources. However, with the same intensity of wind resources, the share of replacement also varies in a certain range, which is affected by the first factor - the coincidence of the consumption schedule with the schedule of wind resource supply.

Based on the above studies, it was decided to take into account simultaneously two factors (coincidence of the consumption schedule with the schedule of wind resource and wind intensity). In this case, in calculating the criterion r_j from formula (27), the calculation of the average hourly values of the power of the WPP will be changed, which should be recalculated in p.u. relative to the average daily power of the consumer according to the formula:

$$K_{N_{WPP_i}} = \frac{N_{WPP_i}}{\bar{N}_{consum}}. \tag{33}$$

Figure 8. The approximating dependences of r_j on $E_{expelled_j}$ according to the data of WDC of the village of Novikovo and the settlement of Ust-Kamchatsk

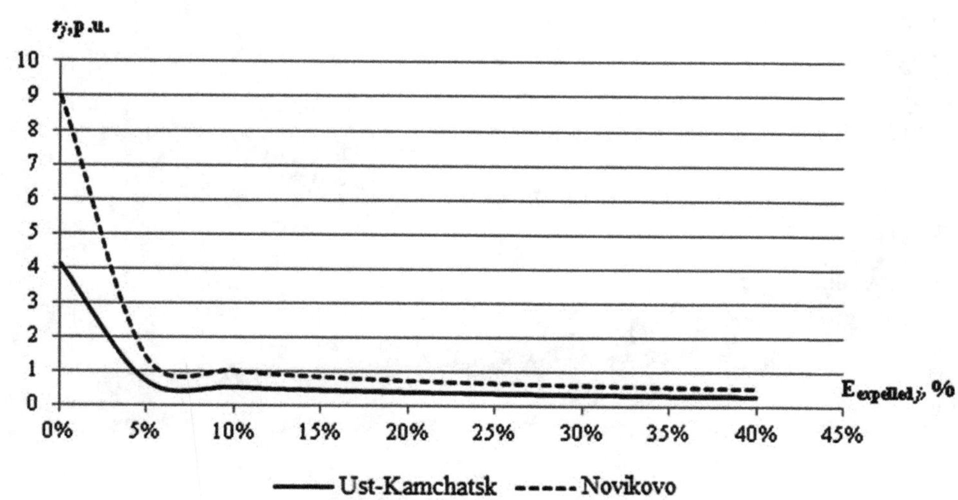

Figure 9. The empirical and approximating dependence of $E_{expelled_j}$ on CUF_j according to the data of WDC of the village of Novikovo, Sakhalin Region

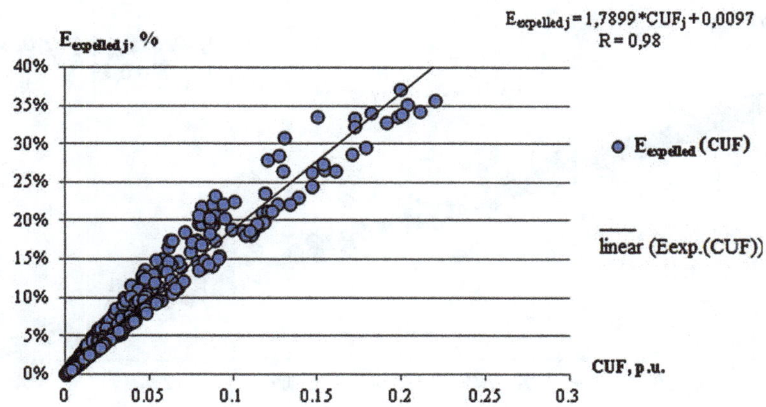

According to the above-described data of the WDC in the settlement of Ust-Kamchatsk (Kamchatka Krai) and the village of Novikovo (Sakhalin Oblast), calculations of r_j were made, the results of which were used to construct empirical tabulated characteristics $E_{expelled}^{DPP}$ on r_j (see Figure 10 and Figure 11) and approximating their dependences with a good correlation dependence:

- For WDC in the settlement of Ust-Kamchatsk: $r = -1,0216 \cdot E_{expelled} + 1,0114$ with $R = 0,99$;
- For WDC of the village of Novikovo: $r = -1,0026 \cdot E_{expelled} + 1,015$ with $R = 0,98$.

Figure 10. The empirical and approximating dependence of r_j on $E_{expelledj}^{DPP}$ according to the data of WDC in the settlement of Ust-Kamchatsk, the Kamchatka Krai

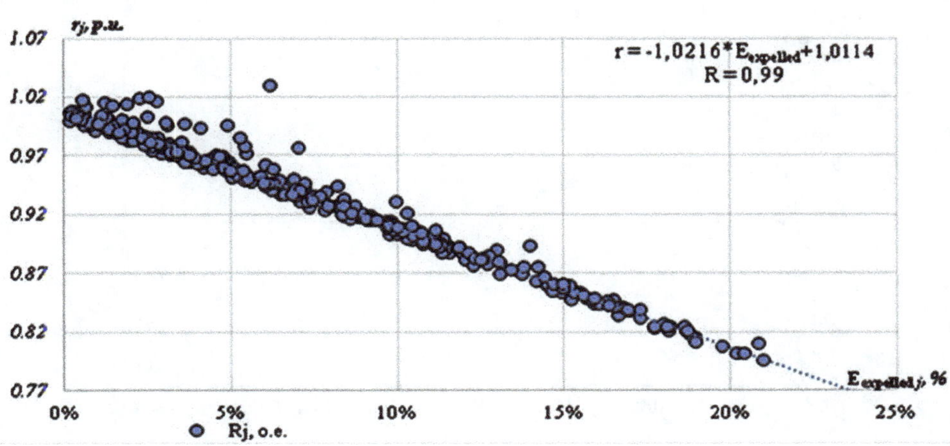

Figure 11. The empirical and approximating dependence of r_j on $E^{DPP}_{expelledj}$ according to the data of WDC of the village of Novikovo, Sakhalin Region

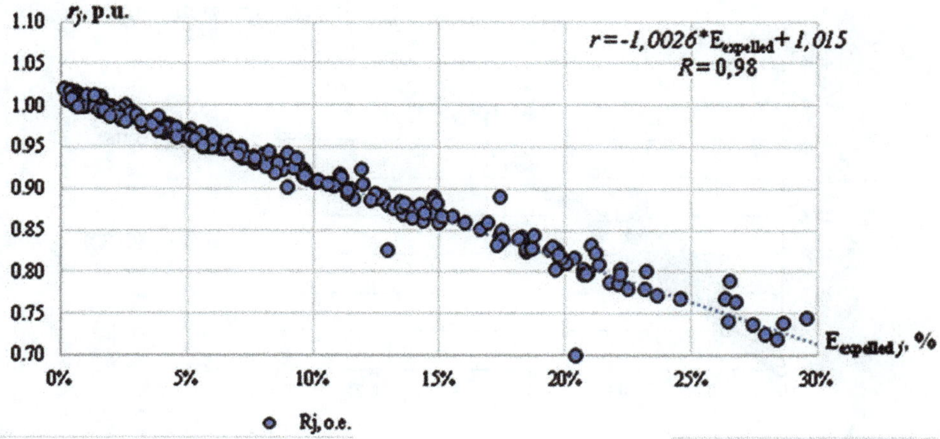

The analysis of the obtained results revealed that the range of the $E_{expelled}$ variation for the same values of r was significantly reduced. For example, at $r = 0,1$, according to the data of the WDC of the settlement of Ust-Kamchatsk, without taking into account the intensity of wind resources, the share of replacement of the DPP energy output changes in the range of 10% -25%, and taking into account the intensity at $r = 0,82$, the share of replacement of the DPP energy output changes in the range of 17% - 19%. Moreover, it can be seen that the approximating dependences of r_j on $E^{DPP}_{expelledj}$ according to the data of two WDC practically coincide (see Figure 12).

Figure 12. The approximating dependences of r_j on $E^{DPP}_{expelledj}$ according to the data of WDC in the settlement of Ust-Kamchatsk (Kamchatka Krai) and the village of Novikovo (Sakhalin Region)

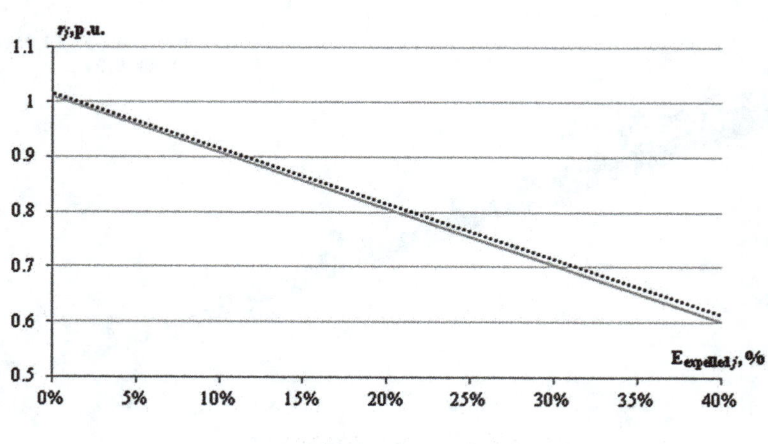

Influence of the Consist of the Existing DPP and the Minimum Load Limitation of the DG on the Displacement of the DPP Energy

Calculation of the power balance, performed in the assessment of the WDC, provides for the minimum load at which the DG can operate. Usually in the DG's passport the minimum load is declared from 20 to 30% of its rated power ($N_{min} = \left(0,2 \div 0,3\right) \cdot N_{rated}$). In case of actual operation, DGs can work with limiting the minimum load to $\left(0,4 \div 0,5\right) \cdot N_{rated}$ (for example, DG-72 is still in operation) and the possibility of replacing the generation of a DPP by a WPP will be different. In addition, in the operating mode below $0,5 \cdot N_{rated}$ of DG, fuel consumption can be higher, depending on the specific fuel consumption of the DG and the duration of its operation in this mode.

The effect of the limitation on the minimum load of the DG (which is taken from 20 to 50% of the unit DG power for the existing WDC «Ust-Kamchatsk») is analyzed for the share of substitution of the energy of the DPP. The calculation is performed according to the condition of the power balance of the local power system in accordance with (18) for the existing equipment of WPP. The results of calculations are presented in Table 11.

The analysis of the obtained results did not reveal the effect of the DG minimum load limitation on the share of penetration in the DPP energy output, since the installed capacity of the WPP is small and amounts to 16% of the maximum load of the system P_{max} .

It is analyzed how the share of the DPP energy output will change when the penetration power of the WPP changes from 16 to 100% of P_{max} with a limitation on the DPP minimum load of $0,2 \cdot N_{rated}$. The results of the calculation are presented in Table 12.

The analysis of the obtained results revealed that the displaced power of the DPP depends on the number of WTGs, that is, on the share of the installed capacity of the WPP from the maximum of the load curve. It was analyzed the change in the share of penetration in the DPP energy output by changing the limit on the minimum load of the DG from $0,2 \cdot N_{rated}$ to N_{rated} of DG with the installed capacity of the WPP about 100% of P_{max} , as well as the share of underutilized energy due to the limitation of the WPP operation in order to minimize the share of free energy in the local power system (see Table 13).

Table 11. Influence of the DG minimum load limitation on the penetration in the DPP energy output for the operate WDC «Ust-Kamchatsk»

Limitation on N_{DG}, %	Penetration of Power N_{penetr}, %
20	8,8
30	8,8
40	8,8
50	8,8

Table 12. The influence of the installed capacity of the WPP on the penetration in the DPP energy output of the WDC «Ust-Kamchatsk»

Number of WTG Komai KWT300 (Installed Capacity of WPP, kW)	Share of Penetration, %
3 WTG (900 kW)	8,8
4 WTG (1200 kW)	11,73
5 WTG (1500 kW)	14,65
10 WTG (3000 kW)	26,65
17 WTG (5100 kW)	36,71
18 WTG (5400 kW)	37,76

Table 13. Influence of the minimum load limitation of the DG on the penetration in the DPP energy output of the WPP with the installed capacity comparable to the maximum load, as well as the share of underutilized WPP energy

Limitation on N_{DPP}, %		Penetration of Power N_{penetr}, %	Share of Underutilized Energy Due to the Limitation of the WPP Operation, %
18 WTG	20	37.76	1.52
	30	37.2	2.93
	40	36.6	4.47
	50	35.9	6.08
	60	35.3	7.75
	70	34.6	9.5
	80	33.8	11.3
	90	33.1	13.2
	100	32.3	15.2

The analysis of the obtained results revealed that there was a decrease in the share of displacement of the DPP energy output from the load curve by 5.4%. Such a small effect for the example in question is explained by the presence of a large number of low-power DGs ($N_{DG}^{\min} < P_{cons}^{\min}$). Also, as the limitation on the DG minimum load increases due to the deterioration of its characteristics, the share of underutilized WPP energy can increase to 15.2%. Therefore, in order to investigate the effect of DG on the penetration of WPPs, it was decided to consider DG of greater unit capacity.

It is analyzed how the share of penetration in DPP energy output for different DG will change in the power range from 1000 to 2500 kW with a minimum load limit in the range from 20% to 50% with a share of WPP 16% of $P_{\max}$.

The results of calculations for the DG with an installed capacity of 2500 kW are given in Table 14.

As seen from the table, the share of WPP (3 WTGs, 900 kW) as a part of the WDC is small and the increase in the DG minimum load does not practically affect the decrease in the share of penetration in the DPP energy output (the decline is from 8.8 to 8.15%).

Table 14. The effect of the minimum load limitation of the DG (with an installed capacity of 2500 kW) on the penetration in the DPP energy output of the operating WPP «Ust-Kamchatsk»

Limitation on N_{DPP}, %	**Share of Penetration in DPP Energy (N_{DG} = 2500 kW, 3 pcs.)**
20	8,8
30	8,77
40	8,62
50	8,15

The influence of the DG installed capacity in the range from 1000 to 2500 kW with a WPP fraction of about 100% of $P_{\max}$ on the share of penetration in the DPP energy output and the share of underutilized WPP energy due to the limitation of its operation due to the reduction of the wind wheel torque with the aim of minimizing the free energy share $E_{WPP}^{underutilized}$ was analyzed (see Table 15).

According to the data in Table 15, empirical dependences of the share of WPP penetration (ΔE_{WPP}) from the DG installed capacity (N_{DG}) were constructed and the approximating $\Delta E_{WPP}\left(N_{DG}\right)$ dependences with a correlation coefficient of about 0.99 have been obtained (see Figure 13 and Table 16).

Also, according to the Table 16, empirical dependences of the share of underutilized energy due to the torque reduction of the wind wheel of WTG as a part of the WPP ($\Delta E_{WPP}^{underutilized}$) from the DG installed capacity (N_{DG}) and the approximating dependences $\Delta E_{WPP}^{underutilized}\left(N_{DG}\right)$ with a correlation coefficient of about 0.99 were constructed (see Figure 14 and Table 17).

The analysis of the obtained results revealed that with an increase in the DG unit installed capacity and the DG minimum load limitation, the share of WPP penetration decreases, as well as the increase in the share of underutilized WPP energy.

Table 15. The effect of the DG minimum load limitation for different DG variants on the share of penetration in the DPP energy output of the WPP with the installed capacity comparable to the maximum consumption, as well as the share of underutilized WPP energy

Limitation on N_{DPP}, %	**N_{DG} = 2500 kW, 3 pcs.**		**N_{DG} = 2000 kW, 3 pcs.**		**N_{DG} = 1500 kW, 4 pcs.**	
	Share of Penetration in DPP Energy, %	**Share of WPP Underutilized Energy Due to the Limitation of Its Operation, %**	**Share of Penetration in DPP Energy, %**	**Share of WPP Underutilized Energy Due to the Limitation of Its Operation, %**	**Share of Penetration in DPP Energy, %**	**Share of WPP Underutilized Energy Due to the Limitation of Its Operation, %**
20	35,1	8,2	35,95	6,1	36,75	4,1
30	32,8	13,92	34,24	10,4	35,5	7,12
40	30,2	20,52	32,34	15,17	34,2	10,4
50	27,1	28,26	30,22	20,52	32,8	13,92

Figure 13. Dependence of the share of WPP penetration on the DG unit installed capacity and the DG minimum load limitation

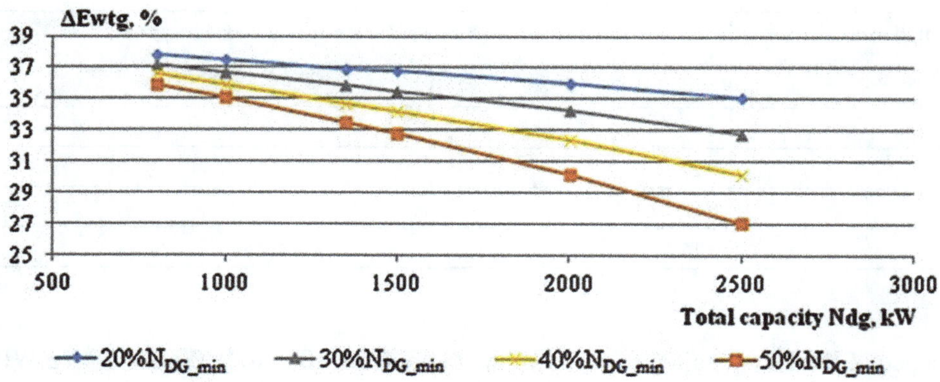

Table 16. Empirical dependences of the share of WPP penetration on the DG unit installed capacity and the DG minimum load limitation

Minimum DG Load (in % of the DG Installed Capacity)	Dependence $\Delta E_{WPP}\left(N_{DG}\right)$
20	$\Delta E_{WPP} = -0,0016 \cdot N_{DG} + 39,045$
30	$\Delta E_{WPP} = -0,0026 \cdot N_{DG} + 39,31$
40	$\Delta E_{WPP} = -0,0038 \cdot N_{DG} + 39,72$
50	$\Delta E_{WPP} = -0,0052 \cdot N_{DG} + 40,365$

Influence of the Actual Fuel Consumption Accounting on Energy Efficiency Indicators of WDC

At the feasibility study for the integration of WTG into the EC, an estimate of the fuel economy at the DPP, as a rule, is carried out at constant specific fuel consumption b_{spec}, which corresponds to the DPP rate power and is specified in the manufacturer's passport data.

According to data of the planned WDC on Popov Island (Vladivostok City District) it assessed the DPP work proportion in the reduced load zone. For this purpose, the DPP work repeatability characteristics with different capacities taking into account and without taking into account the integration of WPP into the EC structure it was built (Deryugina et al, 2017). For the considered local consumer integration of WPP into the existing EC on the basis of a DPP leads to increase in repeatability of DPP operation with a reduced power (less than ½ of rated capacity) for 5,3% that influences increase in fuel consumption during this period of operation.

Figure 14. Dependence of the share of WPP underutilized energy depending on the DG unit installed capacity and the DG minimum load limitation

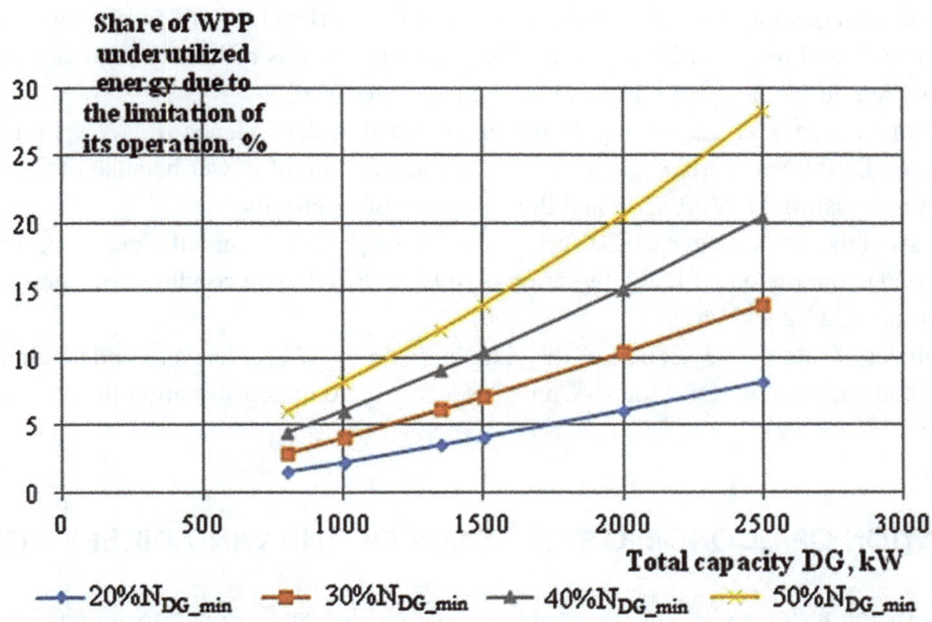

Table 17. Empirical dependences of the share of WPP underutilized energy depending on the DG unit installed capacity and the DG minimum load limitation

Minimum DG Load (in % of the DG Installed Capacity)	Dependence $\Delta E_{WPP}^{underutilized}\left(N_{DG}\right)$
20	$\Delta E_{WPP}^{underutilized} = 0,0039 \cdot N_{DG} - 1,743$
30	$\Delta E_{WPP}^{underutilized} = 0,0065 \cdot N_{DG} - 2,403$
40	$\Delta E_{WPP}^{underutilized} = 0,0094 \cdot N_{DG} - 3,401$
50	$\Delta E_{WPP}^{underutilized} = 0,013 \cdot N_{DG} - 4,985$

The analysis of the obtained results revealed that the most part of a season DPP operates in the mode with the increased fuel consumption. Using actual data on the energy output of DPP on Popov Island and fuel consumption on it has been obtained the approximating dependence of specific fuel consumption on the operating capacity of DPP:

$$b_{spec} = 864,82 \cdot \left(N_{DPP}\right)^{-0.208} \quad (34)$$

where b_{spec} - specific consumption of diesel fuel on DPP, g/kWh, N_{DPP} - operating capacity of DPP, kW.

The annual fuel consumption calculated on a formula (34) differs for 0,6% from actual that allows to verify this formula and to use in further calculations for the analysis of influence of accounting of the actual fuel consumption on power efficiency of the integrated WPP.

Research of influence of accounting of the approximating dependence $b_{spec}(N_{DPP})$ on annual fuel consumption on DPP it was carried out according to the equation of power balance of the local power supply system consisting of WPP, DPP and the consumer of electricity.

On the basis of the carried-out calculations in table 18 are provided: annual fuel consumption on DPP (B_{dpp}, TPA), fuel economy on DPP at integration of WPP (ΔB, TPA) on condition of a constant specific fuel consumption (223,2 g/kWh).

For the object of interest neglection of the characteristic $b_{spec}(N_{DPP})$ leads to underestimation of the planned fuel consumption on DPP for 13% and, respectively, to underestimation the cost of electricity of WDC.

CALCULATION OF ECONOMIC EFFICIENCY OF THE WIND-DIESEL COMPLEX

The analysis of the influence of various factors on the economic calculation (the cost of diesel fuel (thous.rub./t), the error in design output estimation, the specific consumption of diesel fuel at the DPP (g/kWh), etc.) is given for the example of the operating WPP «Ust-Kamchatsk». For this WDC, simple and integral criteria of economic efficiency are calculated: NPV, IRR, profitability index, simple payback period and DPP.

Leading Economic Indicators for Project Evaluation

The purpose of calculating the economic efficiency of the WDC is to determine the following leading economic indicators: net present value (NPV), internal rate of return (IRR), profitability index (PI), simple payback period & discounted payback period (DPP). The criteria for determining the optimal variant of WPP as a part of WDC are: the maximum of the NPV, the maximum of IRR and PI (PI > 1), a minimum of a simple and discounted payback periods.

Net present value - present value of future cash flows of the investment project, calculated by the formula:

Table 18. Annual fuel consumption of DPP, taking into account (without taking into account) the characteristics of $b_{spec}(N_{DPP})$

Composition of the EC	Calculation at b_{spec}=const (223,2 g/kWh)		Calculation for the Characteristic $b_{spec}(N_{DPP})$	
DPP	B_{dpp}, TPA	815,7	B_{dpp}, TPA	935,0
	ΔB, TPA	0	ΔB, TPA	0
DPP + WPP	B_{dpp}, TPA	557,2	B_{dpp}, TPA	612,0
	ΔB, TPA	258,5	ΔB, TPA	323,0

$$NPV = \sum_{i=1}^{n} \frac{CF_i}{\left(1+r\right)^i} - \sum_{i=0}^{n} \frac{IC_i}{\left(1+r\right)^i}, \quad (35)$$

where CF_i - cash flow for the i-th period (month, year), IC_i – invested capital for the i-th period, r – discount rate.

The cash flow is determined by the total effect from the WPP operation. In other words, this is the difference from the WPP revenue from the sale of electricity and the costs associated with the WPP technological process, as well as various types of taxation of WPP.

With a NPV > 0, it is considered that the project will be cost-effective.

Internal rate of return (*IRR*) - this is the interest rate at which the NPV is zero and is calculated by the formula:

$$NPV = \sum_{i=1}^{n} \frac{CF_i}{\left(1+IRR\right)^i} - \sum_{i=0}^{n} \frac{IC_i}{\left(1+IRR\right)^i} = 0 \quad (36)$$

At an *IRR* rate higher than the *WACC* discount rate (see below), it is considered that the project will be cost-effective.

- **Simple Payback Period:** Duration of the period from the initial moment to moment payback. Refers to indicators that do not involve the concept of discounting.
- **Discounted Payback Period:** Duration of the period from the initial moment to moment with discounting payback, i.e. payback period of investments in the present value: $DPP = n,$ at which

$$\sum_{i=1}^{n} \frac{CF_i}{\left(1+r\right)^i} \geq \sum_{i=0}^{n} \frac{IC_i}{\left(1+r\right)^i}, \quad (37)$$

where CF_i - cash flow for the i-th period, IC_i – invested capital for the i-th period, *r* – discount rate.

Most often in the calculation of investment projects, the discount rate is defined as the weighted average cost of capital (*WACC*), which takes into account the cost of equity and the cost of debt and is determined by the formula:

$$r = WACC = \frac{E}{V} \cdot R_e + \frac{D}{V} \cdot R_d \cdot \left(1 - T_c\right), \quad (38)$$

where R_e – cost of equity, E – market value of the firm's equity, D - market value of the firm's debt, V - total market value of the firm's financing ($V = E + D$), R_d – cost of debt, T_c – corporate tax rate.

Determination of CAPEX for the WPP Construction

The cost of wind power equipment depends on a number of factors: the WTG brand, the unit installed capacity, the hub height, the IEC class of WTG, the country of manufacture, etc. This category of costs is determined by the manufacturer and is individual for each WTG. This type of cost can be determined by asking the quotation from the WTG manufacturer.

The design and survey work (R&D) includes:

- Engineering & geodetic works;
- Engineering & geological work;
- Design works;
- Work on the examination of design and survey documentation, etc.

To determine the R&D cost, it is necessary to obtain commercial offers from various companies engaged in engineering surveys. Depending on the type and number of WTG, the R&D may become more expensive due to the increase in the number of drilling points for engineering and geological work, etc.

The construction work includes the following stages of construction:

1. Preparatory stage;
2. Preparation and construction of technological driveways;
3. Construction of the installation site;
4. Preparation and construction of a foundation for WTG;
5. WTG installation.

The cost of delivery of WTG, electrical equipment, special equipment depends on various factors: the country manufacturing the WTG, the remoteness of the WPP construction site, the development of transport routes and logistics, etc. Finally, the value of the delivery cost can be determined after inquiries to the forwarding companies involved in international and domestic cargo transportation.

Table 19 presents the parameters of CAPEX for the construction of 1 kW of installed capacity of the WPP «Ust-Kamchatsk», obtained as a result of statistical processing of the actual financial model.

Table 19. Unit cost parameters of CAPEX for the construction of 1 kW of installed capacity of the WPP «Ust-Kamchatsk» in 2014

Manufacturer	Unit Price for WTG, €/kW	Unit Cost of Construction and Installation Work (Including Transportation Costs), ₽/kW	Unit Cost of R&D and Commissioning Works, ₽/kW	Unit Cost of Electrical Equipment and Its Delivery, ₽/kW
Komai (Japan)	2 565 (1 € = 47 ₽ at the time of WTG ordering)	56 227	22 823	40 812

Calculation of Economic Indicators

Calculation of the indicators "profit-loss" is performed on the basis of the following categories of costs calculated according to specific indicators: R&D, construction and installation works, the cost of WTG, electrical equipment and their delivery.

The WPP energy output (E_{WPP}, ths. kWh) for all 20 years is taken from the data for the long-term average year and is taken from the previous power balance calculations or can be determined using a simplified formula:

$$E_{WPP} = N_{WPP}^{installed} \cdot CUF \cdot 8760, \tag{39}$$

where $N_{WPP}^{installed}$ - installed capacity of WPP (kW), CUF - average annual capacity utilization factor (p.u.), 8760 - number of hours per year.

The losses for auxiliaries equal to 8% of E_{wpp} (E_{aux}, ths. kWh). Economy of diesel fuel DPP (в т.) is calculated by the formula:

$$\Delta B = (E_{WPP} - E_{aux}) \cdot b_{spec}, \tag{40}$$

where b_{spec} – specific consumption of diesel fuel on DPP, g/kWh.

Accordingly, the value of the save diesel fuel (or revenue) is calculated by the formula:

$$Revenue = \Delta B \cdot P_{fuel}, \tag{41}$$

where P_{fuel} - price for diesel fuel per 1 ton (ths.₽/ton).

Price for diesel fuel for the next years is increased proportional to the change in inflation for petroleum products.

Combined costs ($Costs_{comb}$) are consisted of service operation, salary, depreciation cost and miscellaneous expenditures. Expenditures for service WPP and salary for the next years is increased proportional to the change in annual average inflation and multiplying coefficient expenditures for WPP.

Profit before tax is calculated by formula:

$$P = Revenue - Costs_{comb}, \tag{42}$$

If profit for covered period has been got, net profit (P_{net}) is calculated by formula:

$$P_{net} = P - tax_{p}, \tag{43}$$

where tax_p - profit tax per year.

Cash flow statement indirect method is calculated by the formula:

$$CF = P_{net} + Depreciation_{cost} - CAPEX_i + VAT_{rev}, \quad (44)$$

where $CAPEX_i$ – CAPEX per year, VAT_{rev} – reverse charge VAT.

Discount factor (*DF*) is calculated by the formula:

$$DF_i = \left(\frac{1}{1 + WACC_i}\right)^i, \quad (45)$$

where *i* – number of covered period (month, year), *WACC* – the discount rate.

Discounted cash flow (*DCF*) per year is calculated by the formula:

$$DCF_i = CF_i \cdot DF_i \quad (46)$$

Criteria of economic efficiency are calculated by the formulas (35)-(37) for the next benchmark data:

1. Characteristic of the WPP: total capacity of 900 kW (3 WTG). According to the manufacturer expenditures for service WTG Komai is taken equal 23000 €/per year.

CUF is calculated by the formula:

$$CUF = \frac{E_{WTG}}{N_{WPP} \cdot h} = \frac{498.3 \cdot 10^3}{300 \cdot 6480} = 0.26$$

where E_{WTG} - electricity output at WTG from 04.02.2016 to 31.10.2016 (see item 3.2), N_{WPP} - total capacity WTG, *h* - number of hours from 04.02.2016 to 31.10.2016.

2. Specific consumption of diesel fuel on DPP b_{spec} = 375 g/ kWh. Diesel fuel price (for 1 ton) is equal 38000 ₽/ton.;
3. Annual average inflation and inflation for petroleum products are provided in tables 20-21 ("Unifed scenario", 2014).

Table 20. Annual average inflation for petroleum products

Year	-1 (2014 Year)	0	1	2	3	4	5	6	7	8	9
Inflation, %	5.8	4.9	3.8	3.5	6.6	5.4	4.4	4.7	5.0	4.8	3.8
Year	10	11	12	13	14	15	16	17	18	19	20
Inflation, %	3.0	2.8	2.4	2.2	2.1	1.9	1.9	1.9	1.9	1.9	1.9

Table 21. Annual average inflation

Year	-1 (2014 Year)	0	1	2	3	4	5	6	7	8	9
Inflation, %	6.2	5,7	6.0	6.3	6.1	5.6	5.1	4.9	4.8	4.4	3.9
Year	10	11	12	13	14	15	16	17	18	19	20
Inflation, %	3.8	3.7	3.3	3.0	2.8	2.7	2.7	2.7	2.7	2.7	2.7

4. Multiplying coefficient expenditures for WPP are equaled: 1.2 to the beginning of 11 year of operation WPP; 1.5 to the beginning of 16 year of operation WPP
5. The discount rate (WACC) is equaled 12%;
6. The value of salary without including social tax for Kamchatka region is equaled 70 ths.rub./month [22]. Operating personnel at DPP will be servicing WPP (with a operating personnel increase of 2 people).
7. The miscellaneous expenditures are equaled 10% from other types of expenditure.

Assumption: depreciable life - 10 year (accelerated amortization). Useful lifetime of WPP is 20 years. The project is implemented at its sole cost and expense. *VAT* from *CAPEX* is returned in first year of operation WPP.

Calculation is carried out by the formulas (35)-(46). The results of the calculations are provided in Table 22. The analysis of the results revealed that project is economically unviable. Nevertheless project had been realized because the manufacturer of Komai WTG provided WTG free of charge.

Influencing factors being influenced leading economic indicators (increase in energy output by 25%, see item 3.3, decrease in CAPEX by 30%, increase in price of fuel by 30%, gratuitous assignment WPP) had been considered. The results of the calculations are provided in Table 22 and shown in the Figure 15.

CONCLUSION

The different factors influencing on energy efficiency and economic indicators of a WPP has been analyzed. Recommendations for accounting different factors have been given (where was it possible).

Table 22. The summary tabulation on project economics calculating and impact of various alternatives

Indicators	NPV, Ths. Rub.	IRR, %	Simple Payback Period, Year	Discounted Payback Period, Year	PI
According to the initial data	-44445,1	8,93	9,0	>20 years	0,81
Gratuitous assignment WPP	58308.2	19,28	6,0	9,0	1,5
Decrease in CAPEX by 30%	24096.44	14,23	7,0	14,0	1,15
Increase in price of fuel by 30%	18808.4	13,18	7,0	16,0	1,08
Increase in energy output by 25%	8403.2	12,54	8,0	18,0	1,04

Figure 15. Dependence of the discounted cash flow on the influencing factors

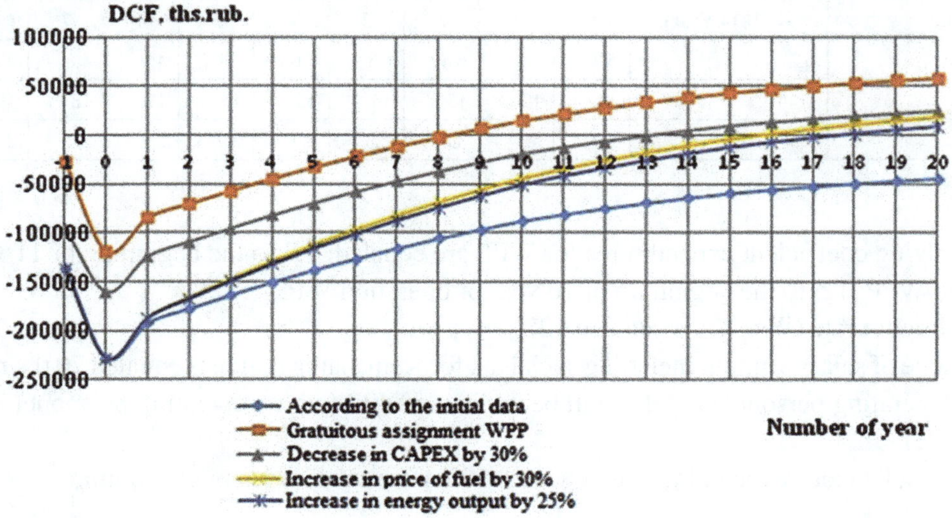

Value of the theoretical energy output substantially is depended on select WSP and the averaging interval of the observational series. Accordingly the WSP model should be developed with averaging interval equal the averaging interval of the observational series.

As a result of the calculations, it was revealed that the incorrectly chosen calculation year may lead to an error in estimation the design WTG energy output to 25%. This issue becomes especially urgent in places where the long-term variation of wind speed can be up to 20%, which may lead to an error in estimation the WTG energy output of up to 35% if the calculation year is incorrect.

With a significant deviation of the air density from the standard value, correction of the WTG power curve should be performed.

As a result of the conducted studies, it was revealed that the share of energy displacements from the DPP depends significantly on the coincidence of the load curve with the curve of wind resource supply taking into account its intensity. An empirical dependence is obtained, which allows estimating the share of energy displacement from the DPP, depending on the value of the chosen criterion «*r*», which takes into account the coincidence of the load curve with the curve of wind resource supply taking into account its intensity.

The effect of the DG minimum load limitation in LES on the share of WPP penetration is noticeable with the share of WPP comparable to the load curve maximum value. The influence of the DG installed capacity and its minimum load limitation on the fraction of the WPP penetration has also been revealed.

The feasibility study taking into account the characteristic $b_{spec}(N_{DPP})$ it is recommended to carry out only when at integration of WPP in EC on the basis of DPP there is an significant increase in proportion of annual DPP operation with a reduced power.

The results found here are applicable for planning future WDC that are subject to similar atmospheric, climatic and topographic conditions, mix of power plant etc.

FUTURE RESEARCH DIRECTIONS

The research data have been done only for existing WDC in the settlement of Ust-Kamchatsk, Novikovo and projected WDC on Popova Island. It is given recommendation only to WDC located on the coastal regions of the FEFD. To give general recommendations for accounting influencing factors it will be necessary have similar research for other atmospheric, climatic and topographic conditions.

Based on research the software package for designing WDC will be planned.

REFERENCES

Afonin, V.S., & Deryugina, G.V. (2012). System properties of hybrid energy complexes based on renewable energy sources. *Energy Security and Energy Efficiency*, 20-27;

Alihodgina, N. V., & Deryugina, G. V. (2017). Investigation of the influence of various environmental factors on the vertical wind profile in the peninsula Kamchatka. *Materials of the XLVII International Scientific and Practical Conference with Elements of the Higher School "Fedorov Readings 2017"*.

Deryugina, G.V., Karpov, N.D., & Shestopalova, T.A. (2017). Investigation of factors and mathematical models influencing the design perfomance of energy efficiency of wind-diesel complexes. *Bulletin of the Kyrgyz-Russian Slavic University*, 44-48.

Deryugina, G.V., Malinin, N.K., & Pugachev, R.V. (2012). *Basic characteristics of the wind*. Publishing house MPEI.

Deryugina, G.V., Malinin, N.K., & Pugachev, R.V. (2016). *Wind Power*. Moscow: Publishing house MPEI.

Deryugina, G.V., & Tyagunov, M.G. (2012). Hybrid energy complexes based on renewable energy sources. *Bulletin of the Kyrgyz-Russian Slavic University*, 11-17.

Gevorgian, V., Touryan, K., Bezrukikh, P., Bezrukikh, P., Jr., & Karghiev, V. (1999). *Wind-Diesel Hybrid Systems for Russia's Northern Territories*. Paper presented at Windpower '99, Burlington, VT.

Gsänger, S., & Denisov, R. (2017). *Perspectives of the wind energy market in Russia*. Retrieved from http://library.fes.de/pdf-files

IRENA. (2017). *REmap 2030 Renewable Energy Prospects for Russian Federation*. Working paper, IRENA.

Katiraei, F., & Abbey, C. (2007). Diesel plant sizing and performance analysis of a remote wind-diesel microgrid. *IEEE Power Engineering Society General Meeting*. 10.1109/PES.2007.386275

Kovalenko, E.A., & Tyagunov, M.G. (2014). The use of hybrid energy complexes based on renewable energy sources in distributed energy. *Energetik (journal)*, 25-27.

MEASNET. (2016). *Evaluation of site-specific wind conditions. Version 2*. Author.

Nikolaev, V. G., Ganaga, S. V., & Kudryashov, Yu. I. (2008). National Cadastre of wind power resources of Russia and methodical bases of their definition. Moscow: Publishing "ATMOGRAF".

Pacific Northwest Laboratory. (1987). Wind Energy Resource Atlas of the United States. Richland, WA: 99352.DOE/CH 10094– 4, Author.

Reliable Prognosis. (n.d.). Retrieved from http://rp5.ru

Ryhlov, A.B. (2011). Regularities in the variation of the average wind speed with altitude in the surface layer of the atmosphere at southeast European Russia for solving the problems of wind energy. *Scientific Notes of the Russian State Hydrometeorological University*, 89-99.

Scott, G. W., Wilreker, V. F., & Shaltens, R. K. (1984). Wind turbine generator interaction with diesel generators on an isolated power system. *IEEE Transactions on Power Apparatus and Systems, PAS-103*(5), 933–937. doi:10.1109/TPAS.1984.318695

Stiller, P. H., Scott, G. W., & Shaltens, R. K. (1983). Measured effect of wind generation on the fuel consumption of an isolated diesel power system. *IEEE Winter Power Meeting*.

Troen, I., & Lundtang Petersen, E. (1989). *European Wind Atlas*. Roskilde: Risø National Laboratory.

Unified scenario conditions of PJSC «RAO Energy Systems of the East» for 2014-2039 years, order of 10.10.2013 N°191.

Vaskov, A. G., Deryugina, G. V., Tyagunov, M. G., & Chernov, D. A. (2012). Certificate of registration of the data basis N°2012620870. Specialized database (SDB) «Wind shear profile». National Research University "Moscow Power Engineering Institute".

Wegley, H.L., Ramsdell, J.V., & Orgill, M.M. (1980). *Siting Handbook for Small Wind Energy Conversion Systems*. Battelle: US DOE.

KEY TERMS AND DEFINITIONS

Aerological Station: A surface location from which upper-air observations are made.

Air Density: The mass per unit volume of Earth's atmosphere.

Capacity Utilization Factor: A coefficient equal to the ratio of the energy output to the installed capacity for a define time interval.

Distributed Power System: A complex of power plants located near electric power consumers and unconnected with unified energy system.

Specific Fuel Consumption: A value equal to the ratio of the fuel rate to the power produced.

Turbulence Intensity: A value equal to the ratio of the root-mean-square deviation to the average wind speed for specified period (usually 10 minute).

Wind Power Plant: A power plant converting wind energy into electrical energy.

Wind Shear Profile: A wind velocity variation depending on height.

Wind-Diesel Complex: A power plant consisting of wind power plant and diesel power plant.

Wind-Measuring Complex: A temporary construction, which is a vertically arranged mast, with a set of measuring equipment installed on it.

Chapter 11
Strategies for Protection Systems of Wind Turbines With Doubly Fed Induction Generator:
Control Strategies and Techniques for Fault Ride Through of Doubly Fed Induction Wind Generator

Dmitry Ilyin
Moscow Power Engineering Institute, Russia

Tatiana Shestopalova
Moscow Power Engineering Institute, Russia

Alexey Vaskov
Moscow Power Engineering Institute, Russia

Aung Ko
Moscow Power Engineering Institute, Russia

ABSTRACT

Doubly fed induction generator (DFIG) is a widely spread technology in modern wind turbines (WT) due to its capability to operate with variable speed, partial scale power converter, and ability to control active and reactive power independently. The main drawback of DFIG is its complicated protection systems. In the chapter, several strategies for DFIG protection are reviewed, and the authors provide a conclusion about their advantages. Penetration of renewable energy sources (in particular, wind power) have a huge impact on power systems; thus, wind turbines should be considered as conventional generation units in terms of frequency and voltage regulation. Modern grid codes require WT stay connected during grid fault and be capable to provide appropriate grid support. Therefore, it is important to implement a DFIG protection system that could meet grid code requirements.

DOI: 10.4018/978-1-5225-9179-5.ch011

INTRODUCTION

Importance of proper protection scheme for DFIG cannot be overestimated. Because the high sensitivity of DFIG to grid disturbances protection circuit should be implemented on each wind turbine. In the first section modern state of wind power generation analyzed and also provided modern grid code requirements. Wind power penetration status shows, that on local scale wind power plants (WPPs) have significant power, so during the fault, they should not be disconnected. Even more, wind turbines should provide support for grid during the fault. Thus, in the second section different approaches for protection strategies proposed and analyzed.

BACKGROUND

In the last decade the penetration of the wind power into the power system has large-scale, and tripping several wind turbines (WTs) during voltage dips can further risk the stability of power system, and as worst case a system collapse (Hansen, Michalke, Sorensen, Lund, & Iov, 2007). So modern grid code require from wind turbine (WT) to act like conventional power generation unit and provide the capability of active and reactive power control, and more important, have an ability to ride through low voltage case.

Doubly fed induction generator (DFIG) has many advantages in terms of independent control of active and reactive power control in the range wider in comparison with the permanent magnet synchronous generator (PMSG). The main drawback of WTs with doubly fed induction generator is their vulnerability to grid disturbances: even small voltage sag can cause WT disconnection and even worse consequences. Protection circuit should be implemented properly to provide WTs ability to stay connected during the fault and support grid with reactive power injection.

This problem of DFIG was raised since DFIG found their application in WT industry. Earlier protection schemes provided a capability for WTs stay connected, but they were unable to support grid with reactive power injection.

To reduce excessive currents in the rotor during the fault, some new fault-control solutions for the rotor side converter (RSC) of DFIG have been developed. Stator transient impact on the DFIG fault ride through (FRT) capability is considered by means of introducing differential terms of stator flux into rotor voltage feed forward compensation terms (Hu, Sun, He, Zhao, 2006), and by the feed-forward of the faulty stator voltage (Liang, Qiao, Harley, 2010). Moreover, the RSC can be protected by counteracting zero and negative sequences of the stator flux (Xiang, Ran, Tavner, Yang, 2006) or by the feed-forward of the measured stator currents as a reference for the rotor currents controllers (Lima et al., 2008). Nevertheless, these techniques are not suitable for severe voltage sags. To keep DFIG connected to the grid during fault and post-fault condition without additional circuits in (Rathi, Mohan, 2005) proposed a method, where the controller is designed using H_{∞} technique and μ-analysis. However, the method is complicated and need a lot of computational resources and time. Authors in (Zhou, He, Sun, 2009; Santos-Martin, Rodriguez-Amenedo, Arnaltes, 2009; Shang, Hu, 2012), pay attention on improving performance during unsymmetrical fault case.

Nevertheless, the optimal strategy is not defined, because protection circuits developed during a long time, with different technologies and cost of power electronics. Different papers focused on one aspect of DFIG behavior during fault but do not consider others.

MAIN FOCUS OF THE CHAPTER

Wind Power Penetration and Grid Codes

The share of renewables in energy consumption continues to grow globally with some technologies developing very rapidly. Despite impressive uptake in sources such as photo-voltaic (PV) panels and wind, growth in renewables had difficulty keeping up with rapidly rising demand.

In the year 2017 there was another record-setting development for renewable energy, characterised by the largest increase in renewable power capacity as well as by falling costs, increases in investment and advances in enabling technologies. Many developments during this year affected the deployment of renewable energy, including the lowest ever bids for solar and wind power in tenders in several countries around the world.

Renewable power generating capacity set its largest annual increase ever in 2017, with an estimated 178 GW installed worldwide, raising total capacity by almost 9% over 2016. Solar PV led the way, accounting for nearly 55% of newly installed renewable power capacity. Wind and hydropower accounted for most of the remaining renewable capacity additions, contributing more than 29% and 11%, respectively. Overall, renewable energy accounted for an estimated 70% of net additions to global power capacity in 2017, up from 63% in 2016. By year's end, global total renewable power capacity was around 2195 GW – enough to supply an estimated 26.5% of global electricity.

A number of countries and regions integrated much higher shares of PV and wind into their energy systems as instantaneous shares of total demand for short periods during 2017. They include, for example, South Australia, which generated more than 100% of load from wind power alone and 44% of load from solar PV alone on two separate occasions; Germany (66% of load from wind and solar power combined); the US state of Texas (54% of load from wind power alone); and Ireland – 60% of load from wind power alone (Renewable Energy Policy Network for the 21st Century [REN21], 2018).

Wind power had a relatively medium year compared with 2015 and 2016, but still saw its third strongest 12-month period, with more than 52 GW added globally in 2017. Cumulative capacity increased nearly 11%, to around 539 GW.

As can be seen from REN21 annual 2017 report (REN21, 2018), renewable energy penetration has significant scale and tendency to future development clear. Most countries have various renewable policies and roadmaps for 2030 and 2050 development strategy. This means that renewable technology will replace traditional fossil fuel based generation. Several countries governmental regulations already limit coal and nuclear power plant construction and some plants are decommissioned.

This development affects a power grid and grid operators should be prepared for renewable penetration. Renewable sources cannot be considered anymore as experimental technology on small scales, in some cases operators have to deal with the significant rated power of wind power plant or PV plant, which can be compared with traditional power plants. Therefore, these renewable power plants should be operated in the same way as conventional power plants: they should participate in frequency and voltage regulation; support grid during faults. In support of this conclusion, the renewable power percentage in the local time in the local area can be considered, such as Texas, Southern Australia or Germany. In these cases, renewable can replace traditional generation with significant percentage, and if a renewable operation is not prepared well, renewable failure can cause significant grid troubles.

Therefore, the importance of appropriate grid codes for renewable energy, and, in particular, for wind power generation could not be overestimated. As wind power plants provide the same amount of power as conventional power plants, they should meet the same requirements for active and reactive power control, as well as frequency and voltage regulation, and more important, their behavior during grid fault should provide maximum grid support (e.g. reactive power injection during voltage sags).

Modern grid codes require that wind power plants (WPPs) must be capable of operating continuously within the voltage and frequency variation limits encountered in normal operating conditions. In addition, they should remain in operation in case of frequency deviations outside the normal operating limits for a specified time and in some cases with a specific reactive power output. With the ability to remain connected to the grid for a wider frequency range, WPPs support the system during abnormal operating conditions and allow for a fast system frequency restoration. Wind turbines (WTs) must be designed appropriately, as frequency deviations can lead to generator windings overheating, insulation degradation, and power electronic devices damage.

The frequency ranges required by the various grid codes are presented in Table 1 and Table 2 (Sourkounis & Tourou, 2013). In the green frequency ranges, the wind turbines must stay connected and operate continuously with full power.

In the white ranges, they must remain connected at least for the minimum specified time, usually at a lower power output and support the grid during frequency restoration.

In many cases, the active power reduction must be controlled proportionally with the frequency deviation from the nominal. In the extreme red frequency ranges, wind turbines are allowed to disconnect from the grid.

Large frequency ranges are allowed in isolated systems with weak interconnections where the system stability is more vulnerable and sensitive to disturbances compared to large interconnected systems (e.g., UCTE).

Active power control is the ability of wind power plants to implement active power output regulation within a defined level and ramp rate (in the case of TSOs commands for active power curtailment). The aim of these requirements to provide a stable system frequency, to prevent overloading of transmission lines and to decrease the effect of the wind speed variations, which influence on wind turbines power production (Sourkounis & Tourou, 2013). The ability of wind turbines to control their active power is also important for transient dynamic stability during faults. Often, active power generation can be reduced temporarily by the control system in case of a voltage dip (Ackermann, 2005). This allows increasing reactive power output without exceeding the converters operating current. After the fault is cleared, a fast transition to normal active power generation is essential to ensure the stability of the grid.

The voltage levels in power systems must be maintained on a constant level because equipment of the utility and consumers are designed to operate at specific voltage levels. Recent versions of grid codes demand from WPPs to contribute to voltage regulation in the system, as conventional power plants do. They must have the ability to inject or absorb reactive power in order to influence the voltage level at the point of common coupling (PCC).

The low voltage ride through (LVRT) is the most important requirement regarding WPP operation. It has been recently presented in the grid codes and essential for a stable and reliable operation of power supply networks, especially in regions with high wind power penetration.

Faults in the grid can cause large voltage dips across wide regions and some generation units can be lost. In the past, during grid disturbances and low grid voltages, the WTs and WPPs were allowed to disconnect from the grid. Nevertheless, if there is a large amount of wind generation in the network,

Table 1. Required frequency range of operation in different grid codes

f, Hz	Great Britain	Nordic	Scotland	Romania	Ireland	Denmark
53.5			**Disconnection**			
53			**within 1 s**			
52.5		3 min. power reduction				
52		possible at any level				
51.5		30 min, small	2% active power	Continuous with 20%	60 min	15 min 60% - 100%
51		P reduction	reduction per Hz	max power losses		
50.5		Continous				
50		Rated power				
49.5						5 hrs 90%-100% power
49	Max 5% linear		Max 5% linear			
48.5	P reduction	30 min, linear reduction	P reduction		60 min	30 min 90%-100% power
48		from 100% to 85%				
47.5						3 min 80%-100% power
47	20 s		20 s	20 s, 80% power	20 s	5 hrs 90%-100% power
46.5			**Disconnection**			
46			**within 1 s**			

the simultaneous disconnection of wind generating units and WPPs can cause larger voltage dip and eventually collapse of voltage in the affected region (Sourkounis & Tourou, 2013).

Wind Turbine LVRT European Grid Code Requirements

Recent grid codes require WPPs to stay connected and support the grid during and after a fault. They must withstand voltage dips of a certain percentage or the specified time duration, as shown in the LVRT voltage-time profiles of Figure 1. Disconnection is prohibited above the borderline. Below the borderline WTs may not contribute to the grid and they can be tripped by circuit breakers.

Above European grid code in terms of wind turbine stability were observed. Nevertheless, USA and China are the two leading countries in wind generation development. In both countries, the rapid deployment of wind power has posed new challenges in maintaining grid stability and reliability in large-scale electric power system operations. China, in particular, has suffered severe integration issues because technical standards for wind power plants (WPPs) lagged behind the surge in installations. The lack of

Table 2. Continuation of Table 1

f, Hz	France	Spain	Cyprus	Germany	Belgium	Italy	Poland	Sweden
53.5								
53								
52.5								
52								
51.5	20 s, power set by TSO				Time by			30 min at reduced
51	15 min, power set by TSO		60 min	30 min	agreement			power output
50.5	60 min, 90% power							Nominal power
50								
49.5								Nominal power
49	5 hrs, 90% power						30 min, 90% power	
48.5	3 min, 90% power			30 min			20min, 85% power	30 min <5% power
48	3 min, 85% power			20 min	Time by		10 min, 20% power	reduction
47.5	20 s, 80% power	3 s	60 min	10 min	agreement			
47			5 s					
46.5								
46								

Figure 1. LVRT grid code requirements for wind turbines

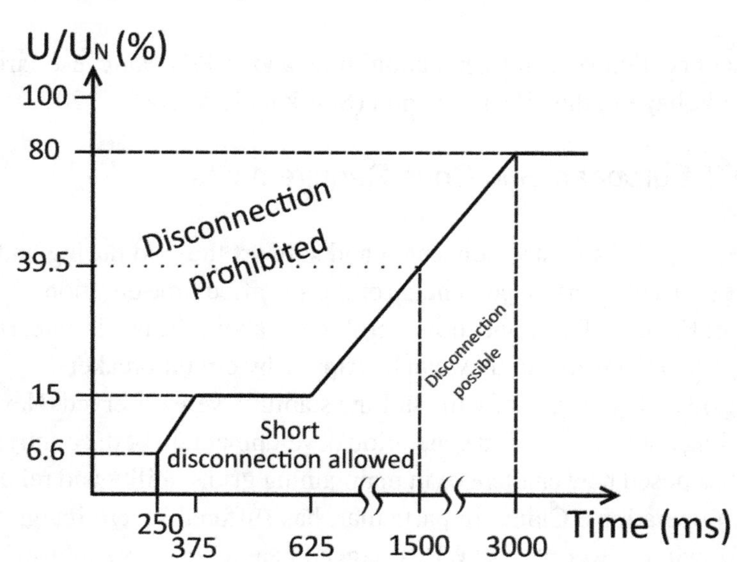

appropriate standardization caused many WPP developers in the industry to maximize the power rating using cheaper technologies without concern for power quality (Wu, Sun, & Du, 2014). Without proper LVRT capability, many wind turbines should be disconnected from the grid during voltage dips (Rebecca & Busby, 2012). As mentioned above this disconnection can cause a significant impact on power system stability due to large scales of WPPs.

China's standards require a wind turbine to remain connected when the voltage is reduced to 0.2 p.u. for up to 0.625 seconds, when US standards have different requirements. In terms of active power support, China's national standard requires a restoration speed of 10%/second of the rated power. In the United States, Independent System Operator–New England (ISO-NE – US regional system operator, who provides local grid codes) suggests a restoration time of 0.5 seconds.

Relative to dynamic reactive power support, China's national standard requires a reactive current response time not longer than 75 ms and proposes explicit technical requirements and a calculation formula for the dynamic reactive current. In the United States, ISO-NE does not have requirements for the corresponding time of the dynamic reactive current, but it proposes qualitative requirements or dynamic reactive support (David, Eduard, Tian, Mackay, Weisheng, 2016).

When the voltage at the point of common connection drops to 20% of the rated voltage, WTs should ensure continuous operation with no disconnection for 0.625 seconds. When the voltage at the point of common connection (PCC) is able to recover to 90% of the rated voltage within 2 seconds after dropping, generators should ensure continuous operation with no disconnection.

FERC Order No. 661 (US Federal Energy Regulatory Commission) requires WPPs to have LVRT capability down to 15% of the rated line voltage for 0.625 second. The plant must also be able to operate continuously at 90% of the rated line voltage, measured at the high-voltage side of the WPP's substation transformer. FERC Order No. 661A amended the LVRT standard, requiring WPPs to have ride through low-voltage capability for events down to a zero voltage level for clearing times up to a maximum of nine cycles. ERCOT and ISO-NE also require WTs to remain connected through low-voltage fault conditions of at least nine cycles at 0 V.

As discussed above, wind power generation penetration in electrical grids worldwide cannot be overestimated. Wind power takes a big role in generation, regardless of geographical factors or governmental energy policy. In China, USA, India and European countries wind power generation develops in its own but similar way. So should change the grid code, and it also aimed at the same goals in different parts of the world.

Two main parts of these requirements are normal operation and fault operation. Wind turbines cannot be 'passive' components in grid power balancing process and disconnect in case of each grid disturbance. As WPPs have significant generated power in the local area, full disconnection of these WPPs in case of grid disturbance could lead to very unpleasant consequences, even if initial disturbance was not critical. This way, WT today considered as valuable unit for power control process and grid support. All existing WT should be modified for these requirements if they do not meet them, and the newly installed WT should be already constructed in the way to fulfil modern standards.

As WT, equipped with Double Fed Induction Generator have important advantages in terms of its variable speed and power control capability, this potentially attractive technology should also meet grid code requirements, or, in other words, DFIG should have an appropriate protection circuit. Actually, WT with DFIG share the significant amount of wind generation market, so appropriate protection circuit is essential for them, and different approaches were developed over the past decade. In the next section, different techniques for protection circuit implementation will be considered, and the best approach in Author's opinion will be proposed.

Doubly Fed Induction Generator Protection Systems for Low Voltage Ride Through

Introduction in DFIG Technology

The presence of power electronics inside modern wind turbines provides enhanced potential for control capabilities, and the main trend of WTs/WPPs concepts develops towards the variable speed operation where the grid connection is partly or entirely performed through power electronic devices. One important advantage of variable speed wind turbines is the ability to control independently their active and reactive power and thus to assist the power system. Two variable speed wind turbine concepts have a solid dominance in the market today. One of them is the variable speed WT with a partial-scale power converter, known as the doubly fed induction generator (DFIG) concept. The other is the variable speed WT with a full-scale power converter and a synchronous generator with permanent magnets (PMSG). These two variable speed WTs compete against each other on the market with their pros and cons.

The main strong feature of the DFIG concept is that only a percentage of the power produced in the generator has to pass through the power converter. Typically, this is only 20%–30% of generator power, compared with full power converter for PMSG wind turbine concept. Thus, it has a cost advantage compared with the full power converter. However, since the stator in a DFIG has a direct connection to the grid, and the rotor interacts through a partial-scale power converter, DFIG is more sensitive to grid disturbances compared with the full-scale power converter concept. The DFIG requires therefore advanced protection systems under grid fault conditions. Without this protection, a DFIG WT is not able to stay grid connected during a grid fault, as high transient currents can damage the power converter device. In the next section, processes in stator and rotor circuit are briefly described.

Power Control in DFIG

Power converters are usually controlled with vector control techniques. Vector control allows separate control of both active and reactive power. With the aid of a rotating reference frame based on an AC flux or voltage currents projected on this rotating frame. Such projections are usually referred as the d and q components of their respective currents. With a proper choice of reference frames, the AC currents appear as DC quantities in the steady state.

For flux-based rotating frames for the rotor-side converter (RSC) of DFIG, changes in the q component will lead to active power changes, while changes in the d component will lead to reactive power changes. In voltage-based rotating frames (and thus 90° ahead of flux-based frames) the effect is the opposite. In the work of Hansen et al. (2007) the converter control is modelled on a generic level, as illustrated in Figure 2.

The rotor-side converter (RSC) operates in the stator flux reference frame, while the grid-side converter (GSC) operates in the stator voltage reference frame. The q-axis current of the RSC is used to control the active power, while the d-axis current is used for reactive power control. The d-axis current of the GSC is used to control the DC link voltage to a constant level, while the q-axis current is used for reactive power control.

Figure 2. Control signals to RSC and GSC

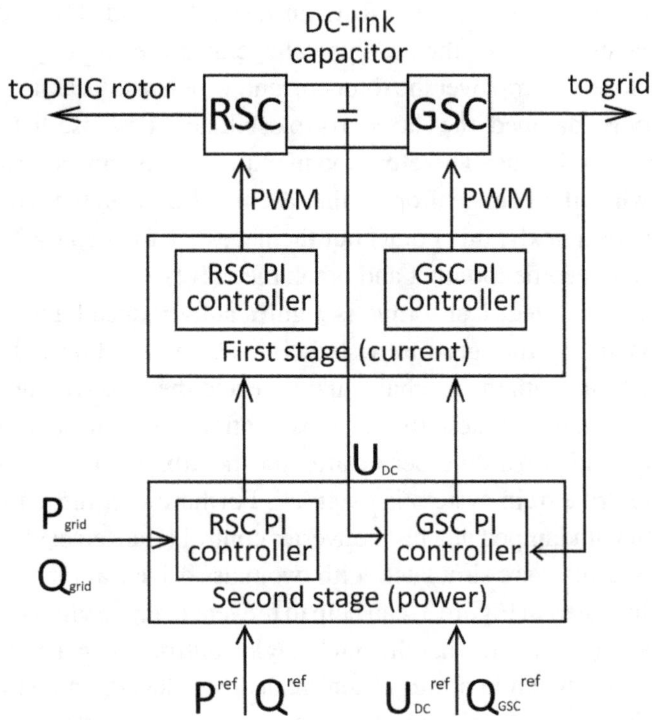

In normal operation, the purpose of the RSC is to control independently the active and reactive power on the grid, while the GSC has to maintain the DC link voltage at a set value regardless of the magnitude and direction of the rotor power and to guarantee converter operation with zero reactive power flow through GSC. The reference for the active power defined by the maximum power tracking point look-up table as a function of the optimal generator speed. The reference for the reactive power of the RSC can be set certain value according to dispatch commands. In normal operation, the reactive power reference for the GSC is usually set to zero. GSC exchanges with the grid only active power, and therefore the injection of reactive power from the DFIG to the grid is done only through the stator (Hansen et al., 2007).

Processes in DFIG Power Converter During Fault

The control performance of the DFIG is great under normal grid conditions. Adjustment of active and reactive power possible with variable speed WT technique, but also by the aid of electrical generator control. The main drawback of the DFIG is usually the fact that large disturbances can lead to large fault currents in the stator, due to the direct connection stator with the grid. Every disturbance in the grid will be transmitted to the stator. Because of the magnetic coupling between the stator and rotor, the stator disturbances is further transmitted to the rotor. High voltages are induced in the rotor windings that can cause excessive currents in the rotor.

As the stator is directly connected to the grid, power converter has only partial control over the generator during grid faults. It can reach its limits quickly and loses the control of the generator. The results are both high rotor currents and voltages during grid faults. The surge following a fault includes an excessive power from the rotor terminals towards the converter. As the grid voltage drops at the fault moment, the GSC is not able to transfer the power from the RSC further to the grid. The additional energy goes for charging the DC bus capacitor and thus the DC bus voltage increases rapidly.

During the fault, RSC loses control over the rotor current, which results as uncontrolled rotor currents. Moreover, if the DFIG poorly damped, the flux starts to oscillate. These oscillations can cause problems when the voltage returns after the sag. Therefore the maximum rotor current can be much higher when the voltage returns than when the voltage drops if the machine flux is in the wrong "direction". As DC-link voltage increases rather quickly, this points out the necessity to keep track of the DC-link voltage so that the voltage does not become too high and break the valves.

Under these conditions, the electrical torque is significantly reduced and therefore the drive train system acts like a torsion spring that gets untwisted (Hansen et al., 2007). Due to the torsion spring characteristic of the drive train, both the mechanical torque and the aerodynamical torque, as well as the generator speed, start to oscillate. As these torsional oscillations may influence the converter operation during the grid disturbance and also a short period after the disturbance, their modelling by using at least a two-mass model for the drive train system is essential. Furthermore, these torsional oscillations can even be excited and become undamped at a fast converter control (Hansen et al., 2007). In next sections, all proposed protection systems were simulated with two mass drive train model.

Pitch angle control, illustrated in Figure 3, can help to prevent overspeeding both in normal operations and during grid faults, owing to the fact that the pitch angle controls the generator speed directly. In the case of overspeeding, the aerodynamic power is automatically reduced, while the speed is controlled to its rated value. On the other hand, the rate of change pitch angle limited and relatively slow. If a wind turbine rotor has enough inertia, it cannot be accelerated instantly, but pitch control action could be too slow to prevent overspeeding.

Therefore, it is important to implement suitable protection system of DFIG converter, which should help to avoid the high currents and the uncontrollable energy flow through the RSC to DC-link. Taking into account aforementioned grid code regulations for WT, which are require WT stay connected during the fault and support grid with reactive power injection it can be concluded, that protection circuit should provide following functions:

1. Prevent uncontrollable action of DFIG, which could lead to extremely undesirable effects in the grid. Under the term of uncontrollable behavior we suggest overcurrent and overvoltages in GSC, mechanical oscillations in power train and rotor overspeeding;
2. Keep WT connected to the grid;
3. Ensure ability to support grid during and after the failure with reactive power injection.

Figure 3. Pitch angle control

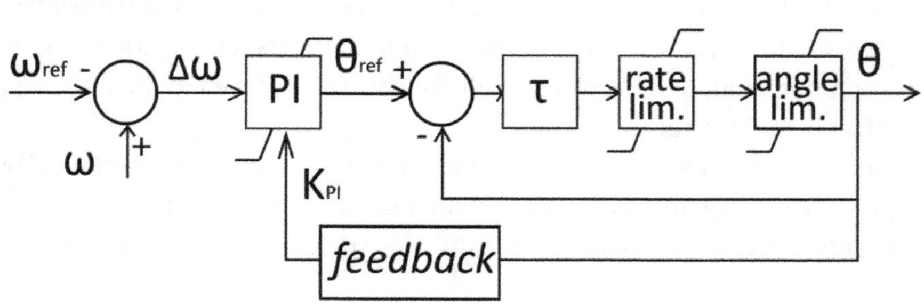

SOLUTIONS AND RECOMMENDATIONS

In the next section, different approaches for implementation of these tasks will be provided and analyzed. Various approaches aimed for FRT improvement, but they achieve these goals with different methods, for example, crowbar limits the rotor current and helps to dissipate excessive energy flow to RSC, DC chopper dissipates excessive energy in DC link. As an action of DC chopper and crowbar aimed at different parts of the converter, both protection systems are important. Other strategies use different topologies of resistance connected from the stator side or passive resistance network; and strategy with series antiparallel thyristors disconnects stator side from the grid for short period of time, to reduce any consequences of transient processes.

Crowbar Approach in the DFIG Rotor Side

DFIG protection system monitors different signals such as the rotor current and the DC link voltage. If one of these monitored signals exceeds a threshold value, the protection triggered. The simplest protection method is to short circuit the rotor via, so-called, crowbar resistance and deactivate rotor converter control. Crowbar system used in modern WT based on a three-phase series shunt resistance controlled via power electronics switches. The crowbar system activated during the high current in the rotor windings or excessive voltage on the DC capacitor, which can appear after a short circuit fault in the grid. The crowbar protection is specific to DFIG and illustrated in Figure 4. It is coupled via the slip rings to the rotor side of the generator.

Following steps involved during the activation and deactivation of the crowbar system performed with the following steps:

1. Disconnection rotor windings from the RSC;
2. Three-phase resistance insertion in series to the rotor windings;
3. Crowbar system disconnection from the rotor windings;
4. RSC reconnection to the rotor windings.

Figure 4. Crowbar in rotor circuit

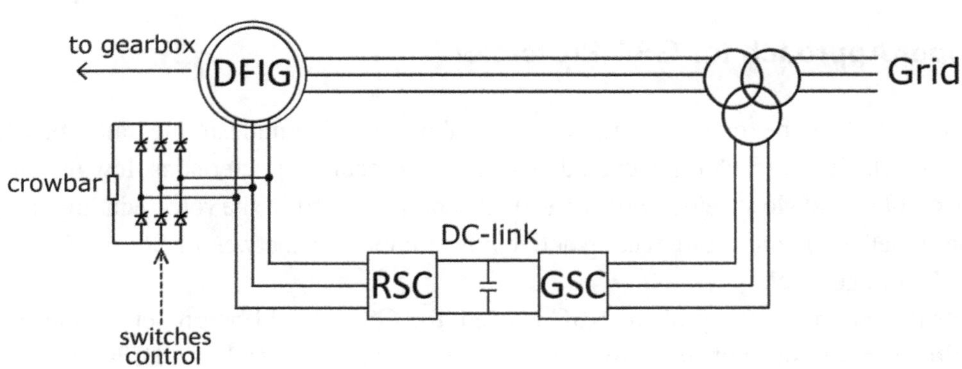

These actions will help to prevent the high rotor currents and DC-link overvoltage. The magnitude of the voltage in the rotor circuit defined by the crowbar resistance. The crowbar also acts as a sink for active power, consuming active power and prevent WT overspeeding. When crowbar is activated, the generator works as an induction generator with increased rotor resistance. In this case, as the RSC is deactivated and bypassed, the controllability of active and reactive power gets lost during fault detection. This implies that the magnetization of the generator, which in normal operation is done over the rotor circuit by the RSC, must in the case of a fault be done over the stator. Since the GSC is not directly coupled to the generator windings, it stays connected to the grid.

As crowbar requires the disconnection of the RSC during a fault, this means that reactive power supply during the fault is impossible, if special control schemes are not implemented. Grid codes require for grid support, so in work of Hansen et al. (2007) control scheme for reactive power supply proposed.

Coordinated Reactive Power Control by RSC and GSC

The voltage grid support capacity of a DFIG depends on the quantity of reactive power injection. At a severe grid fault, if the generator is not tripped, the DFIG wind turbine has to continue its operation with a short-circuited generator, trying to sustain grid connection. Overspeeding of the wind turbine could be partially prevented by the pitch control, but also requires additional protection scheme. The RSC is blocked at the moment when the crowbar is triggered by either an overcurrent in the rotor circuit or an overvoltage in the DC link. The DFIG grid support capability is strongly reduced. Its capability to control independently the active and reactive power is lost when it is actually most needed, namely during and shortly after the fault.

The voltage can be controlled by either the RSC or the GSC or by both of them. In the work of Hansen et al. (2007) voltage control strategy described, where the reactive power injection is performed by both converters in a coordinated way. The idea is that the RSC is used as the default reactive power source, while the GSC is used as a supplementary reactive power source during the RSC blocking.

In the case of less severe grid faults or reactive power unbalance in the system, and when the crowbar is not triggered, the RSC and the GSC have the same tasks as in normal operation. RSC is not blocked and therefore can still independently supply active and reactive power to the grid. On the other hand, in the case of a severe fault, when the crowbar is triggered, the RSC is blocked, its controllability is lost and therefore its task to support the grid with reactive power is taken over by the GSC. The GSC does not block at a grid fault and should provide reactive power as long as the RSC is blocked.

DC Chopper Approach for GSC Protection

Besides high currents in the rotor windings described above, another undesirable transient is the voltage at the DC-link, which can reach high level, depends on two main characteristics: low residual terminal voltage during fault and slow velocity of the RSC disconnection from the rotor winding after fault detection. These facts will contribute making active power unbalance between RSC and GSC higher and thus, DC-link voltage level.

Whatever the control strategy of the RSC is used, the GSC should be properly controlled and the DC-link voltage should be kept stable to realize the ride-through control of a DFIG, if possible. During fault and RSC protection procedure, the DC-link voltage is likely to fluctuate. In addition, the rotor current control might be affected by the DC-link voltage fluctuation when the DFIG is running back to the normal condition.

The authors of Sun, Chen and Blaabjerg (2005) had shown in their research that power electronic converters might be kept in operation during a non-serious voltage dip. For this purpose, DC-link voltage should be maintained stable and fluctuation of the GSC current during a grid fault should be limited, so that the DC-link capacitor could be protected and the adequate voltage on the rotor could be provided.

Most of modern DFIG wind turbines equipped with a crowbar system and sometimes with a DC chopper system in order to be able to keep connected during the fault. Chopper systems have been largely used for industry in the application where the motor operation has a regenerative cycle (for example, systems for elevation). In WTs, the chopper is used to dissipate the unbalance of active power between RSC and GSC. A chopper circuit is additional resistance which can be connected to the DC-link, as shown in Figure 5, with a similar function to that of the rotor side crowbar in order to reduce the DC-link voltage (Takahashi, Tamura, Futami, Kimura, & Idle, 2006; Okedu, Muyeen, Takahashi, & Tamura, 2010). The chopper facilitates a voltage-raising action from the converter terminals, enabling a fast recovery of the DC-link voltage. The protective device in this scheme is a simple chopper circuit and a resistance. The pulse signal to trigger the IGBT is activated when exceeds, and thus, the chopper is turned on and the energy is dissipated by the internal resistance.

The authors of Salles et al. (2009) investigated various protection systems in combination with DC chopper under various fault conditions. First of all, there were considered 3 different cases of control strategies RSC and GSC (Table 3).

Depending on the fault location WT affected in a different manner, so protection system performs different as well. Even more, the difference between various protection schemes smaller in case of a fault at WT terminals. When the fault occurs at the wind power plant point of common coupling (PCC) the injection of active and reactive power is limited due to low terminal voltages. The highest value of terminal voltage during the fault is when the control strategy of the Case B applied. The lowest consumption of reactive power after the fault clearance is also performed by the control of the Case B. The DC-link voltage has reached around 2 pu for all the three control strategies and the RSC currents has reached around 3 pu. These control strategies are used as reference results, without protection schemes. Three different protection schemes were tested by Salles et al. (2009).

In first scenario, a fault occurred at buses connected through the transmission line to wind power plant PCC and cleared after 150 ms. For Case D (Table 4), after the fault clearance, the RSC is not immediately restarted, and the voltage during this period is around 0.9 pu. The RSC is restarted 50 ms after the fault clearance and finally, the voltage returns to the nominal value. Considering that the period of overcurrent on the rotor winding and overvoltage on the DC-link last for a very short time, the RSC

Figure 5. DC chopper in DC-link circuit

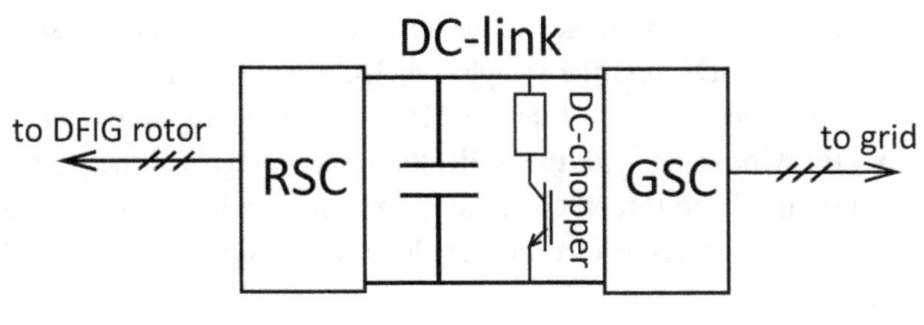

Table 3. Control strategies for RSC and GSC without protection

Case	RSC Control	GSC Control
A	switch control from constant power factor to constant U	as normal
B	as in case A	control injection of reactive power
C	optimum power flow control	as in case B

Table 4. Control strategies for RSC and GSC with protection

Case	Protection	RSC Control	GSC Control
D	Crowbar, removed after 200 ms	RSC blocked when crowbar activated	GSC controls reactive power injection during fault
E	Crowbar, removed after 100 ms	as in case D	as in case D
F	Crowbar, removed after 100 ms and DC chopper	case E with DC chopper	DC chopper activated when DC voltage reaches threshold

can be restarted in a period shorter than 200 ms. Thus, in Case E, the RSC has restarted 100 ms after the fault occurrence, after this period the voltage level rises, even during the fault. Remarkable, that DC-chopper protection does not influence the terminal voltage during the fault, it is identical to Case F. The DC-chopper decreases the overvoltage at DC-link and dissipates the unbalance active power between RSC and GSC.

In cases D, E and F reactive power injected into the grid during the fault. After 100 ms, Cases E and F have the RSC restarted and then are able to inject more reactive power before the fault clearance. This fact increases the voltage level at the wind farm terminals and reduces the reactive power injection smoothly after the fault clearance. Case D consumes reactive power until the RSC is restarted 50 ms after the fault clearance. Evaluation of different deactivation times for the crowbar system has shown that the shorter this time is the better the voltage support during the fault. The use of the DC-chopper turns possible the reactive power injection during the fault. In this case, also, the control of reactive power by RSC and GSC is recommended and the strategy implemented in the Case F should be used.

Series Dynamic Resistors in the Rotor and Stator Circuit

Another approach for limiting excessive current in the rotor is a series dynamic resistor (SDR), proposed by Yang et al. (2010), which connected in series with the rotor circuit. It controlled by a power electronic switches and in normal operation, the switch is closed and the resistor bypassed; during fault conditions, the switch is opened and the resistor is connected in series to the rotor winding (see Figure 6).

The difference between the SDR and the crowbar or DC-link chopper is its topology. The latter is shunt-connected and control the voltage, while the SDR has the distinct advantage of controlling the current magnitude directly. Moreover, with the SDR, the high voltage will be shared by the resistance because of the series topology. Therefore, the induced overvoltage may not lead to the loss of converter control. Therefore, it not only controls the rotor overvoltage which could cause the RSC to lose control,

Figure 6. DFIG rotor equivalent circuit with crowbar, DC chopper and series dynamic resistor

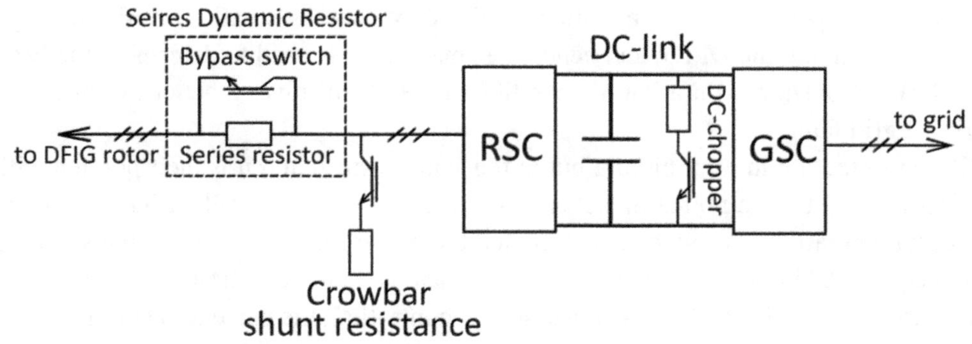

but, more significantly, limits high rotor current. In addition, the limited current can reduce the charging current of the DC-link capacitor, hence avoiding DC overvoltage.

The performance of the crowbar and the SDR protection schemes are compared by Yang et al. (2010). The reactive power, electrical torque and rotor speed of the DFIG system are simulated during the fault.

Both of the two strategies experience the reactive power and electrical torque fluctuations during the fault. However, with crowbar protection, fluctuations are much larger. The authors of Yang et al. (2010) showed with experimental results, that with the RSC connected with the SDR protection scheme, reactive power does not absorbed. On the other side, with crowbar protection, DFIG absorbs reactive power, up to 0.2 p.u. Therefore, in terms of grid voltage recovery, the SDR protection has a significant advantage.

The reactive power and electrical torque ripple are larger with SDR protection compared to crowbar protection. This is due to the higher resistance in the rotor winding and DFIG control system performance during faults, which should be explored more detailed. However, it is clear that the peak torque that occurs when crowbar switches, is significantly higher than that for the SDR. This leads to the large torque fluctuation when the crowbar is engaged. The SDR reduces the rotor overspeeding more effectively than the crowbar circuit. SDR has a much smaller impact than the crowbar, especially during switching off. Improper crowbar switching strategy (without the coordination of controller reference setting (Morren and H. de Haan, 2005) can cause frequent switching, which affects fault recovery.

The authors of Okedu et al. (2012) developed further the idea of series braking resistors and investigated its performance in comparison with another protection scheme based on supplementary rotor current (SRD) controller.

A small size SBDR can be inserted in series with the stator circuit of the DFIG through the control of power electronic switches to balance the active power, which eventually improves the wind generator stability during a grid fault and is less expensive than other techniques such as superconducting fault current limiter (Park, Sung, 2010), passive resistance network (Yan, Venkataramanan, & Wang, 2009) or series antiparallel thyristor (Petersson, Lundberg, & Thiringer, 2005). Two schemes, which work in combination with the DC-link chopper and braking resistor, are proposed by Okedu et al. (2012) to limit the rotor current of the DFIG during a grid fault. The first scheme considers coordinated control of the d and q axis RSC rotor currents via RSC controller. The second scheme conceives small value of SDBR in the stator circuit. Both schemes can limit the rotor current of the DFIG during a grid fault, and Scheme 1 is cheaper but it can only limit the rotor current, while Scheme 2 can also stabilize the DC-link voltage and other variables of the DFIG, like the rotor speed, terminal voltage, etc. In Scheme

2, the effect of the magnitude of the SDBR is also investigated by Okedu et al. (2012) along its insertion time and duration of operation. The simulation results show that the smaller SDBR value gives better performance in stabilizing the DFIG, and hence the small value is used to determine the best insertion time of the SDBR. The shorter duration of operation of the SDBR gives a better response of the DFIG system during a grid fault.

If SDBR connected not in rotor circuit, but in the stator instead and it is incorporated with the DC-link protection scheme, this combination helps to improve further the overall DFIG performance during a fault. In normal operation, the switch is closed and the resistor is bypassed, but the switch opens and the resistor is connected in series to the stator circuit during a fault condition.

As in the case of SDR, SDBR has the advantage of controlling the current magnitude. Other advantages of SDBR topology the same as in case of SDR. The SBDR can also balance the active power of the DFIG, that way improving the DFIG wind generator stability. Another important advantage of SDBR that it will increase the generator output, which in turn, reduces its speed acceleration during a voltage dip, providing overspeed control. It will improve the post-fault DFIG recovery and the entire WPP, because the SDBR controls and improves the rotor speed acceleration during a grid fault.

The distinctive merit of SDBR is based on the fact that its effect is related to current magnitude rather than voltage magnitude. SDBR connection at the stator side is shown in Figure 7 (Causebrook, Atkinson, & Jack, 2007). When the SDBR is connected to the stator of the DFIG instead of rotor side, it will increase the mechanical power extracted from the drive train, thus reducing its speed excursion. Since mechanical torque is proportional to the square of the stator voltage of the DFIG, the effect would enhance the post-fault recovery of the DFIG.

The SDBR control performed by inserting a resistor in the stator during a fault, thus the terminal voltage increases, mitigating the destabilizing condition of electrical torque and power. The bypass switch for the SDBR is normally closed, but when a voltage dip below 0.9 pu occurs due to a grid fault, it opens to allow current to pass through a small series resistance. When the voltage is recovered above a certain specified level, the bypass switch closes and the stator circuit restores to its normal state. The resistor should be selected according to its temperature limit and the maximum energy which can be dissipated during the short period.

Figure 7. Example of DFIG protection schemes connection: crowbar, DC chopper, SDBR in the stator side

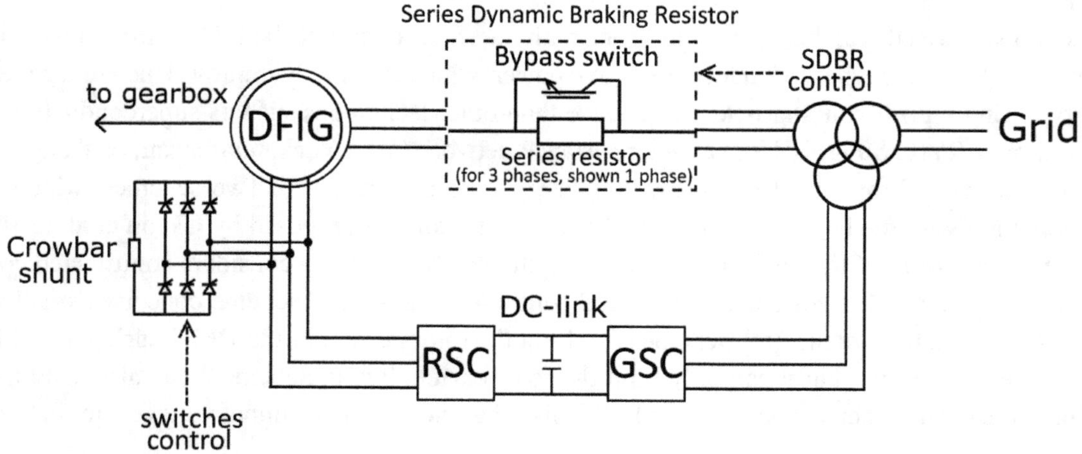

During a grid fault, the SDBR mitigates acceleration of WT. This effect is a result of the additional power, some of which is exported into the grid and the remainder is dissipated in the SDBR resistor (Yang et al., 2010). Dissipated energy can be optimized by changing the switch-out time. The switch-in time, on the other hand, should be as short as possible to maximize its speed limitation effect. Two control strategies by using a bypass switch and a circuit breaker are investigated by Okedu et al. (2012) to show the effect of the SDBR magnitude, along with its insertion time and duration of operation, on the stability of the wind generator. Different values of SDBR resistance, 0.05, 0.1, and 0.15 pu, are used by Okedu et al. (2012). According to the simulation results, the small SDBR resistance of 0.05 pu gives a better response.

Better responses of the terminal voltage and rotor speed of the DFIG were achieved Okedu et al. (2012) when the SDBR is connected at the stator during the grid fault. The effects of the insertion time and duration of operation of the SDBR on the performance of DFIG are investigated in Okedu et al. (2012): the quicker the insertion time and the shorter the duration of operation of the SDBR are, the better responses can be obtained in the rotor speed and terminal voltage of the DFIG during the grid fault.

HIGH CURRENT SWITCHES AND ANTIPARALLEL THYRISTORS

Another proposed technique is to design DFIG converters in such a way, that they could withstand high currents during the fault (Petersson, Lundberg, & Thiringer, 2005). The idea is that only the insulated gate bipolar transistor (IGBT) modules should be designed for a higher current, while the rotor winding and the converter (cooling, etc.) should still only be designed according to the slip power. This means that the converter can briefly (for a few seconds) handle a higher current and thereby stay connected to the grid longer without any crowbar action. Still, this system will have high fault currents from the stator during the voltage sag.

Therefore, the power electronic converter only needs to handle a fraction (20%–30%) of the total power. Since the losses in the power electronic converter depend on the current through the valves, it is important to have a stator-to-rotor turns ratio of the DFIG that minimizes the rotor current. For example, if the stator-to-rotor turns ratio n_s/n_r is 0.4, the rotor current is approximately 0.4 times smaller than the stator current if the magnetizing current is neglected. Moreover, if the slip s of the DFIG is 30%, the rotor voltage will be approximately $Vr = [s/(ns/nr)]Vs = (0.3/0.4)Vs = 0.75Vs$, i.e. 75% of the stator voltage, which leaves room for a dynamic control reserve.

However, the maximum rotor current might actually be much higher, when the voltage returns than at the voltage drop. The authors of Petersson et al., (2005) showed, that the maximum rotor current can be twice as high when the voltage returns. This means that the valves have to be even more over-dimensioned. In order to avoid these high currents when the voltage returns, authors of Petersson et al., (2005) proposed antiparallel thyristors (Figure 8) connected in series with the stator in order to achieve a quick disconnection of the stator circuit. By interrupting the stator circuit, the flux oscillation will also be interrupted. As soon as the flux is interrupted, it is possible to remagnetize the DFIG quickly through the rotor converter and connect the stator circuit to the grid again. The converter need not be disconnected from the rotor circuit since the valves of the converter are over-dimensioned. Moreover, if required, the grid-side converter may provide the grid with reactive power during the sag if the capacity of the DC-link capacitor is large enough or designed accordingly.

Figure 8. Antiparallel thyristors in the stator side

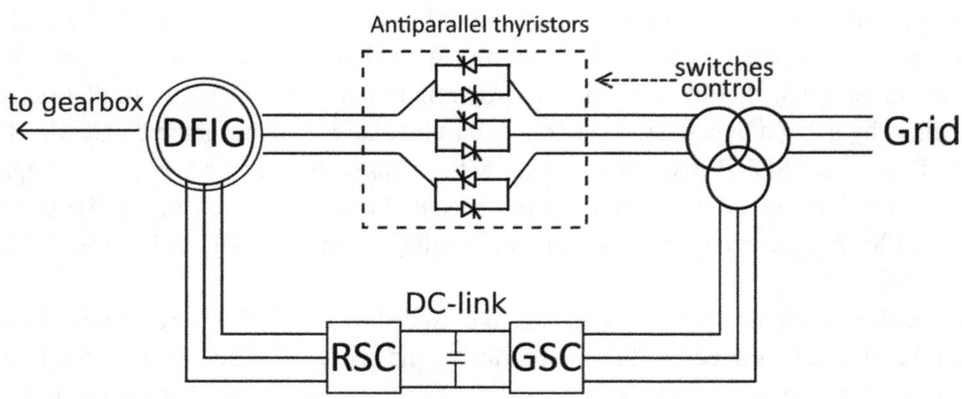

The antiparallel thyristor switch can disconnect the stator within a half cycle, i.e. in 10 ms (Dittrich and Stoev, 2003; Bollen, 2002). However, option that is more preferable would be to have antiparallel IGBTs. Thus, the IGBT modules of the RSC are designed for a higher current rating in order to withstand voltage sags. Moreover, in order to avoid the possible higher rotor currents when the voltage returns, the stator circuit is disconnected from the grid. Then, after a disconnection, the rotor current controller remagnetizes the DFIG, and the DFIG can be synchronized to the grid as soon as the voltage has returned to an acceptable, predefined level. Of course, for IGBT modules that can handle higher currents temporarily, the stator circuit can be disconnected less often owing to voltage sags.

A faster disconnection from the grid could also help to limit the DC-link voltage and possibly also limit the maximum rotor current due to voltage sags even further. Moreover, if required, the GSC may provide the grid with active or reactive power during the sag if the capacity of the DC-link capacitor is large enough.

FUTURE RESEARCH DIRECTIONS

In chapter main approaches for DFIG protection circuit implementation were analyzed and proposed. Nevertheless, there are some pending issues, related to protection systems. First of all, considered protection circuits were simulated by different researches under various conditions. It is important to perform computer simulation or real-time simulation with the same configuration of equipment and various fault scenarios. For example, it is important to investigate the influence of symmetrical and unsymmetrical fault events on different protection circuits. Also, very important to observe all possible DFIG parameters and compare results, because in some research articles protection scheme eliminate voltage and current deviations, but no clue provided how torque or speed changes during this experiment.

For each strategy, it is important to define boundary conditions: what type of fault it can withstand. It will help for DFIG customer or user to decide what protection strategy best for him. If WT DFIG operates in the local microgrid, during the fault it should be capable to replace reserve power generation unit (e.g. diesel generation unit) for the time, this unit starts-up. So for protection circuit requirements will be less severe. On the other hand, protection schemes with maximum performance should be implemented for crucial WTs.

CONCLUSION

Crowbar protection scheme became a standard approach for modern WT with DFIG. Despite its simple, inexpensive and reliable design, crowbar has a significant drawback, because it blocks RSC during the fault. In that way, WT should operate in power grids, because it doesn't meet modern grid code requirements for grid support during the fault.

Nevertheless, a solution for improvement LVRT of the DFIG with crowbar proposed by Hansen et al. (2007). In that solution focus attracted by control algorithm which allows coordinating reactive power control by RSC and GSC. In normal conditions, GSC does not inject reactive power into the grid, reactive power supply performed by the stator circuit. During the fault, RSC is blocked and its ability to control reactive power lost. Reactive power support task is taken over by the GSC.

In addition to GSC reactive power control scheme, it is important to damp possible torsional excitations in the drive train, excited due to the grid fault. Because pitch control damps the slow frequency deviations in the rotor speed, damping controller acts like a more rapid unit and helps to minimize the grid fault effect on both the mechanical and the electrical side of the turbine. Damping controller has an impact on the RSC by providing the set-point for active power.

This improvement of converters control circuit allows WTs to support the grid during the fault and meet grid code requirements. But for this purpose GSC should not be blocked or damaged. When the fault occurs and crowbar activated, DC-link voltage fluctuates. It is important to keep DC-link voltage stable and limit the variation of the GSC current. DC-link capacitor defines the level of voltage sag that converter could withstand, and with higher capacitance overvoltages have a smaller value. But on the other hand, excessive capacitance lead to the higher magnitude of oscillations in electrical torque, rotor and stator currents. Therefore it is important to eliminate DC voltage fluctuations. DC chopper technique was used in electrical machines for this purpose, and in WT chopper is used to dissipate unbalanced active power, flowing between RSC and GSC. For maximum effectiveness of the DC chopper and crowbar systems, which operate together, they should be adjusted properly. Studies performed by Okedu et al. (2012) shown, that with shorter crowbar deactivation time the voltage support is better. The use of the DC-chopper turns possible the reactive power injection during the fault because the risk of GSC deactivation due to overvoltages eliminated.

With SDR protective strategy dynamic resistors in the rotor circuit limit the rotor overcurrent. The difference with a crowbar is the topology of SDR. Both crowbar and SDR experience the reactive power and electrical torque fluctuations during the fault, but with crowbar protection fluctuations magnitude is much larger (Yang et al., 2010). SDR can be connected in the rotor side, but SDBR with a small resistance connected in series with the stator circuit. Unlike crowbar, SDR and SDBR control current magnitude rather than voltage magnitude. Peak torque that occurs when crowbar switches, is significantly higher than that for the SDR commutation, so SDR reduces rotor overspeeding more efficiently than a crowbar.

The effects of the insertion time and duration of operation of the SDBR on the performance of DFIG are investigated by Okedu et al. (2012): the quicker the insertion time and the shorter the duration of operation of the SDBR are, the better responses can be obtained in the rotor speed and terminal voltage of the DFIG during the grid fault.

Last analyzed technique based on RSC and GSC switch design. The authors of Petersson et al., (2005) it proposed, that over-dimensioning IGBT in some degree allows DFIG withstand not severe grid disturbances without designing generator windings and cooling system for higher currents. Power converters can briefly withstand high currents and stay connected without any crowbar action. This does not mean, that such DFIG need not crowbar at all. Protection circuit such as a crowbar, DC chopper or SDBR must be implemented, but they will act only during the severe grid fault, when small voltage dips could ride through with high current switches. Another advantage of this approach, that if power converters not blocked, DFIG can supply active and reactive power without any interruptions.

In addition, the authors of Petersson et al. (2005) proposed a protection strategy, based on the anti-parallel thyristors connected in series with the stator. This strategy can ensure quick disconnection of the stator circuit if current too high. By interrupting the stator circuit, the flux oscillation, caused by a fault, will also be interrupted. As soon as the flux is interrupted, it is possible to remagnetize the DFIG quickly through the rotor converter and connect the stator circuit to the grid again.

REFERENCES

Ackermann, T. (Ed.). (2005). *Wind Power in Power Systems*. Stockholm: John Wiley & Sons. doi:10.1002/0470012684

Bollen, M. H. (2002). Understanding Power Quality Problems: Voltage Sags and Interruptions. *Proceedings of IEEE*.

Causebrook, A., Atkinson, D. J., & Jack, A. G. (2007). Fault ride through of large wind farms using series dynamic braking resistors. *IEEE Transactions on Power Systems*, *22*(3), 966–975. doi:10.1109/TPWRS.2007.901658

David, W. G., Eduard, M., Tian, T., & Mackay, M., Weisheng, W. (2016). *Comparison of Standards and Technical Requirements of Grid Connected Wind Power Plants in China and the United States*. Technical Report NREL.

Dittrich, A., & Stoev, A. (2003). Grid voltage fault proof doubly-fed induction generator system. *Proceedings of 10th European Conference on Power Electronics and Applications*.

Hansen, A. D., Michalke, G., Sorense, P., Lund, T., & Iov, F. (2007). Co-ordinated voltage control of DFIG wind turbines in uninterrupted operation during grid faults. *Wind Energy (Chichester, England)*, *10*(1), 51–68. doi:10.1002/we.207

Hu, J., Sun, D., He, Y., & Zhao, R. (2006). Modeling and control of DFIG energy generation system under grid voltage dip. *Dianli Xitong Zidonghua*, *30*, 21–26.

Liang, J., Qiao, W., & Harley, R. G. (2010). Feed-forward transient current control for low-voltage ride-through enhancement of DFIG wind turbines. *IEEE Transactions on Energy Conversion*, *25*(3), 836–843. doi:10.1109/TEC.2010.2048033

Lima, K., Luna, A., Rodriguez, P., Watanabe, E., Teodorescu, R., & Blaabjerg, F. (2008). *Doubly Fed Induction Generator Control Under Voltage Sags*. Paper presented at the IEEE Energy2030 Conference, Atlanta, GA. 10.1109/ENERGY.2008.4781064

Morren, J., & de Haan, H. (2005). Ridethrough of wind turbines with doubly-fed induction generator during a voltage dip. *IEEE Transactions on Energy Conversion*, *20*(2), 435–441. doi:10.1109/TEC.2005.845526

Okedu, K. E., Muyeen, S. M., Takahashi, R., & Tamura, J. (2010). Comparative study between two protection schemes for DFIG-based wind generator. *IEEE Transactions on Power Energy*, *12*, 138–152.

Okedu, K. E., Muyeen, S. M., Takahashi, R., & Tamura, J. (2012). Wind Farms Fault Ride Through Using DFIG With New Protection Scheme. *IEEE Transactions on Sustainable Energy*, *3*(2), 242–254. doi:10.1109/TSTE.2011.2175756

Park, W., & Sung, B. C. (2010). The effect of SFCL on electric power grid with wind turbine generation system. *IEEE Transactions on Applied Superconductivity*, *20*(3), 1177–1181. doi:10.1109/TASC.2010.2040918

Petersson, A., Lundberg, S., & Thiringer, T. (2005). A DFIG wind turbine ride through system influence on energy production. *Wind Energy (Chichester, England)*, *8*(3), 251–263. doi:10.1002/we.157

Rathi, M. R., & Mohan, N. (2005). *A novel robust low voltage and fault ride through for wind turbine application operating in weak grids*. Paper presented at the IEEE Industrial Electronics Society Annual Conference. 10.1109/IECON.2005.1569295

Rebecca, L. B. (2012). *Wind Power: The Industry Grows Up*. Tulsa, OK: PennWell.

Renewable Energy Policy Network for the 21st Century. (2018). *Renewables 2018, Global status report*. Paris, France: REN21.

Salles, M. B., Cardoso, J. R., Grilo, A. P., Rahmann, C., & Hameyer, K. (2009). Control Strategies of Doubly Fed Induction Generators to Support Grid Voltage. *IEEE Transactions on Power Energy*, *135*, 1551–1556.

Santos-Martin, D., Rodriguez-Amenedo, J., & Arnaltes, S. (2009). Providing ride-through capability to a doubly fed induction generator under unbalanced voltage dips. *IEEE Transactions on Power Electronics*, *24*(7), 1747–1757. doi:10.1109/TPEL.2009.2016965

Shang, L., & Hu, J. (2012). Sliding-mode-based direct power control of grid-connected wind-turbine driven doubly fed induction generators under unbalanced grid voltage conditions. *IEEE Transactions on Energy Conversion*, *27*(2), 362–373. doi:10.1109/TEC.2011.2180389

Sourkounis, C., & Tourou, P. (2012). *Grid Code Requirements for Wind Power Integration in Europe*. Paper presented at the Power Options for the Eastern Mediterranean Region, Limassol, Cyprus.

Sun, T., Chen, Z., & Blaabjerg, F. (2005). Transient stability of DFIG wind turbines at an external short-circuit fault. *Wind Energy (Chichester, England)*, *8*(3), 345–360. doi:10.1002/we.164

Takahashi, R., Tamura, J., Futami, M., Kimura, M., & Idle, K. (2006). A new control method for wind energy conversion system using doubly fed synchronous generators. *IEEE Transactions on Power Energy*, *126*(2), 225–235. doi:10.1541/ieejpes.126.225

Wu, Z., Sun, H., & Du, Y. (2014). A Large Amount of Idle Capacity Under Rapid Expansion: Policy Analysis on the Dilemma of Wind Power Utilization in China. *Renewable & Sustainable Energy Reviews*, *32*, 271–277. doi:10.1016/j.rser.2014.01.022

Xiang, D., Ran, L., Tavner, P., & Yang, S. (2006). Control of a doubly fed induction generator in a wind turbine during grid fault ride-through. *IEEE Transactions on Energy Conversion*, *21*(3), 652–662. doi:10.1109/TEC.2006.875783

Yan, X., Venkataramanan, G., & Wang, Y. (2009). *Grid fault tolerant operation of DFIG wind turbine generator using a passive resistance network*. Paper presented at the Energy Conversion Congress and Exposition (IEEEECCE), San Jose, CA.

Yang, J., Fletcher, E., & O'Reilly, J. (2010). A series dynamic resistor based converter protection schemes for doubly fed induction generator during various fault conditions. *IEEE Transactions on Energy Conversion*, *25*(2), 422–432. doi:10.1109/TEC.2009.2037970

Zhou, P., He, Y., & Sun, D. (2009). Improved direct power control of a DFIG-based wind turbine during network unbalance. *IEEE Transactions on Power Electronics*, *24*(11), 2465–2474. doi:10.1109/TPEL.2009.2032188

KEY TERMS AND DEFINITIONS

Crowbar: A shunt resistance, connected into rotor circuit, and protecting RSC from overcurrent.

DFIG: Doubly fed induction generator is a wound rotor induction generator in which rotor windings powered by variable frequency, provided by power converter via slip rings.

GSC: Grid side converter is an invertor, connected from the grid side. It has the general task of maintaining constant voltage on the DC side.

LVRT: Low voltage ride through is a capability of generation unit to withstand low voltage conditions.

Power Converter: A power electronics device, consisted of two inverters in back-to-back configuration for rectifying one frequency f_1 and then inverting DC into frequency f_2 . During this frequency conversion process power transferred as well.

RSC: Rotor side converter is an invertor, connected to the DFIG rotor. The main task is to regulate active and reactive output from DFIG.

SDRB: Series dynamic braking resistors is a protection device that can be connected to the rotor or stator side and damps excessive power.

Chapter 12
Energy and Water Saving on the Pumping Stations of Karshi Main Canal

Vladimir Aleksandrovich Khokhlov
Turin Polytechnic University in Tashkent, Uzbekistan

Aleksandr Vasilevich Khokhlov
Research-and-Production Enterprise "Vodopodyomnik" ("Water Lifting"), Uzbekistan

Janna Olegovna Titova
Tashkent State Technical University, Uzbekistan

Azamat Ilham Ugli Kurbonov
Tashkent State Technical University, Uzbekistan

ABSTRACT

This chapter describes the advanced technologies of energy and water saving on the Uzbekistan largest pumping stations of Karshi Main Canal. The unique Karshi pumping stations stage consumes more than 4% of the total electricity generated in Uzbekistan. Such a significant consumption of electricity by irrigative pumping stations, gives the problem of energy conservation to them a status of special importance and relevance. Advanced technologies of energy saving and non-cavitation work are scientifically substantiated and developed. The results of carried out field tests are described. The scientific results of the research are recommended to allow efficient use of water and energy resources and to ensure reliable operation of the power equipment of pumping stations in the rural branches of economy.

DOI: 10.4018/978-1-5225-9179-5.ch012

INTRODUCTION

It is difficult to overestimate the importance of pumping stations in the rural and energy sectors of the economy of Uzbekistan.

The pumping stations are among the main electric power consumers in the Republic of Uzbekistan. From about 55 billion kWh/year the electric power developed in Uzbekistan approximately 8 billion kWh/year or 16% are spent for covering the demands of more than 1500 state irrigative pumping stations. All of them are now under the jurisdiction of the Ministry of Agriculture and Water Resources of the Republic of Uzbekistan. At these pumping stations about 5000 pump units are established. Their total power is about 3.8 million kW and the general water delivery is about 7000 m^3/s.

In Uzbekistan 2.4 million hectares out of total 4.3 million hectares of irrigated lands now are irrigated by means of pumping stations. Annually about 55 billion m^3 of irrigating water are pumped. This volume consists of 24 billion m^3 pumped by the first lifting, and 31 billion m^3 by the subsequent liftings of pumping stations stages.

It is enough to say that only one Karshi pumping stations stage, consisting of seven pumping stations, consumes more than 4% of the total electricity generated in the republic.

The Karshi Main Canal (KMC) is the largest irrigation facility providing water for 364 thousand hectares of Karshi steppe land. The total length of the canal is 176 km. The water is taken from the Amu Darya River and is risen to a height of 132 m using a stage of seven pumping stations. The operation of pumping stations is carried out by the Department of Operation of the Karshi Main Canal. In addition, there are other seven small auxiliary pumping stations in the KMC.

The total installed capacity of pumping stations in the Karshi stage is more than 580 thousand kW. The volume of pumped irrigation water is about 4.5 billion m^3 per year. If we consider the total volume of water pumped by all stations of the stage, then the whole flow rate is 25.5 billion m^3 per year. The energy consumption of the stage reaches 2.2 billion kWh per year.

Such a significant consumption of electricity by pumping stations, gives the problem of energy conservation to them a status of special importance and relevance.

A distinctive feature of the Karshi stage of pumping stations is the use of the largest axial pumps. The supply of one pump is about 40 m^3 / s. This is the largest flow rate of pumps in the Commonwealth of Independent States.

If saving water and energy resources, there is a need to improve the mechanism for saving irrigation water and electricity consumption at pumping stations.

Besides, the difference between the pumping stations of the Karshi stage and the vast majority of pumping stations in Uzbekistan is that they operate all year round. In the autumn-winter period, pumping stations of the Karshi stage are switched to fill the Talimarjan reservoir, creating a supply of water for the next irrigation season. This schedule of the stage influences very much on the choice of operating modes of pumping stations and on the type of current and capital repairs.

BACKGROUND

The western half of the Kashkadarya region of Uzbekistan is occupied by a vast sand-bordered plain with large reserves of irrigable land called Karshi steppe.

The Karshi steppe occupies the plain of South-Western Uzbekistan, with a total area of 13.6 thousand km^2. The Karshi steppe in the north and northeast runs along the foot of the Terek-Baba and Zirabulak Mountains; in the west, approximately along the route of the Amu-Bukhara Canal to Lake Karabulak; in the south-west and south along the border between Turkmenistan and Uzbekistan, and in the south-east and east about along the route of the Karshi Main Canal to the city of Karshi and further north (skirting from the west of Mount Kongurtau), almost along the meridian to the Zirabulak mountains.

On the territory of the Karshi Steppe, due to the relief, there are the Karshi Steppe itself in the north, the Nishan Steppe in the south and the irrigated oasis on the Kashkadarya River.

In the southwest, the steppe passes into vast areas of sand. This is the southern tip of the Kyzylkum – the sandy desert of Sundukli, used as pasture land for Karakul sheep.

The Karshi Steppe is characterized by a very short cold winter with little atmospheric precipitations and a long hot summer. Spring is relatively unstable, with a small amount of precipitations per year. Autumn is warm, mostly dry. In the Karshi steppe, there is intense wind activity in the form of wind and dust storms, blowing mainly from the north and north-west.

A distinctive feature of the entire southwestern part of Uzbekistan is the huge disproportion between the available water and land resources. This disproportion is vividly illustrated by the following examples: the head flow of surface water resources per hectare of rural land in the Karshi and Surkhandarya regions is 1.6 thousand m^3, in the Zarafshan valley – 2.5÷2.7 thousand m^3. While in Tashkent, Syrdarya regions and the Fergana valley, the figure reaches 11÷16 thousand m^3, Khorezm region and Karakalpakstan – 14.7 thousand m^3.

However, the limited own surface water resources that the Kashkadarya region possesses are formed outside the Karshi steppe.

The main source of irrigation is the Kashkadarya River. It begins in the western part of the Gissar Range, at an altitude of about 1,300 m. The river flows through the wide intermountain valley to the city of Karshi. Here, the waters of Kashkadarya and its tributaries are intensively sorted out for irrigation, some of them return to the river again in the form of discharges from irrigation fields.

Due to its low capacity, the river brings its waters here mainly during the flood period. Therefore, in the lower reaches of the Kashkadarya River, only a probably irrigated grain farm could exist on a very small area, based on the flood runoff.

The idea of developing the Karshi steppe is rooted in the past. Russian scientist G.L. Arendarenko established the fact of using the waters of the Zarafshan River to irrigate the lands of the Karshi Steppe. Even in ancient times, for these purposes, the Monas canal was stretched for 300 miles from the Zarafshan River. The canal went through a tunnel 1200 meters long. The width of the canal along the bottom in the tunnel was five meters, and its depth after leaving the tunnel was at least four meters.

This message interested Uzbek archeologists and engineers. Further studies have shown that the current Eski-Ankhor canal is laid along the ancient Monas canal. There is a legend that even 500 years ago every spring the farmers of the Karshi steppe came to the village of Djam (Samarkand region), where, together with the local residents, they cleaned the Eski-Ankhor canal, after which they organized a holiday. However, as a result of the internecine wars between the local khanates, the Eski-Ankhor canal was destroyed.

In the projects of engineers of the middle of the XIX century there are interesting proposals on turning the flow of the Amudarya River to where it is most needed. The beginning of wide irrigation of new lands in the republics of Central Asia, including the Karshi steppe, was laid in the twenties of the last century. Equipment, various machines and materials were sent to Karshi Steppe. Specialists of water and agriculture arrived. They scientifically resolved the problems related to improving water use and developing virgin massifs in the Karshi Steppe.

In 1960, the Institute "Sredazgiprovodkhlokopok" compiled the Technical and Economic Report on machine irrigation of the lower reaches of Kashkadarya by the waters of the Amudarya River." It was supposed to develop possible options for machine irrigation of the Karshi steppe, determine the composition of hydraulic structures and parameters of the unique equipment of pumping stations, decide their power supply schemes. This ultimately determined the scope of works, their cost, technical and economic feasibility of the planned activities in irrigation of the Karshi steppe by the waters of the Amudarya River.

As a result of the studies, it was recommended to make KMC along the shortest route from the Kizilayak terminal section on the Amudarya River and primarily include 200 thousand hectares of land located in the downstream part of the Kashkadarya river into irrigation zone of the canal. At the Kizilayak site, the construction of a temporary non-damming intake hub was planned.

In the Technical and Economic Report it was shown that within the Karshi steppe there are about 1 million hectares of land suitable for agriculture, which can be irrigated only with the help of a machine water lift. Water supply to these areas was to be provided by the Karshi Main Canal. Calculations of the TER revealed that the land reclamation of the Karshi steppe on the basis of the machine water lift in terms of the volume of specific capital investments and other technical and economic indicators had a rather high efficiency.

In the 70s of the last century, all the conditions necessary for the development of the Karshi steppe coincided:

- Plants have mastered the products of high-performance pumping and power equipment;
- Relatively cheap electricity appeared in Uzbekistan;
- Equipment was created for the manufacture and installation of a unique metal pipeline;
- Powerful construction equipment and technologies were created for the construction of perfect irrigation and drainage systems.

In March-April 1973, test launches and tests of pumping units began at the central part of the KMC. The tests were successful. On June, 1973, the Amu-Darya water, lifted by pumping stations, filled the Mirishkor canal and was put into the open spaces of the Karshi steppe. In 1973, the first harvest was grown on the irrigated lands.

From 1976, the introduction of the Kizilayak hydroelectric complex was planned. However, the question of its construction has remained unresolved to present day.

This is due to the identification of serious deficiencies in the design of the site, mainly in solving the problem of control of sediment during the concentrated selection of large amounts of water from the river. In this connection, it was necessary to expand the temporary free-flow water intake into the Karshi Main Canal to a throughput of 200 m^3 / s. This flow rate is sufficient for irrigation of the upper zone of the Karshi steppe, if there is a regulating Talimarjan reservoir in the KMC system and for extracting large amounts of water from the river throughout the year.

An increase was made in the cross section of the canal from the Amudarya River and the expansion of the section of the main canal to the pumping station No. 1. This section plays the role of a precipitation reservoir. However, due to the lack of confidence that it will be possible to constantly maintain a stable operational state of the existing canal, the designers developed a version of an improved water intake. Its was located at a more stable point on the right bank of the Amudarya River. A new supply canal with a length of 1.5 km runs in the rocks of the Cape Pulizindan.

A two-chamber precipitation reservoir with a length of more than 2 km and a bottom width of 60 m is created behind the inlet canal. At the end of each chamber, a regulator is provided. Removal of sediment is produced by suction units with storage in the surrounding area.

The practice of operating the existing temporary dam-free water intake into the Karshi Main Canal showed that the withdrawal of water from the Amudarya River at Cape Pulizindan is connected with great difficulties. In this part of the river, intensive processes determine the large variability of the depths. This fact jeopardizes the possibility of guaranteed large amounts of water from the river, especially in the low-flow period.

TECHNICAL REVIEW OF THE CONTEMPORARY STATE OF THE KARSHI MAIN CANAL

The Karshi Main Canal is divided into two parts: the head part which is the machine canal and the working part which is the gravity canal. At the junction of these parts, the Talimarjan water reservoir with a capacity of 1.5 billion m^3 is located. The head part of the KMC is designed to pass a normal water flow of 175 m^3 / s, and the working part is calculated at 360 m^3 / s.

The head part of the KMC and six first pumping stations are located in the territory of Turkmenistan. The seventh dam pumping station is located on the territory of Uzbekistan.

The first six pumping stations of the KMC stage are equipped with axial vertical pumps. Each pump operates on an independent discharge pipe with a diameter of 3640 mm. In the end part, the pressure pipelines are combined with a siphon type outlet pipe from monolithic reinforced concrete. The cross section of the siphon outlet is rectangular. From the inside it has a steel lining.

Normal delivery of pumping stations is 175 m^3 / s, forced - 195 m^3 / s. During the operation of all pumping units installed in the stage, including the backup ones, it is possible to deliver at 234 m^3 / s.

At the first station, three pumps OV-11-260 and three new pumps 300 VO-37 / 26C with a geometrical pressure of 17 m and a flow rate of 39 ÷ 41 m^3 / s are installed. At the stations from the second to the sixth, the OV-10-260 pumps are installed with a geometric pressure of 24 m and a flow of about 35 m3 / s. A total of 36 aggregates were installed - 6 at each station, five of which are working units and one is backup. The pumps are driven by electric motors GVS 375 / 130-24 or GDSM 375 / 105-24 with a power of 12.5 MW each.

The impellers of the pumps have a diameter of 2.6 m. Initially, the pumps were made rotary-blade. The flow and head of the pumps could be changed during their operation. However, literally after several years of operation, the drive system, which allows the blade angle to change, began to fail, and moreover, serious accidents occurred associated with the separation of the blades during pump unit operation. They destroyed the pump casing, which led to the flooding of pumping stations. Therefore, the blades of the pumps were rigidly welded, as a result of which the pumps became unregulated.

The Talimarjan reservoir's facilities include pump station No. 7, earth dams No. 1 and No. 2, a water outlet and a bypass canal.

The pumping station No. 7, designed to fill the Talimarjan reservoir, has a lower flow rate than other pumping stations of the stage of Karshi Main Canal. The pumping station No. 7 operates during about six months a year, and the rest of the time, the water completely passes through a bypass canal.

Pumping station No. 7 pumps only the volume of water that goes to the accumulation, the rest of the flow enters the working part of the canal with direct flow rate from pump station No. 6 through the bypass canal. The normal throughput capacity of the bypass canal is 175 m^3 / s, and the supply of pumping station No. 7 is 142 m^3 / s.

At the pumping station No. 7, centrifugal nine pumps are installed. Seven main pumps 2400VR-25/25 equipped with 8 MW synchronic VDS 375-89-32 electric motors each, as well as two 1600V-10-40 changeable pumps with VDS 325-59-24 electric motors capacity of 4.4 MW each. Two steel pressure pipelines of the pumping station with a diameter of 3640 mm are laid in a concrete gallery.

The Talimarjan reservoir with a capacity of 1.5 billion m^3 accumulates water supplied through the head of the canal in the autumn-winter period and allows accumulating the water evenly throughout the year. During the vegetative period it supplies additional water to the working part of the canal in accordance with the schedule of water consumption.

Technical data on the equipment of pumping stations of the Karshi Main Canal are presented in Table 1. The technical data of the actually supplied water and the consumed electric power of the pump stations of the Karshi Main Canal are presented in Table 2.

Different pumps and pumping stations, when supplying a certain water flow (for example, 1000m^3), consume different amounts of electricity. It depends on such factors as the brand of the pump and the electric motor, as well as their service life, geometric head, diameter of the pipeline and its service life, number of pumps working in parallel or in series for one line of the discharge pipe, quality of the pumped water, etc.

Comparison of actual and standard values of energy consumption will give an objective assessment of the technical quality of operation, will allow to judge quantitatively about the used reserves or the existing technological omissions, i.e. will be a mechanism to control the energy consumption of pumps and pumping stations and stimulate energy savings.

The actual energy consumption in annual terms of pumping stations No. 1 to 7 of the Karshi Main Canal for 2013 is presented in Table 3. Comparison of the actual energy consumption of pumping stations N°N° 1 ÷ 7 KMC is shown in Figure 1.

MAIN FOCUS OF THE CHAPTER

During the operation of the stage, many problems arise in the operation of pumping stations of the Karshi Main Canal.

First, pumping units for a long period of their operation have significantly worn out. For the sake of safety, all units were transferred from turning poles to blades, that is, all vanes were welded to the pump impeller. As a result, the maneuverability in controlling the operation of the stage was completely disrupted. There was nothing to synchronize the operating modes of pumping stations and intermediate canals between them. Constant switching on and off of the units for the purpose of synchronization of the modes aggravated the situation, since each start and stop is associated with a water hammer and significantly reduces the reliability of both the units and the pressure pipelines.

Secondly, the pressure pipelines themselves, due to abrasive wear, have thinned and become partially unsuitable for operation.

Table 1. Technical data of hours of operation of units of pumping stations of the Karshi Main Canal, hours

	Pump	1973-2008y.	2009 y.	2010 y.	2011 y.	2012 y.	2013 y.	Total
PSN°1	1	180058-08	3844-02	3495-07	6422-06	4687-00	5086-00	203592-23
	2	187557-01	5652-35	6288-48	5829-45	4388-00	4787-00	214503-09
	3	191434-00	7586-05	3156-13	6390-15	6820-00	6344-00	221730-33
	4	188296-06	6037-17	7538-22	4016-50	5204-00	6685-00	217777-35
	5	193856-30	6604-27	6762-42	5351-20	6095-000	5390-00	224059-59
	6	177557-06	6024-20	6570-35	5217-15	6090-00	6121-00	207580-16
	Σ	**1118758-51**	**35 748-46**	**33811-47**	**33227-31**	**33284-00**	**34413-00**	**1289243-55**
PSN°2	1	191450-11	6841-02	5728-59	5840-55	4888-000	6222-00	220971-07
	2	190679-43	6704-45	5828-34	5813-53	5556-00	5802-00	220384-55
	3	206280-27	6223-04	4196-00	7059-02	6485-00	5092-00	235335-33
	4	194100-44	7470-42	5846-14	5159-45	4994-00	5369-00	222940-25
	5	188697-04	4986-39	5750-01	6287-17	5420-00	5251-00	216392-01
	6	171400-31	4297-48	4828-50	3124-20	5934-00	7109-00	196694-29
	Σ	**1142608-40**	**36 524-00**	**32178-38**	**33285-12**	**33277-00**	**34845-00**	**1312718-30**
PSN°3	1	183692-26	6263-49	3206-44	5769-10	5685-00	6020-00	210637-09
	2	200093-32	6227-51	4609-09	6204-07	6938-00	5260-00	229332-39
	3	204580-22	5087-08	5517-05	6259-25	5103-00	7353-00	233900-00
	4	183333-56	6651-15	6032-10	5201-55	5733-00	4964-00	211916-16
	5	189841-09	6174-23	6852-34	4712-50	4580-00	5844-00	218004-56
	6	169445-18	5903-41	5783-53	5015-50	5359-00	5113-00	196620-42
	Σ	**1130968-43**	**36 308-07**	**32001-35**	**33144-17**	**33398-00**	**34554-00**	**1300411-42**
PSN°4	1	178076-18	5557-49	3382-55	6408-40	4696-00	6781-00	206902-42
	2	209747-13	5737-04	6386-30	4966-30	5568-00	5413-00	237818-17
	3	195125-13	5485-57	3800-58	4186-20	7015-00	5134-00	220747-28
	4	192340-34	6130-12	5434-52	5936-30	5247-00	6654-00	221743-08
	5	181112-22	5642-02	5287-17	5509-51	5103-00	4915-00	207569-32
	6	176656-24	6322-55	4843-13	5417-15	5067-00	5169-00	203475-47
	Σ	**1133058-04**	**34 875-59**	**31135-45**	**32420-06**	**32696-00**	**34066-00**	**1298256-54**
PSN°5	1	129955-44	4968-35	5050-01	4835-53	4387-00	1634-00	150831-13
	2	122509-10	3577-07	1392-37	3157-47	2385-00	5625-00	138646-41
	3	156138-32	1985-40	3923-13	4741-56	4606-00	4609-00	176004-21
	4	153896-37	5136-15	5200-21	5678-05	3719-00	3983-00	177613-18
	5	137027-01	5205-00	4792-00	4316-20	4237-00	3791-00	159368-21
	6	130137-42	6130-47	2784-15	4057-50	2773-00	6322-00	152205-34
	Σ	**829664-46**	**27 003-24**	**23142-27**	**26787-51**	**22107-00**	**25964-00**	**954669-28**
PSN°6	1	133737-49	6599-24	3671-36	5251-55	4079-00	2921-00	156260-44
	2	131442-14	4206-50	3149-45	4880-50	3561-00	4762-00	152002-39
	3	172513-52	2425-45	3288-07	1627-20	3348-00	3292-00	186495-04
	4	146615-25	3559-57	3987-04	5276-20	3359-00	4837-00	167634-46
	5	116729-52	4929-14	3912-58	4612-50	5494-00	5817-00	141495-54
	6	118414-07	4892-44	4604-39	4362-0	2959-00	3508-00	138740-30
	Σ	**819453-19**	**26 613-54**	**22614-09**	**26011-15**	**22800-00**	**25137-00**	**942629-37**

	Pu mp	1973-2008y.	2009 y.	2010 y.	2011 y.	2012 y.	2013 y.	Total
PSN°7	1	52172-02	1815-07	1203-30	2876-15	1513-00	1319-00	60898-54
	2	33628-01	2274-29	1517-00	2495-53	1421-00	1434-00	42770-23
	3	63564-00	3284-32	3183-05	2706-25	1669-00	2894-00	77301-02
	4	67779-00	2464-25	1123-10	3036-52	2477-00	1984-00	78864-27
	5	52171-05	194-00	18-50	703-20	312-00	745-00	54144-15
	6	39936-29	927-02	747-45	1186-35	1356-00	394-00	44547-51
	7	48398-57	2201-55	1914-30	2733-30	559-00	1495-00	57302-52
	8	53094-22	1554-12	2260-35	1829-0	1480-00	2867-00	63085-09
	9	54129-09	2212-32	882-21	2364-05	2129-00	2053-00	63770-07
Total		**464873**	**16 928**	**12850**	**19931**	**12916**	**15185**	**542685**

The most adverse effects occur during transients associated with a sudden emergency stop of the main pumping units. This gives the largest water hammer in the pressure pipeline and the highest speed of the pump units reverse rotation. For example, on the largest in Uzbekistan Karshi pumping stations stage for the previous 5 years annually about twenty unscheduled electricity outages happened. Each of these outages led to significant losses from occurring transient and water hammer.

For example, electricity outage on January 23, 2013 happened and all 29 working pump units with capacity of 12.5 MW each were turned out with the damage of large amount. The worst effect of such events is the lack of irrigating water during the growing season, leading to significant losses from reduced agricultural yields.

After the implementation of the authors' recommendations and upgrading the power system equipment for its safe operation the number of outages at stage in 2015 reduced from twenty to four per year

Thirdly, since the construction of the dam planned on the Amudarya river in the water intake section of the Karshi stage was not carried out, all the years of operation water with sediments directly from the river came into the Karshi Main Canal. As a result, incredible efforts are necessary to clean the supply canal to the first pumping station.

It is known that axial pumps settled on a KMC are very sensitive to changes in the water level in the downstream station. Even a slight drop in water level causes cavitation. Specialists have proposed a new mode of operation of the pumping station stage. It consisted of a number of technological measures to raise the water level in the pumping station intake to the mark, which ensures stable non-cavitation operation of the pumps, which contributed to the elimination of emergencies and an increase in the pump-overmaintenance period. The new technology has greatly simplified the process of starting the unit, reduced the total height of the stage, and, consequently, reduced the consumption of electricity by the pumps.

One of the most unreliable objects at Karshi pumping stations was a complex with a pressure pools. The installed vacuum breakdown valves of the electropneumatic type not only did not ensure the tightness of the siphons, but also often refused to work. As a result, when the pump units stopped, the pumps began to rotate in the turbine mode, which is extremely harmful for pumps and electric motors. Often it reached the reversal speed, which was associated with serious accidents. It was proposed to remove all complicated electrical and pneumatic equipment, and instead of that to use the flow itself. Aeration

Table 2. Actual data of supplied water and energy consumed by KMC pumping stations for 2013

Name of Pumping Station	Water Supplied, m3	Electricity Expended, kWh
Pumping station Nº 1	4 588 753 550	343 290 760
Pumping station Nº 2	4 487 949 686	420 102 363
Pumping station Nº 3	4 382 891 526	421 423 958
Pumping station Nº 4	4 329 021 026	391 783 359
Pumping station Nº 5	3 295 677 088	283 618 497
Pumping station Nº 6	3 176 782 576	291 294 324
Pumping station Nº 7	1 303 120 000	93 378 745
Pumping station Nº M-1	35 104 320	3 417 867
Pumping station Nº M-2-1	126 122 400	13 846 896
Pumping station Nº M-2-2	110 910 600	13 214 928
Pumping station Nº M-3-1	35 858 920	20 558 400
Pumping station Nº M-3-2	314 906 220	17 660 640
Pumping station Nº M-1-A	8 493 552	583 576
Pumping station Nº M-4-A	1 257 552	465 948
Auxiliary	1 800 000	28 002 274
Total:	**26 518 649 016**	**2 342 642 535**

Figure 1. Comparison of the actual energy consumption of pumping stations NºNº 1 ÷ 7 KMK for 2013, kWh

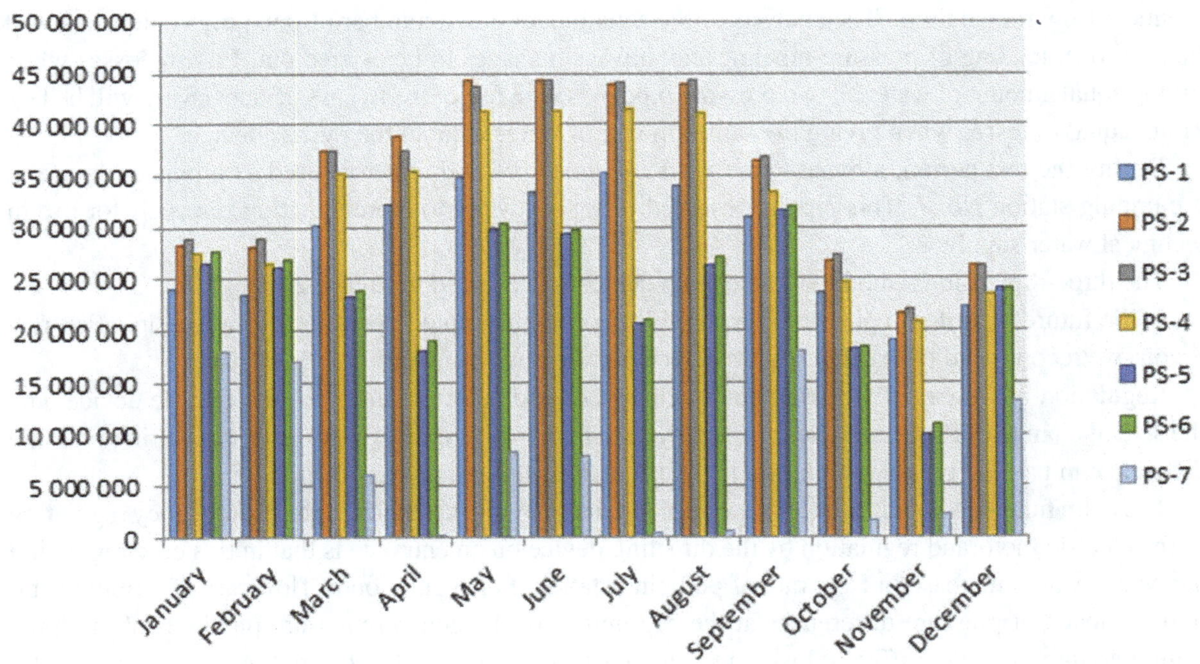

pipes appeared at the pumping stations, which ensured a smooth breakdown of the vacuum in the siphon structure.

Currently the blades of the impellers of the pumps are rigidly welded, as a result of which the pumps became unregulated. This leads to a serious problem of lack of synchronism of the stage. Due to the non-synchronous operation of pumping stations, it is impossible to maintain constant water levels in the intakes, which leads to cavitation and the forced shutdown of one of the station's pumps. The number of on-offs comes up to several times a day, which, as is known, is very harmful for the pump units.

Thus, the problem of regulation of modes of all pumping stations, has become most acute for the Karshi stage and must be solved by all means.

For this, it is promising to apply electromechanical methods of controlling the rotational speed of the pump units, for example, using electric current frequency converters for adjustable electric drives. Analysis of methods for regulating of pumping station operation modes shows that electromechanical methods are mainly considered.

At pumping stations with long pressure pipelines, with large hydraulic losses of energy, in the opinion of the authors, attention should be paid to the energy-hydraulic regulation of the operating modes, including the increasing of the supply of the pumping station.

Research and Production Enterprise "Water Lifting" proposes to use hydraulic methods for regulating the operating modes of pumping stations.

The most effective technology offered by authors is regulation of operating modes relating to pumping stations equipped with two or more parallel long lines of pressure pipeline. At the level of the inventions, an energy-saving bypass device (Patent UZ No. 2373) and energy-saving method of regulation of operating modes by means of bypass device (Patent UZ No. 4380) have been developed for pumping stations.

It was established that at the pumping stations, equipped with two and more long pressure pipelines located in parallel, energetically favorable modes are with uniform loading of pressure pipelines, i.e. at identical flow rate in them. The analysis of the actual modes on a number of pumping stations showed that uniform loading of pressure pipelines not always manages to be carried out. Energy losses when giving equal amount of water on two pressure pipelines with use of the bypass device always will be less or are equal to losses when giving the same amount of water without the bypass device.

During the test period, a bypass pipe with a diameter of 600 mm was used as a bypass device at a pumping station No. 7. This pipe is designed to supply water to a precipitation reservoir for use in technical water supply.

The three-dimensional model of the bypass device is presented in Figure 2.

In the future, in order to preserve the possibility of uninterrupted operation of the precipitation reservoir, we recommend to install a bypass device between two pipelines D = 3640 mm.

Regulation of flow rate is carried out by change of hydraulic resistance of the throttle device, and the size of change of flow rate is determined by indications differential manometer, measuring pressure difference in parallel pressure pipelines in an installation site of a bypass device.

The main fundamental difference of a way of regulation with bypass device from frequency regulation of the electric motorand regulation by the directing device on an entrance is that in this case regulation happens towards increase in flow rate of pumping station. For regulation of flow rate of pumping station existence of pressure differences at the beginnings of the general pressure pipelines of pumping station arising owing to different losses in pipelines is used. At the closed throttle devices when water transfer on a bypass device won't be carried out from one general pressure pipeline in other, each pump unit will work for the pressure pipeline. Total flow rate of pumping station will be equal to the sum of

Figure 2. Three-dimensional model of a bypass device with a throttle device

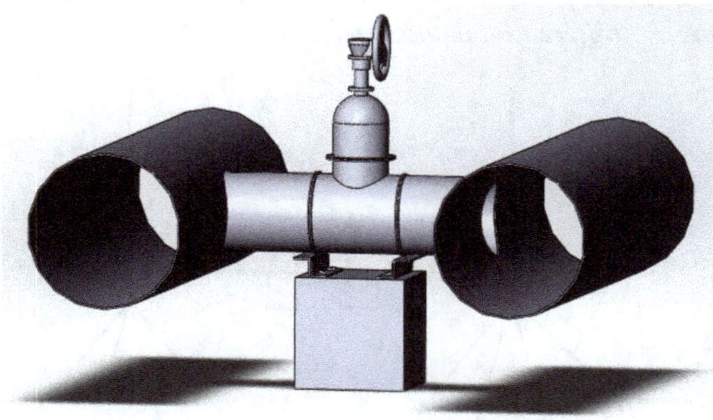

water supply in each pressure pipeline. If completely opened throttle device, the pipeline characteristic will change, and, therefore, working points of pumps will change.

Delivery of pump units at completely opened or closed throttle device will be the range of regulation of flow rate. The more the difference in water delivery in pressure pipelines (at the closed throttle devices on pressure pipelines), the more regulation range is.

It is reasonable to carry out a bypass device at the junctions of individual pressure pipeline with a common pipeline. This is explained by the fact that in this case the range of delivery regulation and the economy of regulation will be largest due to the fact that the flow rate is proportional to the first degree of water velocity in the pipeline, and the loss to resistance is proportional to the velocity`s second degree.

The way of regulation the operating modes of the pumping station is as follows.

The pumping station (Figure 3) consists of centrifugal pump units 1 with stiff-band impellers operating in parallel. Individual lines of pressure pipelines 2 of the pump units 1 are connected in two (or more) common pressure pipelines 3 and 3-a, each of the working pumps is equipped with a disc lock or a valve.

The common pressure pipelines 3 and 3-a are connected to each other by a bypass device 5, which is supplied with a cylindrical throttle device 6. The pressure difference between the inlet and the outlet of the water from the throttle device is measured by a differential pressure gage 7.

Determine the number of pumping units 1, which is necessary and sufficient to ensure a given water supply. The supply of these pumps 1 when working together and the open throttle device 6 on the bypass device 5 must be greater than the predetermined supply, but less than the supply of the pumps working only to the individual pressure pipeline 2.

Here it is the comparison of the energy losses $h_l = \lambda \frac{l}{d}\frac{V^2}{2g}$ for each variant:

$h_{l\ 1} + h_{l\ 2}$ and $2h_{l\ 1-2}$

It is necessary to prove that the energy losses at supplying an equal amount of water along two pressure pipelines will be less or equal to the losses when the same amount of water is supplied without bypass device i.e.:

Figure 3. The scheme of connection of pressure pipelines with the bypass device: 1 – pump units; 2 – individual pressure pipelines; 3 and 3-a – the general pressure pipelines; 4 – butterfly valves; 5 – bypass device; 6 – throttle device; 7 – differential manometer

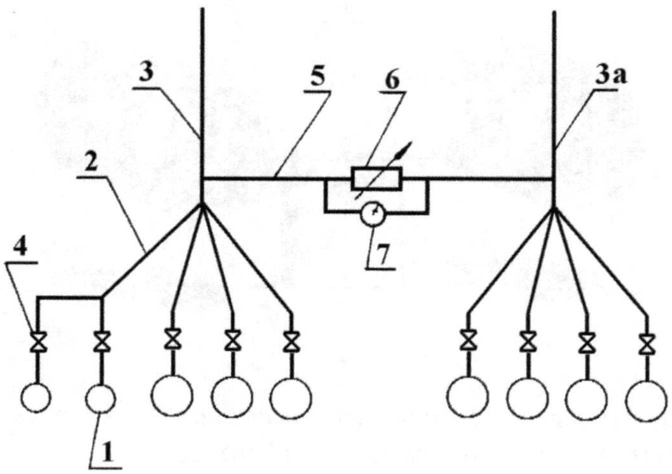

$$h_{l\,1} + h_{l\,2} \geq 2h_{l\,1-2} \qquad (1)$$

The constant $\lambda \frac{l}{d \cdot 2g}$ is denoted by *a*, then:

$$o\left(V_1^2 + V_2^2\right) \geq 2a \frac{\left(V_1 + V_2\right)^2}{4}$$

or:

$$2\left(V_1^2 + V_2^2\right) \geq \left(V_1 + V_2\right)^2 \qquad (2)$$

Then transforming the expression (2) gives:

$$\left(V_1 - V_2\right)^2 \geq 0 \qquad (3)$$

Since any number raised to a square is positive or equal to zero, then the assumption that $h_{l\,1} + h_{l\,2} \geq 2h_{l\,1-2}$ one can assume to be proven.

Thus, at various deliveries in pipelines total losses of energy on overcoming of friction forces in pressure pipelines at the closed throttle devices will be always greater, than losses at pumping of the same consumption of water at bypass device, i.e. at completely open throttle devices.

And since the energy losses in the pipeline are reduced, the characteristic of the pipeline with bypass device will pass below the analogous characteristic of the pipeline without such bypass device. Therefore, the pumps will shift their operating points "to the right", in the direction of increasing the flow rate. Delivery regulation is performed by changing the hydraulic resistance of the throttle device, i.e. by changing the degree of its opening.

The pumping units 1 are put into operation. They must ensure a given supply when the throttle device 6 on the bypass device 5 is open. The pumping station supply will become more than initial one. To reduce the supply, the throttle device 6 is being closed softly and the supply decreases. If necessary, the throttle device is closed completely. The magnitude of the change in flow is judged by the readings of the differential pressure gage 7 by measuring the pressure difference in the common pipelines 3 and 3-a at the point of their connection by the bypass device 5.

Flow rate change at a set size according to indications differential manometer carry out on the following dependence:

$$\Delta Q = Q_{\max}\sqrt{\frac{H_c - H_{\text{int}}}{H_c - H_o}} \tag{4}$$

where:

ΔQ, $Q_{\max}$ – change of flow rate of pumping station required and greatest possible;

- H_c, H_{int}, H_o – indications differential manometer at the closed, intermediate and open throttle device.

When the centrifugal pump units 1 are switched on and the butterfly valves 4 are opened, water begins to flow into the individual pressure pipes 2 and through them into the common pressure pipeline 3, which joins the individual pressure pipelines. In this case, a group of pumping units as well as one of them can deliver water into the pressure pipeline 3.

When the throttle device 6 is closed, large frictional losses of the head occur along the common pipeline 3. With the throttle device 6 open, due to the head difference at the beginning of the pressure pipelines 3 and 3-a, water through the bypass device 5 begins to flow into the pipeline 3-a until the pressures in the pipelines 3 and 3-a equalize. This will lead to a reduction of frictional losses when water is moving in pressure pipelines.

As can be seen from the mathematical analysis with a decrease in the speed of water movement, there is a decrease in head frictional losses along the pipeline, which leads to an increase in the delivery of pump units, i.e. to the extension of the pumping station supply control range.

With a closed throttle device 6, a different number of pump units 1 operates on each of common pressure pipelines 3 or 3-a, and the heads at the beginning of these pipelines are different. Under the effect of the resulting pressure difference with the open throttle device 6 of the bypass device 5 water starts flowing from the pressure pipeline (for example 3) where the greater number of units operates into another pressure pipeline 3-a.

By means of the throttle device 6 it is possible to regulate the water supply from the total water supply of two pipelines with the closed throttle device 6 up to the water supply with the fully opened throttling device 6 when the heads at the beginning of the common pressure pipelines 3 and 3-a are maximally equalized due to a gradual change in the coefficient of hydraulic resistance of the throttle device.

Figure 4 shows the performance of the pump 2400VR-25/25. Position 1 corresponds to the operating characteristic of three pumps working in parallel on the pressure pipeline, whose characteristic corresponds to position 2.

If there is a bypass device with a throttle device at this station, when the latter is closed, the operating point of the pump will be point, which corresponds to a supply of 73.0 m^3/s. When the throttle device is fully open, the water velocity in the pressure pipelines decreases almost twice, and consequently, the energy losses decrease by a factor of 4. The pipeline characteristic will be expressed in this case by position 3, and therefore the operating point of the pumps' joint operation will move to point, which corresponds to the supply equal to 88.4 m^3/s. Thus, the delivery control range for such connection of pumping units will be within 73.0÷88.4 m^3/s.

A distinctive feature of the proposed method for regulating the operating modes of a pumping station with stiff-block pump units is the ability to regulate the supply in the direction of increasing it with a given number of operating pump units, and the efficiency of the pumping station will also increase. This is explained by the fact that with any supply equalizing by means of bypass device the average head, which is required for the operation of pumping units, is reduced, due to the decrease in the amount of energy loss in the pressure pipelines.

Figure 5 graphically depicts the comparison of the effectiveness of water supply regulation by the proposed method as compared to the control by means of the guide device at the inlet to the pump. Position 1 is a characteristic of the power consumption of a pumping station with a guide device at the inlet. Position 2 is a characteristic of the power consumption of the same pumping station when bypass device is used.

Figure 4. Method of regulation of operating modes by means of bypass device

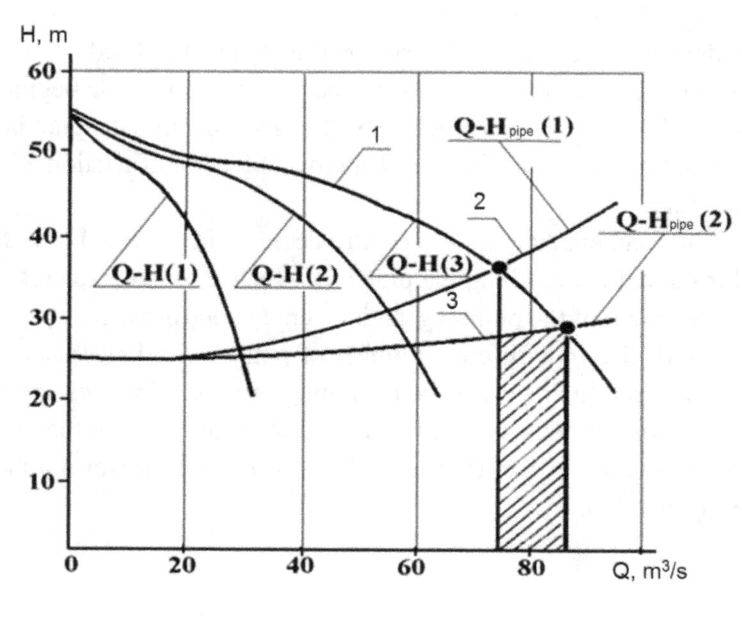

Figure 5. Comparison of methods for controlling the water supply of the pumping station: 1 - the characteristic of the power consumption of the pumping station with the guide device at the inlet, 2 - the same when using the bypass device

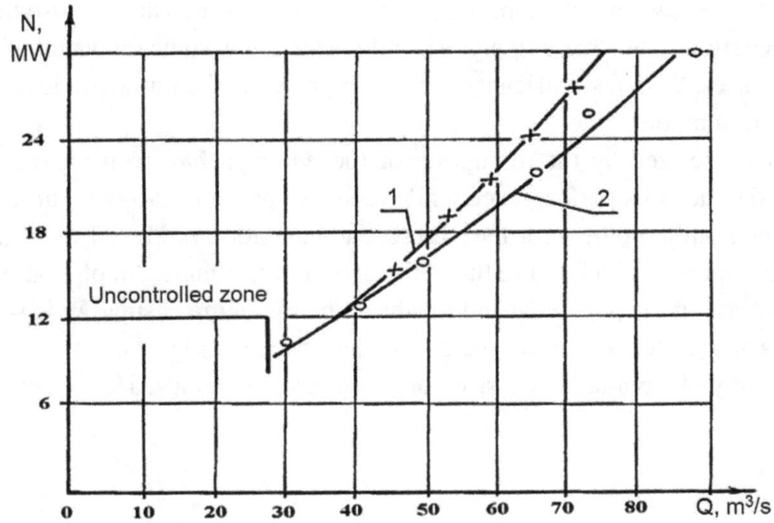

Comparison of graphical dependencies shows a decrease in power consumption by an average of 7% with control by means of bypass device as compared to the regulation by means of the guide device at the pump inlet.

Calculations carried out showed that the increase in the pump station's water supply can be from 3.3 to 28.2%, the reduction in specific energy consumption – from 3.0 to 8.3% compared to the operation of the pumping station without bypass device.

FIELD TESTS OF OPERATING MODES OF THE PUMPING STATION NO. 7 OF KMC

In the course of theoretical studies, a method was developed to control the operating modes of a pump station that operates with a bypass device in accordance with the water supply schedule, which minimizes the power consumed by the pump station.

For the reliability of the obtained results, it was necessary to check with the field tests at the existing pumping station the correctness of the developed calculation program, determine the technical condition of the pressure pipeline, the effectiveness of the proposed energy-saving measures, the allowable pumps operation modes without cavitation.

At the seventh pumping station, centrifugal pumps are installed. Two steel pressure pipelines of the pumping station with a diameter of 3640 mm. Moreover, the first pipe is connected to two variable pumps and three main, and the second pipe - four main pumps.

The length of each pipeline line is 156 m. The pumps are separated from the pipeline threads by butterfly valves: the main pumps are with valves ZD-280-75, and the exchangeable pumps with valves with ZD-220-72.

At the end of the common pipelines, additional valves are installed, designed to ensure the non-cavitation operation of pumps when operating on an unfilled or partially unfilled reservoir.

On the factory characteristics of pumps 2400VR-25/25 and 1600V-10-40 installed at pump station No. 7 KMC, it can be seen that when the operating point is shifted towards an increase in supply (decrease in pressure), the cavitation indicators will significantly deteriorate, since the value of the required cavitation stock Δh increases. With a significant decrease in pressure, the pump mode can be characterized as the second cavitation mode.

This mode is characterized by the disruption of the pump. It has been proven that a centrifugal pump significantly degrades its performance while reducing pressure. So with an increase in flow 1.4 times higher than the nominal pump got into the second cavitation mode. At the same time there was an unbalance of the impeller, which led to the deformation of the shaft. An obligatory phenomenon in this case was an increase in the gap between the wheel and the pump casing, which caused an increase in volumetric losses and a decrease in the energy indices of the aggregates.

In practice, the study of cavitation is carried out by indirect methods. These include:

1. Acoustic;
2. Vibrating;
3. Measurement of pressure pulsations;
4. Measurement of parameters of aggregates.

The norms of vibration are the norms V, D, J. The norms V, D, J on the vibration of the pumps of large pumping units installed on rigid and heavy foundations (pumps OPV, V, FV) are presented in Figure 6.

Figure 6. Norms V, D, J on the vibration of pumps of large pumping units of type V

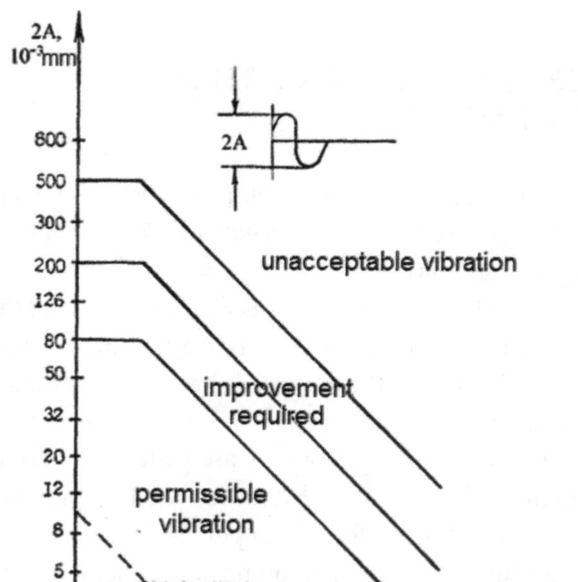

According to these standards, for a type V pump in the speed range from 0 to 1000 min^{-1}, the double vibration amplitude should not exceed 0.5 mm.

The norm on the vibration of the pressure pipe – the double amplitude of vibration should not exceed the value of 0.0005·D, where D is the diameter of the pipeline. The norms on vibration were taken by us as a measure of the cavitation state of the aggregate operating at reduced heads.

PROGRAM AND METHODS OF FIELD TESTS, EQUIPMENT USED AND MEASUREMENT ACCURACY

Field tests were made in order to:

- Identify the actual cavitation vibration state of pumping units operating at reduced heads;
- Verification of the correctness of the developed methodology for calculating the norms of energy consumption;
- Determine the allowable mode of operation of the pump unit at reduced head, in terms of cavitation and vibration of the unit and the pressure pipe;
- Determine the effectiveness of the bypass device of the supplied water from one pipeline to another in order to reduce energy consumption and reduce hydraulic shock in the pipelines.

The tests included the measurement of the following values:

- Levels of water in the downstream and upstream;
- Water pressure in the discharge port of the pump;
- Water pressure on the common pressure pipelines;
- Water pressure at the point of bypass device;
- Measurement of water flow rate in the common pressure pipelines;
- Vibration displacement of the guide bearing assembly and pressure pipeline.

In the course of field tests of pumping station No. 7 KMC, the following parameters were determined: pressure and flow rate of the pump, pressure and pressure pulsation at the outlet from the pump and in the pressure pipe, vibration on the pump bearing cover.

The primary speed converter is installed at the point of the average speed of the straight section of the pressure pipeline of the pumping station No. 7 KMC. To convert the pressure difference into an electrical signal and transmit it to the dispatch console, a small-sized measuring transducer with magnetic flux.

WATER FLOW RATE MEASUREMENT IN PIPELINES WITH LARGE DIAMETERS

The method is based on measuring the velocity of water at the point of the section of the pipeline, in which the local velocity is equal to the average velocity in a given section. It was developed the design of a primary differential pressure transducer into dynamic head, a small-sized measuring transducer having a current signal and a software for converting the current signal of a differential pressure gauge into a flow rate and a supply value of water.

Several pumping models of electromagnetic and ultrasonic flowmeters were previously installed at pumping stations in Uzbekistan. The experience of their work showed that at pumping stations with the operation of powerful high-voltage main equipment, there are significant magnetic fields that have a negative impact on the correctness of the flowmeter's readings, since the flowmeters output is extremely small.

Therefore, proposed a hydraulic method for determining the flow rate and accounting for the flow of the supplied liquid, based on measuring the velocity of water at one point in the section of the pipeline, in which the local velocity is equal to the average velocity in a given section. The product of this velocity by the area of the cross section is equal to the volume flow rate. This method of measurement has obtained the official status of determining the volumetric flow rate of water by measuring the velocity at one point in the section of a cylindrical tube.

The advantage of this method is that it can be used to measure the flow rate of liquids at pumping stations with any diameters, including the largest diameters of the pipelines.

To measure the flow rate, it was designed the structure of a primary converter which is a pressure tube, installed at a point of medium velocity which is adopted equal to 0.24 times the radius of inner wall of the pipeline (Figure 7). The design of the primary converter has to be selected depending on the diameter of the pipeline, local flow rate, measurement range, overpressure and properties of the medium to be measured (density, aggressiveness, presence of solids, etc.).

A small-sized measuring transducer with a magnetic flux compensation and a current output signal is used to convert the pressure difference into an electrical signal. Output signals: current output 4-20 mA, pulse (user defined pulse weight), RS232, USB, ModBUS RTU, HART protocols for connecting pressure sensors is supported. A printed circuit board of a microelectronic device has been developed, which allows receiving and processing a current signal at a distance of up to 1000 meters. In addition, it is possible to transfer the flow rate data through a GSM sim-card.

Special software is developed to convert the output signal in the computer to the flow rate of the pumped liquid. The limit of the permissible error in measuring the flow rate with a confidence probability of 0.95 was 2.2%.

THE MAIN RESULTS OF TESTING THE OPERATION OF PUMPS FOR REDUCED HEADS

According to the test results, the vibration and pressure characteristics of pumps 2400VR-25/25 and 1600V-10-40 were built (Figure 8). Vibration characteristics built in the entire range of delivery. The same figure shows the cavitation characteristics (Δh-Q), which are taken from the factory characteristics. The analysis of the dependences of the vibration and the corresponding cavitation characteristics indicates their identity; therefore, it can be concluded that the main cause of the vibration of the pump unit are cavitation phenomena occurring in the pump. On the same graph, the allowable rate of double amplitude of vibration, which is $2A = 0.5$ mm, is given.

The analysis of the dependences presented in Figure 8, allows to conclude that when the pump 1600V-10-40 is working for a pressure less than 14 m, and for a pump 2400VR-25/25 for a pressure less than 19.5 m, the magnitude of the double amplitude of vibration will exceed the norm, and further even a slight decrease head pressure can cause a sharp increase in double amplitude vibration.

Field tests conclusions:

Figure 7. Flowmeter assembly for pump stations with large pipeline diameters: 1 - large diameter pipeline, 2 - flowmeter primary transducer, 3 - hydraulic signal, 4 - pressure difference converter of hydrodynamic heads into a unified current signal, 5 - electrical signal, 6 - microelectronic flow measuring device, 7 - computer

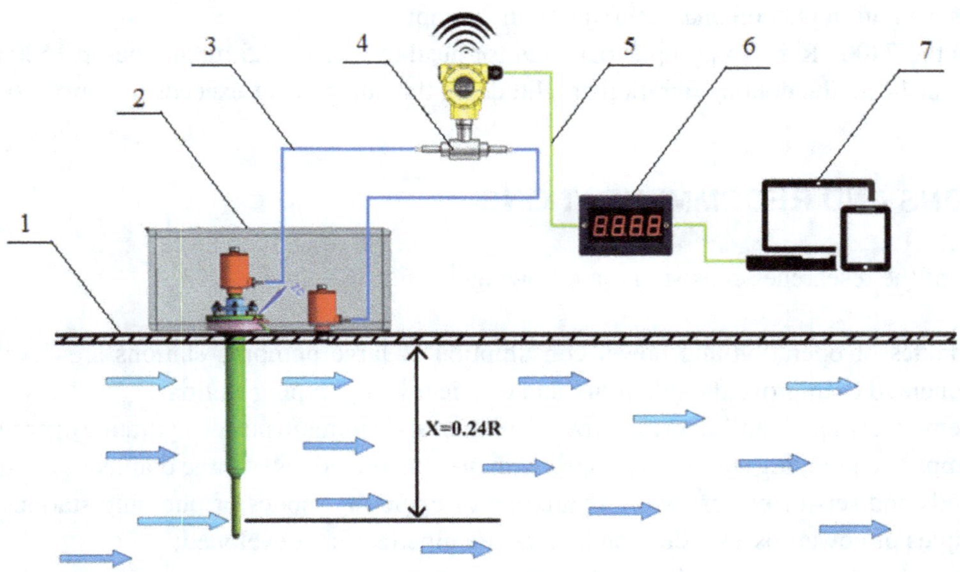

Figure 8. Pressure, cavitation and vibration characteristics of the main units 2400VR-25/25 and changeable units 1600V-10-40 NS No. 7 KMC according to test results

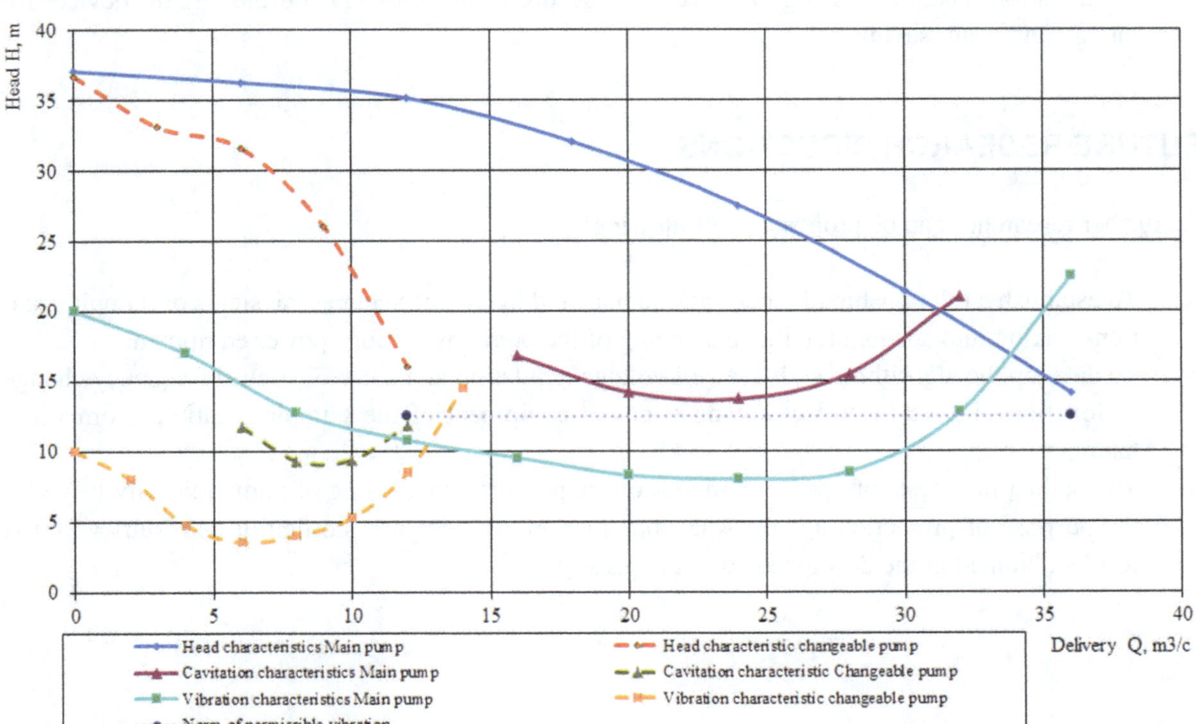

1. It was confirmed that centrifugal pumps significantly impair their cavitation characteristics when operating at reduced heads;
2. The identity of the dependences of the magnitude of the double amplitude of vibration on the pump cover and its cavitation indices allows us to conclude that the main cause of pump equipment vibration is cavitation phenomena occurring in the pump;
3. When the 2400VR-25/25 pump is operated for head less than 19.5 m, and pump 1600V-10-40 is less than 14 m, the double vibration amplitude on the pump cover exceeds the norm (0.5 mm).

SOLUTIONS AND RECOMMENDATIONS

The results of the researches consist in the following:

1. The modes of operation and power consumption of large pumping stations are developed and implemented to improve the reliability and efficiency of pumping stations;
2. Mathematical models and algorithms of calculation and optimum power-hydraulic operating modes of pumps and pumping stations equipped with pressure pipelines of large diameters are developed;
3. Methods and sets of power saving regulation of operating modes of pumping stations based on principles of power losses reduction in pressure pipelines are developed;
4. The use of bypass device at the pressure pipelines for energy loss reduction and operating modes regulation is recommended;
5. The scientific results of the research are recommended to allow efficient use of water and energy resources and to ensure reliable operation of the power equipment of pumping stations in the events of cavitation. The practical significance of the results of the work is in developing the devices for energy and water saving.

FUTURE RESEARCH DIRECTIONS

As further researches can be proposed the following:

- To establish a relationship of electromechanical and hydro-mechanical transients of pumping stations, taking into account the life and ageing of the pumping stations power equipment;
- To develop the algorithms with a law of voltage regulation and a law of control frequency change in function of time to protect the main power equipment of the pumping stations from water hammer;
- To conduct field tests of operating modes of the proportional closing of pump butterfly valve for the purpose of protection against water hammer and to compare the field tests results with the results obtained in the developed software package.

CONCLUSION

Energy efficient methods and means of regulation of steady modes of operation of pumps and pumping stations are implemented at a large pumping stations of Karshi Main Canal. Among them pumping stations No 1÷7 of Karshi stage of Directorate of Exploitation of Karshi Main Canal (Certificate given by the Ministry of Agriculture and Water Resources of the Republic of Uzbekistan N° 04/31-700 on June 28, 2016). Using the results of the researches allowed to reduce the electricity consumption of the pumping stations by an average of 4%.

ACKNOWLEDGMENT

Research is supporting by the Ministry of Innovation Development of the Republic of Uzbekistan [grant number БВ-Атех-2018-10 (2018-2020)].

REFERENCES

Allayev, K.R., Khokhlov, V.A., Sytdykov, R.A., & Titova J.O. (2013). *Elektromekhanicheskiye i gidromekhanicheskiye protsessy v gidroenergeticheskikh ustanovkakh* [Electromechanical and hydro-mechanical processes in hydropower plants]. Tashkent, Fan va texnologiya.

Anderson, J. L., Lowry, M. V., & Thomte, J. C. (2001). Hydraulic and water quality modeling of distribution systems: What are the trends in the U.S. and Canada? In *Proceedings of the AWWA Annual Conference*. Denver, CO: American Water Works Association (AWWA).

Cookson, T., Lang, N., & Thornton, E. (2008). Adjustable speed drives applied to large AC induction motor and pump systems. In *Proceedings of the Twenty-fourth International Pump Users Symposium* (pp. 75-80). Houston, TX: GA Valves. Retrieved January 20, 2016, from http://www.gavalves.co.uk

Garcia-Hernandez, A., Wilcox, M., & Moore, T. (2010). Hydraulic modelling and simulation of pumping systems. *Proceedings of the Twenty-sixth International Pump Users Symposium*, 81-88.

Kamalov, T.S. (2014). *Chastotno-reguliruyemyy elektroprivod nasosnykh stantsiy sistem mashinnogo orosheniya* [Frequency-regulated electric drive of pumping stations of machine irrigation systems]. Tashkent, Fan.

Khashimov, A.A. & Abidov, K.G. (2012). *Energoeffektivnyye sposoby samozapuska elektroprivodov nasosnykh stantsiy* [Energy-efficient ways to self-start electric drives of pumping stations]. Tashkent, Fan va texnologiya.

Khokhlov, A.V., Khokhlov, V.A., Titova, J.O. (2014). *Rezhimy raboty nasosnykh stantsiy Dzhizakskogo kaskada* [Operating modes of Djizak pumping stations stage]. Tashkent, Fan va texnologiya.

Khokhlov, A. V., Khokhlov, V. A., & Titova, J. O. (2015). *Rezhimy raboty nasosnykh stantsiy Karshinskogo kaskada* [Operating modes of Karshi pumping stations stage]. Tashkent: Navruz.

Khokhlov, V.A., Khokhlov, A.V. & Titova, J.O. (2012). *Neustanovivshiyesya rezhimy raboty nasosnykh stantsiy s dlinnymi truboprovodami* [Unsteady operating modes of pumping stations with long pipelines]. Tashkent, Fan va texnologiya.

Larsen, T. (2012). *Water hammer in pumped sewer mains.* Aalborg, Denmark: University of Aalborg, Department of Civil Engineering.

Leznov, B. S. (2006). *Energosberezhenie i reguliruemyy privod v nasosnykh i vozdushnykh ustanovkakh* [Energy saving and adjustable drive for pump and blower installations]. Moscow: Energoatomizdat.

Li, Ch., Ma, Sh., Liu, H., Wu, H., Gao, P., & Yang, F. (2016). Research on scheme of pumping energy-saving based on computer control technology. *MATEC Web of Conferences.* Retrieved March 21, 2017, from http://www.matec-conferences.org

Main Directorate of Pumping Stations Energy and Communications of the Ministry of Ministry of Agriculture and Water Resources of the Republic of Uzbekistan. (2006). *Technical Report.* Tashkent.

Republic of Uzbekistan, Large pumping stations. (2005). *Brief review of technical condition.* Tashkent: State Inspectorate for Control and Supervision over the Technical Condition and Safety of Large and Especially Important Water Facilities under the Cabinet of Ministers of the Republic of Uzbekistan.

Sipos Aktorik. (n.d.). Retrieved January 20, 2016, from http://www.sipos.de

Transient processes of pumping stations. Calculations of hydraulic shock in pressure pipelines. (n.d.). Manual for VSN 33-2.2.12-87 “Reclamation systems and facilities: Pumping stations - Design standards”. Retrieved January 29, 2016, from http://library.fsetan.ru/doc/perehodnyie-protsessyi-nasosnyih-stantsij-raschetyi-gidravlicheskogo-udara-v-napornyih-truboprovodah-posobie-k-vsn-33-2212-87-meliorativnyie-sistemyi-i-sooruzheniya-nasosnyie-stantsii-normyi-proektirovaniya

Valve, V.-M., & Corp, M. (n.d.). Retrieved January 20, 2016, from http://www//valmatic.com

Vissarionov, V. I., & Trusov, I. S. (1986). Matematicheskiye modeli perekhodnykh protsessov v nasosnoy stantsii pri samozapuske elektrodvigatelya posle kratkovremennogo pereryva pitaniya [Mathematical models of transient processes in a pumping station with a self-starting electric motor after a short-time power interruption]. *JournalEnergetika, 11*, 115–119.

APPENDIX

Table 3. Actual energy consumption in annual terms of pumping stations No 1 to 7 of the Karshi cascade for 2013, kWh

Name / Month	Pumping Station N° 1	Pumping Station N° 2	Pumping Station N° 3	Pumping Station N° 4	Pumping Station N° 5	Pumping Station N° 6	Pumping Station N° 7	Total
January	24 155 900	28 401 872	28 894 886	27 434 990	26 529 494	27 756 404	17 927 996	**181 101 542**
February	23 503 538	28 143 944	28 970 492	26 518 841	26 002 846	26 937 011	17 039 449	**177 116 121**
March	30 175 430	37 338 208	37 319 502	35 193 679	23 195 445	23 827 657	6 080 760	**193 130 681**
April	32 065 053	38 788 140	37 336 678	35 490 606	17 972 965	18 935 958	172 588	**180 761 988**
May	34 737 606	44 285 530	43 729 782	41 275 971	29 787 274	30 385 152	8 243 881	**232 445 196**
June	33 281 672	44 225 087	44 274 470	41 246 823	29 363 297	29 767 541	7 880 506	**230 039 396**
July	35 143 066	43 944 673	44 136 142	41 442 438	20 717 615	21 139 578	382 115	**206 905 627**
August	34 081 815	43 890 850	44 290 762	41 073 666	26 232 953	27 053 725	583 866	**217 207 637**
September	31 010 308	36 365 993	36 860 458	33 343 159	31 483 128	32 020 117	18 071 813	**219 154 976**
October	23 738 371	26 688 536	27 379 100	24 592 547	18 308 579	18 455 548	1 587 417	**140 750 098**
November	19 116 345	21 707 404	21 978 062	20 747 810	10 000 365	10 881 178	2 261 768	**106 692 932**
December	22 281 656	26 322 126	26 253 624	23 422 829	24 024 536	24 134 455	13 146 586	**159 585 812**
Total	**343290760**	**420102363**	**421423958**	**391783359**	**283618497**	**291294324**	**93378745**	**2 244 892 006**

Chapter 13
Improving the Power Quality of Rural Consumers by Means of Electricity Cost Adjustment

Alexander Vinogradov
Federal Scientific Agroengineering Centre VIM, Russia

Maxim Borodin
Orel State Agrarian University, Russia

Vadim Bolshev
Federal Scientific Agroengineering Centre VIM, Russia

Natalya Makhiyanova
Orel State Agrarian University, Russia

Nadezhda Hruntovich
Sukhoi State Technical University of Gomel, Belarus

ABSTRACT

Improving power quality of agricultural consumers by adjusting the cost of electricity is considered. The chapter lists the reasons for the output of power quality indicators beyond the standard values. There is the assessment of the quality of electricity in the Orel region. Distribution functions for distortion of power quality indicators are constructed. The relevance of the adjustment of the cost of electricity depending on its quality is justified. The mathematical model for determining the cost of consumed electric energy depending on its quality is developed in chapter. The chapter gives the method allowing to adjust the cost of consumed electric energy depending on its quality and source of distortion. In the chapter, there is the mechanism of carrying out calculations for the consumed and transferred electric power considering change of electric power cost depending on its quality. The feasibility study of the proposed methods and means of improving power quality is realized.

DOI: 10.4018/978-1-5225-9179-5.ch013

INTRODUCTION

Modern agriculture requires qualitative electricity. Power quality (PQ) is a factor influencing energy efficiency directly. In the Russian Federation there are a lot of normative documents which regulate the requirements to PQ but in practice they not always work efficiently. At the same time the necessity to maintain PQ within normative values is becoming a more vital issue in terms of applying modern electrical equipment and automation systems which are sensitive to the deviation of every power quality indicator (PQI). Power quality inconsistency with the values set by normative documents leads to negative economic consequences because of decrease of productivity of factories, non-working equipment, undersupply of the product, its damage, the increase of electricity loss and other negative consequences. Thus, the aim to increase the quality of electric power by means of correcting electricity cost is important.

The culprit of PQ deviation can be a power supply company as well as a consumer. It is why the technical and economic decisions for efficient maintaining PQ are related to a consumer as well as to a power grid company. It demands extra financial investments both sides will do reluctantly. The use of the discount and premium system existing in Russia does not give the guarantee of constant and widespread provision of necessary PQ level as it is used occasionally and for a limited number of objects.

The lack of the technique allowing to realize the technical and economic mechanism of stimulating consumers and power grid companies to maintain PQ on the level of normative documents leads to the decrease of PQ as they are not economically interested in maintaining power quality on the level of normative documents. There are also no developed methods and means which allow to realize the technique of correcting electricity cost in practice. The realization of developed methods and technical means will allow to increase PQ at the expense of doing the immediate correction of electricity cost depending on its quality and thus stimulating consumers and power grid companies to maintain PQ on the level of normative documents. Nowadays there is also no the mechanism calculating bills for electricity which takes into account the change of the cost of the consumed electricity depending on its quality.

BACKGROUND

Power quality along with power supply reliability is an important characteristic of power supply systems (Vinogradov et al., 2018; Bachmanova, 2016; Kuzmichev et al., 2010; Lance, 2010). The power quality should be understood as a set of power characteristics under which the electrical equipment is able to perform its functions (Kartashev et al.,2006; Kartashev, 20014; Leshchinskaya & Naumov, 2008). It has a significant impact on the efficiency of electric receivers as well as on the techno-economic indicators of electrical networks (Suhas et al., 2015; Vinogradova & Skrobov, 2013; Spiganovic, 2004).

According to Fangxing et al. (2010), Spiganovic & Zakharov (2007) and Vinogradov & Borodin (2013) in the context of modern methods and means of providing PQ the electric power should be considered as a product on the one hand and as a physical concept, on the other hand:

- The electric energy as a product have to satisfy to certain quality, requirements of the market and differ from other types of energy in special consumer properties: coincidence in time of production process, transportation and consumption; dependence of PQ characteristics on processes of its consumption; the impossibility of storage and return of poor-quality electric power;

- The electric energy as a physical concept is the ability of the electromagnetic field to perform work under the influence of applied voltage in the technological process of its production, transmission, distribution and consumption.

A number of normative documents are developed for the most effective coordination of actions between the consumer and the power supply organization for providing the power quality in the conditions of operation of power supply systems (PPS) in the Russian Federation. So, according to Grazhdanskij kodeks RF (2000) the power supply organizations are obliged to deliver electric energy to consumers with the quality met requirements of normative documents and power supply contracts. Federal'nyj zakon Rossijskoj Federacii N°35 FZ "Ob Ehlektroehnergetike" (n.d.) establishes the responsibility of energy suppliers and energy supply organizations to consumers for power supply reliability and the quality of supplied electrical energy in accordance with the technical standards and other compulsory requirements.

Norms of power quality in terms of their permissible values in the power supply system of general purposes with single – phase and three-phase alternating current with a frequency of 50 Hz at the connection points of power lines or consumers in Russian federation are established with Interstate Council for Standardization, Metrology and Certification (n.d.). The requirements to the quality of electric energy have changed with the entry into force of GOST 32144-2013 corresponding to the European regional standard EN 50160:2010 "Voltage characteristics of electricity supplied by public distribution networks". At the same time the level of responsibility of power supply organizations for the power quality reduces [Golikov & Vinogradov, 2017; Vinogradov, Borodin & Bolshev, 2014; ZHelezko, 2002). In addition, the permissible intervals of power quality indicators and the time of their deviation were increased, which contributed to the reduction of PQ level in PSS, thereby reducing the energy efficiency of both consumers and power supply systems (Borodin, 2013). Despite the fact that there are a large number of regulatory documents in the Russian Federation which establish requirements to the quality of electric energy, they do not always work effectively in practice (Vinogradov & Borodin, 2012). At the same time, the necessity of maintenance normalized PQ is becoming the more important task in connection with the use of modern electrical equipment and automation systems which are sensitive to the deviation of each power quality indicator (Mandatova et al.,2014; Yasin, 2016; Siano, 2014).

Borodin & Uryupin (2018), Dobrusin (2004) and Vinogradov et al. (2015) point that power quality indicators moving beyond the limits provided by normative documents worsen the operation conditions of electric equipment and also lead to different types of damage. The damage caused by the deterioration of the power quality refers to all forms of negative consequences appearing in the operation of power supply systems, consumers and electric energy receivers. There are two types of damage: electrical and technological according to Perova (2007). Electrotechnical damage is caused by increase of the energy losses, reduction of electric equipment life, sudden reversible and irreversible failures of technology. Technological damage is caused by undersupply of energy and defective products, failures of electrical equipment operation leading to a violation of production technology.

The power quality can be enhanced by applying of organizational, technical and economic activities (Popov, 2010; Gheorghe, 2005; Santarius, 2015).

Organizational activities include:

- Observance of conditions of the contract directed on power quality assurance between consumers and power supply organizations;
- Certification of power quality;

- Creating a structure of power quality management in power supply organizations;
- Development and application of normative documents and regulations defining the norms of power quality and the conditions of their provision; etc.

Technical activities improving power quality include the following steps: measurements; PQI determination; calculations of modes; PQ analysis and identification the reasons of PQ reduction; the choice of method of PQ assurance, PQ controlling etc. Technical devices are often used: synchronous compensators, devices of series compensation, balanced converters, no-break power supplies, capacitor banks, harmonic filters etc. The above-mentioned activities and technical devices do not guarantee the constant and universal provision of the required level of power quality in practice, because they are carried out occasionally and on a limited number of objects (Borodin & Vinogradov, 2018; Rusdy & Liben, 2010). Therefore, it is necessary to apply techno-economic activities for power quality assurance. Economic activities include the creation of effective economic incentive mechanisms both for the consumer and for the energy supply organization to maintain and control power quality at the level of normative documents. It is necessary to develop a methodology to realize the economic mechanism allowing to adjust the electricity cost on its quality, and also technical methods and means to implement this technique.

THE MAIN FOCUS OF THE CHAPTER: ISSUES, CONTROVERSIES, PROBLEMS

In order to assess PQ compliance with GOST 32144-2013 requirements, measurements have been made in Orel region at 100 points of balance delineation between a consumer and a power supply system and at 100 points of common connection during winter and summer. The measurements have been carried out with measuring instruments of PQI "Resource-UF2". The measuring period in each case was seven days. The need for measurements of power quality indicators at 2 points is due to the fact that the modes of operation in common connection differ from the modes of operation of the consumer. This is because each consumer makes an actual contribution to PQI changes in the points of general connection. The control points of measuring of the power quality indicators are shown in Figure 1.

According to Figure 1: A1-A3 - border of balance delineation between a consumer and a power supply system; QS1 – circuit breaker; T1 – transformer 10/0,4 kV; M1 – point of common connection.

The results of power quality assessment at points of common connection and at points of balance delineation between a consumer and a power supplying organization are presented in Figure 2.

Figure 1. The control points of power quality indicators measuring

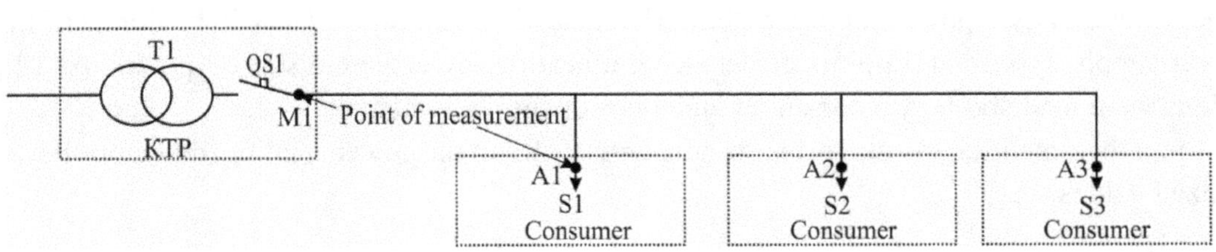

Figure 2. The results of power quality measurements in the Orel region: a - at points of balance delineation between a consumer and a power supplying organization; b – at points of common connection

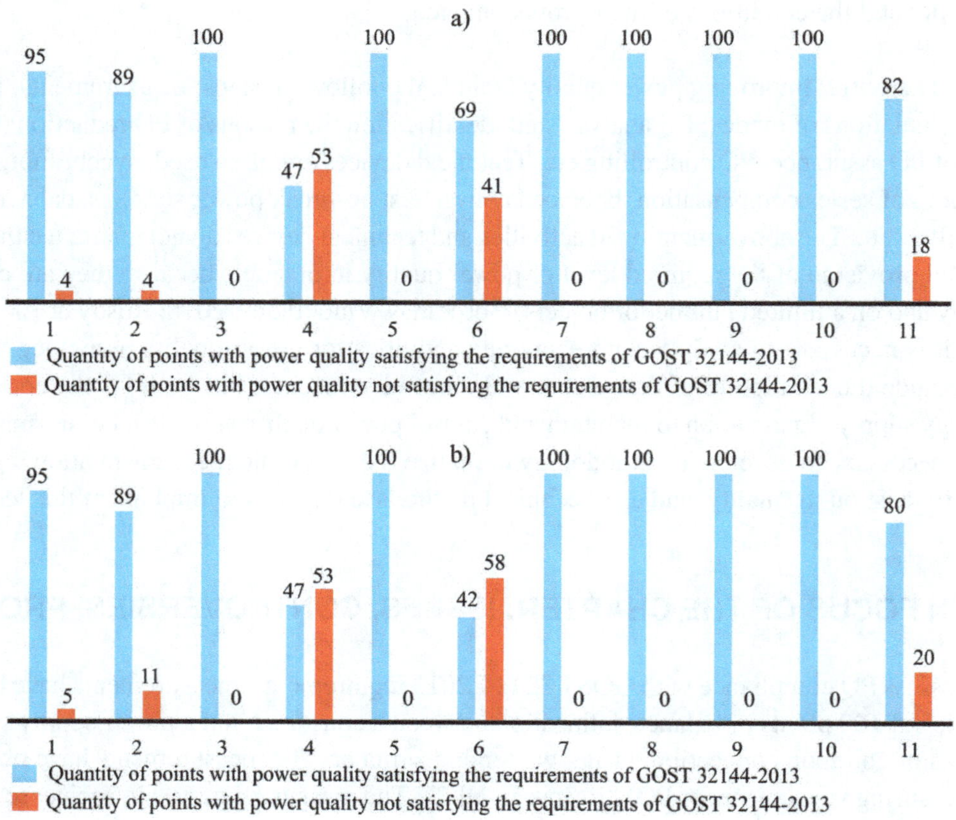

According to Figure 2: 1 – Steady-state voltage deviation; 2 – Range of voltage changes; 3 – Total harmonic distortion of voltage; 4 – Relative level of voltage n-harmonic component; 5 – Negative sequence voltage unbalance ratio; 6 - Zero-sequence voltage unbalance ratio; 7 – Frequency deviation; 8 – Duration of voltage dip; 9 – Impulse voltage; 10 – voltage swell ratio; 11 – Dose of flicker.

The results of the evaluation conclude that the power quality indicators have not satisfy the requirements of GOST 32144-2013 at 81 points of balance delineation and at 70 points of common connection in the normal mode. The analysis allowed to determine most likely perpetrators of the power quality distortion. The energy supply system made the distortion of power quality indicators at 2 points of common connection, the consumers did at 66 points. There are at 2 points the distortions were made both by a consumer and by a power supply system. The distortion of power quality indicators was made by the power supply system at 2 points of balance delineation, by consumers at 67 points. At 12 points distortions were made both by a consumer and a power supply system.

For the above-mentioned categories of consumers, the main power quality indicators beyond the normalized values are:

- **Houses:** The scale of voltage changes (the deviation range is -15,6%...+16.5%); the steady-state voltage deviation (a deviation range is -15,4…+14.6%); relative level of voltage n-harmonic component (harmonics 3, 6, 8, 12, 14, 15, 16, 18, 20, 21, 22, 24, 26, 27); Zero-sequence voltage unbalance ratio (deviation range is 2.01…4,52); short-term dose of flicker (deviation range of 1,45…5,3); long dose flicker (deviation range is 1.1…3.9);

- **Administrative Buildings:** The scale of voltage changes (the deviation range is -7,8%...+15.6%); the steady-state voltage deviation (a deviation range is -7,6%...+14.9%); Relative level of voltage n-harmonic component (harmonics 3, 6, 8, 10, 12, 14, 15, 16, 18, 20, 21, 22, 24, 26, 27); Zero-sequence voltage unbalance ratio (deviation range is 2.02…5,55); short-term dose of flicker (deviation range is 1.41…5,2); long time flicker (deviation range is 1.5…3.5);
- **Agricultural Enterprises:** The scale of voltage changes (the deviation range is -10,6%...+10.8%); the steady-state voltage deviation (a deviation range is -7,2%...+9.4%); Relative level of voltage n-harmonic component (harmonics 3, 6, 8, 12, 15, 16, 18, 20, 21); Zero-sequence voltage unbalance ratio (deviation range is 2.05…4,36); short-term dose of flicker (deviation range is 2.5…2,65), long flicker (deviation range is 1,1…2,3).

Statistical studies have shown that more than 90% of consumers get PQ beyond the limits set by regulatory document requirements. The current situation is connected both with the slow rate of recovery of the electric grid economy and with the lack of investment program financing of power supply companies. Maintaining of the power quality at the level of normative documents requires additional capital investments. The both sides of energy market are not willing on it because of the lack of an effective method of adjusting electricity cost depending on its quality and technical tools to adjust electricity cost in online mode.

The use of the existing system of discounts and bonuses have not guarantee a permanent and universal provision of the required level of power quality as it have been used occasionally and on a limited number of objects. Therefore, the problem of power quality provision is more justified by the lack of an effective method of adjusting electricity cost depending on its quality and the technical methods and means that allow to implement this technique.

DECISIONS AND RECOMMENDATIONS

For solving abovementioned problems, a method of correcting the cost of consumed electricity depending on its quality was developed. The block diagram of this method is illustrated in Figure 3. The proposed method allows to increase PQ due to the economic stimulation of consumers and power supply organizations in maintaining the normative power quality. The adjustment of electricity cost is supposed to carry out on border of balance delineation between a consumer and a power supply system (Borodin & Vinogradov, 2013). The corresponding function must be integrated into the energy meter.

The developed technique is a technical and economic mechanism to stimulate both consumers and suppliers of electricity in terms of maintaining the normative power quality. If the consumer makes PQ distortions he is forced to pay for electric energy at a higher price. If the electrical energy coming to the consumer does not comply with regulations due to the power supply organization fault the consumer pays less.

The cost of the consumed electric power considering adjustment is offered to define by the formula:

Figure 3. Block diagram of the functional unit for determining the cost of consumed electricity depending on its quality

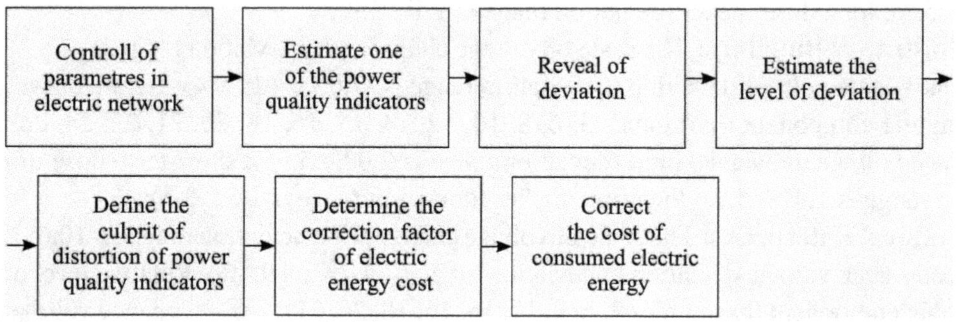

$$C_{CEE} = T_{IN} \cdot \int_{t_1}^{t_2} \kappa_{Pn} \cdot P \cdot dt + T_{IN} \cdot \int_{t_2}^{t_3} \kappa_{Pn} \cdot P \cdot dt + ... + T_{IN} \cdot \int_{t_i}^{t_j} \kappa_{Pn} \cdot P \cdot dt \tag{1}$$

where:

- κ_{Pn} is n-correction factor for a particular level of PQI deviation at a certain point;
- P is power energy consumed at a certain point, kW;
- t_1–t_3 is time interval when the power quality indicators are recognized to be contrary to the requirements of GOST 32144-2013;
- t_i, t_j is the number of intervals t when the PQI is recognized to be contrary to the requirements of GOST 32144-2013.

At the same time, the correction factor for a certain level of deviation of power quality indicators in the time interval when the PQI is recognized as non-compliant with the normative document requirements is determined as specified in Formula 2:

$$\kappa_{Pn} = \frac{\sum_{i=1}^{n} K_{iPPS} + \sum_{j=1}^{m} K_{jPPC}}{n+m} \tag{2}$$

where:

- κ_{PPSO} is the amount of correction coefficients to the cost of the electric power, fair for the power supply organization;
- $\kappa_{PPS!}$ is the amount of correction coefficients to the cost of the electric power, fair for consumers;
- m is the number of K_j values;
- n is the number of K_i values.

In reality, changes in the PQI values occur constantly, which, however, does not affect the use of the technique that allows to adjust electricity cost in online mode. It is why the culprit of PQ distortion and the actual contribution to the PQI change should be determined. The analysis of known approaches to determining the actual contribution have allowed the authors of the studies conducted in Kartashev et al. (2006) to develop a method for determining the actual contribution of the measurement results. This method will be used in determining the culprit of PQ distortion. There is an example of the proposed method in relation to the statistical studies of the power quality in the Orel region. The test results calculate electricity cost per day if the electricity cost excluding correction factors for all measurements is 2.7 rubles per 1 kW h.

Based on the results, a histogram for 21 points shown in Figure 4 has been constructed. In accordance with the results of the calculations without considering the correction factors at 100 points of common connection the electric energy cost has amounted to 703,8 thousand per day while with the correction factors it has amounted to 788,5 thousand rubles per day. It follows that the additional payment for electric energy is 84.7 thousand RUB. At 100 points of balance declination the cost of consumed electric energy per day has amounted to 13.9 thousand rubles. Taking into account the correction factors the cost of consumed electric energy per day has amounted to 14.8 thousand rubles. Additional payment for electricity is 0.9 thousand rubles.

Adjustment of electricity cost in the proposed technique is supposed to be performed by means of correction coefficients. It is necessary to know the statistical characteristics of the power quality indicators for the feasibility study [Borodin & Vinogradov, 2014). PQI characteristics shall be determined by applying the probabilistic-statistical method (Gmurman, 1999). It is primarily due to the random nature of the PQ indicators change. The change intervals are determined for indicators change: the scope of the voltage change (separately for phases A, B, C), the steady-state voltage deviation (separately for phases A, B, C), zero-sequence voltage unbalance ratio. The other PQI outputs has not been significant. The output probability of PQI beyond the norms is determined by the formula:

Figure 4. Electricity cost without and with correction factors for the consumer

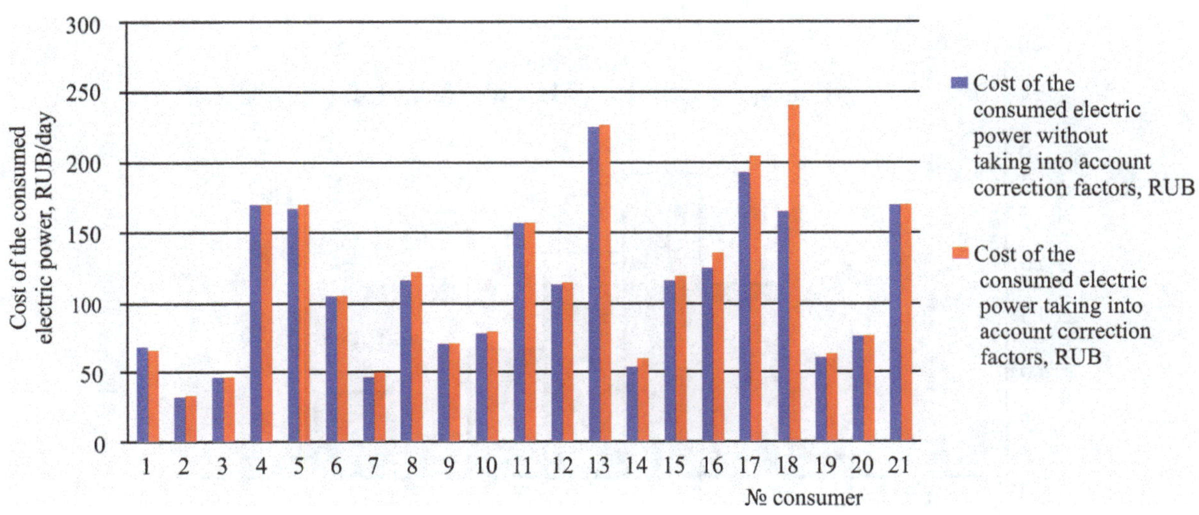

$$P(\kappa_i) = \frac{m}{n} \cdot 100\% \tag{3}$$

where:

- $P(\kappa_i)$ is the probability of occurrence of the k_i-th value of PQ indicators;
- m is the appearance of the k_i values of the PQ indicators in the interval;
- n is number of measurements;
- k_i is the case when m enters the specified interval $k_{i1} < k_i \leq k_{i2}$.

The distribution of PQI values is characterized by a distribution function. The distribution functions for the PQ indicator "Scope of voltage change" has been constructed on the basis of the obtained data and is presented in Figure 5 and Figure 6.

The values of the mathematical expectation and mean-square deviation of PQI have been determined. Mathematical expectation is determined by the formula:

$$MX = \sum_i X_i \cdot P_i \tag{4}$$

where:

- X_I is the average interval of PQI changing;
- P_I is the number of PQI deviations included in the change interval.

The mathematical expectation for the PQI "Zero-sequence voltage unbalance ratio" is defined:

Figure 5. Distribution function of phase A of the PQI "Scope of voltage change" at the point of common connection

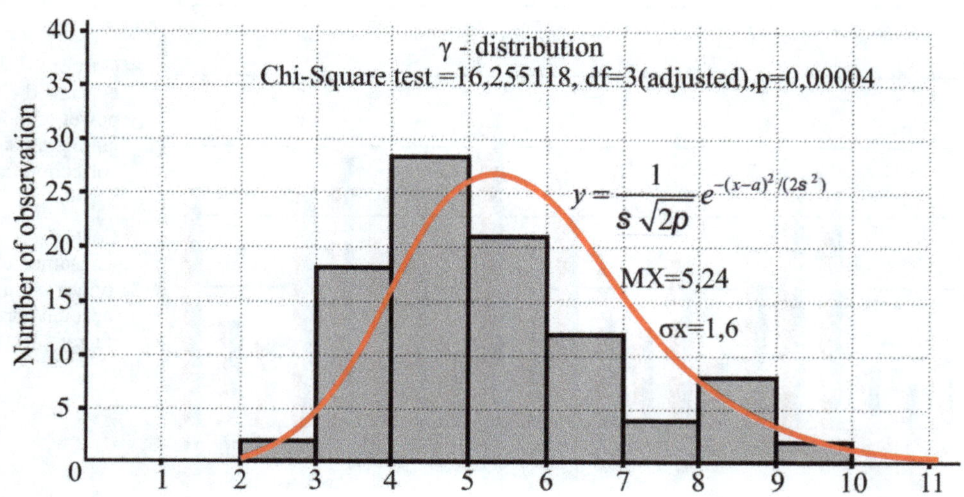

Figure 6. Distribution function of phase A of the PQI "Scope of voltage change" in the points of balance declination of a consumer and a power supply organization

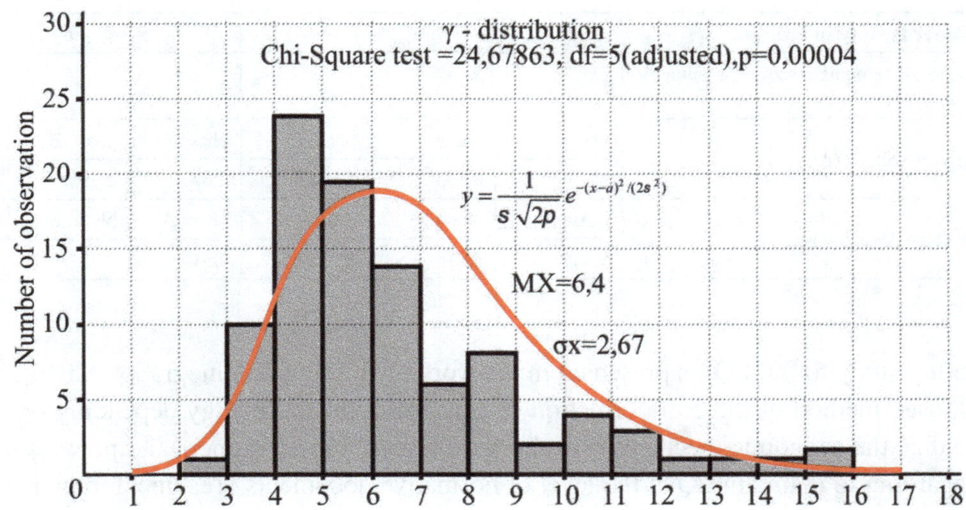

$$MX = \sum_i K_{0Ui} \cdot P_{K_{0Ui}} \tag{5}$$

The standard deviation is determined by the formula:

$$\sigma_x = \sqrt{DX} = \sqrt{\sum_i (K_i - MX)^2 \cdot P_i} \tag{6}$$

The standard deviation for the PQI "Zero-sequence voltage unbalance ratio," is determined by:

$$\sigma_{xK_{0U}} = \sqrt{\sum_i (K_{0Ui} - MX)^2 \cdot P_{K_{0Ui}}} \tag{7}$$

The numerical characteristics for all studied power quality indicators have been calculated, the results are summarized in Table 1 and Table 2.

Table 1. Numerical characteristic of power quality indicators in common connection

Power Quality Indicators	MX			σ_x		
Unbalanced factor of zero phase sequence voltage ratio (K_{0U})	6,61			4,49		
Scope of change of voltage (ΔU_t)	phase A	phase B	phase C	phase A	phase B	phase C
	5,24	5,58	5,7	1,6	1,97	1,94
Steady-state voltage deviation (ΔU)	phase A	phase B	phase C	phase A	phase B	phase C
	4,84	4,995	5,26	1, 35	1,92	1,88

Table 2. Numerical characteristic of power quality indicators in the points of balance declination of a consumer and a power supply system

Power Quality Indicators	MX			σ_x		
Unbalanced factor of zero phase sequence voltage ratio (K_{0U})	5,64			3,4		
Scope of change of voltage (ΔU_t)	phase A	phase B	phase C	phase A	phase B	phase C
	6,4	6,595	6,615	2,67	2,79	2,77
Steady-state voltage deviation (ΔU)	phase A	phase B	phase C	phase A	phase B	phase C
	5,725	5,815	5,965	2,51	2,72	2,54

A feasibility study for all PQI is presented in the work of Borodin & Vinogradov (2014).

The proposed method of the cost correction of consumed electric energy depending on its quality allows to realize the economic mechanism of the stimulating both consumers and power supply organizations regarding to maintain PQ at the level of normative documents. Technical implementation of this technique is carried out using energy meters with unit of PQI control. The evaluation of electricity cost considering the proposed methodology shows that in most cases the cost of electricity increases for consumers (Vinogradov & Borodin, 2013a).

In accordance with the block diagram presented in figure 3 "Method of measuring the electricity consumption and the adjustment of its value based on the quality and source of the distortions" is developed. This method allows to automatically produce a record of the number of consumed electric energy and to determine its value depending on power quality and distortion source. The main point of the proposed method is shown in figure 7. The scheme works as follows: when a consumer gets the electric energy a certain amount of this energy is transmitted in the circuit from the source to the receiver. The voltage sensor 7 and current sensor 6 control the electrical parameters of the transmitted electric energy. The signals from its outputs enter the input of the analog-to-digital converter 8, which converts the analog signal into digital form. Signals from its output go to the inputs of blocks BKKNOP9, BKUON10, BKKNNNP11, BCOC69, BKDPN70, BKKIN71, BKRIN129, BKDF130, BKKISKN131, BKKVP189, BKKGSN190 and BKIPEE229. They control PQI compliance to requirements of GOST 32144-2013 and quantity of the consumed electric energy. If PQI distortion appears there is a signal on the output of BKKNOP9. It fed to the input unit BKIV233, which analyzes the time interval of PQI distortion according to GOST 32144-2013.

If at least one of the power quality indicators goes beyond the normative values with an interval of time not less than the value specified in GOST 32144-2013 the signal from the output of the block BKIV233 enters the input blocks BKUI13 and BKII12.They control the level of distortion and the source of distortion. Suppose that a distortion source is the electric supply system, the distortion of the power quality is caused with change of the unbalanced factor of voltage of negative sequence, the level of distortion of power quality indicators - level 1 (the level is determined based on the deflection of indicator of the quality of electricity from regulated and set for each power quality indicator) (figure 7 b). Signal is supplied to the unit 1 level 22 from the Unit 13 BCUI, and from the block BKII12 the signal arrives on the block ES19 because the source of distortion is the electric supply system. Then the signal is sent to the element AND38 from the blocks ES19 and 1 level 22. The signal is supplied to the PC unit 56 from the output of element 38 AND, unit 56 sets the reduction factor to the cost of the consumed electric energy.

Figure 7. The method of the cost adjustment of consumed electric energy depending on its quality, source of distortion and time of distortion of power quality indicators

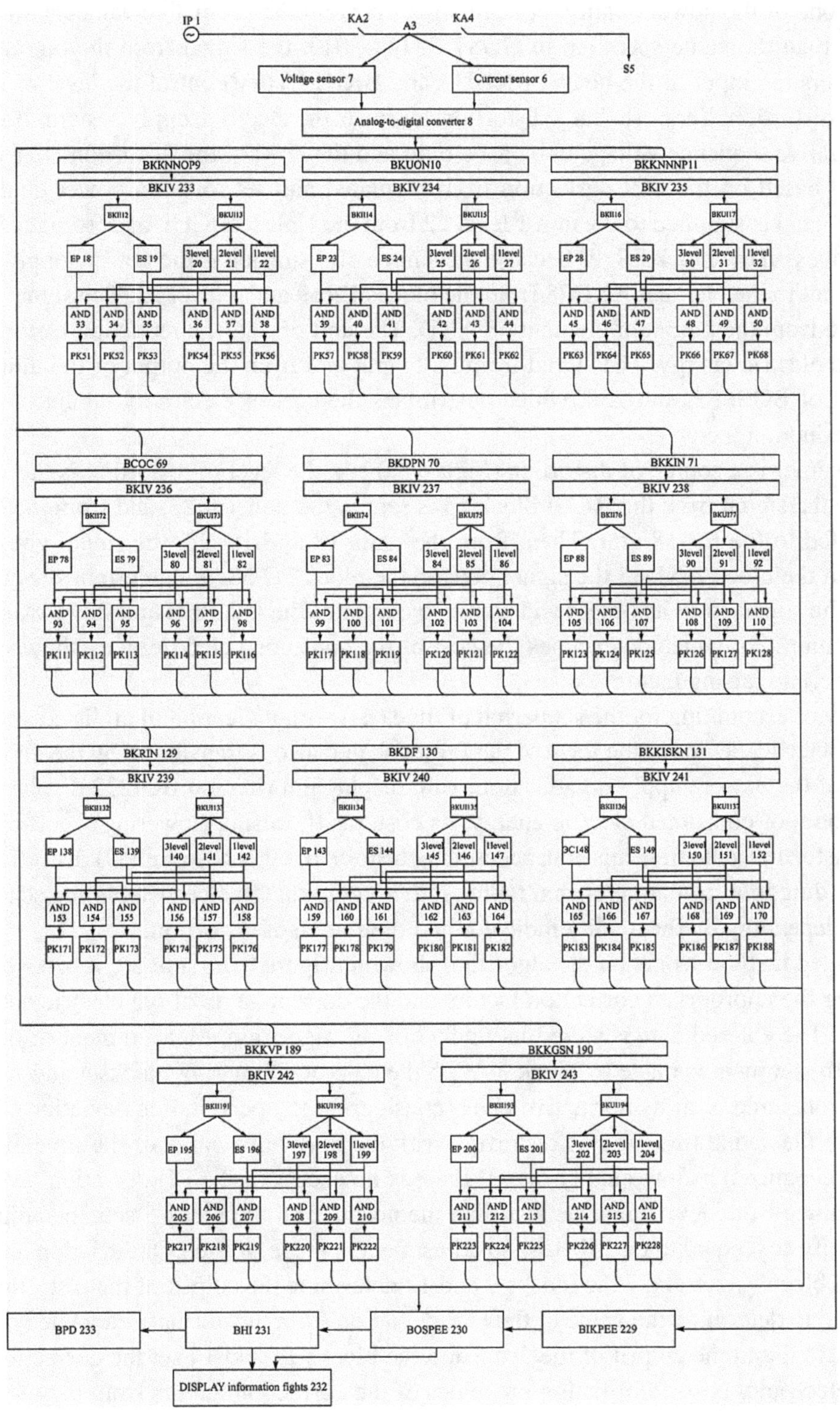

The signal from its output and from the output unit BIKPEE 229 hits BOSPEE input unit 230, which determines the cost of electricity considering the appropriate reduction factor.

If at least one of the power quality indicators goes beyond the normative values with an interval of time not less than the value specified in GOST 32144-2013, the signal from the output of the block BKIV233 enters the input of the blocks BKUI13 and BKII12.They control the level of distortion and the source of distortion. Suppose that a distortion source is the electric supply system, it is caused with change of negative sequence voltage unbalance ratio and the level of the distortion is level 1 (the level is determined based on the PQI deflection from regulated and set for each power quality indicator) (Figure 8b). Signal is supplied to the unit 1 level 22 from the Unit 13 BCUI, and from the block BKII12 the signal arrives on the block ER 19 because the source of distortion is the power supply system. Then the signal is sent to the element AND38 from the blocks ER19 and 1 level 22. The signal is supplied to the PC unit 56 from the output of element 38 AND. The unit 56 sets the reduction factor to the cost of the consumed electric energy. The signal from its output and from the output of the unit BIKPEE229 gets the input of BOSPEE unit 230, which determines the cost of electricity taking into account the appropriate reduction factor.

If the consumer is a source of distortion (Figure 8a) and the level of distortion is the same as in the previous case the signal from the BKUI block 13 is sent to the unit 1ur.22, and from the BKII block12 the signal is fed to the EP 18 unit. Then, from the units 18 and 1ur.22, the signal goes to the block AND35. From the block AND35 the signal goes to the block PC 53, which establishes the increasing coefficient. The signal from its output and from the output of the BIKPEE unit 229 hit on the inputs of the BOSPEE unit 230, which determines the cost of the consumed electricity as shown in Formula 1 with an appropriate raising factor.

The history of accounting for the set period of time (Δt) with indication of all listed sizes is stored in BHI231of a metering device in the form of the Protocol, and also is transferred by means of BPD233 to the database of the power supply organization. Unit display information BOI232 displays information about the amount of consumed electric energy, its cost, tariff, raising (lowering) coefficient, the form and level of distortion, data transmission, accounting history. In other cases of PQI distortion the scheme operates according to a similar algorithm taking into account that, the corresponding scheme elements are triggered depending on the quality indicator and the source of distortion.

The presented method works on the algorithm shown in figure 9. First of all, it is necessary to enter data, including the appropriate correction factors and the current value of the electric energy tariff for the consumer. The entered data is stored in block 2 of the algorithm. Measurement of the load power consumed at the moment is made by block 3. With the help of comparing blocks 4 and 5 the measured PQI value is compared with its normative ones (considering the permissible deviation). The blocks 6 and 7 compare the actual time of PQI distortion with the normative value of the time after which the electricity is recognized as low-quality one. If there is no excess of the PQ deviation level beyond the normalized value or the deviation time less than the normalized the block 8 sets the adjustment coefficient electricity cost equal to one, that is, no adjustment is made. If the PQI deviation above the norm is fixed by the blocks 4,5,6 and 7 the blocks 9 and 12 determine the culprit of the distortion. Blocks 10 and 13 are used to determine the value of the PQI deviation from the normalized value (ΔPQI). Taking into account ΔPQI and the culprit of the distortion the blocks 11 and 14 set the correction factors that adjusted the electricity cost. Information on values of the correction factors from the blocks 8, 11 and 14 is fed to unit 15. This unit calculates electricity cost taking into account correction factors, notably

Figure 8. Diagram of output signals when: A - Energy supply system is the source of distortion; B - Consumer is the source of distortion

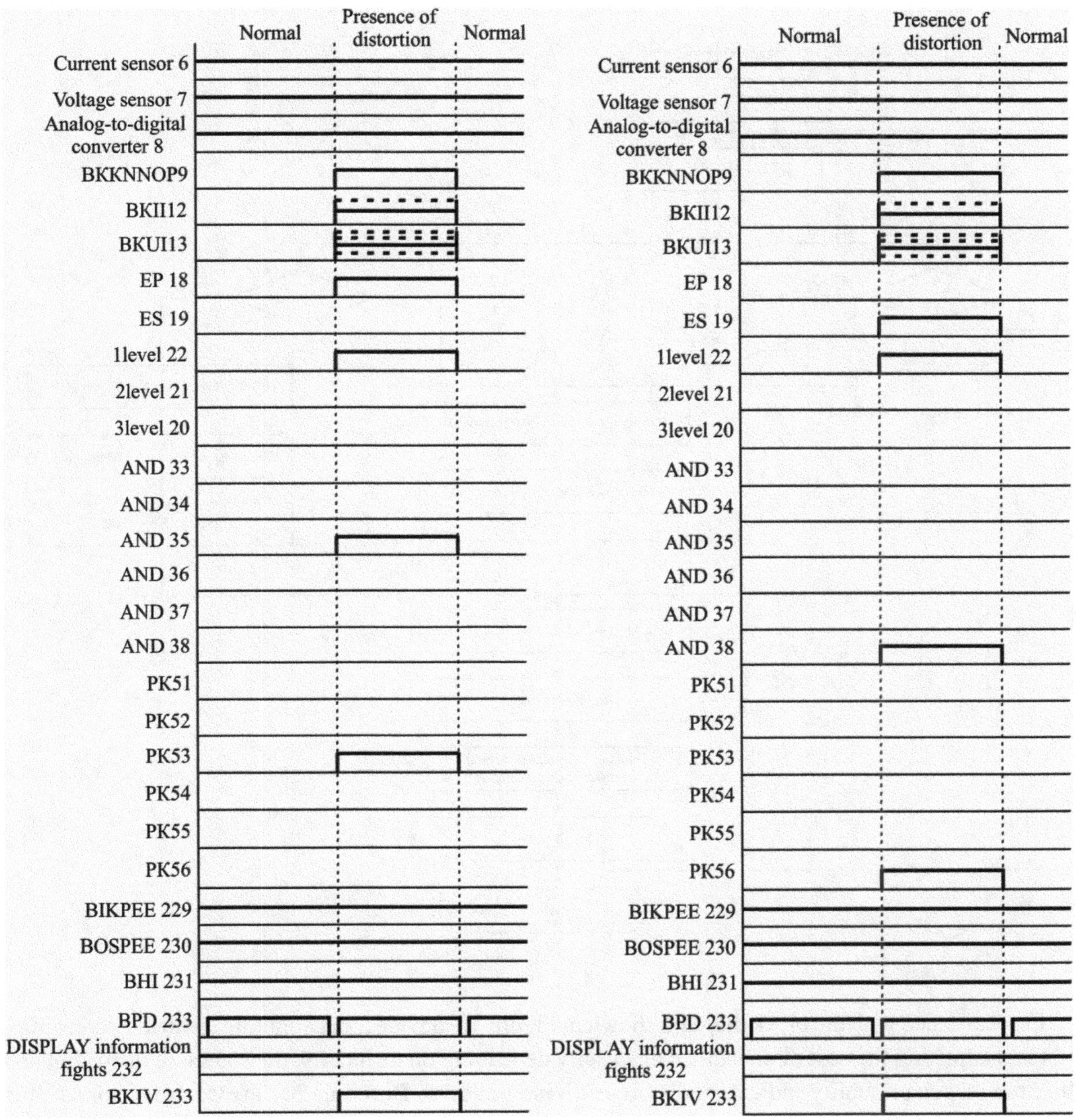

the adjustment is made for the entire specified time interval of measurements. It is rational to enter the measurement time interval equal to a month so electricity cost will be determined directly for the accounting period (monthly). The preservation of the obtained results in the form of Protocol is implemented by the block 16. This Protocol is transmitted through communication channels to the information and measurement center of the power supply organization. Unit 18 displays information on the cost of consumed electricity on the information panel of the meter.

Figure 9. The algorithm of the functional unit, which allows to adjust the cost of consumed electric energy in case of non-compliance with its specified parameters

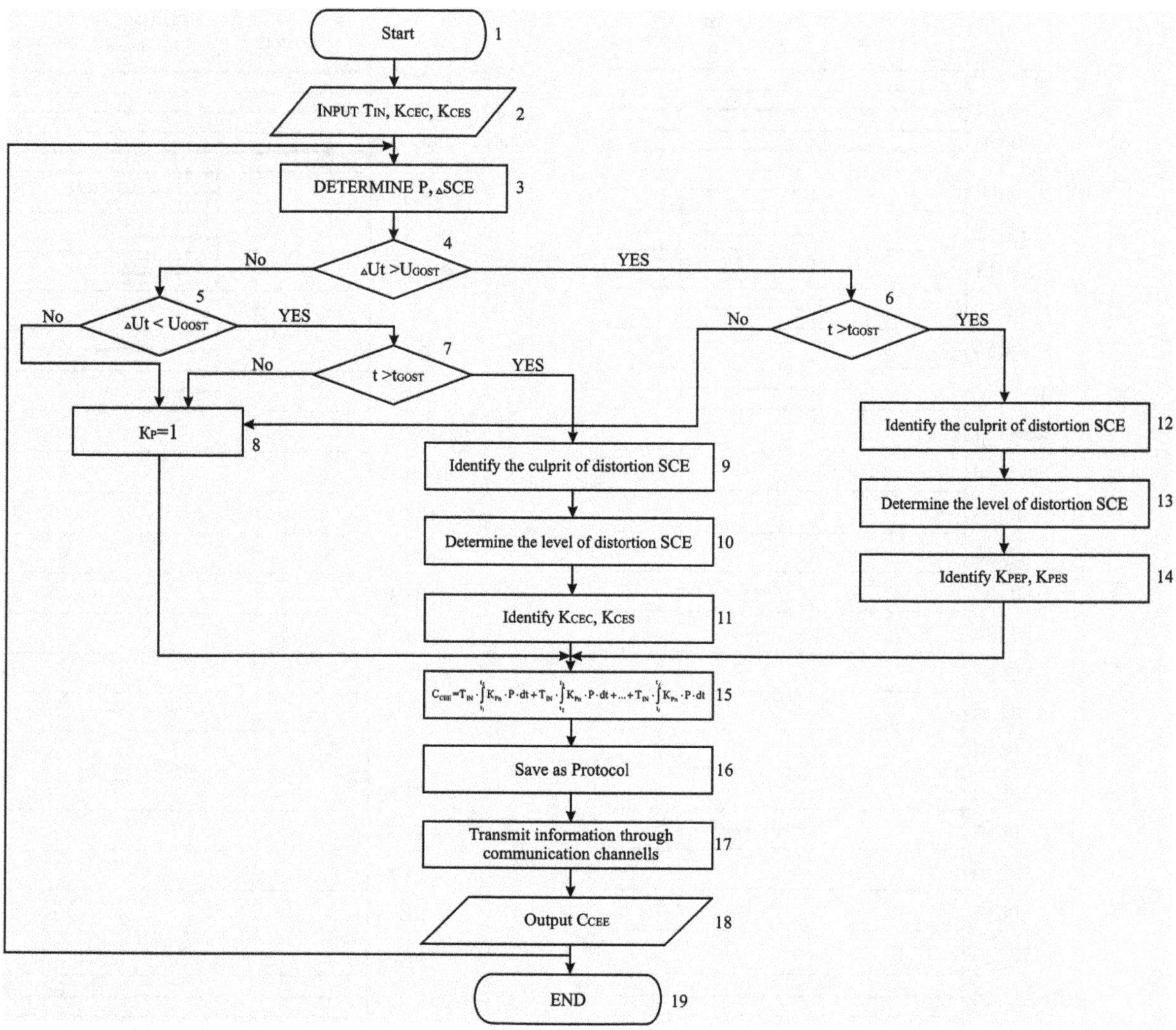

In accordance with the block diagram shown in Figure 3, a device for accounting electricity consumption and adjusting the cost of consumed electricity depending on its quality, the source of distortion and the time of power quality indicators distortion (Vinogradov & Borodin, 2013b) were developed. It is shown in Figure 10.

The device contains a current input terminal 1 and voltage input terminal 2, which are connected to the matching inputs of the energy meter 3 with embedded pulse sensor, store element and communication element. Voltage input clamp 2 is connected to the input of the Converter of the AC voltage to DC voltage 4 (CACDC), the output of which is connected to the information input of analog-to-digital Converter 5 (ADC), the information output of which is connected to the information input of the register 6, the output of which is connected to the input of the digital store element 7, the output of which is connected to the information input of the impulse counter 8, the output of which is connected to the information

Figure 10. The block diagram of the device for accounting electricity consumption and adjusting the cost of consumed electricity depending on its quality, the source of distortion and the time of power quality indicators distortion

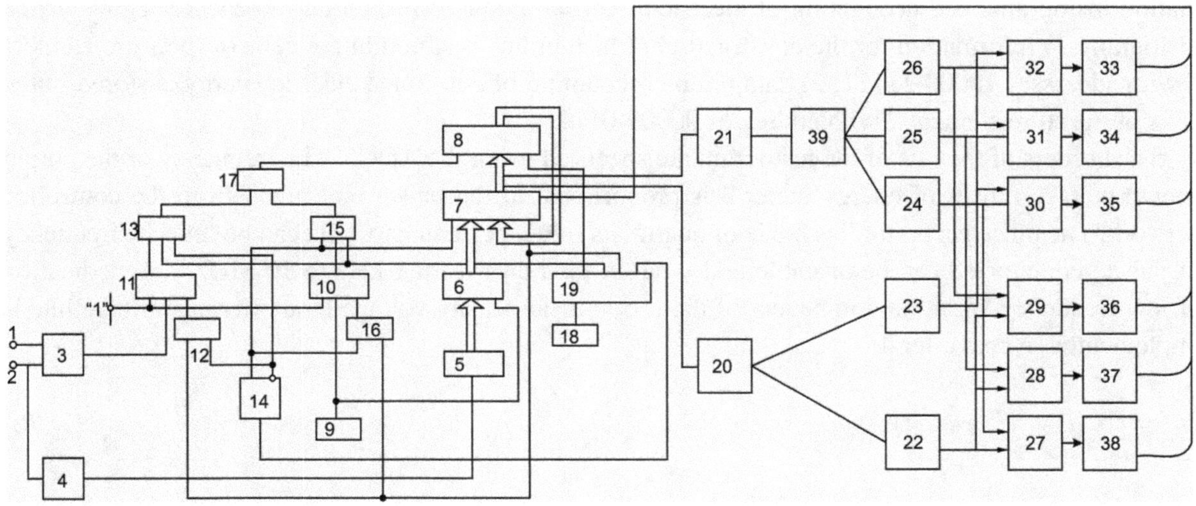

input of store element 7. The output of the pulse generator 9 is connected to the control input of the write register 6 and the clock input of the first D-trigger 10, an information input of which is united with an information input of the second D-trigger 11 and connected to the bus of single voltage. the output of energy meter 3 is connected to the clock input of the second D-trigger 11, the zero-balance input of which is connected to the output of the second AND element 12, the second input of which is combined with a second input of the fourth AND element 13 connected to the inverse output of the SR-trigger 14. The direct output of the SR-trigger 14 is connected to the input of store element 7 and the joint second inputs of the third 15 and the first AND element 16. The output of the last one is connected to the zero-balance input of the first D- trigger 10, the direct output of which is connected to the first input of the third AND element 15, the output of which is connected to the first input OR element 17, the second input of which is connected through the fourth AND element 13 to the direct output of the second D- trigger 11. The output of the OR element 17 is connected to clock input 8. The output of the pulse generator 18 is connected to the clock input of the pulse allocation 19, the first output of which is connected to the input of the recording control 8, the second one is to the combined third inputs of the third 15 and the fourth 13 AND element 3, the third one is to the input of the recording STORE element 7 and the first combined inputs of the first 16 and the second AND element 12, the fourth one is to the clock input of the SR-trigger 14. The control unit of the source distortion BKUI20 and the control unit distortion BKII 21 inputs, which are connected with the information output of the store element 7, block outs BCII 20 which are connected appropriately to the inputs of the fixing unit of the fact of the PQI distortion by consumer ER22 and the fixing unit of the fact of the PQI distortion by power supply system ES23. The output of the block BCUI21 is connected to the unit of determining the time of PQI distortion 39, the outputs of which are connected to the control units of the distortion 1 level 26, 2 level 25, 3 level 24, the outputs are connected respectively to the inputs of AND elements, the outputs of which are connected respectively with the information inputs of the blocks, introducing a multiplying factor 33, 34, 35, or the reducing factor 36, 37, 38, the outputs of which are connected with the information input of the digital block of the STORE element 7.

In the store element 7 of the analyzer accumulates the information for the construction of PQI deviation histograms and accounting of electricity consumption corresponding to each category of the histogram: 1) information for the construction of histograms is stored in the cells of the store element 7 with addresses 10000-11111; 2) data on the accounting of consumed electric energy is stored in the cells of the store element 7 with addresses 00000-01111.

The borders of the bits of the histogram are specified using the ADC5. The frequency of the output square voltage pulses of energy meter 3 is proportional to the power consumed from the controlled network. The pulse generator 9 is made of quartz; its frequency equal to 1 Hz can be set. The frequency of pulse generator 18 can be unstable and is set in the range from 1 kHz to 10 MHz. The dedication of low-frequency variation component of the root – mean-square voltage in alternating voltage line is implemented in converter 4:

$$U_4 = \delta U_t(t) = U(t) - U_n(t) \tag{8}$$

where:

- $U(t)$ is the root-mean-square voltage at the current time t;
- $U(t)$ is the nominal value of the voltage in the network.

The voltage U4 from the output of the PPNP 4 is applied to the information input of the ADC5 and converted into a binary code, for example, 0010.

This code is applied to the low order of CPU address input 7.

Secondly incoming pulse implements the i-th selection of voltage deviation, and the same time the binary unit is recorded into the first D-trigger, the a-d converter input key is entered in register 6 (for example 0010 key).

The pulses coming with variable frequency from the output of the energy meter 3 is transferred to the second D-trigger 11 at bit one state when calculating each j-th quantum of consumed electric energy.

The frequency of the pulse generator 18 is 3-5 orders of magnitude higher than the pulse frequency of the GIB9 and energy meter 3. The output pulses from the of the pulse generator 18 is fed to the input RI19, the outputs of which continuously scans the control signals. SR-trigger 14 changes its state to the opposite at the end of each cycle RI 19 that contribute to the alternate removal of information from the first 10 and second 11 D-triggers

Consider the next work cycle of RI 19.

The first output impulse of RI19 records the information $n_{2(i-1}$ accumulated as a result of the previous statistical analysis of voltage deviations in the discharge of the histogram number 0010 from the cell CBP7 with the address 10010 (at bit one state of the SR-trigger 14) in SI8.

The elements 13 and 15 are gated by the pulse from the second output RI19. The D-trigger is bit one state if another sample of voltage deviation has taken place before the work cycle of RI 19. At the same time, the pulse from the second output of RI19 passes sequentially through the elements AND15 and OR17 to the clock input SI 8, increasing its content $n_{2(i-1)}$ by one, which becomes equal:

$$n_{2i} = n_{2(i-1)} + 1 \quad (9)$$

The pulse from the third output of RI19, passing through the element AND16, sets the D-trigger 10 to zero state as well as acting on the input of the recording control of the CPU 7 returns twice increased n_{2i} information to its cell with the address 10010.

The zero condition of SR-trigger is set by the fourth pulse output of RI19.

Appearing pulse in the next cycle of operation at the first output RI19 enters from the cell with the address CBP 7 00010 (SR-trigger 14 is now in zero state) to SI 8 information $m_{2(j-1)}$, which is accumulated as a result of the previous accounting for power consumption at voltage deviations corresponding to the category of the histogram number 0010.

Pulse from the second output RI19 strobes the element AND 13. D-trigger is at bit one state if another pulse at the output of CE3 have appeared before the work cycle of RI19. At the same time, the pulse from the second output of RC 19 passes sequentially through the elements AND13 and OR17 to the clock input of SI 8, increasing its content $m_{2(j-1)}$ by one, which becomes equal:

$$m_{2j} = m_{2(j-1)} + 1 \quad (10)$$

The pulse from the third output RI19 passing through the element AND12 sets the D-trigger 11 to zero state and also acts on the input of the recording control of the CPU7, returning m_{2j} to its cell with the address 00010.

The pulse from the fourth output RI19 SR-trigger 14 sets towards the single condition etc.

The information n_{ki} or m_{kj} (where k is the key of ADC output 5 suitable to the number of histogram order) returns by pulse from the third output PI19 to the matching cells 7 from the 8 without changes (without increasing by one) in those cycles when D-trigger 10 and 11 are in zero condition.

The histogram of voltage deviation is plotted using the cell content store element 7 with address 10000-11111 after measurements continuing the time T (day, week, month). Ordinate of k – digit is determined by formula:

$$P_{:} = n_k / N = n_k / T \times f_9 \quad (11)$$

where:

- n_k is k-th cell content in store element 7 with address 10000-11111;
- N is total cell content in store element 7 with address 10000-11111;
- T is time of measurement, sec;
- f_9 – frequency of pulse generator 9.

The quantity of consumed electric energy matching the k-th digit of voltage deviation histogram is defined with the cell content with address 00000-01111:

$$W_k = C_3 \times m_k \quad (12)$$

where:

- C_3 – cost of output pulse of energy meter, kW*h/pulse;
- m_k – k-th cell content in store element 7 with address 00000-01111.

The accordance of power quality indicators to GOST demands and quantity of consumed electric energy are controlled using the total cell content of CBP with matching address 00000-01111. The PQI compliance to standard is tested with histogram. The signal entered the input 21 and 39 appears in the presence of distortion of power quality indicators. The units 21 and 39 control the level and the source of distortion. Suppose that the source of distortion is power energy system, the distortion is the presence of a steady voltage deviation and the level of PQI distortion is the 1st. In this case from the unit BKYU 21 the signal enters the unit 39 where the distortion time of power quality indicators is defined. Then the signal enters the unit 1 level 26, the signal from the unit BKKI 20 enter the unit 23. After this the signal from the units 23 and 1 level 26 enters the AND element 32. The output signal of AND element enters the unit 33, set the reducing factor of electricity cost. Its output signal enters the unit CBP defining the cost of consumed energy with matching reducing factor. If the source of distortion is the consumer at the 1st level of distortion of the same type as in the previous case a signal goes from the block 21 to the block 39 where the time of PQI distortion is determined. Then the signal goes to the block 1level 26. A signal is supplied from block 20 BCII to the block ER 22. Next, the signal of the blocks ER22 and 1level.26 goes to the unit AND29. From the block AND29 the signal is sent to the block PC36, which sets the increasing coefficient. Signal from its output goes to the input of block 7 of store element, which determines the electricity cost with matching increasing factor.

The cost of the consumed electricity is proposed to be determined by the Formula 1. If there is no distortion of power quality K_p=1.

The history of accounting for a given period of time (Δt) with the indication of all the listed values of SE3 is stored in the memory of the metering device in the Protocol form and is transmitted by means of the information transmission unit to the database of power supply organization. The energy meter 3 displays information about the amount of consumed electric energy, its cost, tariff, correction factor, the type, level and time of PQI distortion, data transmission, accounting history.

Thus, this device allows automatically accounting the amount of consumed electric energy and determining its cost depending on its quality, source and time of the distortion of power quality indicators.

A feasibility study of the proposed method is presented further. It is offered to implement the proposed method with the help of the Resource-E4 counter, a retrofitted embedded software unit implementing operational adjustment of the cost of electricity depending on its quality. The list of components used for the implementation of this method is shown in Table 3.

Table 3. Calculation of the cost of components

N°	Name	Cost, Thousand RUB
1	Energy meter Resource -E4	35,63
2	Program unit realizing the correction	4
3	Total	39,63

4 thousand RUB includes payment of the patent and materials to create the device.

Capital investments are determined by the formula:

$$\kappa = C + I + T \tag{13}$$

where:

- C is the expense of components, rubles;
- I is installation expense, RUB (5 ... 10% of C);
- T is transportation expense, rubles (5% of C).

The transportation expense for their delivery is:

$$T = 5\% \times C \tag{14}$$

$$T = 0,05 \times 39,63 = 1,98 \ thousand \ RUB$$

The expense of installation of the system is:

$$I = 10\% \times C \tag{15}$$

$$I = 0,1 \times 36,42 = 3,64 \ thousand \ RUB$$

Total capital investments is:

$$\kappa = 39,63 + 1,98 + 3,64 = 45,25 \ thousand \ RUB$$

The operation cost of the device that implements the cost adjustment of consumed electric energy depending on its quality is determined by the formula:

$$E = E_{AD} + E_{RM} + E_{other} \tag{16}$$

where:

- E_{AD} is accrued depreciation, RUB;
- E_{RM} is the repair and maintenance expense of the system, RUB;
- E_{other} is other direct expense, RUB.

Accrued depreciation is determined by the formula:

$$E_{AD} = R_{AC} \times \kappa \quad (17)$$

where R_{AC} is the rate of accrued depreciation (accept 4% from the calculation of the service life of the device 25):

$$E_{AD} = 0,04 \times 45,25 = 1,81 \ thousand \ RUB$$

The expense of repair and maintenance of the device is calculated by the formula:

$$E_{RM} = R_{RM} \times \kappa \quad (18)$$

where R_{RM} is the rate of deductions for technical inspection and technical repair of the device (take 5% as maintenance of the device is to systematically perform work: cleaning the body and the front panel once per month), monitoring the correctness of the work (once per month), checking the reliability of the connection to the meter measuring or interface circuits (once for 2 years).

The expense of repair and maintenance is:

$$E_{RM} = 0,05 \times 45,25 = \ 2,26 \ thousand \ RUB$$

Other expense is determined depending on the value of all operating costs:

$$E_{other} = R_{other} \times (E_{AC} + E_{RM}) \quad (19)$$

where R_{other} is the deductions for other expenses (5%).

Other expense is:

$$E_{other} = 0,05 \times \left(1,81 + 2,26\right) \ = \ 0,2 \ thousand \ RUB$$

Total operating expense is:

$$E = 1,81 + 2,26 \ + 0,2 = \ 4,27 \ thousand \ RUB$$

To determine the economic efficiency of the device implementation, it is proposed to use the following procedure.

Step 1: Accept annual consumption limits ranging from 122880 kW•h to 1228800 kW•h, as the annual electricity consumption for different groups of consumers varies in a wide range. The specified range is proposed to divide into 10 parts in increments of 122880.

Step 2: Consider two options. In the first version the culprit of power quality deterioration is a power supply organization, in the second one the culprit is a consumer.

Step 3: Calculate the cost of electricity consumption (initial tariff T_{init} = 2,7 rubles per 1 kW h) excluding and taking into account the correction factors.

Step 4: Determine the saving on electricity payment by the formula:

$$S_E = C_{withoutCF} - C_{withCF} \tag{20}$$

where:

- $C_{withoutCF}$ is the electricity cost excluding correction factors for the year, RUB;
- C_{withCF} is the electricity cost taking into account the correction factors for the year, RUB.

Step 5: Determine the annual savings by the formula:

$$S_{E_{annual}} = S_E - E \tag{21}$$

Step 6: Determine the payback period of capital investments by the formula:

$$T_{DPP} = \frac{-\ln(1 - R_D \cdot K / I_N)}{\ln(1 + R_D)} \tag{22}$$

where:

- T_{DPP} is discounted payback period, year;
- R_D is discount rate (Bank refinancing rate, take equal to 0.08), in shares;
- I_N is the value of annual net income from the implementation of equipment, RUB per year.

The results of calculations are summarized in Table 4 and Table 5.

For the accepted ranges of electricity consumption, the use of methods and means of adjusting the cost of consumed electric energy depending on its quality is economically profitable with a short payback period at different correction factors. If electricity consumption is less than 122880 kWh per year payback period may be reduced to zero. If the consumption is less than 1581 kWh per year it may be negative. A schedule of changes based on the results of calculations in the payback period depending on the volume of consumed electric energy and the correction factor is shown in Figure 11 for the case when the culprit of PQ distortion is a power supply organization and in Figure 12 for the case when the culprit of PQ distortion is a consumer.

The recommended payback period for energy during 7 years is achieved with annual consumption for both a power supply system and a consumer:

2501 kWh ($К_п = 0$)

5002 kWh ($К_п = 0{,}5$)

Table 4. Economic efficiency when a power supply organization is the source of power quality distortion

Consumed Electric Energy, kWh	The Electricity Cost Excluding Correction Factors Per Year, RUB	The Electricity Cost Considering Correction Factors Per Year, RUB	Savings Under Electricity Cost, RUB	Annual Savings, RUB	Payback Period, Year
Correction factor K_P=0					
122880	331776	0	331776	327506	0,165798
245760	663552	0	663552	659282	0,082362
368640	995328	0	995328	991058	0,05479
491520	1327104	0	1327104	1322834	0,041048
614400	1658880	0	1658880	1654610	0,032817
737280	1990656	0	1990656	1986386	0,027336
860160	2322432	0	2322432	2318162	0,023424
983040	2654208	0	2654208	2649938	0,020491
1105920	2985984	0	2985984	2981714	0,018211
1228800	3317760	0	3317760	3313490	0,016388
Correction factor K_P=0,5					
122880	331776	165888	165888	161618	0,335977
245760	663552	331776	331776	327506	0,165798
368640	995328	497664	497664	493394	0,110054
491520	1327104	663552	663552	659282	0,082362
614400	1658880	829440	829440	825170	0,065805
737280	1990656	995328	995328	991058	0,05479
860160	2322432	1161216	1161216	1156946	0,046934
983040	2654208	1327104	1327104	1322834	0,041048
1105920	2985984	1492992	1492992	1488722	0,036474
1228800	3317760	1658880	1658880	1654610	0,032817
Correction factor K_P=0,8					
122880	331776	265420,8	66355,2	62085,2	0,874605
245760	663552	530841,6	132710,4	128440,4	0,422764
368640	995328	796262,4	199065,6	194795,6	0,278754
491520	1327104	1061683	265420,8	261150,8	0,207926
614400	1658880	1327104	331776	327506	0,165798
737280	1990656	1592525	398131,2	393861,2	0,137866
860160	2322432	1857946	464486,4	460216,4	0,117988
983040	2654208	2123366	530841,6	526571,6	0,10312
1105920	2985984	2388787	597196,8	592926,8	0,09158
1228800	3317760	2654208	663552	659282	0,082362

Table 5. Economic efficiency when a consumer is the source of power quality distortion

Consumed Electric Energy, kWh	The Electricity Cost Excluding Correction Factors Per Year, RUB	The Electricity Cost Considering Correction Factors Per Year, RUB	Savings Under Electricity Cost, RUB	Annual Savings, RUB	Payback Period, Year
Correction factor K_P=2					
122880	331776	663552	331776	327506	0,165798
245760	663552	1327104	663552	659282	0,082362
368640	995328	1990656	995328	991058	0,05479
491520	1327104	2654208	1327104	1322834	0,041048
614400	1658880	3317760	1658880	1654610	0,032817
737280	1990656	3981312	1990656	1986386	0,027336
860160	2322432	4644864	2322432	2318162	0,023424
983040	2654208	5308416	2654208	2649938	0,020491
1105920	2985984	5971968	2985984	2981714	0,018211
1228800	3317760	6635520	3317760	3313490	0,016388
Correction factor K_P=1,5					
122880	331776	497664	165888	161618	0,335977
245760	663552	995328	331776	327506	0,165798
368640	995328	1492992	497664	493394	0,110054
491520	1327104	1990656	663552	659282	0,082362
614400	1658880	2488320	829440	825170	0,065805
737280	1990656	2985984	995328	991058	0,05479
860160	2322432	3483648	1161216	1156946	0,046934
983040	2654208	3981312	1327104	1322834	0,041048
1105920	2985984	4478976	1492992	1488722	0,036474
1228800	3317760	4976640	1658880	1654610	0,032817
Correction factor K_P=1,2					
122880	331776	398131,2	66355,2	62085,2	0,874605
245760	663552	796262,4	132710,4	128440,4	0,422764
368640	995328	1194394	199065,6	194795,6	0,278754
491520	1327104	1592525	265420,8	261150,8	0,207926
614400	1658880	1990656	331776	327506	0,165798
737280	1990656	2388787	398131,2	393861,2	0,137866
860160	2322432	2786918	464486,4	460216,4	0,117988
983040	2654208	3185050	530841,6	526571,6	0,10312
1105920	2985984	3583181	597196,8	592926,8	0,09158
1228800	3317760	3981312	663552	659282	0,082362

17212 kWh ($K_п = 0,8$)

The typical cases arising in the calculations for the consumed and transmitted electric energy with the implementation of the device for electricity cost adjusting depending on its quality:

- If a consumer makes PQ distortions, he has to pay for electric energy at a higher price, what leads to the fact that a power supply organization receive more money. Therefore, the payment for the transferred electric energy in this case will be greater by the amount of overpayment that the consumer made;
- If there is no distortion of power quality the cash flow scheme works according to the order approved by regulatory documents;
- If PQ distortions are made by a power supply organization the consumer have to pay for electricity at a lower price, what leads to the fact that a power supply organization receive less money. Therefore, the payment for the transferred electricity in this case will be less by the amount of underpayment that the consumer made.

Figure 11. Schedule of the payback period change depending on the amount of consumed electric energy and the correction factor in case of electricity quality distortion by a power supply organization

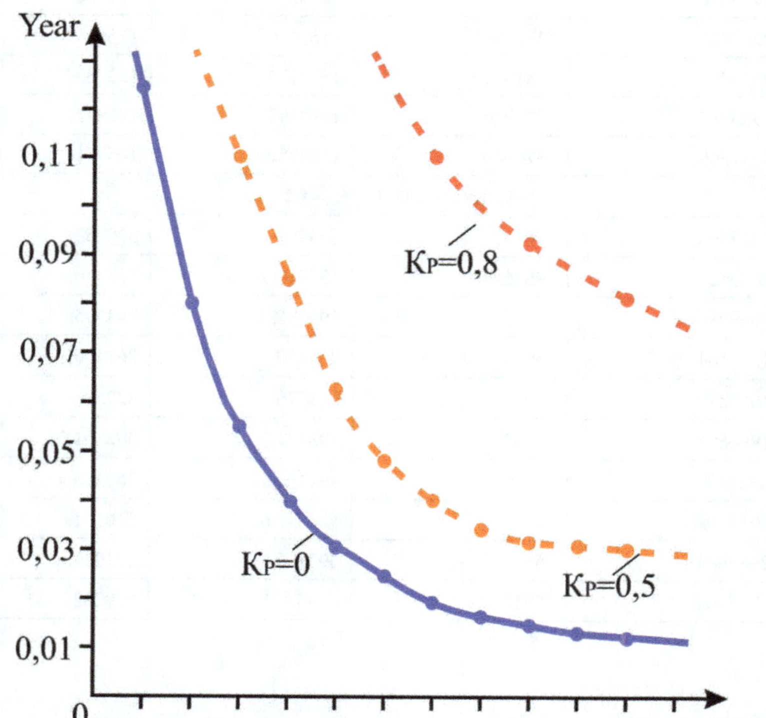

Figure 12. Schedule of the payback period change depending on the amount of consumed electric energy and the correction factor in case of electricity quality distortion by a consumer

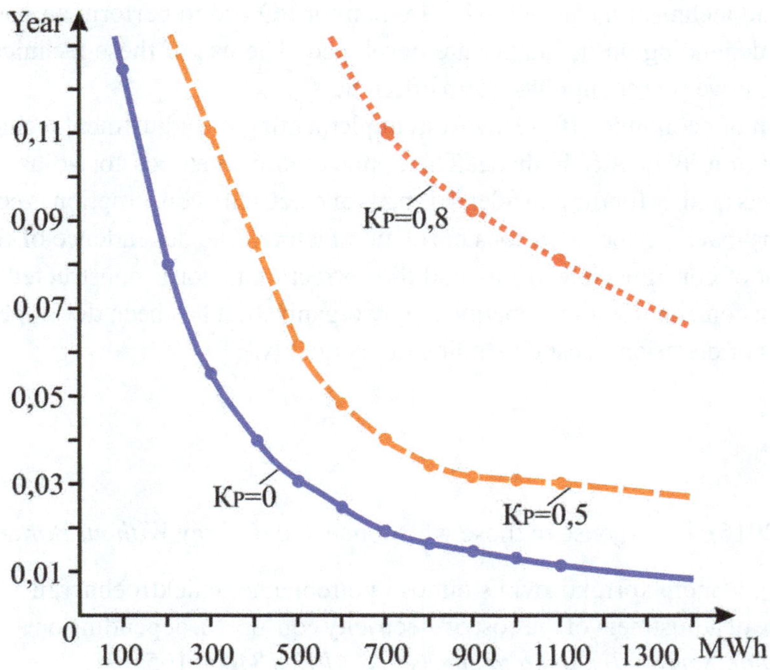

FUTURE RESEARCH DIRECTIONS

The main directions of future research on this topic are the development of technical and software tools for the proposed method of adjusting electricity cost depending on its quality. There should be the development the technical means of PQ managing and preventing PQI distortions by both consumers and power supply organizations. A method of PQ controlling of a particular consumer and a method of group PQ regulation should be developed as well. The developed methods and technical means should provide not only the possibility of adjusting electricity cost depending on its quality, but also automatic PQI correction, that is, the elimination of distortion both on the part of consumer network and on the part of power supply organization network. In addition, it is necessary to develop regulatory documents that establish requirements for means of operational adjustment of the electricity cost depending on its quality and the necessary coefficients.

CONCLUSION

On the basis of the carried out studies, a number of conclusions can be drawn:

1. A mathematical model of the electricity metering system has been developed. The model makes it possible to determine electricity cost depending on its quality. The presented model allowed

to develop a technical and economic mechanism to stimulate both consumers and power supply organization to maintain PQ at the level of normative documents;

2. Algorithms and technical means allowing to monitor PQ and to perform an operative adjustment of electricity depending on its quality are developed. The use of these technical means makes it possible to improve power supply system efficiency;
3. The estimation of economic efficiency from implementing the adjustment technique of electricity cost depending on its quality is made. The application of methods for adjusting electricity cost depending on its quality for the considered ranges of electricity consumption is economically viable with a short payback period at various correction factors. The dependence of the payback period on the amount of consumed electricity and the correction factor is constructed. A mechanism for calculating the consumer with an energy supply organization has been developed. It takes into account changes of electricity cost depending on its quality.

REFERENCES

Bachmanova, K. (2016). In response to those who connected. *Energy Without Borders, 5*(40), 18-19.

Borodin, M. (2013). Princip korrektirovki stoimosti potreblennoj ehlektroehnergii v zavisimosti ot eyo kachestva [Principle of adjustment of the cost of electricity consumed, depending on its quality]. *Naukovij visnik NUBiP Ukraïni. Seriya: Tekhnika ta energetika APK, 184*(1), 165-171.

Borodin, M., & Uryupin, N. (2018). Obosnovanie ehnergoehffektivnogo secheniya linij ehlektroperedach pri otkloneniyah napryazheniya ±5%, ±10% [Rationale energy-efficient section of the power lines with voltage deviations of ±5%, ±10%]. *Vestnik NGIEHI, 2*(81), 20–29.

Borodin, M. & Vinogradov, A. (2013). Korrektirovka stoimosti potreblennoj ehlektroehnergii v zavisimosti ot ee kachestva [Adjustment of the cost of consumed electricity depending on its quality]. *Promyshlennaya ehnergetika, 7*, 12-16.

Borodin, M., & Vinogradov, A. (2014). *Povyshenie ehffektivnosti funkcionirovaniya sistem ehlektrosnabzheniya posredstvom monitoringa kachestva ehlektroehnergii* [Improving of power supply system efficiency by means of power quality monitoring]. Orel, Russia: FGBOU VPO Orel GAU.

Borodin, M., & Vinogradov, A. (2018). Redakciya metodiki korrektirovki stoimosti potreblennoj ehlektroehnergii v zavisimosti ot eyo kachestva i algoritm eyo realizacii v sootvetstvii s gost na kachestvo ehlektroehnergii [A revision of the adjustment method of the electricity cost depending on its quality and the algorithm for its implementation in accordance with the standard on power quality]. *Vestnik NGIEHI, 4*(83), 54–64.

Dobrusin, L. (2004). O sovershenstvovanii tekhnicheskogo regulirovaniya kachestva ehlektricheskoj ehnergii v sistemah ehlektrosnabzheniya obshchego naznacheniya [About improvement of technical regulation of quality of electric energy in General-purpose power supply systems]. *Vesti v ehlektroehnergetike, 1*, 44-50.

Fangxing, L., Wei, Q., Hongbin, S., Hui, W., Jianhui, W., Yan, X., ... Pei, Z. (2010). Smart transmission grid: Vision and framework. *IEEE Transactions on Smart Grid*, *1*(2), 168–177. doi:10.1109/TSG.2010.2053726

Federal'nyj zakon Rossijskoj Federacii N°35 FZ "Ob Ehlektroehnergetike". (n.d.). [Federal law of the Russian Federation N°35 "About Electric Power"]. Available from: http://ivo.garant.ru/#/document/185656/paragraph/539078:1

Gheorghe, S., Tanasa, C., Ene, S., & Mihaescu, M. (2005). Power quality, Energy efficiency and the performance in electricity distribution and supply companies. In *Proceedings of the 18th International Conference and Exhibition on Electricity Distribution: CIRED*. Turin, Italy: IET.

Gmurman, V. (1999). *Teoriya veroyatnostej i matematicheskaya statistika. Ucheb. Posobie dlya vuzov. Izd. 7-e ster* [Probability Theory and mathematical statistics. Studies'. The manual for high schools. Ed. 7's erased]. Moscow, Russia: Vyssh. shk.

Golikov, I., & Vinogradov, A. (2017). *Adaptivnoe avtomaticheskoe regulirovanie napryazheniya v sel'skih ehlektricheskih setyah 0,38 kV* [Adaptive automatic voltage regulation in rural power networks of 0.38 kV]. Orel, Russia: FGBOU VO Orel GAU.

Grazhdanskij kodeks RF. (2000). [The civil code of the Russian Federation]. Moscow, Russia: Publishing house NORM.

Hartungi, R., & Jiang, L. (2010). Investigation of Power Quality in Health Care Facility. *Proceedings of International Conference on Renewable Energies and Power Quality (ICREPQ'10).* 10.24084/repqj08.555

Interstate Council for Standardization, Metrology and Certification. (n.d.). *GOST 32144-2013 EHlektricheskaya ehnergiya. Sovmestimost' tekhnicheskih sredstv ehlektromagnitnaya. Normy kachestva ehlektricheskoj ehnergii v sistemah ehlektrosnabzheniya obshchego naznacheniya* [GOST 32144- 2013 Electric energy. Electromagnetic compatibility of technical means. Norms for power quality in general-purpose power supply systems]. Moscow, Russia: *Standartinform.*

Irwin, L. (2010). Asset management benefits from a wide area power quality monitoring system. In *Proceedings of the 23rd International Conference on Electricity Distribution: CIRED.* Lyon, Italy: IET.

Kartashev, I. (2001). *Upravlenie kachestvom ehlektroehnergii* [Power Quality in Power Supply Systems]. Moscow, Russia: Izdatel'skij dom MEHI.

Kartashev, I., Tul'skij, V., SHamonov, R., SHarov, Y., & Vorob'ev, A. (2006). *Upravlenie kachestvom ehlektroehnergii* [Power quality management]. Moscow, Russia.: Izdatel'skij dom MEHI.

Kuzmichev, A. V., Malyshev, V. V., & Tikhomirov, D. A. (2010). Efficiency of combined pasteurization of milk by UV and IR irradiation. *Light & Engineering*, *5*, 6–9.

Leshchinskaya, T., & Naumov, I. (2008). *EHlektrosnabzhenie sel'skogo hozyajstva* [Power supply of agriculture]. Moscow, Russia: Colossus.

Mandatova, P., Massimiano, M., Verreth, D., & Gonzalez, C. (2014). Network tariff structure for a smart energy system. In *Proceedings of the CIRED Workshop*. Rome, Italy: IET.

Perova, M. (2007). *EHkonomicheskie problemy i perspektivy kachestvennogo ehlektrosnabzheniya sel'skohozyajstvennyh potrebitelej v Rossii* [Economic problems and prospects of qualitative power supply of agricultural consumers in Russia]. Moscow, Russia: INP RAN.

Popov, N. (2010). *Hlektrosnabzhenie. Rabochie rezhimy setej 0,38...10kV: uchebnoe posobie* [Power supply of the networks of 0,38...10kV: working modes]. Kostroma, Russia: KGSHA.

Santarius, P., Krejci, P., Brunclik, Z., Prochazka, K., & Kysnar, F. (2015). Evaluation of power quality in regional distribution networks. In *Proceedings of the 23rd International Conference on Electricity Distribution: CIRED*. Lyon, France: IET.

Siano, P. (2014). Demand response and smart grids – a survey. *Renewable & Sustainable Energy Reviews, 30*, 461–478. doi:10.1016/j.rser.2013.10.022

Spiganovic, A. (2004). Obespecheniya kachestva ehlektroehnergii sistem ehlektrosnabzheniya [Ensure the power quality of power systems]. In *Proceedings of the konferencii «EHlektroehnergetika i ehnergosberegayushchie tekhnologii»*. Lipeck, Russia: LGTU.

Spiganovic, A., & Zakharov, K. (2007). *Vnutrizavodskoe ehlektrosnabzhenie i rezhimy: uchebnik* [In-Plant electricity and modes: tutorial]. Lipetsk, Russia: LGTU.

Suhas, C. D., Nitin, R. L., & Pillai, R. (2015). Power quality monitoring with smart meters. In *Proceedings of the 23rd International Conference on Electricity Distribution: CIRED*. Lyon, France: IET.

Vinogradov, A. & Borodin, M. (2012). EHkspertnaya ocenka popravochnyh koehfficientov k stoimosti potreblennoj ehlektroehnergii v zavisimosti ot eyo kachestva [Expert evaluation of correction factors to the cost of electricity consumption depending on its quality]. *Vesti vysshih uchebnyh zavedenij CHernozem'ya, 3*, 14-19.

Vinogradov, A., & Borodin, M. (2013a). EHkonomicheskaya ehffektivnost' vnedreniya sposobov i sredstv korrektirovki stoimosti potreblennoj ehlektroehnergii v zavisimosti ot ee kachestva [Economic efficiency of introduction of methods and means of adjustment of cost of the consumed electric power depending on its quality]. *Proceedings of the mezhdunarodnoj nauchno-prakticheskoj konferencii. "Osobennosti tekhnicheskogo i tekhnologicheskogo osnashcheniya sovremennogo sel'skohozyajstvennogo proizvodstva.*

Vinogradov, A., & Borodin, M. (2013b). *R.F. Patent No. 134338*. Patent Service of the Russian Federation.

Vinogradov, A., Borodin, M., & Bolshev, V. (2014). Sposob upravleniya kachestvom ehlektricheskoj ehnergii [Method for managing power quality]. *Tekhnika v sel'skom hozyajstve, 4*, 30-31.

Vinogradov, A., Borodin, M., Volchenkov, Y. U., & Peshekhonova, Z. H. (2015). *Sovershenstvovanie deyatel'nosti po ehnergosberezheniyu i po osushchestvleniyu tekhnologicheskih prisoedinenij filiala OAO "MRSK Centra" - "Orelehnergo* [Improvement of the energy efficiency and the implementation of technological connection of the branch of JSC "IDGC of Center" - "Orelenergo"]. Orel, Russia: FGBOU VPO Orel GAU.

Vinogradov, A., Vasiliev, A., Bolshev, V., Semenov, A., & Borodin, M. (2018). Time Factor for Determination of Power Supply System Efficiency of Rural Consumers. In V. Kharchenko & P. Vasant (Eds.), Handbook of Research on Renewable Energy and Electric Resources for Sustainable Rural Development (pp. 394–420). Academic Press. doi:10.4018/978-1-5225-3867-7.ch017

Vinogradova, A., & Skrobov, A. (2013). Ob uvelichenii poter' ehlektroehnergii v LEHP vsledstvie izmenenij v proektirovanii s uchetom uvelicheniya dopustimyh otklonenij napryazheniya soglasno GOST 54149-2010 [On the increase of electric power losses in transmission lines due to changes in design taking into account the increase in permissible voltage deviations in accordance with GOST 54149-2010]. *Vestnik NGIEHI*, *6*(25), 16–19.

Yasin, K. (2016). A survey on smart metering and smart grid communication. *Renewable & Sustainable Energy Reviews*, *57*, 302–318. doi:10.1016/j.rser.2015.12.114

Zhelezko, Y. (2002). O normativnyh dokumentah v oblasti kachestva ehlektroehnergii i uslovij potrebleniya reaktivnoj moshchnosti[About normative documents in the field of power quality and consumption conditions of jet capacity]. *EHlektricheskie stancii, 6*, 18–24.

KEY TERMS AND DEFINITIONS

Point of Common Connection: The network point closest to the specific load of a user, where the loads of other network users are connected.

Power Quality: The conformity degree of electrical energy characteristics at a point in the electrical system with a set of standardized indicators.

Power Supply Consumer: A legal entity or a private person exercising the use of electric energy (capacity) on the basis of a concluded contract.

Power Supply Reliability: The ability of the power supply system to transmit and distribute the required amount of electricity from sources to consumers at standard voltage levels and in accordance with a specified load schedule.

Power Supply System: A set of electrical installations and electrical devices designed to supply electric power to various consumers of electrical networks.

System for Monitoring Power Quality: A set of interrelated organizational and technical measures ensuring the monitoring power quality. Typically, monitoring system uses equipment complexes designed to monitor power quality at various point of power supply system and transfers data through communication channels to the processing center.

Technological Connection: A complex of measures carried out to connect power receivers of natural or legal persons to electric grids.

Chapter 14
Resonant Power System for Electric Transport

Dmitry Strebkov
Federal State Budgetary Scientific Institution, Russia

Alexey Nekrasov
Federal State Budgetary Scientific Institution, Russia

Anton Nekrasov
Federal State Budgetary Scientific Institution, Russia

ABSTRACT

Noncontact electric power supply with the use of cable line is an innovative trend in the development of electric transport. Work to develop such systems is carried out in the USA, South Korea, Germany, Israel, the Russian Federation. The VIESH researchers have developed and patented experimental models of a car and a tramway of the future without accumulators, which receive energy from external energy system through air gap from a single wire cable laid under the surface of road pavement. Noncontact systems for electric power supply to transport are developed at VIESH on the basis of resonant methods of electric power supply through a single-wire cable. This is a basic difference of the technology under consideration from similar ones. This technology makes it possible to realize at the new technological level the need for electric roads with minimal costs. Given that all the world's automakers are working on the production of an electric vehicle for mass production, the need for creating a motorway in the city and on inter-urban autobahns is even more topical.

DOI: 10.4018/978-1-5225-9179-5.ch014

INTRODUCTION

Noncontact electric power supply with the use of cable line is an innovative trend in the development of electric transport. Work to develop such systems is carried out in the USA, South Korea, Germany, Israel, the Russian Federation.

N. Tesla patented the contactless method of electric power supply to rail vehicles from single-conductor cable laid in ground (Tesla, N. 1894, Tesla, N. 1999, Tesla, N. 1978, Tesla N. Lectures 1956, Tesla, N. 2008, Sarbon S., M.Sc. 2013). In his letters dated June 14 and 17, 1905 N. Tesla wrote: «...by the use of a generator [transmitter] of stationary waves and receiving apparatus properly placed and adjusted in any other location, however remote, it is practicable to transmit intelligible signals or to control or actuate at will any one or all of such apparatus for many other important and valuable purposes... [or] other features or property of disturbances of this character.

Electricity is displaced by the transmitter in all directions, equally through the earth and the air; that is true, but energy is expended only at the place where it is collected and used to perform some work.

Although the electrical oscillations would manifest themselves all over the earth, at the surface as well as high in the air, virtually no power would be consumed.

Electromagnetic energy of the transmitter is sent to any place on Earth or its atmosphere where there is a receiver with resonant frequency, tuned to the transmitter's frequency».

BACKGROUND

Resonant Single Trolley and Cable Systems for Supplying Power to Electric Vehicles

The scheme of the resonant system, used to power a mobile mechanism, is shown in Figure 1 (Strebkov, D.S., Nekrasov, A.I., Avramenko, S.V., & Sakharov, D.N. 2002; Strebkov, D.S., Avramenko, S.V., & Nekrasov, A.I. 2000; Strebkov, D.S., Avramenko S.V., Nekrasov A.I., & Roshchin, O.A. 2002; Strebkov D.S., Avramenko S.V., & Nekrasov, A.I. 1999). The power source *1* is connected to the frequency converter *2*, whose output frequency ranges from 1 to 20 kHz. The converter powers the resonant step-up Tesla transformer *3*, containing the low-voltage and high-voltage windings *4*, *5*, mounted on the common core. One terminal of the transformer secondary winding *5* is coupled to the contact system *6*, the second terminal is open or grounded. The trolley *7* is connected to the contact system *6*, whose voltage ranges from 1 to 35 kV. This voltage is applied to the high-voltage winding *9* of the step-down Tesla transformer *10*. Its low-voltage winding *10* is connected to the inverter *12* through the diode unit *11*. The inverter output voltage (with the required frequency) is applied to the traction motor *14* via the control unit *13*; the motor is joined to the driving wheels *15* of the mobile mechanism.

The use of the resonant power supply system enables enhancement of the mechanism efficiency and reliability, as well as reduction of energy loss in the transmission line. The considered method may be used to power a group of mobile vehicles through contact systems of various structures (for example, a grid of conductors). In an alternative version, a thin insulated conductor with a high mechanical strength may be used instead of the conventional contact system.

Freight transportation inside farms may be carried out using ecologically pure electric vehicles powered through a trolley or a flexible single-wire conductor.

Figure 1. Scheme of the resonant system used to power a mobile mechanism

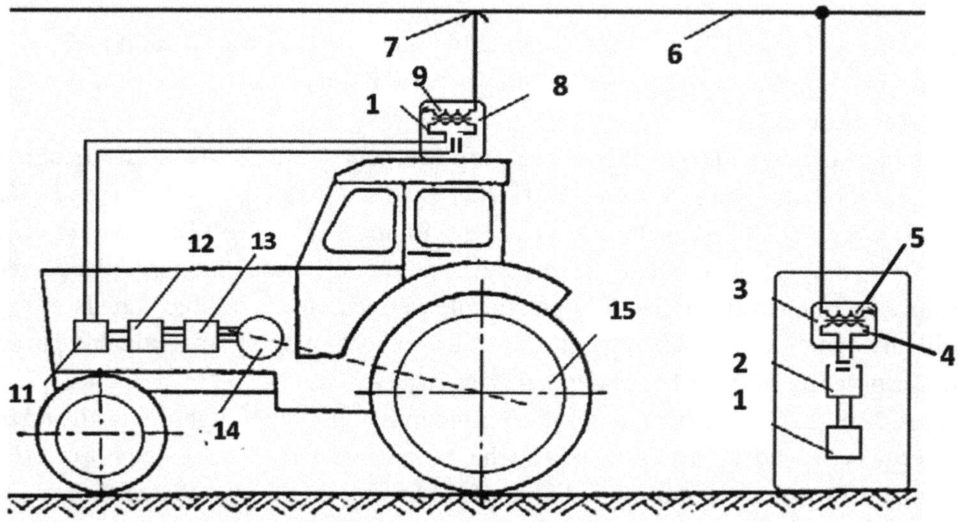

Let us consider resonant electric systems used to supply power to vehicles with insulated wheels. The proposed equipment is characterized by high reliability and low loss; it ensures high maneuverability of vehicles in case of multilane traffic (Strebkov, D.S., Avramenko, S.V., & Nekrasov, A.I. 2000).

In order to supply power to a single-conductor overhead contact system, located above each traffic lane, it is connected, in regular distance intervals, to high-frequency high-voltage generators. The electric energy is delivered in the resonant regime through the contact system and trolleys to vehicles, in order to power the electric drive equipment.

The considered resonant power supply system is illustrated in Figure 2. Each power substation contains a power source *1*, a high-frequency converter *2* and a high-voltage step-up transformer *3*, connected to the contact system *4*. The contact system includes two sets *6* and *7* of single-conductor lines *5* (for both driving directions), located above each highway lane and connected together by means of electric jumpers *8*, *9*.

The block diagram of the vehicle equipment is shown in Figure 3. The apparatus contains the trolley *10* with the telescopic drive *11*, connecting the receiver *12* to the contact system *4*. The receiver *12* is connected to the diode unit *13*, whose output is coupled to the dc-to-ac converter *14*. The converter ensures the required frequency and voltage values and supplies power to the brushless electric motor *16* via the control unit *15*.

A block diagram of a power substation delivering electric energy to the contact system is shown in Figure 4. The high-voltage generator *3* includes a resonant Tesla transformer (whose resonant frequency is within the range from 0.5 to 50 kHz). It consists of the high-voltage winding *17* and the low-voltage winding *18*, which are winded on the common core *19*. The low-voltage winding *18* is connected to the power source *1* through the high-frequency converter *2*. One terminal of the high-voltage winding *17* is insulated, the other terminal is connected to the contact system *4*.

The converted energy of the power source is supplied through the contact system to mobile vehicles. The single-trolley receiver of the vehicle converts the received energy and supplies it to the electric motor. The use of the resonant method for supplying power to vehicles through single-conductor lines

Figure 2. Power supply system for single-trolley electric vehicles

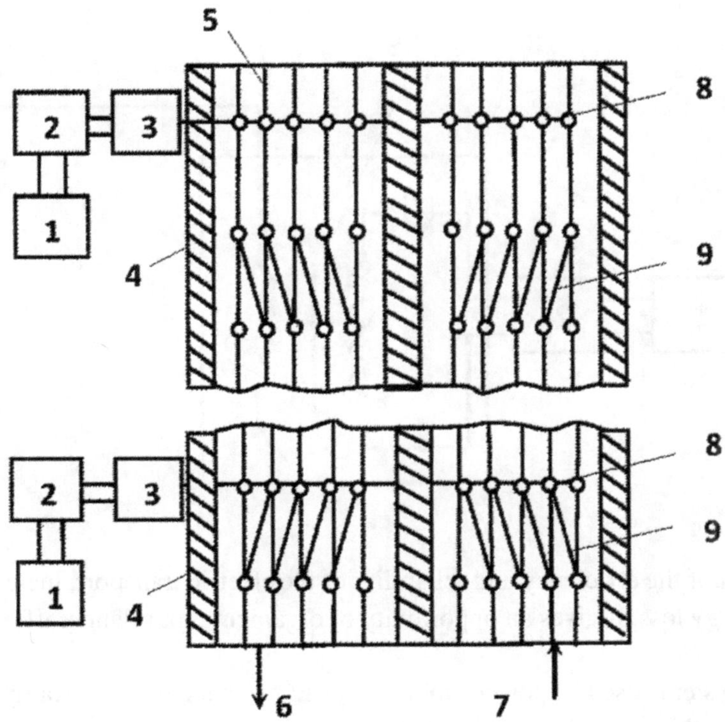

Figure 3. Block diagram of vehicle equipment powered from a single-conductor contact system

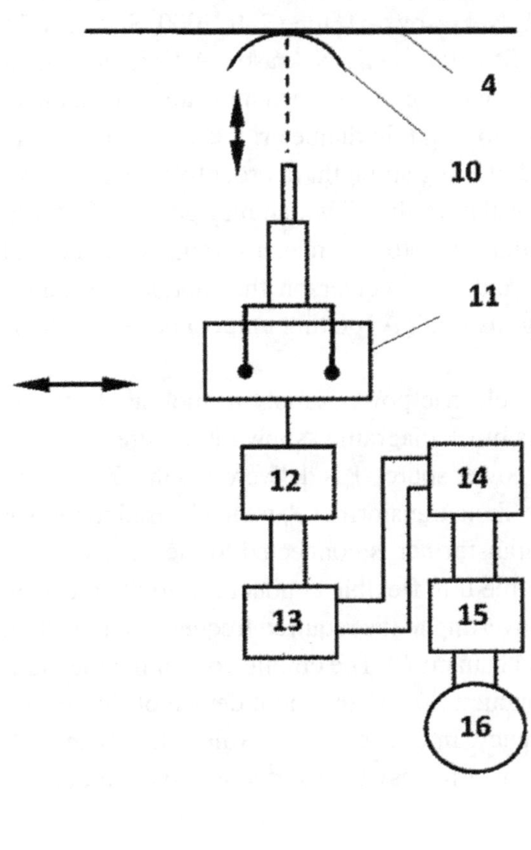

Figure 4. Block diagram of a power substation delivering electric energy to the contact system

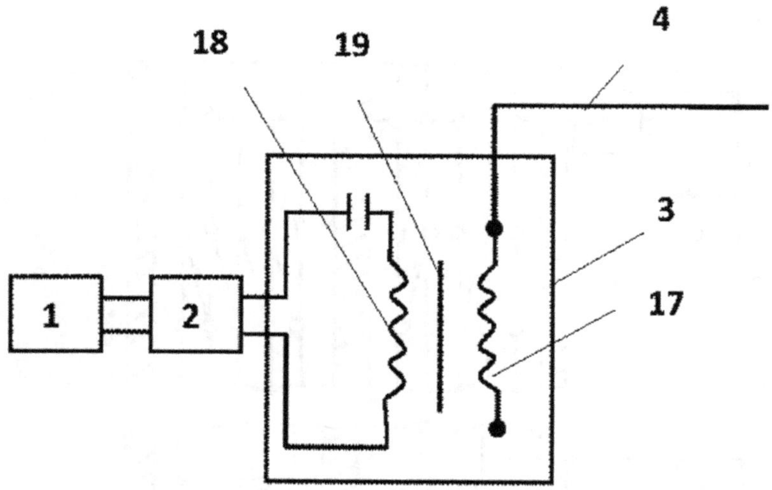

ensures enhancement of the efficiency and reliability of the electric transport, increase of its service life and reduction of energy loss. It gives an opportunity of organizing multilane traffic for electric vehicles along highways.

Let us consider systems used to power mobile electric mechanisms (for example, tractors) through thin flexible wires. In this case, the energy of the power source is supplied through a high-frequency converter is into a flexible conductor. Its end terminal is connected to a receiver located on a mobile electric mechanism. The received energy is delivered through a coupling unit to the electric motor (Strebkov, D.S., Avramenko, S.V., & Nekrasov, A.I. June 7–9, 2000; Strebkov, D.S., Korolev, V.A., Nekrasov, A.I., & Toporkov, V.N. 2006; Strebkov, D.S., Nekrasov, A.I., Juferev, L.Ju, Roshin, O.A. et al, 2007).

In order to increase the service life of the power supply apparatus, the flexible conductor is manufactured of a conductive wire, 0.5 to 5 mm in diameter; the wire material is characterized by high tensile strength, bending strength and twisting strength. In order to reduce the corona loss and ensure electrical safety, the conductor has an insulating sheath which may be manufactured of extra-strong plastic.

In order to enhance maximum stroke and maneuverability of the mobile electric mechanism, the power substation transmitter and the receiver on the mobile mechanism are equipped with towers manufactured of an insulating material. A rotating unit connected to the flexible conductor is installed on the top of each tower.

Versions of the system for electric power supply to mobile electric mechanisms are illustrated in Figures 5 to 11. The apparatus block diagram is shown in Figure 5.

The electric energy of the power source 1 is delivered to the frequency converter *2* (in order to raise the frequency) and then to the Tesla transformer *3*, which is used to increase the operating voltage. The high-voltage winding of the transformer is connected to the transmitter *5* via the cable *4*. The energy is further transmitted through the thin flexible conductor *6* to the receiver *7*, and via the cable *8* to the load-coupling unit *9*. After converting to the required frequency and voltage, the received energy is used to drive the mobile electric mechanism *10*. The circuit, containing the high-voltage winding of the Tesla transformer *3*, the flexible conductor *6* and the input device of the receiver *7*, is tuned to the resonant regime at the operating frequency. In case of its low value (less than 30 kHz), additional tuning high-voltage capacitors and induction coils may be inserted into the circuit.

The structure of the proposed apparatus is shown in Figure 3.6. The power substation contains the energy source *1*, the frequency converter *2* and the step-up high-frequency Tesla transformer *3*. The terminal *11* of the transformer secondary winding is connected to the flexible conductor *6* through the cable *4* and the transmitter *5*. The transmitter rotating unit *14* is mounted on the insulators *13*, located on the top of the tower *12*. It consists of the lower fixed ring *15* and the upper mobile ring *16*, rotating round its vertical axis on the bearing. The rings have good electric contact to each other. The receiver *7*, located on the mobile electric mechanism *10*, has a similar structure with the fixed ring *21*, the mobile ring *17* and the insulator *20*. If the towers *12* and *19* are manufactured of an insulating material, the insulators *13* and *20* may be removed. The thin flexible conductor *6* is connected between the mobile rings *16* and *17*. The receiver fixed ring *21* is connected to the load-coupling unit through the cable *8*. This unit contains two opposing diodes *23* and *24*, whose common terminal is connected to the cable *8*. Other diode terminals are coupled to the capacitor *25* and to the control unit *27* of the mechanism drive *28* (through the diode *26*).

An alternative structure of the rotating unit is shown in Figure 7. The unit *7* contains the reel *29* installed on the frame *30*, joined to the electric motor *31*, which drives the reel *29* via the insulated muff *32*. The flexible conductor *6* is fastened to the reel *29*. The motor *31* is insulated from the reel frame by the insulators *33*. The frame *30* is installed on the insulators *34*, fastened to the mobile platform *35*. The platform *35* can rotate round the vertical axis on the bearing *36*. It is driven by the motor *37* with the reduction gear *38*; the motor is fastened to the tower *19*.

The cable *8* ends with a sliding contact *39* to the frame *30*. The frame *30* and the reel *29* are manufactured of a conductive material. The motor *31* is powered via the cable *40*. The motor *37* is power through the cable *41*. The cables *8*, *40* and *41* pass inside the tower *19* to the mechanism control system. In an alternative structure version, the motor *37* and the reduction gear *38* may be eliminated.

Figure 5. Block diagram of the apparatus for electric power supply to a mobile electric mechanism

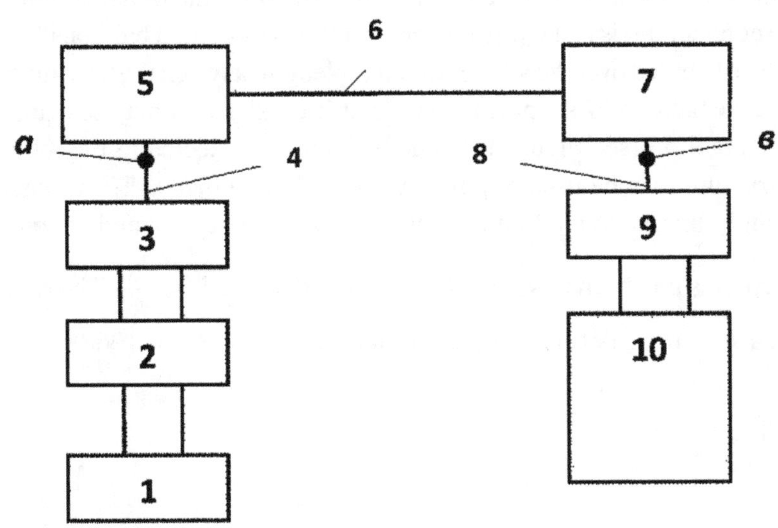

Figure 6. Structure of a transmitter and a receiver with rotating units

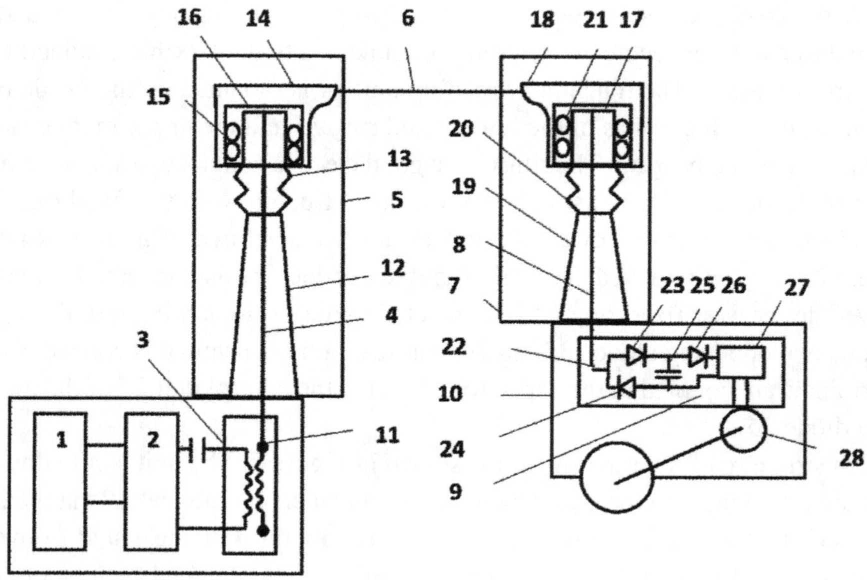

The thin flexible conductor used to transmit the power to the receiver, may be suspended above the ground in several ways. One of them is illustrated in Figure 3.8. The conductor *6*, connecting the transmitter *5* to the mobile mechanism *10* through the rotating units *14* and *18*, is fastened in the middle point *42* to a light balloon *43*. The balloon is filled with a gas whose density is less than the air density. In order to decrease the disturbance caused by wind, the balloon *43* has a shape of a disk.

In Figure 9, the conductor 6, connecting the transmitter *5* to the mechanism *10*, is suspended in three points *44*, *45*, *46* to three gas-filled light balloons *47*, *48* and *49*. The point *45* is in the middle of the conductor length, while the points *44* and *46* are located above the transmitter and receiver towers.

In Figure 10, the conductor *6* is fastened to several extended gas-filled balloons *50*, small in diameter. The fixing points are distributed over the whole conductor length between the transmitter and the receiver.

As an alternative, the balloon *51* supporting the flexible conductor *6* may be equipped with a propeller *52* driven by the motor *53* (see Figure 11). The balloon is connected to the rotating unit *14* with a cable *54*. Similar structure may be used to power various flying objects. To reduce radiation losses, a single-conductor supply line is divided into small shielded sections isolated from other sections. The length l_s of each section is much smaller than the wavelength of λ_0, $l_s = \frac{\lambda_0}{K}$. The radiation loss in a line of length l without a screen are defined by the formula (Semenov, A. A. 1968):

$$P_{irr} = I_{eff} 80\pi^2 \left(\frac{l}{\lambda_0}\right)^2 \quad (1)$$

where:

I_{eff} : The effective current density in the line
l : The length of the line
λ_0 : Wavelength

At the length of the screened section $l_s = \frac{\lambda_0}{K}$ - radiation losses are:

$$P^1_{irr} = \frac{I^2_{eff} 80\pi^2}{K^2} \tag{2}$$

with the length of the unshielded line $l = \lambda_0$:

$$P^2_{irr} = I^2_{eff} 80\pi^2 \tag{3}$$

Ratio $\frac{P^2_{irr}}{P^1_{irr}} = K^2$ i.e. the radiation losses decrease in K^2 time.

Examples for Systems Supplying Power to Mobile Electric Mechanisms

Example 1: The mobile mechanism 10 is designed on the basis of a hybrid diesel-electric tractor. Both the diesel engine power and the electric motor power are equal to 55 kW; any of them may be used to drive the tractor.

Figure 7. Rotating unit with two motors

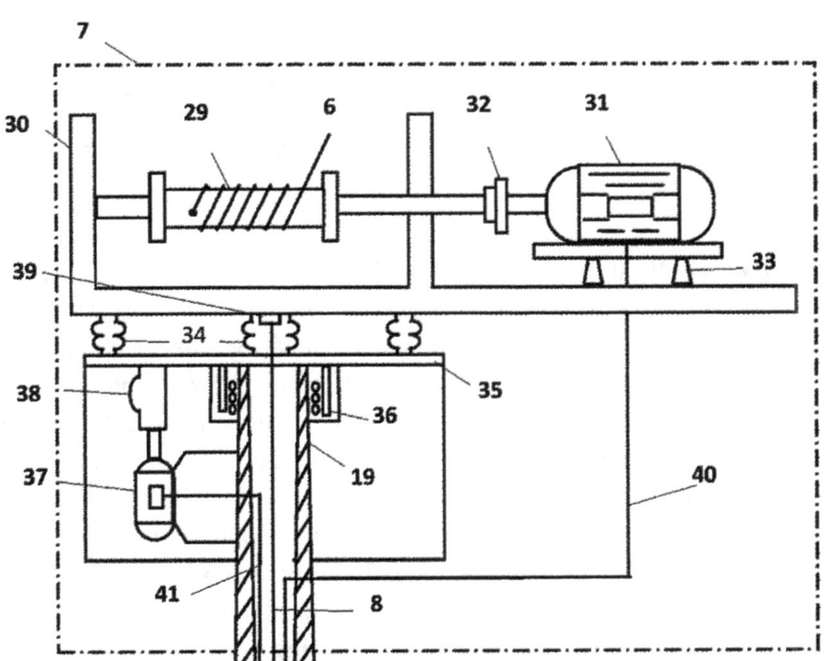

Figure 8. Block diagram of a power supply system with a flexible conductor suspended on a single gas-filled balloon

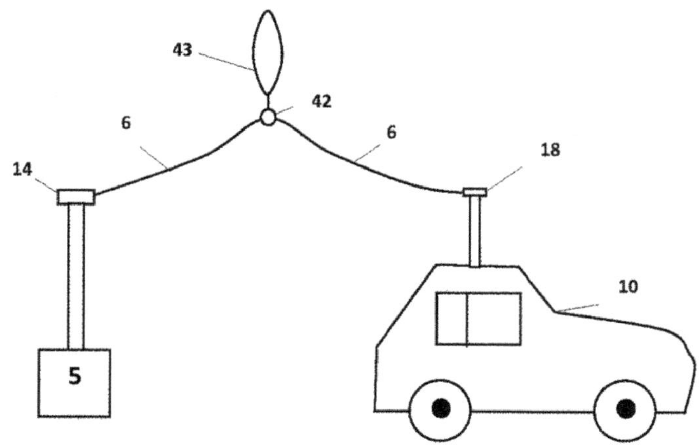

Figure 9. Apparatus for electric power supply with a flexible conductor suspended on three gas-filled balloons

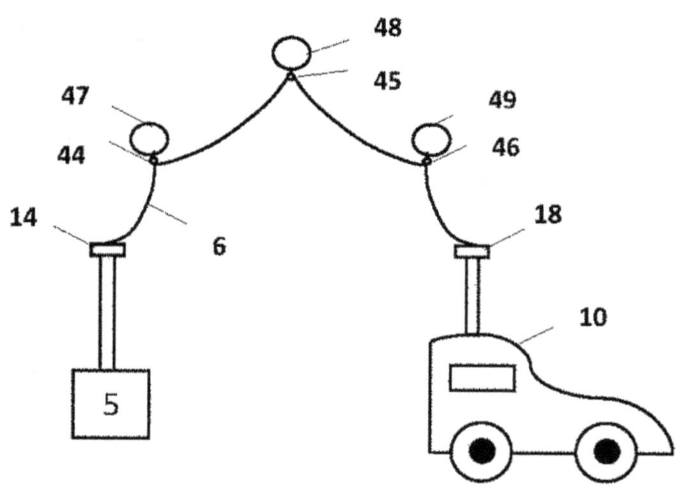

The structure of the power substation corresponds to Figure 3.6. The energy of the three-phase mains 380/220 V is delivered to the frequency converter *2* (whose output frequency is 30 kHz) and further to the Tesla transformer *3* with the voltage transformation ratio 220/400 V. The terminal 11 of the transformer secondary winding is connected through the cable 4 to the transmitter 5. The transmitter is located on the tower *12* (0.15 m in diameter and 10 m in height), which can be folded. The conductor 6 is a wire (0.5 mm in diameter and 1000 m in length), manufactured of alloyed steel.

The structure of the transmitter rotating unit corresponds to Figure 3.7. The reel *29* is 0.2 m in diameter without the conductor *6*, and 0.4 m, when the conductor is winded up; the reel length is equal to 0.2 m. The motor *31*, used to drive the reel *29*, is installed on the insulators *34*; it is connected to the reel shaft through the insulated muff *32*. The motor *31* is powered from the source *1* via the cable *40*. The reel *29* is installed on the frame *30*, which is fastened to the rotating platform *35* of the receiver *7*. The reel

Figure 10. Apparatus for electric power supply with a conductor suspended along its whole length using several gas-filled balloons

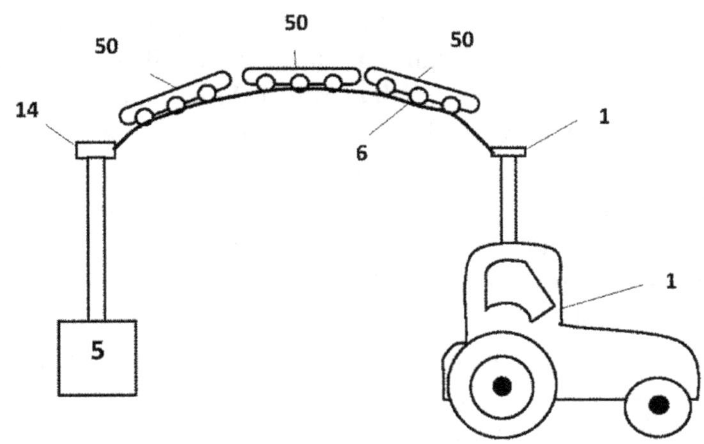

Figure 11. Apparatus for electric power supply using a gas-filled balloon with a propeller driven by a motor

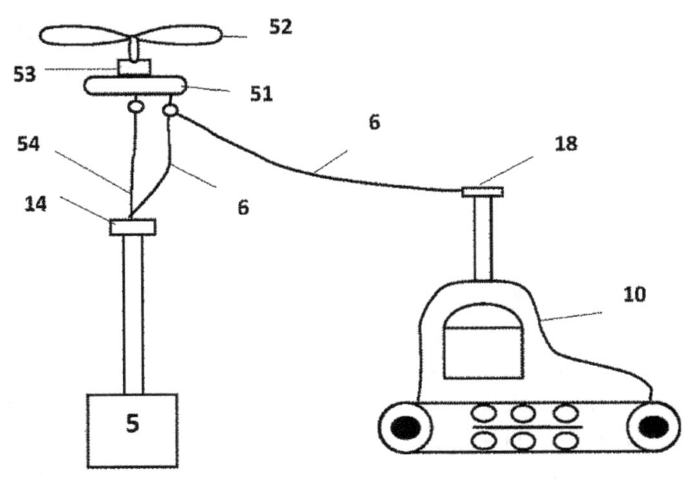

29 is supplied with a sensor of the conductor tension. It is used to control automatically the conductor tension by means of the motor *31*.

The structure of the receiver installed on the tractor *10* corresponds to Figure 3.6. Its tower 19, which is 6 m in height and 1 m in diameter, can be folded, when the tractor is driven by the diesel engine. The brushless motor *28* has a mass of 60 kg. It is driven through the control system *27* supplying the nominal voltage 220 V.

Example 2: The mobile mechanism 10 is connected to the transmitter 5 through the conductor 6 (1 mm in diameter and 500 m in length), which is manufactured of zinc-coated alloyed steel. The conductor mass is 3.1 kg, its tensile strength is equal to 200 kg. A disk-shaped balloon filled with hydrogen is fastened to the conductor middle point (see Figure 3.8). The balloon volume is equal to 5 m^3. In an alternative system version (see Figure 3.9), the conductor 6 is fastened in three points 44, 45 and 46 to the balloons 47, 48 and 49.

Example 3: The mobile mechanism 10 is connected to the rotating unit 14 by means of a conductor 6 (0.5 mm in diameter and 1000 m in length), manufactured of alloyed steel (see Figure 3.10). The conductor 6 is fastened over its whole length to extended gas-filled balloons 50. Each balloon, 10 m in length and 0.25 m in diameter, is filled with hydrogen. The mass of the zinc-coated conductor 6 (0.5 mm in diameter and 1000 m in length) is 166 kg, its tensile strength is equal to 60 kg.

Example 4: The mobile mechanism 10 is connected to the transmitter 5 by means of a zinc-coated conductor 6 (2 mm in diameter and 1000 m in length), manufactured of alloyed steel (see Figure 11). The conductor tensile strength is equal to 750 kg. The conductor is fastened to a kite balloon 51 (50 m^3 in volume) located above the transmitter at a height of 100 m. The balloon may be equipped with a propeller 52 driven by the motor 53, 1 kW in power. The balloon is connected to the tower with an individual cable. The receiver of the mobile mechanism is equipped with a reel (similar to Figure 7), used to wind up the conductor 6.

The considered examples use the resonant method for supplying electric power to a mobile mechanism. The energy is transmitted along the conductor *6*, which represents a guiding system for the electromagnetic waves. At a frequency of 10 kHz, the wavelength is equal to 30 km, that considerably exceeds (by factor of 20 to 30) the length of the conductor 6. At a frequency from 1 to 10 kHz, the radiation loss is low. The loss in the line resistance is also low, because the conductance current in the line is much lower than the current circulating in a conventional double-wire or three-wire line. This allows to reduce the conductor diameter by factor of 10 or 20 and to increase the conductor length up to 1 or 2 km.

Another essential advantage of the proposed method is that the conductor *6* is suspended in the air (instead of being located on the ground). The rotating units and the reel are manufactured of a conductive material and have the same potential, as the conductor. As a result, the mass of the conductor and the reel may be reduced by factor of a hundred or more, compared to a conventional cable used to power mobile mechanisms.

A working model of a mobile electric mechanism, 10 W in power, is shown in Figure 12. It is used to demonstrate the advantages of the single-conductor power supply system which needs no metal-consuming cable reels or accumulator batteries. The experimental sample consists of a power supply unit (including a frequency converter and a step-up high-frequency transformer) and a model of a mobile mechanism powered through a thin insulated single-wire conductor. It contains a step-down transformer, a rectifier and a matching device, used to power a dc motor, driving the vehicle wheels through a reducer. The rectified voltage is also applied to light-emitting diodes used as headlights.

A working model of a mowing machine with similar operating principle is shown in Figure 13; the mechanism is powered via an insulated conductor, fastened on a high support.

Resonant Contactless Method for Powering Electric Vehicles

A contactless method for powering electric vehicles is proposed, which can be used for trolleybuses, electromobiles, electric loaders, trams, electric tractors, electric locomotives and other vehicles.

The known contactless methods for powering electric vehicles by means of an high-frequency electromagnetic induction are accompanied by high loss in the line because of high capacitive and inductive impedance of the wires Therefore, at a frequency of 50 Hz, the maximum length of cable lines does not

Figure 12. Model of a mobile mechanism with a single-conductor power supply system

Figure 13. Model of an electric mowing machine powered through a flexible insulated conductor

exceed 80 km. At high frequencies 25-100 kHz, the self-induction emf arises in the receiver winding and in the cable line. Its polarity is opposite to the supplied substation voltage, so that the transmission efficiency is considerably reduced and power transmission distance is limited by several hundreds meters. The measures used to suppress this phenomenon result in raising the price of the power transmission to an inadmissible level (Rozenfeld, V.E., Staroselskii, N.A. 1975; Strebkov, D.S., Nekrasov, A.I., Trubnikov, V.Z., & Nekrasov, A.A. 2018). The proposed method for supplying power to vehicles is characterized by low loss and ensures high maneuverability of the vehicle. In addition, it allows to power vehicles whose wheels are manufactured of rubber and other insulating or conductive materials in case of multilane driving along a highway. In this case, the electric energy of the source is delivered through a high-voltage high-frequency converter and a single-conductor line to a vehicle via an individual capacitive coupler which is represented by an air-gap capacitor. The coupler capacitance is formed by the insulated line conductor and a thin insulated plate manufactured of a conductive material, located under the bottom of the vehicle. The line conductor is laid inside the roadway covering or in the ground near the surface. The power is supplied using a resonant single-conductor transmission system at a frequency of 0.1 to 100 kHz with the line voltage from 0.5 to 500 kV.

The proposed apparatus for supplying power to electric vehicles is illustrated in Figure 14. The apparatus contains the source of high-frequency electric energy *1*, the transmitter resonant circuit *2*, the step-up transformer *4*, the receiver resonant circuit *3* and the step-down transformer *5*; one of the terminals of the transformer high-voltage winding *8* is connected to the natural capacitor *9*. The transmitter is installed at the substation, the step-up transformer output terminal is connected to the single-conductor line *6*, laid inside the roadway covering *12*. The receiver resonant circuit, the step-down transformer and the natural capacitor are located on the mobile vehicle *10*. The stationary and movable parts of the apparatus are connected to each other through the air capacitor *11*. The conductor *6* represents one of the capacitor electrodes. The second electrode is a capacitive coupler, represented by the plate *13* fastened under the bottom *14* of the vehicle *10*. The air gap *a* between the capacitor electrodes ranges from 0.1 to 1 m. The line conductor *6*, covered with the insulation layer *15*, is located under each driving lane. The capacitive coupler *13* fastened on the insulators *16* is connected to the receiver resonant circuit *3*. Its output *17* is coupled to the vehicle motor *21* through the rectifier *18*, the battery *19* and the control unit *20*. The natural capacitor *9* is represented by a conductive body or by the vehicle casing.

Figure 14. Apparatus for contactless power supply to electric vehicles

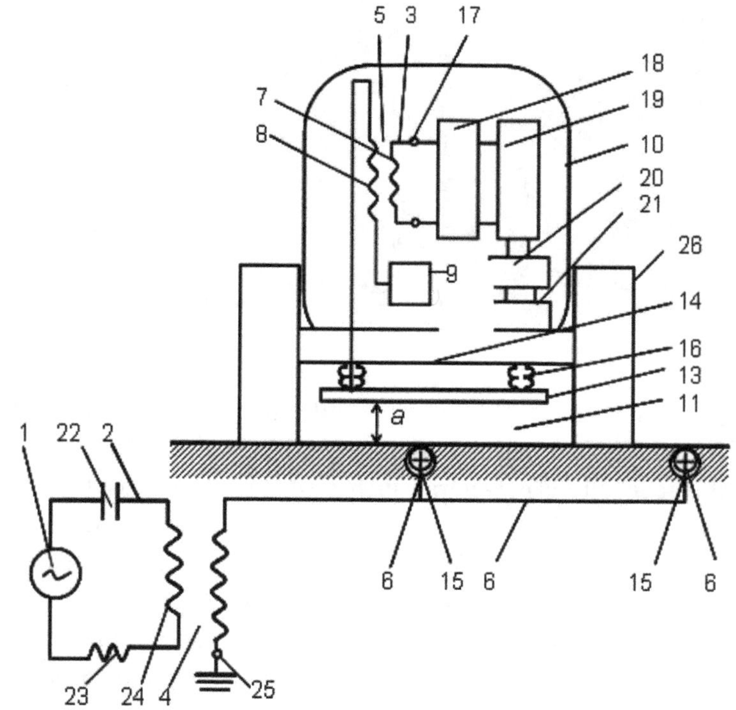

An apparatus version shown in Figure 15 contains a 50-Hz three-phase power source, a three-phase frequency converter and a resonant electric system transmitting power to three vehicles. The power source *1* and the frequency converter *2*, connected to three transmitter resonant circuits *3* and three step-up transformers *5,* are installed at the substation. The transformer output terminals are connected to three single-conductor cable lines *7*, laid inside the roadway covering *8* under each of three traffic lanes. The cables are supplied with additional electric insulation *15*, comprising a layer of quartz sand between the line and the road surface. The sand is fastened with a water-proof hydrophobic impregnant producing monolithic medium, which is intended to reduce the line capacitance and the leakage current to the ground (the current leakage may be particularly high in case of wet ground).

The vehicle equipment includes the capacitive coupler *14*, fastened on the insulators *16* under the vehicle bottom. The air gap between the plate *14* and the line conductor ranges from 0.02 to 1 m. The plate *14* is connected to the step-down high-frequency transformer *6*. One terminal of the transformer high-voltage winding is connected to the natural capacitor *11*, represented by a conductive body or the ground. The transformer low-voltage winding forms the receiver resonant circuit *4*. Its output terminal *17* is coupled through the rectifier *18*, the battery *19* and the control unit *20* to the vehicle motor *21*.

In the example of Figure 15, three traffic lanes are shown, powered from the three-phase power source. The single-conductor lines use the operating voltage of 0.4 to 100 kV at a frequency from 0.1 to 400 kHz. If the length of each line corresponds to the half-wave length, the transmission regime is near to the stationary-wave condition. The voltage loop is located approximately in the middle of the line *7*, the current loops are located at the line ends. The loss in the line conductor is comparatively low. The transmitted energy delivered via the contactless capacitive coupler *14* is efficiently used to power the vehicle motor.

An alternative contactless power supply system is shown in Fig.16. In this case, the single-conductor lines *7* are located above the road covering along each driving lane at a height exceeding the height of the vehicle *12*. The line conductors are fastened to supports *25*, the conductors are insulated from each other by means of insulators *26*. The capacitive couplers *14* are located at a distance *a* from the lines *7*

Figure 15. Apparatus for contactless power supply to vehicles, using a three-phase power source

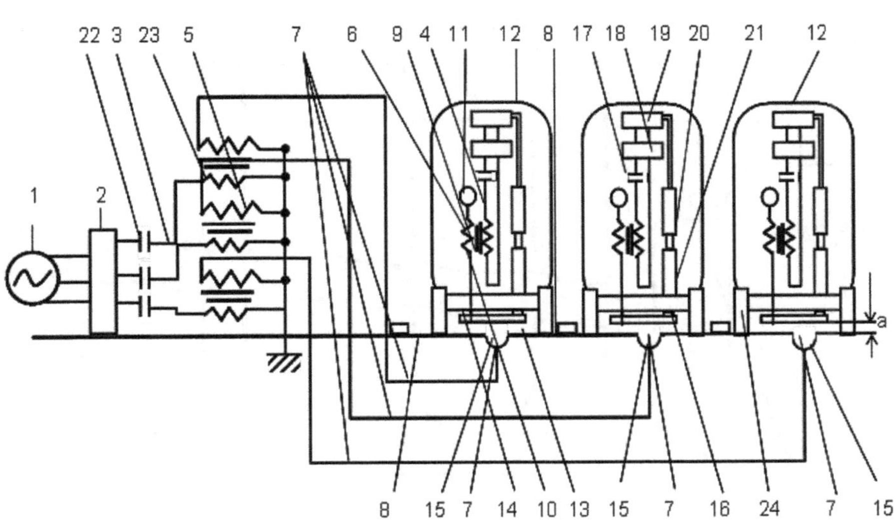

(its value has been specified in the previous examples). The insulated metal plates *11* located under the bottom of the vehicles *12* on the insulators *27* are used as natural capacitors.

In the system shown in Figure 17-18, two conductor segments of the same single-conductor line 30 are located in the road covering 8 under each traffic lane. The electric energy of the high-frequency source *28* is delivered through the resonant circuit *3* (including the capacitor *22* and the low-voltage winding *23* of the transformer *29*) to the zigzag-type line *30* (see Fig.17). The distance *b* between adjacent line segments: *32* and *33*, *34* and *35*, located under each traffic lane, ranges from 0.5 to 5 m. The length of each segment (for example, *32* and *33*) *l* /2 = (0.1 to 100) km, depending on the operating frequency. Additional electric insulation 15 is described on page 57.

Figure 16. Block diagram of a contactless power supply system using overhead lines

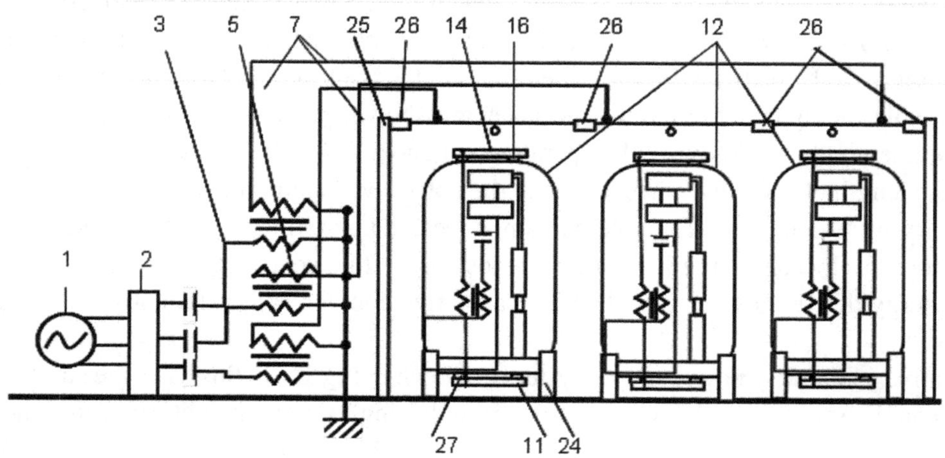

Figure 17. Contactless power supply system with two conductor segments of the same single-conductor line under each traffic row

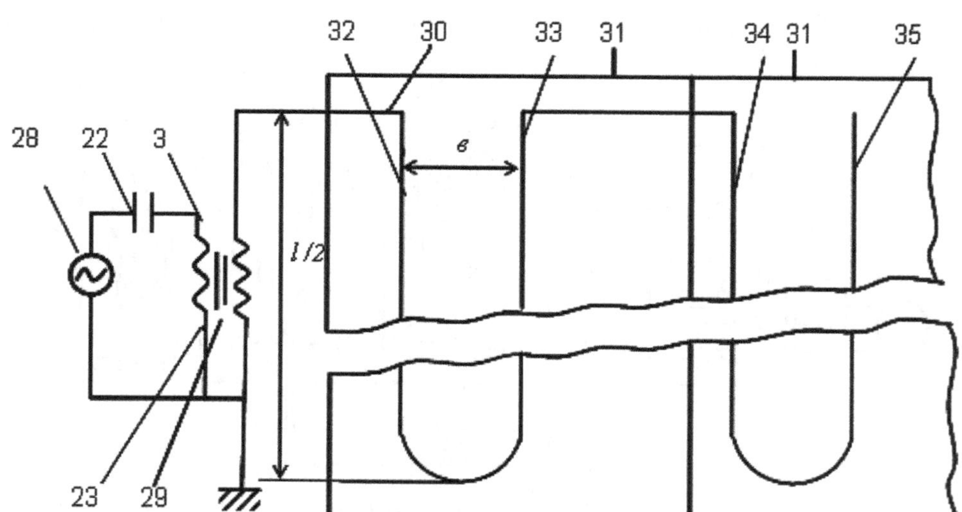

Figure 18. The vehicle with capacitive couplers under the bottom and two line segments

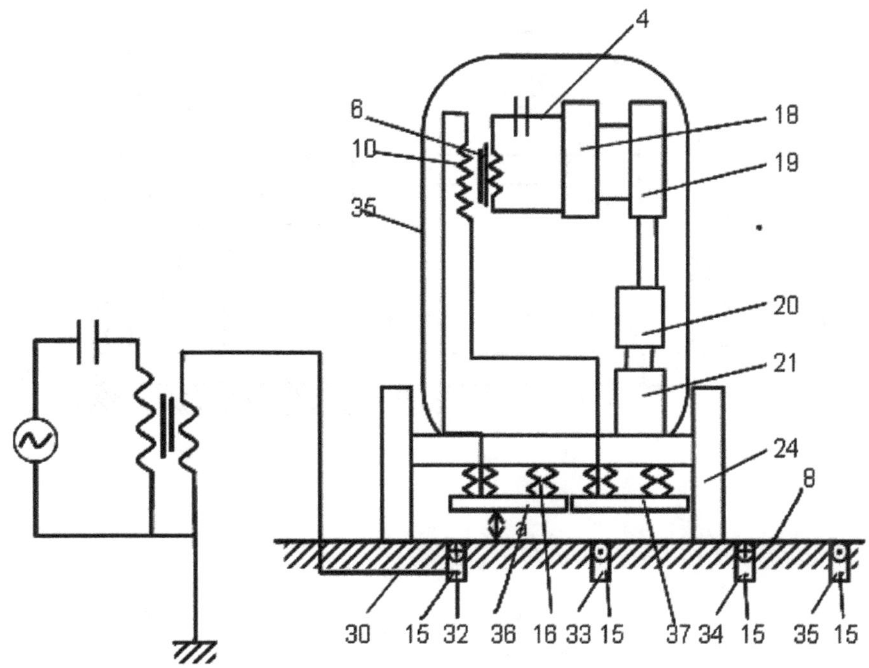

The vehicle *12* has two capacitive couplers *36* and *37*, fastened on the insulators *16* under the bottom; their distance from the road surface *8* is equal to *a* (see Figure 18). The couplers move along the line segments 32 and 33; they are connected to the high-voltage winding 10 of the resonant step-down transformer *6*. The currents in the adjacent segments of the single-conductor line *30* flow in opposite directions, which is shown with the signs + 32 and 33 and ʘ. The potential difference between these segments depends on the distance along the line 30, the operating frequency and the line voltage. This potential difference is applied to the resonant circuit 4 (through the transformer *6*). The electric energy generated in the resonant circuit *4* is delivered through the rectifier *18*, the battery *19* and the control unit *20* to the motor *21*, which drives the vehicle wheels *24*.

A contactless single-conductor power supply system for rail vehicles is shown in Figure 19-20. The single-conductor line 38 is installed above a multiple-carriage rail vehicle (train) 39, having a high length l. The extreme carriages 40 and 41 are equipped with capacitive couplers 42 and 43 (see Figure 19). The air gap between the couplers and the line conductor is a. The couplers 42 and 43 are connected to the high-voltage winding 10 of the resonant step-down transformer 6. The method for supplying power to the vehicle 39 is similar to that illustrated in Figure 19. If the potential difference at the length l is insufficient, both capacitive couplers 42 and 43 are connected to two additional resonant step-down transformers 44 (see Figure 20). In this case, the electric energy is obtained using the full line potential at the place where the vehicle 39 is located. The single-conductor line 38 is fastened to the supports 46 and 47 via the insulators 26. The terminal 9 of the high-voltage transformer winding is connected to the ground through the rails 48.

Figure 19. Contactless single-conductor power supply system for rail vehicles

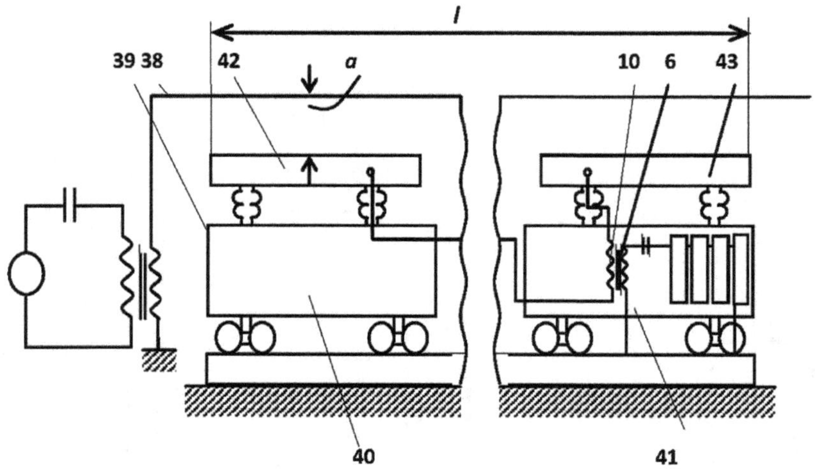

Figure 20. Power supply for rail way vehicle

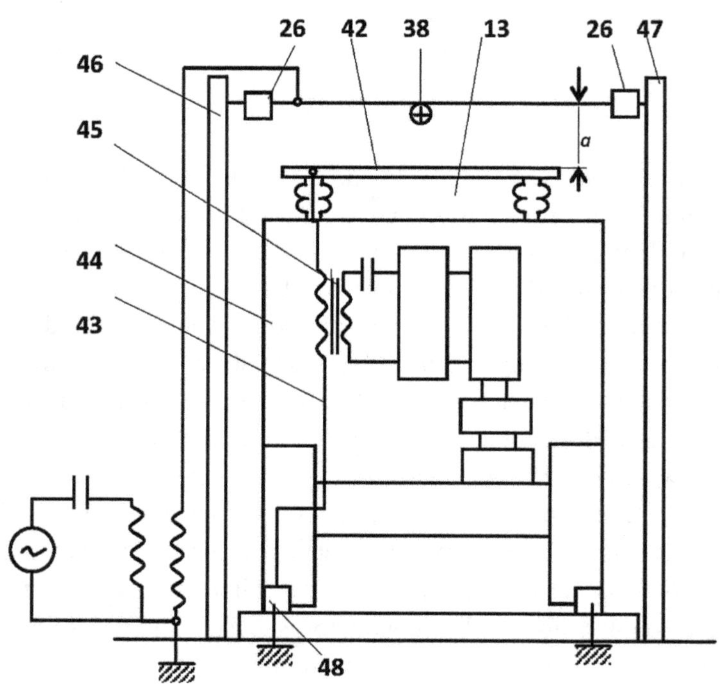

An apparatus for power supply to a rail vehicle using a three-phase resonant system is shown in Figure 21. The power is transmitted through three single-conductor lines *7*, located under each railroad track. The conductors are fastened on the crossties *49* between the rails *48* inside the insulating housing *50*. Three flat capacitive couplers *27* are installed under the bottom of the vehicle *46* on the insulators *14*. The couplers are connected to high-voltage windings *51* of three resonant step-down transformers *52*. When the power source *1* delivers energy to the single-conductor lines *7*, stationary waves appear in the lines, characterized by a phase shift between the current and voltage waves equal to 90°. The potential difference induced in the capacitive couplers generates oscillations in the resonant circuits of the step-down transformers. The low-voltage transformer windings *53* are connected through the capacitors *54* to the three-phase rectifier *55*. The rectifier output is coupled to the drive motor *21* through the battery *19* and the control unit *20*. The free terminals *56*, *57* of the high-voltage and low-voltage windings *51* are connected to the ground through the rails *48*.

An apparatus for power supply to an air-cushion vehicle (or airfoil boat) is shown in Figure 22; the conductive water surface 49 is used as the single-conductor line. The electric energy is transmitted to the vehicle *50* from the high-frequency power source *28* through the resonant transformer *29*, water medium *49* and the capacitive coupler. The coupler is represented by the capacitor with the air gap *51* between the water surface *49* and the flat electrode *52*, located under the vehicle bottom and insulated from it with the dielectric lining *53*. The electrode *52* is connected to the high-voltage winding *10* of the step-down resonant transformer *6*. The other winding terminal *9* is connected to the natural capacitor *54*, manufactured as a metal cover in the upper part of the vehicle. The cover is insulated from the vehicle casing *55* with the dielectric layer *56*. The electric energy is delivered from the low-voltage transformer winding to the motors *21* and *57* through the rectifier *18*, the battery *19* and the control unit *20*. The motors are used to produce the air cushion and to move the vehicle.

Several versions of a conductive lines under the ground surface used to power electric vehicles are illustrated in Figure 23-26.

In Figure 23, the single-conductor line located on the ground surface (or just under the ground surface) in the form of a spiral-shaped insulated conductor 56. The spiral area corresponds to the area intended for the displacement of the vehicle 57.

Figure 21. Apparatus for non-contact supplying power to a rail vehicle using the three-phase lines

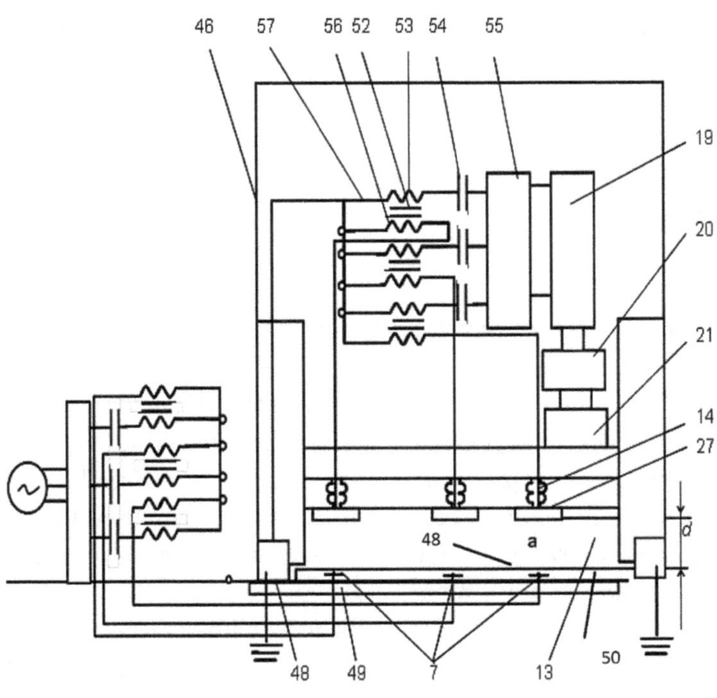

Figure 22. Apparatus for supplying power to an air-cushion vehicle

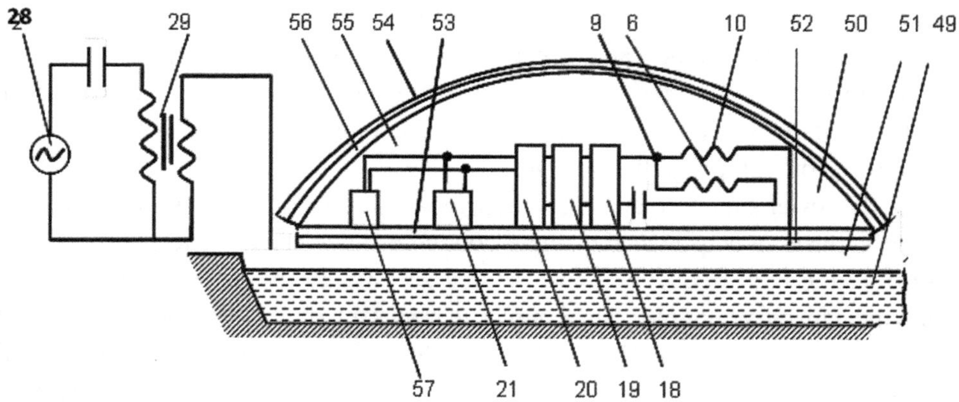

In Figure 24, the conductive surface is manufactured as a grid 58 of a conductive lines located on the ground.

In Figure 25, the conductive surface is represented by a set of single-conductor lines 59, connected in parallel. The line length L corresponds to the length intended for the displacement of the vehicle 57. The distance between the lines is less than the vehicle width b by factor of 3 to 5.

In Figure 26, the conductive surface is manufactured as an insulated metal film 60, located on the ground surface.

In case of using a capacitive coupler, the electric energy, delivered through the air capacitor is proportional to the square of the capacitor voltage. In contrast to double-conductor and three-conductor lines used for power transmission by means of electromagnetic induction, the single-conductor line has low inductive and capacitive impedance components, which allows to transmit electric power using the contactless method with a low loss.

The electric power P_{el}, delivered through the air capacitor C to the vehicle is determined using the following equation:

$$P_{el} = 2\pi f_0 C V^2 \tag{4}$$

where V is the line voltage.

Assuming:

$f_0 = 5$ kHz, $C = 10$ nF and $V = 10$ kV, we obtain $P_{el} = 31.4$ kW

The proposed method for supplying power to vehicles through the underground single-conductor line has considerable advantages. The road is cleared of supports, cables, and wires; it becomes possible to power electric lorries, cars, buses, invalid carriages, as well as store-supply vehicles, etc. The single-conductor line is insulated, therefore, it is more safe than a bare wire used to contact trolleys. Besides that, in case of contactless power supply, the wear and sparking are suppressed, which results in higher service life and reliability of the supply systems.

Models of an electrified railroad and an electric airplane with resonant single-conductor power supply systems are illustrated in Figs. 27. and 28. (Yuferev, L.Yu, Strebkov, D.S., & Roshin, O.A. 2010).

Figure 23. The spiral – shaped insulated conductor connected to Tesla transformer

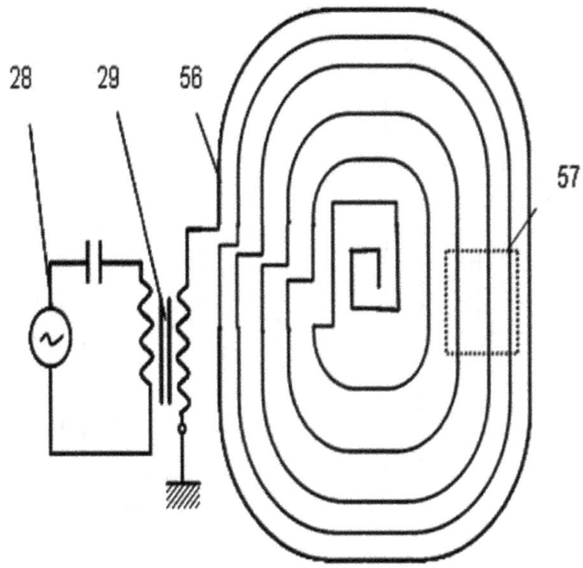

Figure 24. A grid of single-wire conductive lines on the ground surface

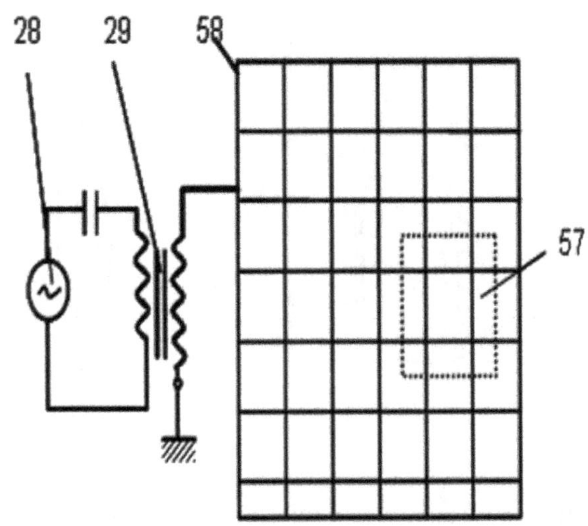

The considered working models demonstrate the efficiency of the high-frequency resonant system in transmitting electric energy from the power source to the consumer through a single conductor (or another conductive medium), using reactive current. They may be used to investigate the new power transmission method at scientific and educational laboratories; they may be also of interest for technicians, developing new electric technologies for various branches of industry.

The VIESH project "Resonant electric power transmission system" has been awarded a Golden Medal of the 5-th Moscow International Workshop of Innovations and Investments (Moscow, All-Russia Exhibition Center, February 15 to 18, 2005).

Figure 25. Parallel connection of single wire conductive fines

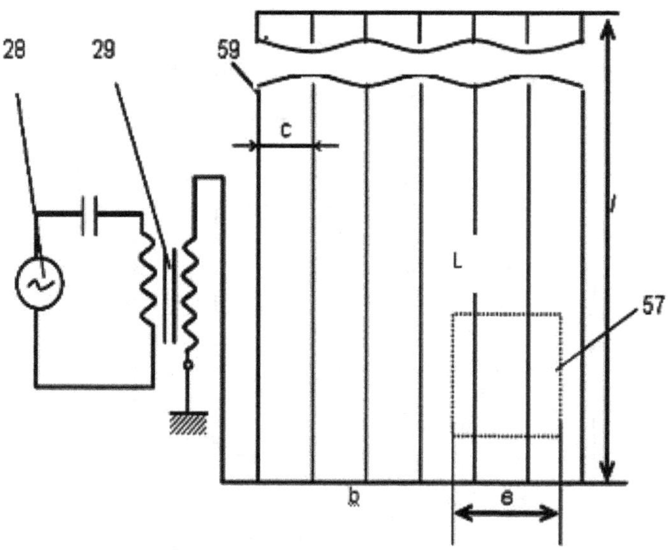

Figure 26. Insulated metal film on the ground surface

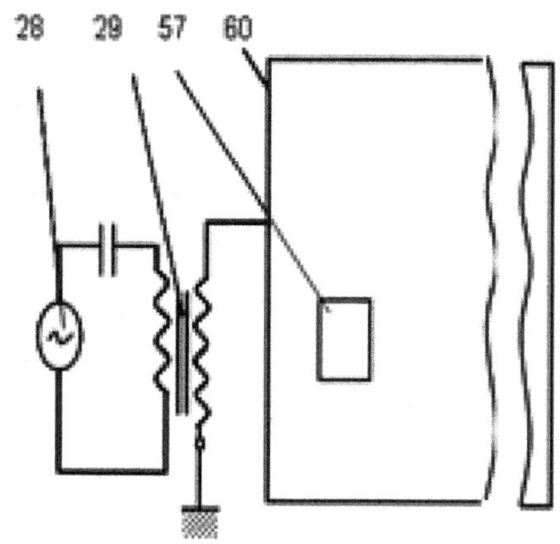

The VIESH project "Mobile electric mechanism with a resonant single-conductor power supply system" has been awarded a Diploma of the 6-th Moscow International Workshop of Innovations and Investments (Moscow, All-Russia Exhibition Center, February 7 to 10, 2006). In modern electric cars, the cost of a lithium-ion battery is almost half the cost of an electric car. Non-contact resonant methods of power supply allow to reduce the battery capacity and weight by 2-3 times by charging the battery during driving and at stops and reduce energy costs per 100 km of mileage and increases the weight of the useful cargo.

The equipment of high-speed tracks of Formula One by contactless power supply systems will allow to organize races of electric vehicles, to reduce injuries and harmful emissions during the competition.

In high-speed rail trains, the main problem limiting the speed of movement of 300 km/h is the sparking and wear contact troll. Non-contact resonant power supply systems for rail trains increase the speed of high-speed trains up to 600 km /h.

Figure 27. Model of an electrified railroad powered by a resonant single-conductor system

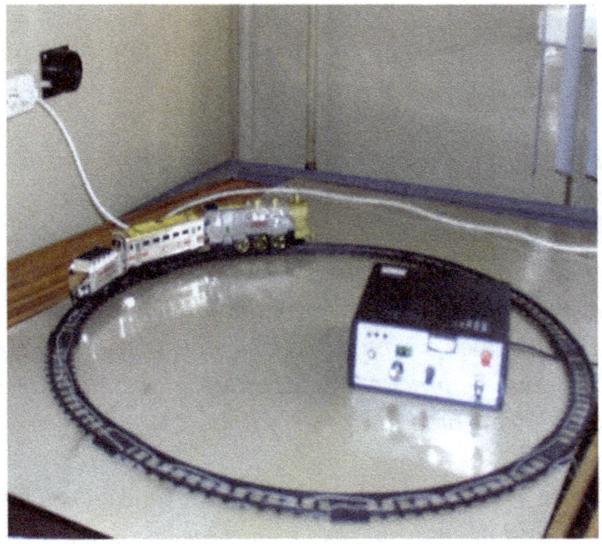

Figure 28. Models of an electric airplane powered by a resonant single-conductor system

Resonant Power Systems for Marine Transport

Sea water is an excellent conductive media for resonant electric power transmission Nikola Tesla wrote (N. Tesla 1914, 1915):

In the near future we shall see a great many new uses of electricity, particularly of vessel at sea. We shall have electrical instruments for preventing collisions, and we shall even be able to disperse fogs by electric force and powerful and penetrating rays. I am hopeful that within the next few years wireless plants will be installed for purpose of illuminating the oceans. The project is perfectly feasible, and if carried out will contribute more than any other provision to the safety of property and human lives at sea. The same plant could also produce stationary electrical waves and enable vessels to get at any time accurate bearings and other valuable practical date without resorting to present means. It could also be used for time signalling and many other purpose of similar nature...

The new principle can also be applied to a submarine, and particularly in connection with control from great elevation, it will afford the most perfect means for coast defence so far devised. But its full possibilities will only be appreciated when the use of certain electrical waves to which the earth is resonantly responsive becomes general. It will then be practicable to despatch a crewless boat or balloon to distances of hundreds of miles, guide it along any chart at wile and release its potential energy at any point desired...

Battleships and submarines will be provided with electric and magnetic feelers so delicate that the approach of anybody under water or in darkness will be detected torpedoes and floating mines are almost in sight which direct themselves automatically and without fail get in fatal contact with the object to be destroyed.

The electromagnetic waves, whose frequency lies in the considered range (0.1 to 100 kHz) are characterized by comparatively low attenuation in the sea-water and in the ground. Therefore, they can be used in a system of underwater or underground transmission of information. The electromagnetic wave attenuation factor for the sea-water is (Nefyodov, V.I. 2002):

$$\alpha (f) = 0.00345\sqrt{f} \text{ (in dB/m)} \quad (5)$$

If the frequency is $f = 100$ Hz, the attenuation at the sea depth of 300 m will is equal to 90 dB.

As shown in our experiments, not only electronic information, but also electric power can be transmitted through the sea-water and through the ground. It means, for example, that a submarine can receive power from water, without raising any antenna device into the air.

A model of an electric launch, powered through the water, has been also developed and tested at the VIESH (see Figure 29-32) (Yuferev, L.Yu, Strebkov, D.S., & Roshin, O.A. 2010). The power supply unit *1* contains a resonant generator with a step-up transformer, whose high-voltage terminal is connected to the transmitter electrode *3* by means of the single-conductor line *2* (see Figure 32). The unit power equals to 100 W, the line voltage is 1 kV. The electrode *3* is located in any point of the insulated water-filled basin *5*. The water medium plays also a part of a single-conductor transmission line. The receiver electrode *6* is connected to a high-voltage terminal of the step-down transformer which forms part of the matching unit *8*, installed on the launch. It is used to drive the screw propeller *10* by means of the dc motor *9*. The matching unit is coupled to a natural capacitor, which may be represented by the mast *12* with a sail, fastened to it. Non-contact power supply of river and sea vessels will reduce emissions from the internal combustion engine and energy costs for the transportation of goods and passengers.

The developed resonant single-conductor system for supplying power to mobile electric mechanisms requires a frequency converter, as well as step-up and step-down high-frequency transformers, which results in additional cost. At the same time, it has several advantages, as compared to conventional power supply methods using three-conductor or four-conductor feeder cables, including high reliability, low energy loss, enhanced mechanism maneuverability, low expenditure of the conductor material and reduced metal consumption.

Figure 29. Diagram of the launch model powered through water medium used as a single-conductor circuit

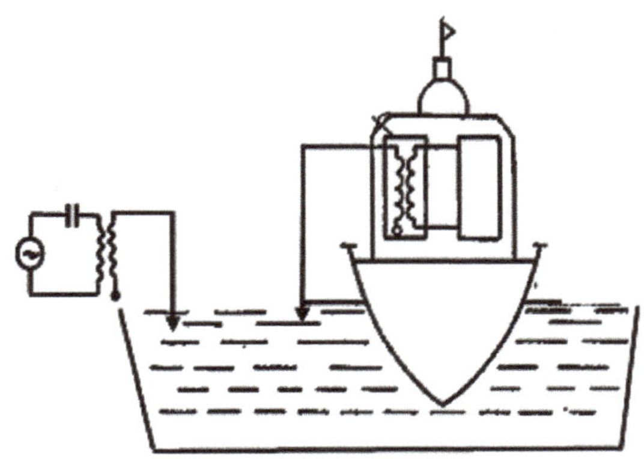

Figure 30. View of the tested model of the boat and the power supply unit

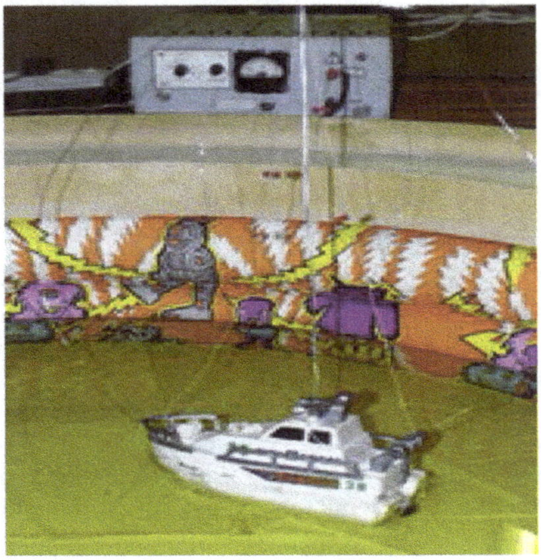

Figure 31. Experimental sample of an electric launch

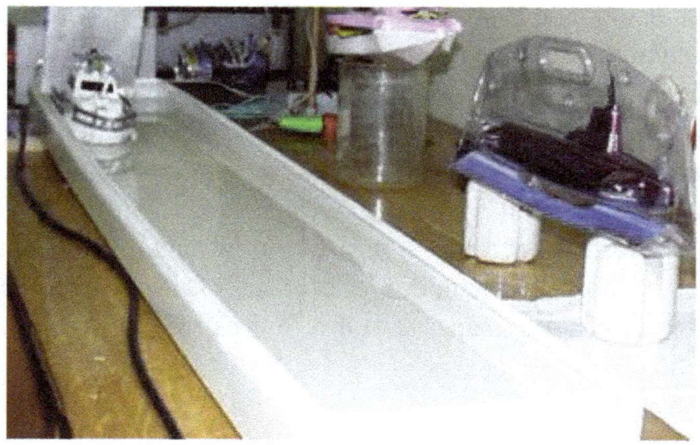

Figure 32. Diagram of the launch power supply circuit, using water medium as a single-wire line
1 – resonant generator; 2, 7 – single –conductor lines; 3 – transmitter electrode; 4 – water; 5 – basin; 6 – receiver electrode; 8 – matching unit; 9 – motor; 10 – propeller; 11 – launch; 12 – natural capacitor.

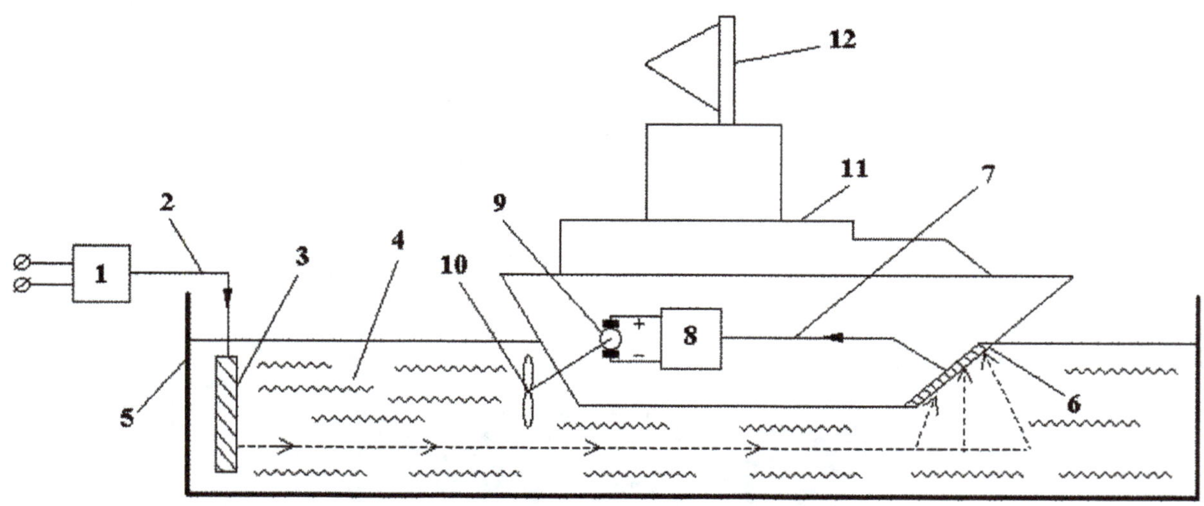

Charge of electric transport is possible all the continuously, which will significantly increase the range of any electric vehicle, both passenger and cargo, and will allow not only to carry out traffic but also to charge them. Planning to 2025 the exponential growth in the number of electric vehicles in the city, the construction of a single-lane and multi-lane electric road becomes an indispensable solution, which can be realized using Tesla technique. In contrast to the schemas and methods practiced in the world by various companies built on overhead lines and electromagnetic induction (Strebkov, D.S., & Nekrasov, A.I. 2018).

The use of high-frequency generators, driven by wind-power or hydropower installations, enables direct power transmission through a single-conductor line without frequency converters. The use of the single-conductor systems based on new low-cost solar plants for supplying power to mobile electric mechanisms will enable considerable reduction in organic fuel consumption.

FUTURE RESEARCH DIRECTIONS AND CONCLUSION

1. Noncontact electric power supply with the use of cable line is an innovative trend in the development of electric transport. Work to develop such systems is carried out in the USA, South Korea, Germany, Israel, the Russian Federation;
2. The VIESH researchers have developed and patented experimental models of a car and a tramway of the future without accumulators, which receive energy from external energy system through air gap from a single wire cable laid under the surface of road pavement. Noncontact systems for electric power supply to transport are developed at VIESH on the basis of resonant methods of electric power supply through a single-wire cable. This is a basic difference of the technology under consideration from similar ones;
3. Electric power from the source is supplied to an electrical receiver through the air gap between the car bottom and road pavement where single-wire line has been laid. When a car is in motion, electric power costs will amount to 1 USD per 100 km of the road. Electric cars price will be reduced twice due to the absence of accumulators and will be less expensive than cars with internal combustion engines. Moreover, ecological problems of large cities and highways will be completely solved. Electric power supply systems without trolleys will increase reliability of tramways and high-speed trains and will allow to use heavy electric trucks on inter-city highways;
4. Our experiments demonstrate that electric energy can be transmitted through the sea water. The working prototype of the motor boat with water drive through the speed reduction gear from the direct current motor has been developed and fabricated. Energy to the motor boat is supplied through the water volume where the boat is, as water plays the role of a single-wire electric power transmission line.

REFERENCES

Nefyodov, V. I. (2002). *Fundamentals of Radio Electronics and Communications*. Moscow: Vysshaya Shkola. (in Russian)

Rozenfeld, V. E., & Staroselskii, N. A. (1975). *High-Frequency Noncontact Electrical Transport*. Moscow: Transport. (in Russian)

Sarbon, S., & Sc, M. (2013). *The unresolved patents of Nikola Tesla. The inventions that Tesla did not protect*. Nikola Tesla Museum.

Semenov, A. A. (1968). *The Theory of Electromagnetic waves*. Moscow State University. (in Russian)

Strebkov, D. S., Avramenko, S. V., & Nekrasov, A. I. (1999). RF Patent 2136515. Method and Apparatus for Supplying Power to Electric Vehicles. *Inventor's Bulletin*, 25. (in Russian)

Strebkov, D. S., Avramenko, S. V., & Nekrasov, A. I. (2000). System for Supplying Electric Power to Mobile Electric Mechanisms [in Russian]. *Drive Techniques*, *4*(26), 32–38.

Strebkov, D. S., Avramenko, S. V., & Nekrasov, A. I. (2000). RF Patent 2158206. Apparatus for Supplying Power to a Mobile Electric Mechanism. *Inventor's Bulletin*, 30. (in Russian)

Strebkov, D. S., Avramenko, S. V., & Nekrasov, A. I. (2000). Single-Trolley System for Supplying Power to Mobile Electric Mechanisms, *Proceedings of the International Scientific-and-Technical Conference "Automation of Agricultural Production"*, 65–66.

Strebkov, D. S., Avramenko, S. V., Nekrasov, A. I., & Roshchin, O. A. (2002). Investigation of Electric Parameters of a Single-Conductor Power System. [in Russian]. *Drive Techniques*, *6*(40), 28–30.

Strebkov, D. S., Korolev, V. A., Nekrasov, A. I., & Toporkov, V. N. (2006). RF Patent 2269854. Cable system for Power Supply of Mobile Agricultural Machine. *Inventor's Bulletin, 1*. (in Russian)

Strebkov, D. S., & Nekrasov, A. I. (2018) Resonant Methods of Generation Transmission and Application of Electrical Energy. Moscow: FSAC VIM. (in Russian)

Strebkov, D. S., Nekrasov, A. I., Avramenko, S. V., & Sakharov, D. N. (2002). Methods for Supplying Electric Power to Mobile Agricultural Vehicles, Electrification of the Agriculture. *Proceedings of the VIESH*, 88, 74–82. (in Russian)

Strebkov, D. S., Nekrasov, A. I., Juferev, L. J., & Roshin, O.A. (2007). RF Patent 2297928. Method and Apparatus for Electric Power Supply of Electric Transport. *Inventor's Bulletin*, 12. (in Russian)

Strebkov, D.S., Nekrasov, A.I., Trubnikov, V.Z., & Nekrasov, A.A. (2018). Single-wire resonant electric power systems for renewable based electric grid. In *Handbook of Research on Renewable Energy and Electric Resources for Sustainable Rural Development.* Academic Press. DOI: . doi:10.4018.978-1-5225-3867-7

Tesla, N. (1894). *Electric Railway System.* US Patent 614942.

Tesla, N. (1914). Science and Discovery are the greater forces which will lead to the consummation of the war. The Sun, Dec. 20. 1914. In *Nicola Tesla Selected works.* Beograd: Academic Press.

Tesla, N. (1915). The wonder world to be created by electricity manufactures record. In Nikola Tesla. Selected works. Beograd: Academic Press.

Tesla, N. (1978). *Colorado Springs Notes 1899 – 1900*. Beograd: Published by Nolit.

Tesla, N. (1999). *The Wonder World to be Created by Electricity. Manufactures. Record, Sept. 9. 1915. In Nicola Tesla. Selected works* (pp. 369, 377, 378). Beograd: Articles.

Tesla, N. (1999). *The Problem of Increasing Human Energy. The Century Illustrated Monthly Magazine. June, 1900. In Nikola Tesla. Selected works* (p. 133). Beograd: Articles.

Tesla, N. (1999). *My Inventions. Electrical Experimenter, February, March, June and October. 1919. In Nikola Tesla. Selected works* (p. 56). Beograd: Articles.

Tesla, N. (2008). *From Colorado to Long island: Research Notes – Colorado Springs 1899-1900 – New York. 1900-1901*. Nikola Tesla Museum.

Tesla, N. L. (1956). *Patents*. Beograd: N. Tesla Museum.

Yuferev, L. Yu., Strebkov, D. S., & Roshin, O. A. (2010). *Experimental Models of Resonant Electric Power Transmission Systems*. Moscow: VIESH. (in Russian)

KEY TERMS AND DEFINITIONS

Electric Transport: A mode of transport using electricity as a source of energy, and a drive motor is used in the drive.

Frequency Converter: A secondary power source that produces an alternating current with a frequency different from the frequency of the source current.

Non-Metallic Electrically Conductive Medium: An electro conductive medium (liquid, water) that does not contain metals.

One Trolley Electric Transport: An electric vehicle that receives food by one trolley.

Power Line: One of the components of the electrical network, a system of power equipment designed to transmit electricity through an electric current.

Resonance Method of Electric Energy Transmission: Transmission of electrical energy through a single-wire line with high-frequency currents in the resonance mode of voltages.

Transformer Tesla: Resonant transformer, producing high voltage high frequency.

Wireless Transmission of Electricity: A method of transmitting electrical energy without the use of conductive elements in the electrical circuit.

Chapter 15
Methods of Reducing the Power Supply Outage Time of Rural Consumers

Alexander Vinogradov
Federal Scientific Agroengineering Centre VIM, Russia

Alina Vinogradova
Federal Scientific Agroengineering Centre VIM, Russia

Alexey Vasiliev
Federal Scientific Agroengineering Centre VIM, Russia

Tatyana Kudinova
Orel State Agrarian University, Russia

Vadim Bolshev
Federal Scientific Agroengineering Centre VIM, Russia

Nikolay Sorokin
Orel State Agrarian University, Russia

Nikolai Hruntovich
Sukhoi State Technical University of Gomel, Belarus

ABSTRACT

The chapter analyzes the power supply restoration time. The analysis shows that this time consists of the time for obtaining information about failure, the time of information recognition about failure, the damage repair time and the time of connection and disconnection harmonization. Each of them, in turn, is also considered in more detail that allows identifying reserves to reduce the power supply outage time. The chapter proposes reducing the power supply restoration time by going to the monitoring system combination of both operating modes and technical condition of power networks. Based on the study results, there are methods allowing to reduce the power supply outage time of rural consumers by means of automating rural electrical networks and improving the operating mode monitoring of power supply systems. The chapter also suggests a set of devices that allows to implement the proposed methods.

DOI: 10.4018/978-1-5225-9179-5.ch015

INTRODUCTION

Statistical analysis shows that the damage causes of rural transmission lines within power supply outage are: operating shortcomings (45%), impact of natural weather phenomena (33%), impact of unauthorized people and organizations (15%), other causes (impact of birds, animals, etc.) (7%). Besides, it is noted that operating shortcomings also include tree falling and short circuits due to the touching of power line wires by the tree crowns. It results from untimely cutting of routes, wire breakage, falling of power transmission line supports, etc. It all leads to damage for both power consumers and electric network companies. At the same time, because of the lack of the operating mode monitoring systems as well as insufficient (or often completely absent) network automation, the power supply outage time for rural electrical networks is quite long (on the average more than 5 hours). During the year the power supply outage time for rural consumers is about 100 hours (OAO «Rosseti», 2013).

To develop methods for reducing the power supply restoration time, it is firstly necessary to investigate the components of the given time. In literary sources this aspect is not paid enough attention to, so it was carried out a series of studies based on statistical data obtained in the operation of electric grid companies and agricultural enterprises.

Evaluation of all time components of power supply outages allows developing methods for reducing them, which include first and foremost the methods for automation electrical networks. It includes methods of the sectionalizing and the backup of power networks.

Another important method of reducing the power supply outage time is to monitor the operating modes of electrical networks and their technical condition. It allows controlling the number and duration of outages in these networks, diagnosing outage causes as well as the voltage level deviation more than the permissible values. The monitoring system also automatically informs the dispatcher about the outages on specific sections of the electrical network considering the number and duration of these outages.

Thus, the purpose of the work is to develop methods for reducing the power supply outage time of rural consumers by means of automating and monitoring the operational modes of rural electrical networks.

To achieve the objective, it is necessary to solve a number of tasks:

1. To analyze the structure of the power supply outage time and identify possible reserves for its reduction;
2. To develop methods and means for reducing the power supply outage time by monitoring the operating modes and the technical condition of electrical networks;
3. To develop methods and means for reducing the power supply outage time by automating electrical networks of rural consumers.

BACKGROUND

The recent power industry faces such challenges as the introduction of distributed power generation, the need to improve the power quality (PQ) and the reliability of power supply, environmental problems with traditional methods of power generation, the desire to reduce the expenses on the maintenance and modernization of power network equipment, and finally, the need to improve the power supply efficiency (Vinogradov et al, 2018). The works of both domestic and foreign scientists are devoted to these problems (Leshchinskaya & Magadeev, 2008; Leshchinskaya & Polyanina, 2005; Perova, 2007;

Stepanov & SHpiganovich, 2015; Trojanowska, 2014). The development of technologies has influenced the improvement of distributed power generation, including renewable energy sources (RES) and energy storage, as well as the development of traditional power supply systems. The main direction of power supply system development is the multiple complication of electrical network management systems. Despite the complication, this leads to an increase in the efficiency of entire power supply systems (PSS). PSS management schemes should consider the many uncertainties in the system that may arise due to the load starting and load shedding in the network in order to respond to them (Momoh, 2009). To solve these problems, many energy companies modernize their generation and distribution system, which leads to the use of Smart Grid (SG).

Smart Grid has improved the traditional concept and functionality of electrical networks by using information technologies to extract these network components from power producers to consumers in order to maximize the efficiency and reliability of PSS. The ideal design of Smart Grid should consider the reliability of the system, the adaptation of components and the prediction of system processes development. Besides, the issues related to the electricity demand-side management, the possibility of end-to-end monitoring, the quality of service, data security, the optimization of maintenance costs, self-restoration and reconstruction of the system should be considered. This, in turn, requires the improvement of monitoring systems for electrical network operating modes and network element technical state as well as the development of system automation devices (CHeremisin et al., 2016; Golikov & Vinogradov, 2017; Vinogradov & Vinogradova, 2016). Many studies both in the industry itself and in the scientific community have been carried out in an attempt to put SG conception into practice. Despite the fact that the achievements are great, there are still problems that need to be solved. And at the same time, although Smart Grid allows to solve many of them, it is a source of new problems such as confidentiality of passed information (Goulden et al.,2014; Ipakchi & Albuyeh, 2009).

The greatest problem of introducing Smart Grids is the absence or presence of only one-way communication between the power supplier and the consumer (Yasin, 2016). It prevents operational interaction between the participants in the power market and does not allow the full implementation of monitoring and management of the power supply system. Bidirectional communication, used as the basis of SG, can be implemented on the basis of the Advanced Metering Infrastructure (AMI) or based on monitoring and management systems not connected with accounting systems. In contrast to the Automated Meter Reading (AMR), it includes several advanced technologies such as Smart Meters (SM), Home Area Networks (HAN), Wide Area Networks (WAN) and Neighborhood Area Networks (NAN) (Fangxing et al., 2010; Siano, 2014).

The development of both options of "smart grid" is urgent especially in rural electrical networks because of some peculiarities connected to their considerable length (Vinogradov & Vinogradova, 2016; Vinogradova, 2014), a sufficiently high degree of wear (OAO «Rosseti», 2013), the lack of trained personnel and necessary automation tools. The low economic efficiency of implementing complex and expensive control systems with small loads into rural networks also make it difficult to use smart grids based on intelligent meters widely. This problem is especially urgent for 0.38 ... 10 kV rural power networks.

Nowadays various authors offer solutions for the creation of monitoring systems for electrical network operating modes (Samarin, Rygalin & SHklyaev, 2012; Vinogradov, Vinogradova & Bolshev, 2017), systems for monitoring network element technical condition (Kostikov, 2011; Vinogradov et al., 2014) 22, 23 and devices for automating 0.38 ... 10 kV power networks (Vinogradov & Vinogradova, 2016). But not always the applications of proposed solutions are sufficiently justified. At the same time, the main task of increasing power supply system efficiency is to reduce maintenance costs while meeting the requirements for minimizing power supply outage time, the power quality inconsistency time with set values and the implementation time of technological connection to electric networks (Vinogradov, Vinogradova & Kuchinov, 2017; Vinogradov et al, 2018). The introduction of monitoring systems and automation equipment for electrical networks can reduce the time of power supply outages and their number though it requires an analysis of all time components in order to draw conclusions about the efficiency of considered monitoring systems or automation systems.

TIME ANALYSIS OF POWER SUPPLY OUTAGES

The power supply outage time can be applied as a basic indicator of the power supply reliability (Vinogradov, Vinogradova & Bolshev, 2017; Vinogradov et al., 2017; Vinogradov et al, 2018). Here it is proposed to determine the power supply outage time taking into account all its components as follows:

$$t_{outage} = t_{restor} + t_{outage\ sched} \tag{1}$$

where t_{restor} is the power supply restoration time; $t_{outage\ sched}$ is the power supply outage time scheduled for maintenance and repair.

Power supply restoration time can be represented as follows:

$$t_{restor.} = t_{obt.infor.} + t_{rec.\ infor.} + t_{repair} + t_{harmonize} \tag{2}$$

where $t_{obt.\ infor.}$ is the time to obtain information; $t_{rec.\ infor.}$ is the time to recognize information; t_{repair} is the time to repair; $t_{harmonize}$ is the time to harmonize connection and disconnection.

The time for obtaining information $t_{obt.\ infor.}$ means the time interval from the moment of the failure occurrence to the moment of obtaining information about it the dispatching service of a power grid company. This time can be determined as follows:

$$t_{obt.\inf or.} = t_{\inf or1} + t_{\inf or2} + t_{\inf or3} \tag{3}$$

where $t_{inf.1}$ is the time interval for obtaining information about the failure by the primary information link (for example, the sensor of the automation or monitoring system); $t_{inf.2}$ is the time interval for obtaining information about failure by the secondary information link, for example, by a person who has noticed a stop of electrical equipment orby the comparing element of the automation or monitoring system; $t_{inf.3}$ is the time interval for obtaining information about the failure by the third information link, for example, by a dispatcher receiving a signal about a failure in the network from the consumer or a monitoring system.

The time for the recognizing the information $t_{rec.infor}$ can be represented as follows:

$$t_{rec.\inf or.} = t_{read.\inf or} + t_{dec.} + t_{search} + t_{report} \tag{4}$$

where $t_{read.infor.}$ is the time interval required to read (recognize) the received information message about the failure in the electrical network; $t_{dec.}$ is the time interval from the moment when a dispatcher makes a decision to the moment a brigade is sent to find a place and a fault type; t_{search} is the time interval required forthe field service team departure to find out a place and a fault type; t_{report} is the time interval required to report a location and a fault type.

The repair time $t_{repair.}$ is the time interval from the moment of sending repair brigade to repair a fault until the moment of the beginning the harmonization of equipment connection. It can be presented in the following form:

$$t_{repair} = t_{repair.prepar.} + t_{repair.reach.} + t_{repair.switch.} + t_{repair.permit.} + t_{repair.work.} + t_{repair.complet.} \tag{5}$$

where $t_{repair.prepar.}$ is the time interval necessary to prepare the repair brigade for departure, including the time for preparation of work permit, equipment, devices and loading on transport; $t_{repair.reach}$ is the time interval necessary for the repair brigade to reach the fault site; $t_{repair.switch}$ is the time interval for necessary switching; $t_{repair.permit}$ is the time interval necessary for permitting a repair team to work; $t_{repair.work}$ is the time interval required to carry out direct repair work; $t_{repair.complet.}$ is the time interval required for the repair brigade to complete the work, to clean a workplace, to exit from a workplace, to document the completion of work, up to the point of time for the harmonization of equipment connection.

The time of connection and disconnection harmonization $t_{harmonize}$ can be presented as follows:

$$t_{harmonize} = t_{inf.transfer.} + t_{pre.connect.} + t_{connect.} + t_{encsure} \tag{6}$$

where $t_{inf.transfer}$ is the time interval required to transfer the information on the need of connection; $t_{pre.connect}$. is the time interval required to prepare connection and documenting a connection; $t_{connect.}$ is the time interval required for equipment connection; t_{ensure} is the time interval required to ensure that the connection is successful.

Actual total restoration time $T_{restor.act.}$ for a certain period of time (e.g., 1 year) is determined by:

$$T_{restor.act.1year} = \sum_{i=1}^{n} t_{restor.i} \tag{7}$$

where $t_{restor.i}$ is the restoration time of the i-th powers supply outage; n is the case quantity of power supply restoration per year.

Estimated total restoration time for a certain period of time (e.g., 1 year) $T_{restor.est.1year}$ might be determined as the following formula:

$$T_{restor.est.1year} = SAIFI \times SAIDI = v_{aver.} \times t_{aver.restor.} \tag{8}$$

where SAIFI (System Average Interruption Frequency Index) characterizes the average number of times per year when consumers lost power; SAIDI (System Average Interruption Duration Index) characterizes the average duration of one power supply outage in the power system per year (we accept the equivalent of the average restoration time); $v_{aver.}$ is the average quantity of failures in the power system during the considered period of time (determined on the basis of statistical data or with account of the probability of failure); $t_{aver.restor.}$ is the average restoration time.

The time of one scheduled power supply outage can be represented as follows:

$$t_{outage.sched.} = t_{maint.sched.} + t_{harm.con.sched.} \tag{9}$$

where $t_{maint.\ sched.}$ is the time interval for scheduled maintenance work; $t_{harm.con.sched}$. is the time interval required for the harmonization of equipment connection after scheduled maintenance.

Scheduled maintenance work time $t_{maint.sched}$:

$$t_{maint.sched.} = t_{maint.switch.} + t_{maint.permit.} + t_{maint.work.} + t_{maint.complet.} \tag{10}$$

where $t_{maint'swich}$ is the time interval for necessary switching; $t_{maint'permit}$ is the time interval necessary to permit a maintenance brigade to work; $t_{maint'work}$ is the time interval required to carry out direct maintenance work; $t_{maint'complet}$ is the time interval required for the maintenance brigade to complete the work, to clean a workplace, to exit from a workplace, to document the completion of work, up to the point of time for the harmonization of equipment connection.

Actual scheduled power supply outage time can be determined as follows:

$$T_{outage\ sched.act.1year} = \sum_{j=1}^{m} t_{outage\ sched.j} \tag{11}$$

where $t_{outage.sched.}$ is the time of the J-th scheduled power supply outage; m is the case quantity of scheduled power supply outage per year.

The estimated total time of scheduled power supply outages $T_{outage.sched.est}$ for a certain period of time (e.g., 1 year) can be determined either by the plan of outages per year or based on the quantity probability of scheduled outages due to the deterioration of equipment conditions. The application of the second variant is possible when companies use the "maintenance by condition" system and the maintenance of a detailed information based on the used equipment and its condition.

In the first case:

$$T_{outage.sched.est.1year} = \sum_{k=1}^{s} t_{outage.sched.est.k} \tag{12}$$

where $t_{outage.sched.est.}$ is the time of the k-th estimated scheduled power supply outage; s is the case quantity of estimated scheduled power supply outage per a year.

In the second case:

$$T_{outage.sched.est.1year} = w_{aver.outage.sched.} \times t_{aver.outage.sched} \tag{13}$$

where $w_{aver.outage.sched.}$ is the average quantity of scheduled power supply outage occurring in the system (it is determined on statistical data basis or taking into account scheduled maintenance probability); $t_{aver.outage.sched}$ is the scheduled power supply outage time.

In the literature the above-mentioned time intervals are not given fully, although the analysis of them allows to identify the potential for shortening the restoration time and, therefore, to reducethe power supply outages of consumers and the damage caused by them.

The work of Vinogradov et al. (2018) presents all studies of each time component of power supply outages. The expected time for obtaining the information about the damage in the electrical network is 1.01 hours, that is, about 60 minutes. It is noted that this time can be reduced by various means, for example, by using the monitoring the network element technical condition and the damage site location. According to Diagnosticheskie resheniya v ehnergetike (2017), telemechanics can reduce this time up to 20 minutes. The time for recognizing informationis about up to three fourths of the power supply restoration time. The mathematical expectation of the time for recognizing the information about electrical network damage is 2.94 hours, that is, approximately 180 minutes (determined by the method of expert assessments, the consistency of experts was 79%). The expected repair time is 1.25 ± 0.1 hours with a probability exceeding 95%. The time of connection and disconnection harmonization is up to 20 minutes, or 0.33 hours. Thus, the power supply restoration time can be, in total, more than 5 hours. But the restoration time, which determines the amount of damage from power supply outages, does not consider the idle time of electrical equipment after the power supply restoration. Nevertheless, more than a half of the electric consumers in agricultural enterprises are controlled via magnetic starters, which might be switched off if the power fails even for the fraction of a second, and subsequent restarting the magnetic starters requires time (Vinogradov & Vinogradova, 2016).

The idle time depends on the amount of electric equipment at the considered enterprise, its location, the sequence of switching on and the necessary pauses between the start-up of equipment in the process lines as well as from the power supply restoration time. In addition, computerized technological processes, where microprocessor equipment is used, require quite a long time to switch on and to prepare to work. Therefore, it is necessary to determine the time required to start up all electrical equipment after the power supply restoration. Vinogradov et al. (2018) investigated it by the method of expert surveys and they assume that the mathematical expectation of electrical equipment idle time reached 1.5 h. The agreement of experts was 62%.

The component analysis of the power supply restoration time shows that, first of all, special attention should be paid to the creation of monitoring systems for the technical condition of the transmission lines. Moreover, a significant number of Russian rural power lines are in bad condition and require replacement (Vinogradova, 2014; OAO «Rosseti», 2013). Komarova & Vinogradov (2017), Semenov, Selezneva & Vinogradov (2015) and Vinogradov & Perkov (2015) also prove it by the failure reasons. The power supply outage time can be reduced by means of a monitoring system of the controlled network. This monitoring system should automatically inform the dispatcher about the outages on specific sections of the electrical network considering the number and duration of these outages and their possible causes.

The damage from power non-supply to consumers depends, first of all, on the facts of outages in the networks. Therefore, the outage sensors of the monitoring system in the electrical network must be located at the input of consumers or at several points of the network, for example at the beginning, middle and the end of the transmission line as well as at the buses of the transformer substation.

The power quality was considered in detail in (Borodin & Vinogradov, 2014; Golikov & Vinogradov, 2017). In these studies, it is justified to install sensors to monitor power quality indicators at customer inputs and create power quality control system based on data received from these sensors. It is also proposed to combine these monitoring systems with advanced metering infrastructure, which allows using the data channels of the system to transmit the necessary data.

SOLUTIONS AND RECOMMENDATIONS

A System for Monitoring the Number and Duration of Power Cutoffs and Voltage Deviations in 0.38 kV Electrical Networks

The system for monitoring the number and duration of power supply outages and voltage deviation in 0.38 kV electrical networks is proposed to build as follows. The monitoring sensors of power quality indicators and sensors for recording the number and duration of outages are installed at the inputs of all consumers. Both types of sensors can be integrated into an electricity meter. Borodin & Vinogradov (2014) proposes the sensor integrated into a meter which allows to monitor power quality and adjust electricity cost depending on its quality. Golikov & Vinogradov (2017) proposes the scheme of the device for monitoring the voltage deviation at customer inputs. In Figure 1 there is the proposed scheme of the sensor for calculating the number and duration of outages.

The sensor circuit (Figure 1) contains the switching device SD1, the voltage sensor VS2, the short-circuit sensor SCS4, the overload sensor OLS5, the Meter 12, electric meter with information processing unit, the NOT3 element, the NOT7 element, the OR6 element, the AND8 element, the element Memory9, the information processing unit IPU10/1, the IPU10/2, the Meter11.

In the normal operation mode there is voltage in the power transmission line supplying a consumer, there is no overload or short circuit in theconsumer internal network. The signal is present at the output of the voltage sensor VS2 and there is no signal at the outputs of the elements SCS4 and OLS5. There is also no signal from the outputs of the sensors OVS13 and UVS14. In this case, the output of NOT7 is signaled to one of the inputs of the AND8 element, and there is no signal at the outputs of the elements NOT3, OR6, AND8, Memory8, Memory11. In this case, the signal from the output of the voltage sensor VS2 is sent to reset the element Memory9. The circuit does not start.

Figure 1. The device for recording the number and duration of outages

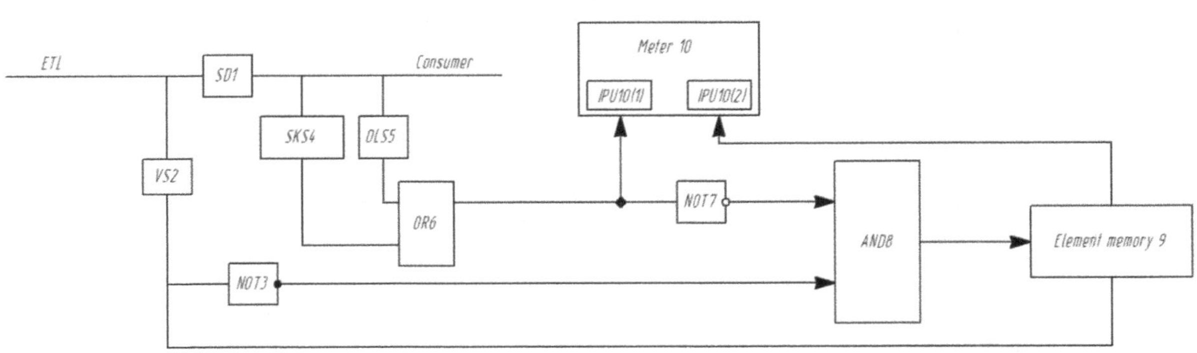

At the moment of failure in the transmission line, the voltage at the consumer input disappears and the signal from the output of the element VS2 disappears. It accordingly leads to the signal appearance at the output of NOT3, which is fed to one of the inputs of the AND8 element.If there are no signals at the sensor outputs the short-circuit current of the SCS 4 and the overload sensor OLS 5, the signal at the output of the NOT 7 element is present and fed to the second input of the AND 8 element. At both inputs AND8 there are signals, hence a signal will appear at its output, which will be memorized by the Memory9 element and transmitted to thespecial information processing unit IPU10, built-in the electric meter or performed by a separate unit. IPU10(1) stores the fact of the outages and its duration. The signal of the outageis transmitted via Automatic Meter Reading (AMR) channel or via another channel to the power company dispatcher. When the voltage in the transmission line is restored, the VS2sensor will detect its presence, the signal from VS2 will be "reset" by the MEMORY9 element, the circuit will return to its original state. The diagram of the signals at the outputs of the elements for this mode is shown in Figure 2.

In case if a short-circuit current appears in the consumer internal network and after its disappearing the voltage is in the power line the circuit will work as follows. At the moment of the short-circuit current appearance at the output of the SCS4 element there will be a signal which will be fed to the input of the OR6 element and from its output to the input of the NOT7 element and also to the information processing unit IPU10(2) that fixes the facts of the short-circuit current and overloads in the internal network of the consumer. At the output of the element NOT7, the signal during the period of the presence of short-circuit current will be absent. At the same time the short-circuit current in the consumer network will cause a failure in the input voltage, as a result, the signal disappears at the output of the sensor VS2, and at the output of the element NOT3 the signal appears and will be fed to one of the inputs of the element AND8. But because of the signal absence at its second input the element AND8 will not

Figure 2. The diagram of signals at the element outputs shown in Figure 1 when the voltage at the consumer input is disconnected

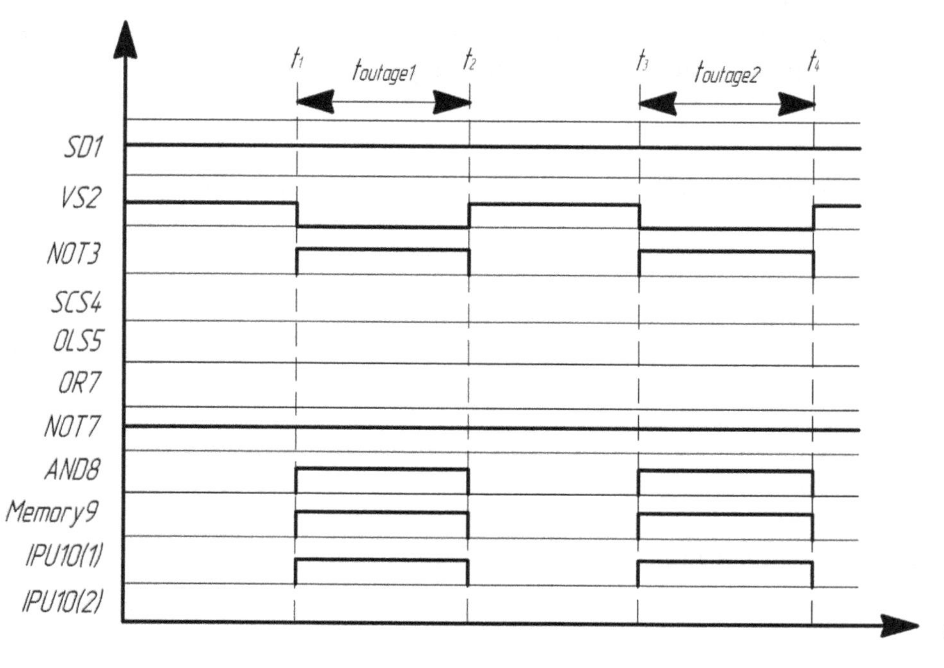

work and the signal at its output will not appear. Once the short-circuit current will disappear because of the switching device disconnection, the signal from the SCS4 output will disappear, the voltage level at the input will return, a signal will appear at the output of the VS2 and the signal at the output of the NOT3 will disappear. The circuit will not start. Thus, in this operation mode the block IPU10(2) will detect the fact of a short circuit in the consumer network without switching off the input voltage.

The diagram of signals at the outputs of the elements for this mode is shown in Figure 3.

In case of a non-selective tripping of the switch installed in the power line supplying the consumer with short circuit current in the internal network the circuit will work as follows. At the moment of the appearance of the short circuit current, at the output of the element SCS4 there will be a signal that will be fed to the input of the OR6 element and from its output to the input of the element NOT7 and to the information processing unit IPU10(2) that fixes the facts of short circuit and overloading in the internal network of the consumer. At the output of the element NOT7 the signal during the period of the short-circuit current will be absent. At the same time a short circuit in the consumer network will cause a voltage drop at the input; as a result the signal will disappear at the output of the VS2 sensor and will appear at the output of the element NOT3 and be fed to one of the inputs of the AND8 element. But because of the absence of a signal at its second input, during the short-circuit current flow the element AND8 will not work and the signal at its output will not appear. After the short-circuit current has disappeared due to the disconnection of the switching device installed in the power line, the signal from the output of the SCS4 will disappear, but due to the nonselective disconnection of the switching device in the line, the voltage at the consumer input will disappear (not recover). Therefore, the signal at the output of the VS2 will not appear and the signal at the output of the NOT3 will not disappear. Thus, signals will be fed to both inputs of AND8, and a signal will appear at its output. This signal, memorized by the element Memory9, will be fed to the input of the block IPU10(1). The fact of the voltage cutting-off at the

Figure 3. The diagram of signals at the element outputs shown in Figure 1 with a short circuit in the internal network of the consumer

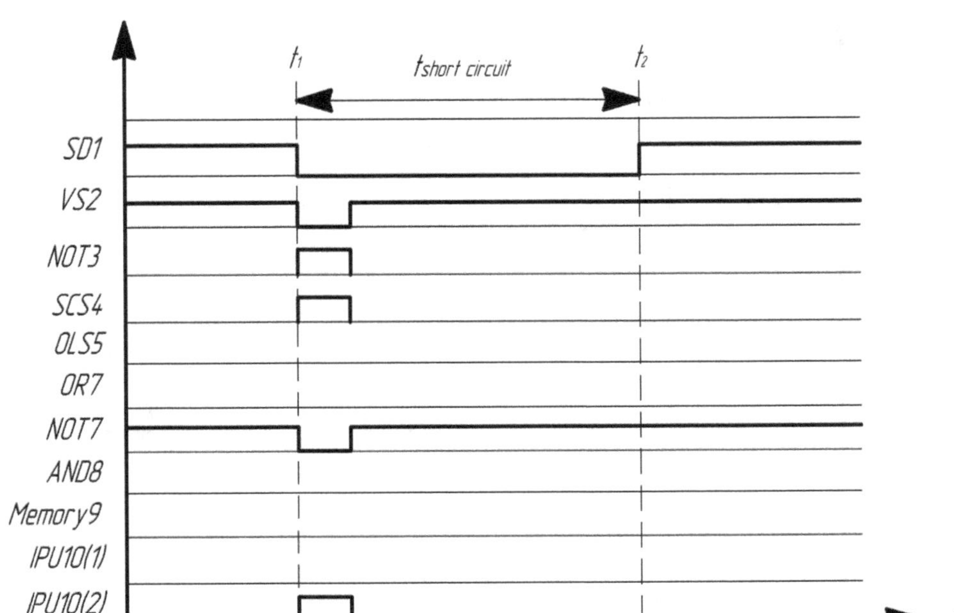

consumer's input and the duration of this cutoff will be fixed by this unit. Also, in this mode of operation, the block IPU10(2) will detect a short circuit in the customer's network. The information about these facts will be stored in the IPU and will be transmitted through the data transmitting channel to the dispatcher.The diagram of the signals at the outputs of the elements for this mode is shown in Figure 4.

If there is an overload in the internal network of the consumer, and in this case, after its disappearance, the voltage in the transmission line does not disappear, the circuit will work as follows. At the moment of the appearance of the overload current at the output of the element OLS5 a signal will appear which will be fed to the input of the element OR6 and from its output to the input of the element NOT7, as well as to the information processing unit IPU10(2), which fixes the facts of the short circuit and overloads in the internal network of the consumer. At the output of the element NOT7 the signal during the presence of the overload current will be absent. At the output of the sensor VS2 the signal won't disappear, but at the output of the element NOT3 it won't appear. Because there is no signal on one of the inputs, the element AND8 will not work and the signal at its output will not appear. After the overload current has disappeared, as the switching device is disconnected, the signal from the output of OLS5 disappears. Thus, in this mode of operation, the IPU10(2) unit will record the fact of an overload in the consumer network without switching-off the input voltage.The diagram of the signals at the outputs of the elements for this mode is shown in Figure 5.

Situation when there is a non-selective disconnection of the switching device installed in the line during an overload in the consumer's network is generally analogous to the situation of non-selective triggering of the switching device during of a short circuit in the consumer network. The difference is that the input signal of the element OR6 will be fed not from the element SCS4 but from the sensor OLS5. In this case, just as in the previously described, the block IPU10(1) will fix a disconnection in the line, and the block IPU10(2) will fix an overload in the consumer network.

Figure 4. Diagram of signals at the element outputs shown in Figure 1 with a short circuit in the customer network and the subsequent non-selective tripping of the switching device in the power line

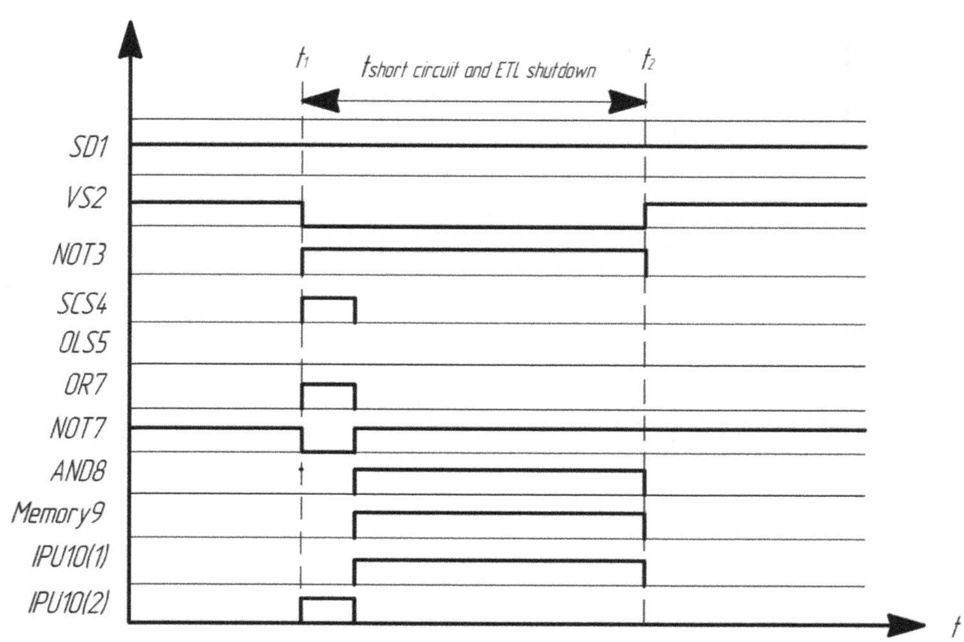

Figure 5. Diagram of signals at the element outputs shown in Figure 1 when the consumer network is overloaded

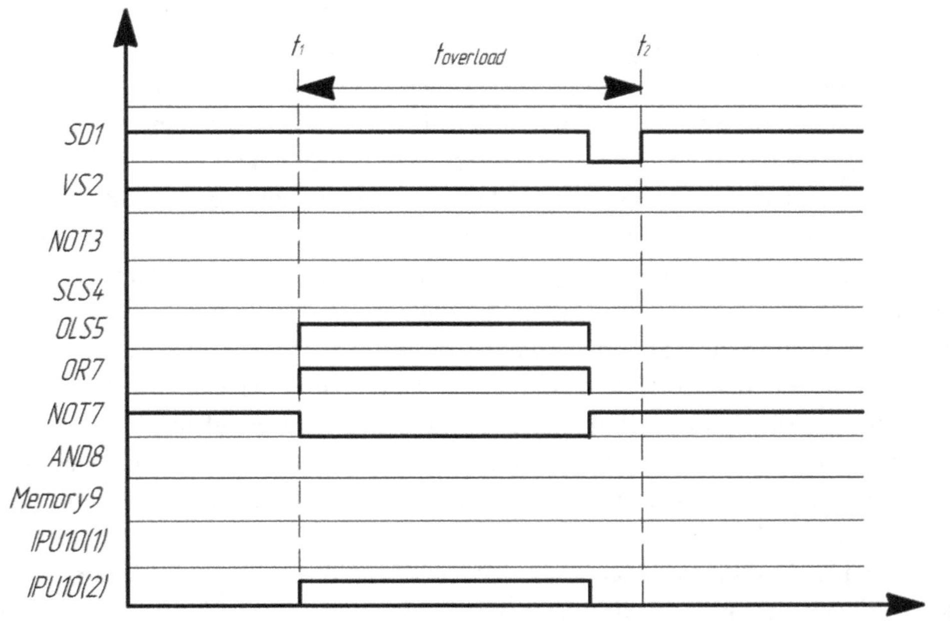

Both the history of accounting the number and the time of outages and the facts of short circuits and overloads in the consumer network are stored in the memory of IPU10(2) in the form of protocols; all these data can be transferred to the dispatching office of the electric grid company. If IPU10(1) and IPU10(2) are built into Meter11, the Meter11 panel can display information on the amount of electricity consumed, its cost, tariff, number and duration of outages for a given period of time (when equipping the meter with the system of power cost adjustments depending on its quality). If IPU is not built into the meter, information on the number and duration of outages is displayed on their information panels. Thus, the worked out device supports automatic calculation of the amount of the consumed power, accounting of the number and duration of power cutoffs, monitoring and recording of emergency situations in the consumer network.

A simplified version of considered device is shown in Figure 6.

Figure 6. The device for recording the number and duration of power supply outages using the position sensor of the switching device

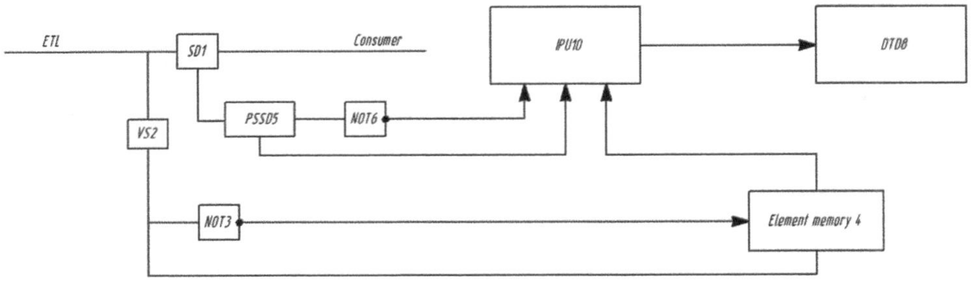

In this variant there the sensors SCS and OLS are absent since in general their functions are performed by the switching device. But the circuit is equipped with a position sensor of the switching device (PSSD), which makes it possible to record the disconnection of this device without specifying the reasons (manual or due to an emergency situation in the consumer network).

In fact, in this case the device reacts only to the presence or absence of voltage at the input of the consumer. In the absence of voltage, a signal about this comes to the IPU and is transmitted to the dispatcher. The on-off state of the SD can also be fixed. This scheme does not allow to detect cases of non-selective operation of switching devices installed in power lines during overloads and short circuits in the consumer network. In case if the consumer switching device is chosen incorrectly, it can lead not only to loss of voltage in the whole supply line, but also to fire-hazardous situations for consumers. Therefore, the first version of the scheme is more rational.

In Figure 8 there is the device allowing to use all the features of the device shown in Figure 1 (the device for recording the number and duration of outages) and monitor the level of voltage deviation. It is much cheaper than monitor allpower quality indicators and in most cases sufficient for a set of functions

The proposed version of the divicecan be applied without its integration into the electric meter and without the AMR channels that makes it possible to reduce the cost of the device option. The overvoltage sensor OVS and the undervoltage sensor OVS give a signal about the voltage deviation higher than the standardized values. The information on the voltage deviation mismatch is transmitted to the IPU block and then to the electric grid company dispatcher using a data transmission device via one of the communication channels (JPS, JPRS, Glonassetc)

The system for monitoring the power quality and power supply reliability can be implemented as shown in Figure 8 (an example is given for the case of using the device shown in Figure 7 i.e. the device for monitoring the number and duration of power outages and voltage deviation DMNDOandVD).

According to the figure TS – transformer substation; DMNDOandVD – the device for monitoring the number and duration of outages and voltage deviation; S1 ... Sn – consumers; EGCD – the electric grid company dispatcher; DATD – Data Acquisition and Transmission Device.

The system works as follows. The information from the DMNDOandVD of each consumer is collected in the DATD and sent to the EGCD. On the basis of this information the dispatcher office makes decisions on the need for the departure of repair teams and voltage regulation. The DMNDOandVD is also installed at the TS and connected to the TS buses. This device monitors the voltage deviation on the buses and the loss of voltage on TS buses, information about it is also transmitted with the help of

Figure 7. The device for monitoring the number and duration of power supply outagesand monitoring the level of voltage deviation at the consumer inputs

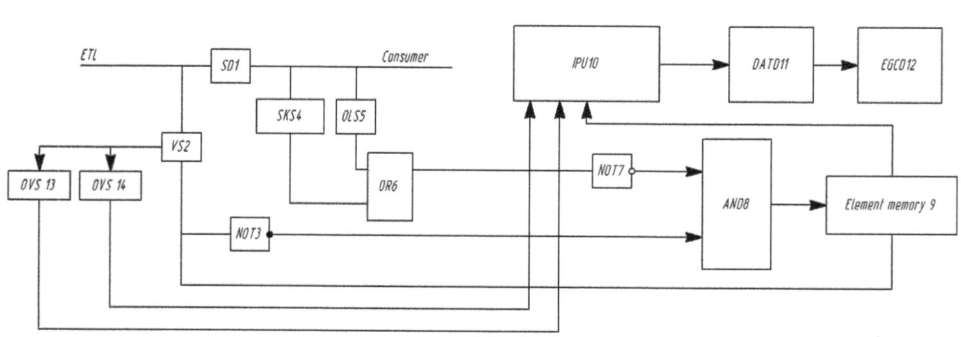

Figure 8. Implementation of a system for monitoring the power quality and power supply reliability on the example of the 0.38 kV outgoing line

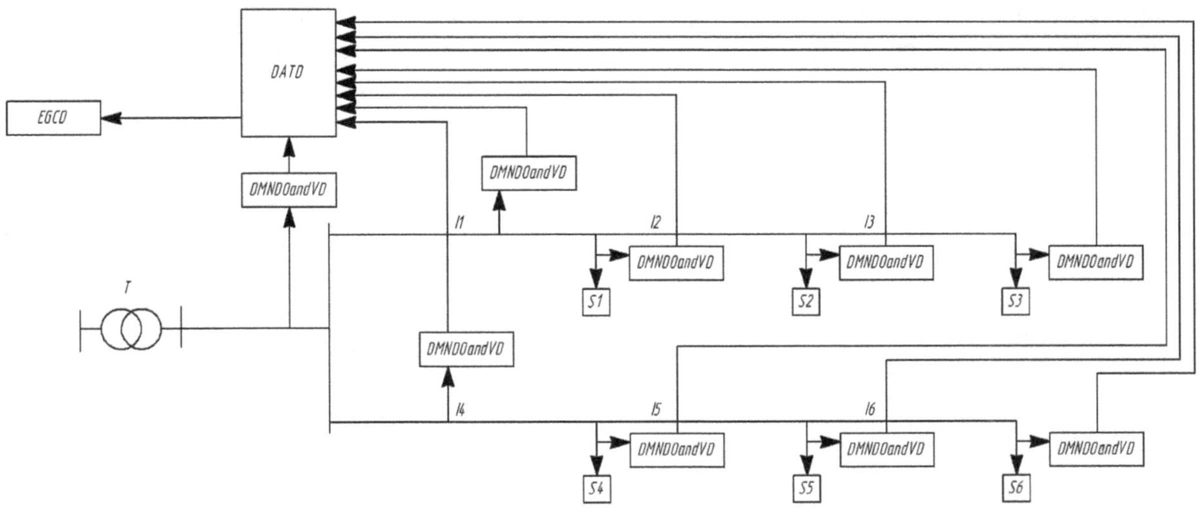

the DATD to the EGCD. The DMNDOandVD installed on the outgoing transmission lines from the TS work similarly. They are connected after the automatic switches of the outgoing transmission lines and monitor the disappearance of the voltage, the voltage deviation at the beginning of the transmission line. If the task of monitoring voltage deviation is not set and power lines do not have a back-up power the outage control can be made cheaper by installing voltage monitoring devices on the TS buses and on the outgoing lines. Then, with the disappearance of the voltage on one of the lines, it would be possible to judge about the interruptions in the power supply of consumers connected to it. With the disappearance of the voltage on the TS buses, one can talk about a break in the power supply for all consumers connected to the TS. But this method is not acceptable if the transmission lines are equipped with means of power line sectionalizing and redundancy as well as it is also necessary to control the level of voltage deviation at the inputs of consumers. Therefore, more functional is a version of the system in which the inputs of all consumers are equipped with the DMNDOandVD. Moreover, the application of this system makes it possible to identify some modes of the network, which are difficult to be detected in other ways. For example, if a wire (of one phase) breaks in the area between consumers S2 and S3, the signal from the DMNDOandVD installed at the consumer S3 will show the presence of a voltage failure, and the DMNDOandVD installed at the consumer S2 will indicate that the voltage is present.

The development of systems for monitoring the power supply reliability in 0.38 kV electrical networks is an urgent task. This solution will make it possible to increase power supply system efficiency as a whole by reducing the power supply restoration time.

Using such systems as indicated by the work of Vinogradov et al. (2017) allow to decrease the power supply restoration time by approximately 2 hours. When determining the economic effect it is necessary to take into account such network characteristics as the reliability indicators of elements as well as their number (for transformers, disconnectors, etc.) and length (for transmission lines). Reliability indicators depend on where the cable or air lines are used (in urban or rural areas).

The economic effect of implementing a monitoring system by reducing the power supply restoration time should be determined as indicated below. The first step is to assess the failure probability in the considered networks during the year. Vinogradov & Perkov (2015) and Semenov, Selezneva & Vinogradov (2015) propose the failure rate for various 0.38 kV network elements. For a 0.38 kV insulated overhead lines (IOL) with a length L_{IOL} and failure rate of 2.7 failures/100 km per year, the probable number of failures per year is:

$$N_{IOL0,38} = \frac{2,7 \times L_{IOL}}{100} \text{ failures per year} \tag{14}$$

For power transformers with number n_{TR} and failure rate 3,5 failures/100 per year:

$$N_{TR} = \frac{3,5 \times n_{TR}}{100} \text{ failures per year} \tag{15}$$

The number of failures per year for other network elements is determined similarly.

The next step is to determine the power undersupply per year for a given number of failures for the various elements of power supply system.

The rate of power undersupply is determined by the formula:

$$W_{undersup.} = N \times T_{av.} \times P_{rat.} \tag{16}$$

where $P_{rat.}$ - estimated power of the load connected to the failed equipment; $T_{av.}$ - the average time for power supply restoration. CHeremisin, CHerkashina & Holod (2016) takes it equal to 5.86 hours.

The reduction of power undersupply is defined as follows:

$$W_{red.} = W_{undersup.} - W_{undersup.1} \tag{17}$$

where $W_{undersup.1}$ – undersupply when implementing a monitoring system taking into account the reduction of the average restoration time by two hours.

In the above sequence, it is possible to assess the economic damage caused by the power undersupply to consumers during power outages.

The Use of Automation Means on the Example of Power Line Sectionalizing and Line Backup

Sectionalizing is a method to improve the power supply reliability of consumers. It consists in dividing the transmission line into sections using a switching device. The use of sectionalizing makes it possible to disconnect only the damaged section of the power line while other consumers are powered by the undamaged section.

The most effective variant is to apply power line sectionalizing together with redundancy. There is a positive experience of the use of sectionalizing points (SP) in electrical networks. Vinogradov & Vinogradova (2016) and Vinogradova, Vinogradov, & Konstantinov (2016) present scheme application of SP with the automatic reclosing (AR) function.

The joint application of power line sectionalizing and line backup in the ring network makes it possible to significantly improve power supply reliability to consumers connected to the power transmission line located between the SP and the AR. In Figure 9 there is scheme of this joint application. At the same time it is possible to significantly reduce the costs of building a redundant power line in the ring network containing the SP, if a responsible consumer is connected to the section 2 of L1 (S7 in Figure 9). This is achieved as follows. A customer is connected to two power points before and after SP2 installed in L1. The AR3 is installed in the scheme in order to to feed the responsible consumer iether from section 2 of L1 or section 3 of L1 depending on where the voltage is located. The AR3 can be installed directly into the consumer input distribution device (IDD) or at the installation site of the SP altough it is a less reliable option.

According to the scheme: T1, T2 – power transformers; Buses1, Buses2 – TS buses 10/0,4 kV; SP1 ... SP4 – sectioning points; AR1, AR2 – the points of automatic reclosing; S1 ... S6 – consumers; SC1 ... SC5 – short circuit points; CP1 ... CP10 - the points at which currents and voltages are controlled during network automation.

Figure 9. The method of the joint application of sectioning and redundancy means

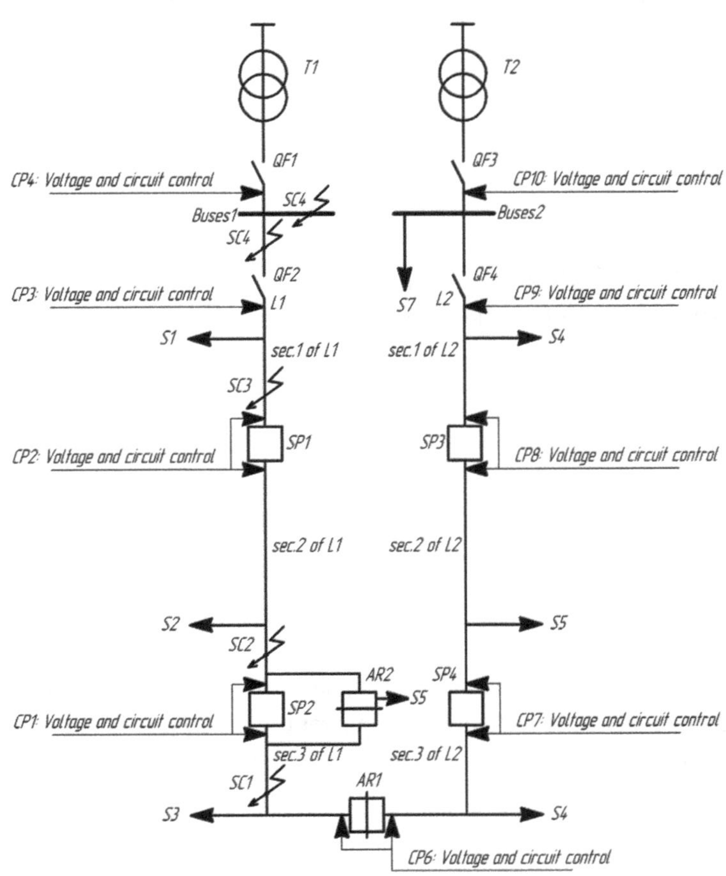

The use of power line sectionalizing and redundancy makes it possible to shorten the power supply restoration time. It is possible to consider it on the example of the sheme shown in Figure 9. Suppose that a damage, for example a short circuit, has occurred in the section 2 of L 1. In this case, if there were no sectionalizing means of the power line, namely, the sectionalizing points (SP1 and SP2) and the redundancy mean (AR2), the whole line L1 and all connected consumers would be disconnected. In the case of equipping the transmission power line with the means of sectionalizing and line backup, the following will happen. If a short circuit occures at the point K2, the SP1 will be turned off due to the action of the relay protection tuned to the short-circuit current. The SP2 will be turned off as well due to the voltage disappearance in the section 2 of L1. Thereby the damaged section of L1 will be isolated. there will be no power supply outage for consumers connected to the section 1 of L1 and accordingly the restoration time for them will be reduced to 0 in comparison with the power line without sectionalizing. Nevertheless power supply outages will be for consumers connected to the section 3 of L1. The power supply restoration time in this case will be equal to the automatic control time of the AR2. Depending on the AR type and its coordination with the operation of the SP1 and the SP2 can range from fractions of a second to seconds. Thus, the restoration time for these consumers will also be close to zero.

The reduction of power supply restoration time is determined similarly for consumers connected to other sections of power lines and for situations where damage occurs in other sections of L1 or L2. Thereby, the power supply restoration time should be defined proceeding from the transmission line equipping by means of automation (first of all by means of sectionalizing and redundancy) taking into account the reliability of this equipment.

Table 1 gives the formulas for determining the power supply restoration time depending on power line equipping by means of sectionalizing and redundancy.

The significance of the damage depending on the power supply restoration time and the specific damage will be:

$$Y = f(t_{rest.}) = y \times t_{rest.} \tag{18}$$

where y is the specific damage from a power supply outage, rubles/h.

The damage reduction for different applications of sectionalizing and redundancy means can be determined as follows:

$$\Delta Y = Y(t_{rest.})_{withoutSPandAR} - Y(t_{rest.})_{withSPandAR} \tag{19}$$

where $Y(t_{rest.})_{withoutSPandAR}$ is the damage from the power supply outage without the means of power line sectionalizing and redundancy; $Y(t_{rest.})_{withSPandAR}$ is the damage from the power supply outage with the means of power line sectionalizing and redundancy which can be determined depending on the situations described in Table 1.

Thus, power line sectionalizing and redundancy makes it possible to increase the power supply reliability of rural consumers. It also allows to decrease power under-supply by preventing the switching-off consumers and by reducing the power supply restoration time after power line damage.

Table 1. Determination of power supply restoration time depending on the equipment of power lines by means of sectionalizing and redundancy

No. p/p	Situation	Calculation Formula	Approximate Restoration Time
1	Damage to power lines without means of sectionalizing and redundancy	$t_{restor.} = t_{obt.\inf or.} + t_{rec.\inf or.} + t_{repair} + t_{harmonize}$ where $t_{obt.\ infor.}$ is the time to obtain information; $t_{rec.\ infor.}$ is the time to recognize information; t_{repair} is the time to repair; $t_{harmonize}$ is the time to harmonize connection and disconnection.	Up to several hours
2	Damage to power lines equipped by means of sectionalizing	• $t_{restor.} = 0$ is for consumers connected before the SP if a section after the SP is damaged; • $t_{restor.} = t_{AR\ of\ SP\ operation\ time}$ is for consumers connected after the SP if there is unstable damage in a section after the SP; • $t_{restor.} = t_{obt.\inf or.} + t_{rec.\inf or.} + t_{repair} + t_{harmonize}$ is for consumers connected after the SP if a section after the SP is damaged; • $t_{restor.} = t_{obt.\inf or.} + t_{rec.\inf or.} + t_{repair} + t_{harmonize}$ is for consumers connected before and after the SP if a section before the SP is damaged;	From zero to several hours depending on the location of the damage and a portion of consumers (connected before and after the SP)
3	Damage to power lines equipped by means of sectionalizing and redundancy	• $t_{restor.} = 0$ is for consumers connected before the SP if a section after the SP is damaged; • $t_{restor.} = t_{AR\ of\ SP\ operation\ time}$ is for consumers connected to the damaged section if there is unstable damage in a section after the SP; • $t_{restor.} = t_{obt.\inf or.} + t_{rec.\inf or.} + t_{repair} + t_{harmonize}$ is for consumers connected to the damaged section and if they do not have an individual power redundancy; • $t_{restor.} = t_{AR\ operation\ time}$ is for consumers connected to the redundant section.	From zero to several hours, depending on the location of the damage and a portion of consumers (connected to the damaged or other section)

The power supply restoration time for consumers should be determined using different formulas depending on the application of power supply sectionalizing and redundancy means, the types of this equipment and the power line section the consumers are connected to. The damage reduction from power supply outages in power lines equipped with means of sectionalizing and redundancy occurs due to reduction of the power supply restoration time. In some cases the outages can be completely prevented and, accordingly, there can be no the damage.

FUTURE RESEARCH DIRECTIONS

The presented paper considers options to improve the power supply reliability of rural consumers by creating systems for monitoring power supply outages and power quality in 0.38 kV electrical networks as well as by power line sectionalizing and redundancy. The devices of sectionalizing and redundancy of electrical networks have been made by the authors and tested on the basis of the branch of the PJSC "Interregional Distribution Network Companies of the Centre" – "Orelenergo" and the tests gave positive results.

Experimental operation of the SP in the Mezenka settlement of the Orel region has already proved its effectiveness. So on November 17, 2015 (3 weeks after connecting the SP) a breakdown occurred in the power line after the SP and the SP was cut off. At the same time, the consumers connected before the SP did not lose power. The power supply restoration time was 45-50 minutes for consumers connected after the SP. During this time, the power under-supply of consumers in the amount of about 20 kW/h was eliminated, the effect of eliminating the power under-supply amounted to more than 2400 rubles. The short-circuit in the power line was unstable and after switching-on the SP the power of the disconnected consumers was restored. The appearance of SP installed in the Mezenka settlement is shown in Figure 10.

It should be noted that in the Mezenka settlement the disconnection of the 0.38 kV transmission line has also been noted a week before the SP started its work. At the same time, all consumers were losing power, since the circuit-breaker was switched off at the beginning of the transmission line. The interruption also amounted to 45-50 minutes (the time before the arrival of the emergency crew of the Orel distribution zone).

Nevertheless, it is necessary to work on the improvement of these devices, since the development of microprocessor technology allows to reduce the dimensions of these devices and to increase their functionality significantly.

As for the developed monitoring systems, there is work to be done further to justify their use and optimize their hardware options taking into account the peculiarities of domestic power supply systems. Of course, the most promising variant is the use of such systems in conjunction with intelligent meters, but in the short term it is rational to use simpler but less expensive systems.

Figure10. Appearance of SP installed in the Mezenka settlement

CONCLUSION

On the basis of the carried out studies a number of conclusions is drawn:

1. Analysis of the time components of power supply outage was performed. It showed that it is possible to reduce the components of the given time by means of rural electrical network automation and the improvement of monitoring of the operation modes of power supply systems;
2. The structural diagram of the system for monitoring the operating modes of power supply systems has been developed. It makes it possible to determine the number and time of outages in 0.38 kV electrical networks and to determine the presence of voltage deviations at the consumers inputs from the standardized value. The application of this system allows to reduce the power supply restoration time, and consequently the power supply outages of rural consumers by approximately 2 hours;
3. The influence of the electrical network equipment by means of sectionalizing and redundancy on the power supply restoration time has been considered. The variant to construct an electrical network with the use of sectionalizing points and automatic reclosing devices, which allows maximizing the power supply reliability for rural consumers, is proposed;
4. The main directions of further research on the topics presented in the work are indicated. They should be aimed at improving the efficiency of the proposed means of electrical network automation and also optimizing the structure and hardware of the developed system for monitoring electrical network operating modes. They should take into account the features of their design and operation as well as financial possibilities to implement monitoring systems.

REFERENCES

Borodin, M. V., & Vinogradov, A. V. (2014). *Povyshenie ehffektivnosti funkcionirovaniya sistem ehlektrosnabzheniya posredstvom monitoringa kachestva ehlektroehnergii* [Improving of power supply system efficiency by means of power quality monitoring]. Orel, Russia: FGBOU VPO Orel GAU.

Cheremisin, N. M., Cherkashina, V. V., & Holod, A. V. (2016). Povyshenie "intellekta" ehlektricheskoj seti za schet avtomatizirovannogo monitoringa parametrov vozdushnyh linij [The "intelligence" increase of an electric network due to automated monitoring of the of overhead line parameters]. *Ehnergetika ta komp'yuterno-integrovani tekhnologiї v APK, 1*(4), 19-22.

Diagnosticheskie resheniya v ehnergetike. (2017). [Diagnostic solutions in power engineering]. Available from: http://dimrus.ru/dilin.html

Fangxing, L., Wei, Q., Hongbin, S., Hui, W., Jianhui, W., Yan, X., ... Pei, Z. (2010). Smart transmission grid: Vision and framework. *IEEE Transactions on Smart Grid, 1*(2), 168–177. doi:10.1109/TSG.2010.2053726

Golikov, I. O., & Vinogradov, A. V. (2017). *Adaptivnoe avtomaticheskoe regulirovanie napryazheniya v sel'skih ehlektricheskih setyah 0,38 kV* [Adaptive automatic voltage regulation in rural power networks of 0.38 kV]. Orel, Russia: FGBOU VO Orel GAU.

Goulden, M., Bedwell, B., Rennick-Egglestone, S., Rodden, T., & Spence, A. (2014). Smart grids, smart users? The role of the user in demand side management Energy. *Energy Research & Social Science, 2*, 21–29. doi:10.1016/j.erss.2014.04.008

Ipakchi, A., & Albuyeh, F. (2009). Grid of the future. *IEEE Power & Energy Magazine, 7*(2), 52–62. doi:10.1109/MPE.2008.931384

Komarova, A. A., & Vinogradov, A. V. (2017). Analiz dinamiki sostoyaniya liniy elektroperedachi 6-20 kV [Analysis of the state dynamics of 6-20 kV transmission lines]. In Energosberezheniye I effektivnost' v tekhnicheskikh sistemakh: Materialy IV Mezhdunarodnoy nauchno-tekhnicheskoy konferentsii studentov, molodykh uchenykh I spetsialistov. Tambov, Russia: Tambov State Technical University.

Kostikov, I. (2011). Sistema monitoringa CAT-1 — povyshenie propusknoj sposobnosti i nadezhnosti v LEHP [CAT-1 monitoring system - increase of capacity and reliability in power lines]. *EHnergetika, 3*(38). Retrieved 26.04.2017, from http://kazenergy.kz/arhiv/38/44.htm

Leshchinskaya, T. B., & Magadeev, E. H. V. (2008). *Metodika vybora optimal'nogo varianta povysheniya nadezhnosti ehlektrosnabzheniya sel'skohozyajstvennyh potrebitelej* [The method of choosing the optimal option for increasing the power supply reliability of agricultural consumers]. Moscow, Russia: FGOU VPO MGAU.

Leshchinskaya, T. B., & Polyanina, I. N. (2005). *Povyshenie ehffektivnosti funkcionirovaniya raspredelitel'nyh setej rajonov s maloj plotnost'yu ehlektricheskih nagruzok (na primere Joshkar-Olinskij setej)* [Efficiency increase of distributive network functioning in the areas with low density of electric loads (on an example of the Yoshkar-Ola network)]. Moscow, Russia: Agrokonsalt.

Momoh, J. A. (2009). Smart grid design for efficient and flexible power networks operation and control. In *Power Systems Conference and Exposition, PSCE '09*. Seattle, WA: IEEE. doi:10.1109/PSCE.2009.4840074

OAO "Rosseti". (2013). Polozheniye OAO «Rosseti» o yedinoy tekhnicheskoy politike v elektrosetevom komplekse [Position of Rosset JSC on a unified technical policy in the power grid complex]. Moscow: OAO «Rosseti».

Perova, M. B. (2007). *EHkonomicheskie problemy i perspektivy kachestvennogo ehlektrosnabzheniya sel'skohozyajstvennyh potrebitelej v Rossii* [Economic problems and prospects of qualitative power supply of agricultural consumers in Russia]. Moscow, Russia: INP RAN.

Samarin, A. V., Rygalin, D. B., & SHklyaev, A. A. (2012). Sovremennye tekhnologii monitoringa vozdushnyh ehlektrosetej LEHP [Modern technologies for monitoring of power transmission lines]. *Estestvennye i tekhnicheskie nauki, 1*.

Semenov, A. E., Selezneva, A. O., & Vinogradov, A. V. (2015). Sravnenie pokazatelej nadezhnosti vozdushnyh i kabel'nyh linij v gorodskoj i sel'skoj mestnosti. In *Proceedings of the VII All-Russian Scientific and Practical Conference: Osnovnye napravleniya razvitiya tekhniki i tekhnologii v APK*. Knyaginino, Russia: NGIEHU.

Siano, P. (2014). Demand response and smart grids – a survey. *Renewable & Sustainable Energy Reviews, 30*, 461–478. doi:10.1016/j.rser.2013.10.022

Stepanov, V.M., & Shpiganovich, A.A. (2015). Opredelenie iskhodnyh dannyh dlya povysheniya ehffektivnosti funkcionirovaniya sistem ehlektrosnabzheniya [Determination of initial data for improving power supply system efficiency]. *Izvestiya Tul'skogo gosudarstvennogo universiteta. Tekhnicheskie nauki, 12*(2).

Trojanowska, M. (2014). Bezpieczeństwo elektroenergetyczne terenów wiejskich [Electricity safety in rural areas]. *Agrotekhnika i ehnergoobespechenie, 1*(1), 468-475.

Vinogradov, A., Vasiliev, A., Bolshev, V., Semenov, A., & Borodin, M. (2018). Time Factor for Determination of Power Supply System Efficiency of Rural Consumers. In V. Kharchenko & P. Vasant (Eds.), Handbook of Research on Renewable Energy and Electric Resources for Sustainable Rural Development (pp. 394–420). Academic Press. doi:10.4018/978-1-5225-3867-7.ch017

Vinogradov, A., Vasiliev, A., Semenov, A., Sinyakov, A., & Bolshev, V. E. (2017). Analiz vremeni pereryvov v energosnabzhenii sel'skikh potrebiteleyi metodov yego snizheniya za schet monitoringa tekhnicheskogo sostoyaniya liniy elektroperedach [Analysis of the power supply outage time for rural consumers and methods of its reducing due to monitoring power line technical condition]. *Bulletin of All-Russian Institute of Electrification of Agriculture*, 2(27), 3–11.

Vinogradov, A. V. (2007). *Sovershenstvovanie avtomaticheskogo rezervirovaniya na dvuhtransformatornyh podstanciyah* [Improvement of automatic backup on two-transformer substations]. Orel, Russia: FGBOU VPO Orel GAU.

Vinogradov, A. V., Chernyshov V. A., Semenov A. E., & Moroz A. O. (2014). Sposob monitoringa tekhnicheskogo sostoyaniya ehlementov vozdushnoj linii ehlektroperedach [Method for monitoring the element technical condition of an overhead power line]. *Agrotekhnika i ehnergoobespechenie, 3*, 64-68.

Vinogradov, A. V., & Perkov, R. A. (2015). Analiz povrezhdaemosti ehlektrooborudovaniya ehlektricheskih setej i obosnovanie meropriyatij po povysheniyu nadezhnosti ehlektrosnabzheniya potrebitelej [Analysis of the damageability of electrical equipment in power networks and the rationale of measures to improve power supply reliability of consumers]. *Vestnik NGIEHI, 12*(55), 12–20.

Vinogradov, A. V., & Vinogradova, A. V. (2016). *Povyshenie nadezhnosti ehlektrosnabzheniya sel'skih potrebitelej posredstvom sekcionirovaniya i rezervirovaniya linij ehlektroperedachi 0,38 kV* [Improving the power supply reliability of rural consumers by the distribution sectionalizing and the backup of 0.38 kV power networks]. Orel, Russia: FGBOU VO Orel GAU.

Vinogradov, A. V., Vinogradova, A. V., & Bolshev, V. E. (2017). Ustroystva I sistema monitoringa nadezhnosti elektrosnabzheniya I otkloneniya napryazheniya v elektricheskikh setyakh 0,38 kV [Devices and monitoring system for power supply reliability and voltage deviation in 0.38 kV electrical networks]. *Vestnik of the Nizhniy Novgorod State Engineering and Economic Institute, 11*(78), 69–82.

Vinogradov, A. V., Vinogradova, A. V., & Kuchinov, A. A. (2017). Vremya osushchestvleniya tekhnologicheskogo prisoyedineniya k elektricheskim setyam kak factor effektivnosti system elektrosnabzheniya [The time of technological connection to electrical networks as a factor of the power supply system efficiency]. *Vestnik of the Nizhniy Novgorod State Engineering and Economic Institute*, *6*(73), 54–60.

Vinogradova, A.V. (2014). Statisticheskaya kharakteristika sel'skikh elektricheskikh setey [Statistical characteristics of rural electrical networks]. *Agrotekhnika I energoobespecheniye, 1*(1), 419-423.

Vinogradova, A.V., Vinogradov, A.V., & Konstantinov, A.V. (2016). Primeneniye sektsionirovaniya liniy elektroperedachi 0,38 kV dlya povysheniya nadezhnosti elektrosnabzheniya sel'skikh potrebiteley [Application of sectionalizing of 0.38 kV transmission lines to improve power supply reliability for rural consumers]. *Teoreticheskiyi nauchno-prakticheskiy zhurnal "Innovatsii v APK: problemyiperspektivy"*, *2*(10), 12-27.

Yasin, K. (2016). A survey on smart metering and smart grid communication. *Renewable & Sustainable Energy Reviews*, *57*, 302–318. doi:10.1016/j.rser.2015.12.114

KEY TERMS AND DEFINITIONS

Point of Common Connection: The network point closest to the specific load of a user, where the loads of other network users are connected.

Power Quality: The conformity degree of electrical energy characteristics at a point in the electrical system with a set of standardized indicators.

Power Supply Consumer: A legal entity or a private person exercising the use of electric energy (capacity) on the basis of a concluded contract.

Power Supply Interruption Time: The duration of the absence of voltage at the customer input, connected with emergency or planned disconnection in the electrical network.

Power Supply Reliability: The ability of the power supply system to transmit and distribute the required amount of electricity from sources to consumers at standard voltage levels and in accordance with a specified load schedule.

Power Supply Restoration Time: The duration of an emergency repair or replacement of a failed element of the power supply system.

System for Monitoring the Electrical Network Operating Modes: A set of interrelated organizational and technical measures ensuring the monitoring of the technical condition of power transmission lines, determining the location and type of damage. Typically, monitoring systems use equipment complexes designed to monitor the technical condition of power lines, determine the location and type of damage, and transfer data through various channels to the processing center.

Compilation of References

Ackermann, T. (Ed.). (2005). *Wind Power in Power Systems*. Stockholm: John Wiley & Sons. doi:10.1002/0470012684

Adadurov, E. A. (2005). Povyshenie ehffektivnosti ispol'zovaniya akkumulyatorov teploty s vozobnovlyaemymi istochnikami ehnergii [Increasing the efficiency of using heat accumulators with renewable energy sources]. Krasnodar: Dis. ... cand. tech. sciences. (in Russian)

Adomavicius, V., Kharchenko, V., Valickas, J., & Gusarov, V. (2013). RES-based microgrids for environmentally friendly energy supply in agriculture. *Proceedings of 5th International Conference TAE 2013, Trends in Agricultural Engineering*, 51-55.

Afonin, V.S., & Deryugina, G.V. (2012). System properties of hybrid energy complexes based on renewable energy sources. *Energy Security and Energy Efficiency*, 20-27;

AFREA. (2014). Energizing Africa: Achievements and Lessons from the Africa Renewable Energy and Access Program (AFREA). The International Bank for Reconstruction and Development/WB Group.

Alekseev, V. V. (1991). Solnechnaya ehnergetika: perspektivy razvitiya [Solar energy: development prospects]. Knowledge. (in Russian)

Alferov, Zh. I., Andreev, V. M., & Rumyantsev, V. D. (2004). Tendencies and prospects for the development of solar photovoltaics. *Physics and Technology of Semiconductors*, *38*(8), 937–948.

Alihodgina, N. V., & Deryugina, G. V. (2017). Investigation of the influence of various environmental factors on the vertical wind profile in the peninsula Kamchatka. *Materials of the XLVII International Scientific and Practical Conference with Elements of the Higher School "Fedorov Readings 2017"*.

Allayev, K.R., Khokhlov, V.A., Sytdykov, R.A., & Titova J.O. (2013). *Elektromekhanicheskiye i gidromekhanicheskiye protsessy v gidroenergeticheskikh ustanovkakh* [Electromechanical and hydro-mechanical processes in hydropower plants]. Tashkent, Fan va texnologiya.

Amerhanov, R. A., Butuzov, V. A., Tomarov, G. V., & Butuzov, V. V. (2008). Jenergeticheskij potencial solnechnoj radiacii i jekonomicheskaja celesoobraznost' primenenija gelioustanovok v Krasnodarskom krae i Jakutii [Energy potential of solar radiation and economic feasibility of solar installations in the Krasnodar Territory and Yakutia]. *Works of the Kuban State Agrarian University*, *1*, 26–32.

Amirat, J., Benbouzid, M. E. H., Bensaker, B., & Wamkeue, R. (2007). Generators for wind energy conversion systems: State of the art and coming attractions. *Journal of Electrical Systems*, *3*(1), 26–38.

Anderson, J. L., Lowry, M. V., & Thomte, J. C. (2001). Hydraulic and water quality modeling of distribution systems: What are the trends in the U.S. and Canada? In *Proceedings of the AWWA Annual Conference*. Denver, CO: American Water Works Association (AWWA).

Angelis-Dimakis, A., Biberacher, M., Dominguez, J., Fiorese, G., Gadocha, S., Gnansounou, E., ... Robba, M. (2011). Methods and tools to evaluate the availability of renewable energy souces. *Renewable & Sustainable Energy Reviews*, *15*(2), 1182–1200. doi:10.1016/j.rser.2010.09.049

Anisimov, O., & Reneva, S. (2006). Permafrost and changing climate: the Russian perspective. *AMBIO: A Journal of the Human Environment*. Retrieved from http://www.bioone.org/doi/abs/10.1579/0044-7447(2006)35%5B169:PACCTR%5D2.0.CO%3B2

Anisimov, O.A., & Lavrov, S.A. (2004). Global warming and melting of permafrost: assessment of risks for production facilities of the fuel and energy sector. *Technologies of Fuel and Energy Complex*, (3), 78-83.

Antonov, O. (2014). *Green Energy in Kazakhstan in 21st Century: Myths, Realities and Prospects*. Retrieved from http://www.kursiv.kz/news/details/kompanii/Zelenaya-energetika-Kazahstana-v-21-veke-mify-realnost

Arán Carrión, J., Espín Estrella, A., Aznar Dols, F., Zamorano Toro, M., Rodríguez, M., & Ramos Ridao, A. (2008). Environmental decision-support systems for evaluating the carrying capacity of land areas: Optimal site selection for grid-connected photovoltaic power plants. *Renewable & Sustainable Energy Reviews*, *12*(9), 2358–2380. doi:10.1016/j.rser.2007.06.011

Aronova, E. S., & Grilihes, V. A. (2006). Metodika rascheta real'noj plotnosti solnechnogo izluchenija pri proektirovanii fotojelektricheskih ustanovok [Method for calculating the real density of solar radiation in the design of photovoltaic installations]. *SPbSU Scientific and Technical Statements*, *6*(1), 62–66.

Ashourian, M. H., Cherati, S. M., Mohd Zin, A. A., Niknam, N., Mokhtar, A. S., & Anwari, M. (2013). Optimal green energy management for island resorts in Malaysia. *Renewable Energy*, *51*, 36–45. doi:10.1016/j.renene.2012.08.056

Atwater, M., & Ball, J. T. (1978). A numerical solar radiation model based on standard meteorogical observation. *Solar Energy*, *21*(3), 163–170. doi:10.1016/0038-092X(78)90018-X

Babaev, B. D. (2015). Resursy vozobnovlyaemyh istochnikov ehnergii Respubliki Dagestan: uchebno-spravochnoe posobie [Resources of renewable energy sources of the Republic of Dagestan: a manual]. Makhachkala: Publishing house "Rainbow". (in Russian)

Babaev, B. D. (2016a). Razrabotka i issledovanie ehnergosistem na osnove vozobnovlyaemyh istochnikov s fazoperekhodnym akkumulirovaniem tepla [Development and investigation of power systems based on renewable sources with phase-transfer heat accumulation]. Moscow: RSL. (in Russian)

Babaev, B. D., & Danilin, V. N. (2002). EHnergoakkumuliruyushchie ustanovki. Uchebnoe posobie [Energy storage facilities. Tutorial]. Makhachkala: CPI of the DSU. (in Russian)

Babaev, B.D. & Khalilullayev, G.M. (2005). *Opisanie termohimicheskogo vzaimodejstviya v mnogokomponentnyh vzaimnyh sistemah v zavisimosti ot temperatury. Teplovoj ehffekt v zavisimosti ot temperatury* [Description of thermochemical interaction in multicomponent reciprocal systems as a function of temperature. Thermal effect in relation to temperature]. Patent of the Russian Federation N° 2005610201. (in Russian)

Babaev, B. D. (2014). Principy teplovogo akkumulirovaniya i ispol'zuemye teploakkumuliruyushchie materialy [Principles of thermal accumulation and used heat storage materials] [in Russian]. *Thermophysics of High Temperatures*, *52*(5), 760–776.

Babaev, B. D. (2016b). Akkumulyatory tepla pri ispol'zovanii vozobnovlyaemyh istochnikov ehnergii. Perspektivnye napravleniya novyh razrabotok [Heat accumulators using renewable energy sources. Perspective directions of new developments]. *Energetik*, *3*, 19–22.

Bachmanova, K. (2016). In response to those who connected. *Energy Without Borders, 5*(40), 18-19.

Baranov V. K. (1976). *Methods of Calculation for the compound parabolic concentrator profiles, Applied Solar Energy*. Moscow: Publishing house of Nauka.

Baskakov, A. P. (1974). *Heating and cooling of metals in the fluidized bed*. Moscow: Metallurgy.

Baskakov, A. P., Dubinin, A. M., & Tuponogov, V. G. (2008). On mechanism of coal steam gasification. *Industrial Power Engineering, 4*, 40–42.

Baskakov, A. P., Volkova, J. V., & Plotnikov, N. S. (2014). Optimum chemical regeneration of the gases burnt in solid oxide fuel cells. *Journal of Engineering Physics and Thermophysics, 87*(4), 763–778. doi:10.100710891-014-1070-9

Baskakov, A. P., & Volkova, Y. V. (2013). *Physicochemical Principles of Thermal Processes*. Moscow: Teplotekhnik.

Beckmann, G., & Gilli, P. (1987). Teplovoe akkumulirovanie ehnergii [Thermal energy storage: Translation from English]. Mir. (in Russian)

Bekirov, E. A. (2011). *Avtonomnye istochniki ehlektropitaniya na baze solnechnyh batarej: uchebnoe posobie* [Autonomous power supplies based on solar panels: a tutorial]. Simferopol: Arial. (in Russian)

Belenov, A. T. Kharchenko, V.V., Rakitov, S.A., Daus, Y.V., Yudaev, I.V. (2016). The Experience of Operation of the Solar Power Plant on the Roof of the Administrative Building in the Town of Kamyshin, Volgograd Oblast. *Applied Solar Energy, 52*(2), 105–108.

Benbouzid, M. E. N., Beltran, B., Mangel, H., & Mamoune, A. (2012). A high-order sliding mode observer for sensorless control of DFIG – based wind turbine. IEEE IECON, 7670 – 7674.

Benkaciali, S., Benkaciali, S., & Gairaa, K. (2012). Comparative study of two models to estimate solar radiation on an inclined surface. *Revue des Energies Renouvelables, 15*(2), 219–228.

Bezrukih, P. P. (2008). Renewable energy sources and reliability of power supply. *Energy Policy, 3*, 3–8.

Blaesser, G., Munro, D. (1995). *Guidelines for the assessment of PV plants. Document B: Analysis and presentation of monitoring data.* Joint Research Centre of the European Commission: Ispra, Varesse, Italy. CL-NA-16-339-EN-C

Blood & Skolof. (2014). *Huge thermal plant opens as solar industry grows*. Associated Press.

BNEF (Bloomberg New Energy Finance). (2016). *Off Grid Solar Market Trends Report 2016.* Retrieved from https://about.bnef.com/white-papers/off-grid-solar-market-trends-report-2016

Bodrova, E., & Solomin, E. (2016). Overview of the problem of wind power plants utilization. *Proceedings of the Russian Scientific Conference with International Participation and the X Scientific Youth School. Renewable Energy Sources*, 16-21.

Bollen, M. H. (2002). Understanding Power Quality Problems: Voltage Sags and Interruptions. *Proceedings of IEEE.*

Borodin, M. & Vinogradov, A. (2013). Korrektirovka stoimosti potreblennoj ehlektroehnergii v zavisimosti ot ee kachestva [Adjustment of the cost of consumed electricity depending on its quality]. *Promyshlennaya ehnergetika, 7*, 12-16.

Borodin, M. (2013). Princip korrektirovki stoimosti potreblennoj ehlektroehnergii v zavisimosti ot eyo kachestva [Principle of adjustment of the cost of electricity consumed, depending on its quality]. *Naukovij visnik NUBiP Ukraïni. Seriya: Tekhnika ta energetika APK, 184*(1), 165-171.

Borodin, M., & Uryupin, N. (2018). Obosnovanie ehnergoehffektivnogo secheniya linij ehlektroperedach pri otkloneniyah napryazheniya ±5%, ±10% [Rationale energy-efficient section of the power lines with voltage deviations of ±5%, ±10%]. *Vestnik NGIEHI, 2*(81), 20–29.

Borodin, M., & Vinogradov, A. (2014). *Povyshenie ehffektivnosti funkcionirovaniya sistem ehlektrosnabzheniya posredstvom monitoringa kachestva ehlektroehnergii* [Improving of power supply system efficiency by means of power quality monitoring]. Orel, Russia: FGBOU VPO Orel GAU.

Borodin, M., & Vinogradov, A. (2018). Redakciya metodiki korrektirovki stoimosti potreblennoj ehlektroehnergii v zavisimosti ot eyo kachestva i algoritm eyo realizacii v sootvetstvii s gost na kachestvo ehlektroehnergii [A revision of the adjustment method of the electricity cost depending on its quality and the algorithm for its implementation in accordance with the standard on power quality]. *Vestnik NGIEHI*, *4*(83), 54–64.

BP. (2017). *Statistical Review of World Energy*. Retrieved from www.bp.com

Budzko, I. A., & Zul', N. M. (1990). *Jelektrosnabzhenie sel'skogo hozjajstva* [Electricity supply of agriculture]. Moscow: Agropromizdat.

Burkett, V. (2011). Global climate change implications for coastal and offshore oil and gas development. *Energy Policy*, *39*(12), 7719–7725. doi:10.1016/j.enpol.2011.09.016

Bushuev, V. V., Zaichenko, V. M., Morgunova, M. O., Soloviev, D. A., & Chernyavsky, A. A. (2017). The potential of renewable energy in the Arctic: new challenges. In *Proceedings of the V International Conference "Renewable Energy: Problems and Prospects" and the X School of Young Scientists "Actual Problems of Development of Renewable Energy Resources" named after Corr. RAS EE Shpilrayna* (vol. 1, pp. 94-99). Makhachkala: FGBUN Institute of Geothermal Problems, Dagestan Scientific Center, Russian Academy of Sciences.

Butuzov, V.V. (2009). Raschetnye znachenija intensivnosti solnechnoj radiacii dlja proektirovanija gelioustanovok [Calculated values of the solar radiation intensity for the design of solar power plants]. *Alternative Energy and Ecology*, *11*, 75–80.

Castillo-Aguilella. (2012). *Non imaging applications of volume diffractive optics*. Department of electrical and computer engineering, the University of Arizona.

Causebrook, A., Atkinson, D. J., & Jack, A. G. (2007). Fault ride through of large wind farms using series dynamic braking resistors. *IEEE Transactions on Power Systems*, *22*(3), 966–975. doi:10.1109/TPWRS.2007.901658

Chade, D., Miklis, T., & Dvorak, D. (2015). Feasibility study of wind-to-hydrogen system for Arctic remote locations - Grimsey island case study. *Renewable Energy*, *76*, 204–211. doi:10.1016/j.renene.2014.11.023

Chen, J., Chen, J., & Gong, C. (2014). On optimizing the aerodynamic load actingon the turbine shaft of PMSG – bared direct-drive wind energy conversion system. *IEEE Transactions on Industrial Electronics*, *61*(8), 4022–4031. doi:10.1109/TIE.2013.2284148

Cheremisin, N. M., Cherkashina, V. V., & Holod, A. V. (2016). Povyshenie "intellekta" ehlektricheskoj seti za schet avtomatizirovannogo monitoringa parametrov vozdushnyh linij [The "intelligence" increase of an electric network due to automated monitoring of the of overhead line parameters]. *Ehnergetika ta komp'yuterno-integrovani tekhnologiï v APK*, *1*(4), 19-22.

Chernyshev, Y.A., Shinyakov, A.I., & Gordeev, K.G. (1991). *Autonomous power supply system*. Patent of the Russian Federation 2035109 IPC H02J 7/35.

Choi, D. W., Byun, S. I., & Cho, Y. H. (2014). A study on the maximum power control method of switched reluctance generator for wind turbine. IEEE Trans. Magnetics, 50(1).

Cookson, T., Lang, N., & Thornton, E. (2008). Adjustable speed drives applied to large AC induction motor and pump systems. In *Proceedings of the Twenty-fourth International Pump Users Symposium* (pp. 75-80). Houston, TX: GA Valves. Retrieved January 20, 2016, from http://www.gavalves.co.uk

Da Rosa, A. (2010). Vozobnovlyaemye istochniki ehnergii. Fiziko-tekhnicheskie osnovy. Uchebnoe posobie [Renewable energy sources. Physical and technical basis. Tutorial]. Intellect. Ed. house of Moscow Power Engineering Institute. (in Russian)

DanSolar. (n.d.). *Danish solar energy supplier.* Retrieved from http://dansolar.dk/en/

Daus, J. V., Kharchenko, V.V., & Yudaev, I.V. (2018). Solar Radiation Intensity Data as Basis for Predicting Functioning Modes of Solar Power Plants. In Handbook of Research on Renewable Energy and Electric Resources for Sustainable Rural Development. Hershey, PA: IGI Global.

Daus, Ju. V. (2016). Opredelenie optimal'nogo ugla naklona panelej solnechnyh jenergoustanovok v uslovijah Rostovskoj oblasti [Determination of the optimal inclination angle of solar energy panels under the conditions of the Rostov region]. *Proceedings of the Higher Educational Institutions. The North Caucasus region. Series: Engineering, 2*(190), 67-71.

Daus, Yu. V., & Yudaev, I. V. (2016). Designing of Software for Determining Optimal Tilt Angle of Photovoltaic Panels. In *Proceedings of International Conference on Education, Management and Applied Social Science (EMASS2016).* Beijing: DEStech Publications, Inc.

Daus, Y. V., Kharchenko, V. V., & Yudaev, I. V. (2016a). Evaluation of Solar Radiation Intensity for the Territory of the Southern Federal District of Russia when Designing Microgrids Based on Renewable Energy Sources. *Applied Solar, 52*(2), 124–129.

Daus, Y. V., Kharchenko, V. V., & Yudaev, I. V. (2016b). *Ocenka potenciala solnechnoj jenergii v zadannoj tochke Juzhnogo federal'nogo okruga* [Estimation of solar energy potential at a given point in the Southern Federal District]. FGBOU VO "Donskoj GAU".

Daus, Y. V., Kharchenko, V. V., & Yudaev, I. V. (2016c). *Poisk optimal'nogo ugla naklona prijomnoj ploshhadki otnositel'no gorizonta v zadannoj tochke Juzhnogo federal'nogo okruga* [Search for the optimal inclination angle of the receiving area relative to the horizon at a given point in the Southern Federal District]. FGBOU VO "Donskoj GAU".

Daus, Yu. V., & Kharchenko, V. V. (2018). Evaluating the applicability of data on total solar-radiation intensity derived from various sources of actinometric information. *Applied Solar Energy, 54*(1), 71–76. doi:10.3103/S0003701X1801005X

Daus, Yu. V., Yudaev, I. V., & Stepanchuk, G. V. (2018). Reducing the costs of paying for consumed electric energy by utilizing solar energy. *Applied Solar Energy, 54*(2), 137–141. doi:10.3103/S0003701X18020056

David, W. G., Eduard, M., Tian, T., & Mackay, M., Weisheng, W. (2016). *Comparison of Standards and Technical Requirements of Grid Connected Wind Power Plants in China and the United States.* Technical Report NREL.

Deryugina, G.V., & Tyagunov, M.G. (2012). Hybrid energy complexes based on renewable energy sources. *Bulletin of the Kyrgyz-Russian Slavic University,* 11-17.

Deryugina, G.V., Karpov, N.D., & Shestopalova, T.A. (2017). Investigation of factors and mathematical models influencing the design perfomance of energy efficiency of wind-diesel complexes. *Bulletin of the Kyrgyz-Russian Slavic University,* 44-48.

Deryugina, G.V., Malinin, N.K., & Pugachev, R.V. (2012). *Basic characteristics of the wind.* Publishing house MPEI.

Deryugina, G.V., Malinin, N.K., & Pugachev, R.V. (2016). *Wind Power.* Moscow: Publishing house MPEI.

Diagnosticheskie resheniya v ehnergetike . (2017). [Diagnostic solutions in power engineering]. Available from: http://dimrus.ru/dilin.html

Díaz-Cuevas, P., Camarillo-Naranjo, J., & Pérez-Alcántara, J. (2018). Relational spatial database and multi-criteria decision methods for selecting optimum locations for photovoltaic power plants in the province of Seville (southern Spain). *Clean Technologies and Environmental Policy*, 1–14.

Dittrich, A., & Stoev, A. (2003). Grid voltage fault proof doubly-fed induction generator system. *Proceedings of 10th European Conference on Power Electronics and Applications.*

Dobrusin, L. (2004). O sovershenstvovanii tekhnicheskogo regulirovaniya kachestva ehlektricheskoj ehnergii v sistemah ehlektrosnabzheniya obshchego naznacheniya [About improvement of technical regulation of quality of electric energy in General-purpose power supply systems]. *Vesti v ehlektroehnergetike, 1*, 44-50.

DualSun. (n.d.). Retrieved from https://dualsun.com/en/

Dubinin, A.M., Shcheklein, S.E., Tuponogov, V.G., & Ershov, M.I. (2017). Mini CHP based on the electrochemical generator and impeded fluidized bed reactor for methane steam reforming. *Alternative Energy and Ecology*, (19-21), 95-105.

Dubinin, A. M., & Shcheklein, S. E. (2015). Coal mini CHPP plants on the basis of a gas generator and electrochemical generator. *Alternative Energy and Ecology*, *7-9*, 60–74.

Dubrovsky, V., & Viestur, U. (1988). *Methane Fermentation of Agricultural Wastes*. Riga: Zinatne. (in Russian)

Duffy, J., & Beckman, U. A. (2013). Osnovy solnechnoj teploehnergetiki [Fundamentals of solar thermal power]. Dolgoprudny: Intelligence. (in Russian)

Earth System Research Labarotory. (2016). *The NCEP / NCAR Reanalysis Project.* Retrieved from http://www.cdc.noaa.gov/cdc/reanalysis

EC JRC. (2011). *Renewable energies in Africa – current knowledge. Scientific and technical Report. The European Commission Joint Research Center*. Luxembourg: Publications Office of the European Union.

Elistratov, V. V., & Denisov, R. S. (2016). Estimation of energy losses in the selection of wind turbine equipment for severe climatic conditions. In *Polar Mechanics Materials of the Third International Scientific Conference* (pp. 896-907). Vladivostok: Far East. Fed. un-t.

Elistratov, V. V. (2017). *Problems and solutions to create effective energy supply systems for consumers in the Arctic with the use of innovative renewable energy technologies. In Arctic-2017 (Arctic: offshore projects and sustainable development of regions) II International Conference* (pp. 1–41). Moscow: System Consulting. Retrieved from https://energy.s-kon.ru/arktika-2017/

Elistratov, V. V., & Aronova, E. S. (2009). Solar photo energy technologies for electric power consumers. *Applied Solar Energy*, *45*(3), 143–147. doi:10.3103/S0003701X09030025

Energy Efficiency and Renewable Energy Consulting. (2016). *Bioenergetics.* Retrieved from http://www.energypartner.kz

Energy Strategy of Russia for the period up to 2030. (2010). *Environmental Consulting.* Author.

Energy-efficient Russia. (2016). *Kazakhstan alternative.* Retrieved from http://energohelp.net

EnergyPartner LLP. (2016). *Solar energy.* Retrieved from http://www.energypartner.kz

ERA Interim Reanalysis System. (n.d.). Retrieved from http://www.era

Essefi, R. M., Souissi, M., & Abdallah, H. H. (2014). Maximum Power Point Tracking Control Using Neural Networks for Stand–Alone Photovoltaic Systems. *International Journal of Modern Nonlinear Theory and Application*, *3*(03), 53–65. doi:10.4236/ijmnta.2014.33008

Fangxing, L., Wei, Q., Hongbin, S., Hui, W., Jianhui, W., Yan, X., ... Pei, Z. (2010). Smart transmission grid: Vision and framework. *IEEE Transactions on Smart Grid*, *1*(2), 168–177. doi:10.1109/TSG.2010.2053726

Farrell, E., Fuhrer, E., Ryan, D., Andersson, F., Huttl, R., & Piussi, P. (2000). European forest ecosystems: Building the future on the legacy of the past. *Forest Ecology and Management*, *132*(1), 5–20. doi:10.1016/S0378-1127(00)00375-3

Fay, G., & Udovyk, N. (2013). Factors influencing success of wind-diesel hybrid systems in remote Alaska communities: Results of an informal survey. *Renewable Energy*, *57*, 554–557. doi:10.1016/j.renene.2013.02.021

Federal'nyj zakon Rossijskoj Federacii N°35 FZ "Ob Ehlektroehnergetike ". (n.d.). [Federal law of the Russian Federation N°35 "About Electric Power"]. Available from: http://ivo.garant.ru/#/document/185656/paragraph/539078:1

Ferroukhi, R. (2016). *Renewable Energy Benefits: Measuring. The Economics*. Abu Dhabi, UAE: IRENA.

Gabderakhmanova, T. S., & Shakun, V. P. (2017). Criteria for the prospects of the territories of the Arctic zone of the Russian Federation for the introduction of solar and wind generation facilities. In *Proceedings of the V International Conference "Renewable Energy: Problems and Prospects" and the X School of Young Scientists "Actual Problems of Development of Renewable Energy Resources" named after Corr. Of the Russian Academy of Sciences EE Shpilrayna* (vol. 1, pp. 114-117). Makhachkala: IP Ovchinnikov (ALEF). Retrieved from http://www.ipgdncran.ru/confayer

Gabderakhmanova, T. S., & Tarasenko, A. B. (2016). On the possibility of using solar energy in the charging infrastructure of electric transport. In *Construction - the formation of the environment of life: a collection of materials of the XIX International Interuniversity Scientific and Practical Conference of students, undergraduates, graduate students and young scientists* (pp. 1025-1027). Moscow: NIU MSSU. Retrieved from http://elibrary.ru/item.asp?id=26275764

Gabderakhmanova, T. S., Kiseleva, S. V., Popel, O. S., & Tarasenko, A. B. (2016). Some aspects of the development of renewable energy in the Arctic zone of the Russian Federation. *Alternative Energy and Ecology*, (19-20), 41-53. Retrieved from http://elibrary.ru/item.asp?id=26935418

Gabderakhmanova, T. S., Director, L. B., Popel, O. S., & Tarasenko, A. B. (2015). Comparative analysis of electrochemical energy storage devices. *Alternative Energy and Ecology*, *23*, 184–195.

Gabderakhmanova, T. S., & Shakun, V. P. (2016). Geoinformation mapping of resources and renewable energy facilities in the Arctic zone of the Russian Federation. In S. V. Kiseleva (Ed.), *Renewable Energy Sources: Proceedings of the All-Russian Scientific Conference with International Participation and the Xth Scientific Youth School: A Compilation* (pp. 348–353). Moscow: University Book. Retrieved from https://lomonosov-msu.ru/eng/event/3654/

Garcia-Hernandez, A., Wilcox, M., & Moore, T. (2010). Hydraulic modelling and simulation of pumping systems. *Proceedings of the Twenty-sixth International Pump Users Symposium*, 81-88.

Gevorgian, V., Touryan, K., Bezrukikh, P., Bezrukikh, P., Jr., & Karghiev, V. (1999). *Wind-Diesel Hybrid Systems for Russia's Northern Territories*. Paper presented at Windpower '99, Burlington, VT.

Gevorkian, P. (2016). *Solar power generation Problems, Solutions, and Monitoring*. Cambridge University Press. doi:10.1017/CBO9781316341414

Gheorghe, S., Tanasa, C., Ene, S., & Mihaescu, M. (2005). Power quality, Energy efficiency and the performance in electricity distribution and supply companies. In *Proceedings of the 18th International Conference and Exhibition on Electricity Distribution: CIRED*. Turin, Italy: IET.

Gidrometeoizdat. (1997). *Scientific and Applied Guide to Russia's Climate (Arctic Region): Solar Radiation*. St. Petersburg: Author.

Gigavat. (n.d.). *All about power plants*. Retrieved from http://www.gigavat.com/

GIS. (2018). *Renewable Energy Sources in Russia*. Retrieving January 15, 2018, from http://gisre.ru/

Gmurman, V. (1999). *Teoriya veroyatnostej i matematicheskaya statistika. Ucheb. Posobie dlya vuzov. Izd. 7-e ster* [Probability Theory and mathematical statistics. Studies'. The manual for high schools. Ed. 7's erased]. Moscow, Russia: Vyssh. shk.

Golikov, I., & Vinogradov, A. (2017). *Adaptivnoe avtomaticheskoe regulirovanie napryazheniya v sel'skih ehlektricheskih setyah 0,38 kV* [Adaptive automatic voltage regulation in rural power networks of 0.38 kV]. Orel, Russia: FGBOU VO Orel GAU.

Gorbunova, T., & Gorbunov, R. (2016). Study of resource opportunities and land-use restrictions on the use of solar energy in the example of the southeastern Crimea. *Proceedings of the Russian Scientific Conference with International Participation and the X Scientific Youth School.*

GOST 28977 - 91 (IEC 904-187). (2004). *Photovoltaic devices. Part 1. Measurements of photoelectric volt-ampere characteristics*. IPK Publishing Standards.

GOST 51597 - 2000. (2000). *Unconventional power engineering. Modules solar photovoltaic. Types and basic parameters*. IPK Publishing Standards.

Goulden, M., Bedwell, B., Rennick-Egglestone, S., Rodden, T., & Spence, A. (2014). Smart grids, smart users? The role of the user in demand side management Energy. *Energy Research & Social Science*, *2*, 21–29. doi:10.1016/j.erss.2014.04.008

Grazhdanskij kodeks RF. (2000). [The civil code of the Russian Federation]. Moscow, Russia: Publishing house NORM.

Gridasov, M. V., Kiseleva, S. V., Nefedova, L. V., Popel, O. S., & Frid, S. E. (2011). Development of the Geoinformation System "Renewable Sources of Russia": Statement of the Problem and Choice of Solution Methods. *Thermal Engineering*, *58*(11), 924–931. doi:10.1134/S0040601511110073

Grigorev, V. A. (2003). Razrabotka akkumulyatorov teploty s zernistym teplonositelem i metody ih rascheta na osnove matematicheskogo modelirovaniya [Development of heat accumulators with a granular coolant and methods for calculating them on the basis of mathematical modeling]. RSL. (in Russian)

Grinevich, G. (1952). *Experience in developing elements of a small wind energy cadastre in Central Asia and Kazakhstan*. Publishing House of Academy of Sciences.

Grishko, V. I., & Misjura, L. I. (Eds.). (2007). *Spravochnik po resursam vozobnovljaemyh istochnikov jenergii Rossii i mestnym vidam topliva. Pokazateli po territorijam: spravochnoe posobie* [A guide to renewable energy resources in Russia and local fuels. Indicators for territories: reference book]. Moscow: IAC Jenergija.

Gsänger, S., & Denisov, R. (2017). *Perspectives of the wind energy market in Russia*. Retrieved from http://library.fes.de/pdf-files

Guba, A. (2006). Digital sensors of illumination MAX44007 and MAX44009 of Maxim Company. *Electronic Components*, *6*, 11–13.

Guemard, C. (2003). Direct solar transmittance and irradiance prediction with broadband models. Part I: Detailed theoretical performance assessment. *Solar Energy*, *74*(5), 355–379. doi:10.1016/S0038-092X(03)00195-6

Gupta, S., & Chandrashekar, M. (1995). A unified approach to modeling photovoltaic powered systems. *Solar Energy, 55*(4), 267–285. doi:10.1016/0038-092X(95)00033-N

Halinen, M., Thomann, O., & Kiviaho, J. (2012). Effect of anode off-gas recycling on reforming of natural gas for solid oxide fuel cell system. *Fuel Cells (Weinheim), 12*(5), 754–760. doi:10.1002/fuce.201200047

Hansen, A. D., Michalke, G., Sorense, P., Lund, T., & Iov, F. (2007). Co-ordinated voltage control of DFIG wind turbines in uninterrupted operation during grid faults. *Wind Energy (Chichester, England), 10*(1), 51–68. doi:10.1002/we.207

Hartungi, R., & Jiang, L. (2010). Investigation of Power Quality in Health Care Facility. *Proceedings of International Conference on Renewable Energies and Power Quality (ICREPQ'10).* 10.24084/repqj08.555

Hoesen, J., & Letendre, S. (2010). Evaluating potential renewable energy resources in Poultney, Vermont: A GIS-based approach to supporting rural community energy planning. *Renewable Energy, 35*(9), 2114–2122.

Hosseini, S. M., Aslani, A., Hafeznia, H., & Naaranoja, M. (2017). Analysis of Energy System in Sweden Based on Time series Forecasting and Regression Analysis. *International Journal of Energy Optimization and Engineering, 6*(3), 97–114. doi:10.4018/IJEOE.2017070105

Hrebenikov, V.V., Petrenko, A.V. (2009). *Mahnitoelektrychnyi liniinyi henerator* [A Magnetic electrical linear generator]. Patent of Ukraine, IPC (2009) H02K 35/00, N° 41883; publ. 10.06.2009. Bul. N° 11.

Hu, J., Sun, D., He, Y., & Zhao, R. (2006). Modeling and control of DFIG energy generation system under grid voltage dip. *Dianli Xitong Zidonghua, 30*, 21–26.

IEA. (2014). *Africa Energy Outlook. A focus on the energy prospects in sub-Saharan Africa.* Paris, France: International Energy Agency.

IEA. (2017). *World Energy Outlook 2017. From Poverty to Prosperity. Special Report.* Paris, France: International Energy Agency.

International Energy Agency. (2014). *Energy Technology Perspectives. Executive Summary.* Retrieved from http://www.iea.org/publications/freepublications/publication/name-51003-en.html

Interstate Council for Standardization, Metrology and Certification. (n.d.). *GOST 32144-2013 EHlektricheskaya ehnergiya. Sovmestimost' tekhnicheskih sredstv ehlektromagnitnaya. Normy kachestva ehlektricheskoj ehnergii v sistemah ehlektrosnabzheniya obshchego naznacheniya* [GOST 32144- 2013 Electric energy. Electromagnetic compatibility of technical means. Norms for power quality in general-purpose power supply systems]. Moscow, Russia: *Standartinform.*

Ipakchi, A., & Albuyeh, F. (2009). Grid of the future. *IEEE Power & Energy Magazine, 7*(2), 52–62. doi:10.1109/MPE.2008.931384

IRENA. (2016). *Renewable energy benefits: measuring the economics.* Abu Dhabi.

IRENA. (2016). *Unlocking Renewable Energy Investment: The Role of Risk Mitigation and Structured Finance.* Abu Dhabi: International Renewable Energy Agency.

IRENA. (2017). *REmap 2030 Renewable Energy Prospects for Russian Federation.* Working paper, IRENA.

IRENA. (2017a). *REthinking Energy 2017: Accelerating the global energy transformation.* Abu Dhabi: International Renewable Energy Agency.

IRENA. (2017b). *Electricity storage and renewables: costs and markets to 2030.* Abu Dhabi: International Renewable Energy Agency.

IRENA. (2017c). *Accelerating Off-grid Renewable Energy: Key Findings and Recommendations from IOREC (2016).* Abu Dhabi: International Renewable Energy Agency.

Irwin, L. (2010). Asset management benefits from a wide area power quality monitoring system. In *Proceedings of the 23rd International Conference on Electricity Distribution: CIRED.* Lyon, Italy: IET.

Iurevych, Gubin, & Dudeck. (2013). Modelling of a Hybrid Solar Panel with Solar Concentration. *Electrical Engineering Research, 1*(2).

Iurevych, O., Gubin, S., & Dudeck, M. (2012). Combined Receiver of Solar Radiation with Holographic Planar Concentrator. *IOP Conference Series: Materials Science and Engineering Physics*, 29. 10.1088/1757-899X/29/1/012016

Jarymbash, D.S., & Daus, Ju.V. (2014). Osobennosti identifikacii intensivnosti solnechnoj radiacii pri proektirovanii solnechnyh jelektrostancij [Features of identifying the solar radiation intensity in the designing solar power plants]. *Power and Electrical Engineering, 1,* 74-78.

Jebaseeli, E. A. E., & Susitra, D. (2010). *Performance Analisis of various configurations of switched reluctance machine for wind energy applications.* IEEE RSTSCC.

Kamalov, T.S. (2014). *Chastotno-reguliruyemyy elektroprivod nasosnykh stantsiy sistem mashinnogo orosheniya* [Frequency-regulated electric drive of pumping stations of machine irrigation systems]. Tashkent, Fan.

Karapetyants, M. K. (1975). *Chemical thermodynamics* (3rd ed.). Moscow: Khimiya.

Kartashev, I. (2001). *Upravlenie kachestvom ehlektroehnergii* [Power Quality in Power Supply Systems]. Moscow, Russia: Izdatel'skij dom MEHI.

Kartashev, I., Tul'skij, V., SHamonov, R., SHarov, Y., & Vorob'ev, A. (2006). *Upravlenie kachestvom ehlektroehnergii* [Power quality management]. Moscow, Russia.: Izdatel'skij dom MEHI.

Katiraei, F., & Abbey, C. (2007). Diesel plant sizing and performance analysis of a remote wind-diesel microgrid. *IEEE Power Engineering Society General Meeting.* 10.1109/PES.2007.386275

Kelion, L. (2017). Record-sized data centre planned inside Arctic Circle. *BBC News.* Retrieved July 13, 2018, from https://www.bbc.com/news/technology-40922048

Kharchenko, V. V., Nikitin, B. A., & Tikhonov, P. V. (2010). Estimation and forecasting of PV cells and modules parameters on the basis of the analysis of interaction of a sunlight with a solar cell material. *Proceedings of 4th International Conference TAE 2010, Trends in Agricultural Engeneering*, 307-310.

Khashimov, A.A. & Abidov, K.G. (2012). *Energoeffektivnyye sposoby samozapuska elektroprivodov nasosnykh stantsiy* [Energy-efficient ways to self-start electric drives of pumping stations]. Tashkent, Fan va texnologiya.

Khokhlov, A.V., Khokhlov, V.A., Titova, J.O. (2014). *Rezhimy raboty nasosnykh stantsiy Dzhizakskogo kaskada* [Operating modes of Djizak pumping stations stage]. Tashkent, Fan va texnologiya.

Khokhlov, V.A., Khokhlov, A.V. & Titova, J.O. (2012). *Neustanovivshiyesya rezhimy raboty nasosnykh stantsiy s dlinnymi truboprovodami* [Unsteady operating modes of pumping stations with long pipelines]. Tashkent, Fan va texnologiya.

Khokhlov, A. V., Khokhlov, V. A., & Titova, J. O. (2015). *Rezhimy raboty nasosnykh stantsiy Karshinskogo kaskada* [Operating modes of Karshi pumping stations stage]. Tashkent: Navruz.

Kiseleva, S., Ermolenko, G., & Popel, O. (2015). *Atlas of renewable energy resources in Russia.* Moscow: D.I. Mendeleyev RCTU.

Kitaeva, M. V. (2015). Solar tracking systems for solar energy. *Collection XV International Scientific and Practical Conference "Modern technology and technology"*, 54-55.

Kokorin, A. O., Karelin, D. V., & Stetsenko, A. V. (2008). *Impact of climate change on the Russian Arctic: analysis and ways to solve the problem*. WWF Russia. Retrieved from http://narfu.ru/upload/medialibrary/2af/vozdeistvie_na_arcticu.pdf

Komarova, A. A., & Vinogradov, A. V. (2017). Analiz dinamiki sostoyaniya liniy elektroperedachi 6-20 kV [Analysis of the state dynamics of 6-20 kV transmission lines]. In Energosberezheniye I effektivnost' v tekhnicheskikh sistemakh: Materialy IV Mezhdunarodnoy nauchno-tekhnicheskoy konferentsii studentov, molodykh uchenykh I spetsialistov. Tambov, Russia: Tambov State Technical University.

Korovin, N. A. (2005). *Fuel Cells and Electrochemical Power Installations*. Moscow: MPEI.

Kostikov, I. (2011). Sistema monitoringa CAT-1 — povyshenie propusknoj sposobnosti i nadezhnosti v LEHP [CAT-1 monitoring system - increase of capacity and reliability in power lines]. *EHnergetika, 3*(38). Retrieved 26.04.2017, from http://kazenergy.kz/arhiv/38/44.htm

Kovalenko, E.A., & Tyagunov, M.G. (2014). The use of hybrid energy complexes based on renewable energy sources in distributed energy. *Energetik (journal)*, 25-27.

Kovalenko, A. S., Morgunova, M. O., & Gribkovskaia, V. (2018). Infrastructural Synergy of the Northern Sea Route in the international context. *Jenergeticheskaja Politika*, *4*, 57–67.

Kozyrskyi, V.V., Kaplun, V.V., Petrenko, A.V. (2007). *Vitroustanovka* [A Wind system]. Patent of Ukraine, IPC (2006) F03D 11/00, N° 23461; publ. 25.05.2007. Bul. N° 7.

Krasovsky, N. (1935). Wind power resources of the USSR and prospects of their use. Wind Power Resources, 1(3).

Kudryashov, V.S., Elman, V.O., & Nesterishin, M.V. (2009). *The power supply system of the spacecraft*. Patent of the Russian Federation 2396666 H02J7/35.

Kumar, A., Sah, B., Singh, A. R., Deng, Y., He, X., Kumar, P., & Bansal, R. C. (2017). A review of multi criteria decision making (MCDM) towards sustainable renewable energy development. *Renewable & Sustainable Energy Reviews*, *69*, 596–609. doi:10.1016/j.rser.2016.11.191

Kuzmichev, A. V., Malyshev, V. V., & Tikhomirov, D. A. (2010). Efficiency of combined pasteurization of milk by UV and IR irradiation. *Light & Engineering*, *5*, 6–9.

Kuznetsov, P.N., & Borisov, A.A. (2017). *A method for selecting electric energy from photovoltaic converter batteries*. Patent of the Russian Federation 2634590.

Kuznetsov, P.N., & Safonov, V.A. (2016). Increase of the efficiency of the photovoltaic station. *Energy Security and Energy Saving, 3*, 26-30.

Kuznetsov, P.N., Yuferev, L.Y. (2017). Increase of efficiency of work of photoelectric converters at consecutive connection. *Bulletin of Agrarian Science of Don, 37*, 15-25.

Kuznetsov, P. N., & Avdeev, D. S. (2017) Development of a laboratory model of the device for coordinating photovoltaic modules. *Abstracts of the International Scientific and Technical Conference "Electrotechnical Complexes and Systems"*, 45-46.

Larsen, P., Goldsmith, S., Smith, O., & Wilson, M. (2008). Estimating future costs for Alaska public infrastructure at risk from climate change. *Environmental Change, 18*(3), 442–457. Retrieved from http://www.sciencedirect.com/science/article/pii/S0959378008000216

Larsen, T. (2012). *Water hammer in pumped sewer mains*. Aalborg, Denmark: University of Aalborg, Department of Civil Engineering.

Lebsir, A., Benbouzid, M. E. N., Bentonnsi, A., & Salah, I. B. (2016). Switched reluctance generator models for wind turbine systems. *ELECTROMOTION*, *23*, 3–9.

Lee, A. H. I., Chen, H. H., & Kang, H. Y. (2009). Multi-criteria decision making on strategic selection of wind farms. *Renewable Energy*, *34*(1), 120–126. doi:10.1016/j.renene.2008.04.013

Leshchinskaya, T. B., & Magadeev, E. H. V. (2008). *Metodika vybora optimal'nogo varianta povysheniya nadezhnosti ehlektrosnabzheniya sel'skohozyajstvennyh potrebitelej* [The method of choosing the optimal option for increasing the power supply reliability of agricultural consumers]. Moscow, Russia: FGOU VPO MGAU.

Leshchinskaya, T. B., & Polyanina, I. N. (2005). *Povyshenie ehffektivnosti funkcionirovaniya raspredelitel'nyh setej rajonov s maloj plotnost'yu ehlektricheskih nagruzok (na primere Joshkar-Olinskij setej)* [Efficiency increase of distributive network functioning in the areas with low density of electric loads (on an example of the Yoshkar-Ola network)]. Moscow, Russia: Agrokonsalt.

Leshchinskaya, T., & Naumov, I. (2008). *EHlektrosnabzhenie sel'skogo hozyajstva* [Power supply of agriculture]. Moscow, Russia: Colossus.

Leznov, B. S. (2006). *Energosberezhenie i reguliruemyy privod v nasosnykh i vozdushnykh ustanovkakh* [Energy saving and adjustable drive for pump and blower installations]. Moscow: Energoatomizdat.

Li, Ch., Ma, Sh., Liu, H., Wu, H., Gao, P., & Yang, F. (2016). Research on scheme of pumping energy-saving based on computer control technology. *MATEC Web of Conferences*. Retrieved March 21, 2017, from http://www.matec-conferences.org

Liang, J., Qiao, W., & Harley, R. G. (2010). Feed-forward transient current control for low-voltage ride-through enhancement of DFIG wind turbines. *IEEE Transactions on Energy Conversion*, *25*(3), 836–843. doi:10.1109/TEC.2010.2048033

Li, H., Huang, W., Huang, F., Hu, P., & Chen, Z. (2013). *Optical analysis and optimization of parabolic dish solar concentrator with a cavity receiver*. Elsevier.

Lima, K., Luna, A., Rodriguez, P., Watanabe, E., Teodorescu, R., & Blaabjerg, F. (2008). *Doubly Fed Induction Generator Control Under Voltage Sags*. Paper presented at the IEEE Energy2030 Conference, Atlanta, GA. 10.1109/ENERGY.2008.4781064

Lisbona, P., & Romeo, L. M. (2008). Enhanced coal gasification heated by unmixed combustion integrated with an hybrid system of SOFC/GT. *International Journal of Hydrogen Energy*, *33*(20), 5755–5764. doi:10.1016/j.ijhydene.2008.06.031

Liu, B. Y. H., & Jordan, R. C. (1961). Daily insolation on surfaces tilted towards the equator. *ASHRAE Journal*, *3*(10), 53–63.

Magomedov, M. M. (2000). Kompleksnaya metodologiya razrabotki fazoperekhodnyh teploakkumuliruyushchih materialov na osnove mnogokomponentnyh system [Complex methodology of development of phase-transfer heat-accumulating materials on the basis of multicomponent systems]. Moscow: RSL. (in Russian)

Main Directorate of Pumping Stations Energy and Communications of the Ministry of Ministry of Agriculture and Water Resources of the Republic of Uzbekistan. (2006). *Technical Report*. Tashkent.

Malczewski, J. (1999). *GIS and multicriteria decision analysis*. New York: John Wiley & Sons Inc.

Mandatova, P., Massimiano, M., Verreth, D., & Gonzalez, C. (2014). Network tariff structure for a smart energy system. In *Proceedings of the CIRED Workshop*. Rome, Italy: IET.

Maudi, B. (2018). A cost optimal resolution for Sub-Saharan Africa powered by 100% renewables in 2030. *Renewable & Sustainable Energy Reviews, 92*, 440–457. doi:10.1016/j.rser.2018.04.110

MCPFE. (2002). *Improved Pan-European Indicators for Sustainable ForestManagement.* Paper presented at the MCPFE Expert Level Meeting, Vienna, Austria.

MEASNET. (2016). *Evaluation of site-specific wind conditions. Version 2.* Author.

Mentis, D., Howells, M., Rogner, H., Korkovelos, A., Arderne, C., Zepeda, E., ... Scholtz, E. (2017). Lighting the World: The first application of an open source, spatial electrification tool (OnSSET) on Sub-Saharan Africa. *Environmental Research Letters, 12*(8), 085003. doi:10.1088/1748-9326/aa7b29

Momoh, J. A. (2009). Smart grid design for efficient and flexible power networks operation and control. In *Power Systems Conference and Exposition, PSCE '09*. Seattle, WA: IEEE. doi:10.1109/PSCE.2009.4840074

Moner-Girona, M., Solano-Peralta, M., Lazopoulou, M., Ackom, E. K., Vallve, X., & Szabó, S. (2018). Electrification of Sub-Saharan Africa through PV/hybrid mini-grids: Reducing the gap between current business models and on-site experience. *Renewable & Sustainable Energy Reviews, 91*, 1148–1161. doi:10.1016/j.rser.2018.04.018

Morgunova, M. O., & Soloviev, D. A. (2016). Energy supply of the Russian Arctic: hydrocarbons or renewable energy sources? *Jenergeticheskaja Politika*, (5), 44-51. Retrieved from http://www.energystrategy.ru/editions/EP.htm

Morgunova, M. O. (2015). *Arctic Offshore Hydrocarbon Resource Development: Past, Present and Vision of the Future.* KTH Royal Institute of Technology.

Morgunova, M. O., & Solovjov, D. A. (2015). Decentralized power supply based on local and renewable energy sources: a case of Russian Arctic. In *International Geographical Union Regional Conference: Geography, Culture and Society for Our Future Earth*. Moscow, Russia: IGU.

Morgunova, M. O., & Solovyev, D. A. (2017). Challenges to overcome: Energy supply for remote consumers in the Russian Arctic. *Journal of Physics: Conference Series, 891*, 012157. doi:10.1088/1742-6596/891/1/012157

Morren, J., & de Haan, H. (2005). Ridethrough of wind turbines with doubly-fed induction generator during a voltage dip. *IEEE Transactions on Energy Conversion, 20*(2), 435–441. doi:10.1109/TEC.2005.845526

Moscow, V. N. B. (Ed.). (1956). *Thermal-physical properties of substances.* Gosenergoizdat.

Mueller, M. A. (2005). *Design and performance of a 20 kW, 100 rpm, switched reluctance generator for a direct drive wind energy converter.* IEEE IEMDC. doi:10.1109/IEMDC.2005.195701

Muhamedyev, R., Kiseleva, S., Gopejenko, V.I., Amirgaliyev, Y., Muhamedyeva, E., Gopejenko, A.V., & Abdoldina, F. (n.d.). Visualization of the renewable energy resources. *Lecture Notes in Computer Science, 9768*, 218.

Muhamedyev, R. (2015). Machine learning methods: An overview. *Computer Modelling and Network Technologies, 19*, 14–29.

Mukhamediev, R., & Mustakaev, R. (2016). *Aggregation of heterogenius data in the decision support system for the use of renewable energy sources.* International Scientific Conference "Informatics and Applied Mathematics", Almaty, Kazakhstan.

Munts, V. A., Volkova, Y. V., Plotnikov, N. S., Dubinin, A. M., Tuponogov, V. G., & Chernishev, V. A. (2015). Studying the characteristics of a 5 kW power installation on solid-oxide fuel cells with steam reforming of natural gas. *Thermal*, *62*(11), 779–784. doi:10.1134/S004060151511004X

Mustafina, R. (2010). The issue of energy security in Kazakhstan regions. *Vestnik PSU*, *2*, 110–116.

NASA Atmospheric Science Data Center. (n.d.). Retrieved from https://eosweb.larc.nasa.gov/sse

Nefedova, L.V., & Soloviev, A.A. (2015). Analysis of risks of development of renewable energy sources in Russia. *Problems of Risk Analysis, 12*(6), 56-63.

Nefedova, L.V., Solovyev, A.A., Shilova, L.A., & Solovyev, D.A. (2016). Risk factors during construction of power plants using renewable energy sources. *Vestnik MGSU*, (12), 79–90. doi:10.22227/1997-0935.2016.12.79-90

Nefyodov, V. I. (2002). *Fundamentals of Radio Electronics and Communications*. Moscow: Vysshaya Shkola. (in Russian)

Nikolaev, V. G., Ganaga, S. V., & Kudryashov, Yu. I. (2008). National Cadastre of wind power resources of Russia and methodical bases of their definition. Moscow: Publishing "ATMOGRAF".

Nilsen, J. (2015). *Her spyler helikopteret bort et tykt lag med is - Tu.no*. Retrieved June 27, 2017, from https://www.tu.no/artikler/her-spyler-helikopteret-bort-et-tykt-lag-med-is/222975

Novakovskij, B. A., Prasolova, A. I., Kiseleva, S. V., & Rafikova, Yu. Yu. (2011). Geoinformacionnye sistemy po vozobnovljaemoj jenergetike [Geoinformation systems for renewable energy]. In *Proceedings of the International Conference INTERKARTO / INTERGIS. Belokuriha: Denpasar.* Retrieved from Official website "Quantum", at: http://npp-kvant.ru

OAO "Rosseti". (2013). Polozheniye OAO «Rosseti» o yedinoy tekhnicheskoy politike v elektrosetevom komplekse [Position of Rosset JSC on a unified technical policy in the power grid complex]. Moscow: OAO «Rosseti».

Okedu, K. E., Muyeen, S. M., Takahashi, R., & Tamura, J. (2010). Comparative study between two protection schemes for DFIG-based wind generator. *IEEE Transactions on Power Energy*, *12*, 138–152.

Okedu, K. E., Muyeen, S. M., Takahashi, R., & Tamura, J. (2012). Wind Farms Fault Ride Through Using DFIG With New Protection Scheme. *IEEE Transactions on Sustainable Energy*, *3*(2), 242–254. doi:10.1109/TSTE.2011.2175756

Orlin, A. S., & Kruglov, M. G. (1983). *Internal Combustion Engines: Theory of Reciprocating and Compound Engines*. Moscow: Mashinostroenie.

Pacific Northwest Laboratory. (1987). Wind Energy Resource Atlas of the United States. Richland, WA: 99352.DOE/CH 10094– 4, Author.

Park, W., & Sung, B. C. (2010). The effect of SFCL on electric power grid with wind turbine generation system. *IEEE Transactions on Applied Superconductivity*, *20*(3), 1177–1181. doi:10.1109/TASC.2010.2040918

Perova, M. (2007). *EHkonomicheskie problemy i perspektivy kachestvennogo ehlektrosnabzheniya sel'skohozyajstvennyh potrebitelej v Rossii* [Economic problems and prospects of qualitative power supply of agricultural consumers in Russia]. Moscow, Russia: INP RAN.

Petersen. (1999). *UNDP and the Government of the Republic of Kazakhstan project "Removing Barriers to Wind Energy Development in Kazakhstan. Wind energy potential of the Dzungar gate and Shelek corridor in Kazakhstan. Brief annotation.* National laboratory RISO.

Petersen. (2011). *The UNDP and the Government of Kazakhstan Project "Elimination of barriers to development of wind energy in Kazakhstan. Wind energy potential of Jungar gates and the Shelek corridor in Kazakhstan. Abstract.* Astana European Bank for Reconstruction and Development. Retrieved from http://ebrd.com

Peters, R., Deja, R., Blum, L., Pennanen, J., Kibiaho, J., & Hakala, T. (2013). Analysis of solid oxide fuel cell system concepts with anode recycling. *International Journal of Hydrogen Energy*, *38*(16), 6809–6820. doi:10.1016/j.ijhydene.2013.03.110

Petersson, A., Lundberg, S., & Thiringer, T. (2005). A DFIG wind turbine ride through system influence on energy production. *Wind Energy (Chichester, England)*, *8*(3), 251–263. doi:10.1002/we.157

Petreuş, D., Fărcaş, C., & Ciocan, I. (2008). Modelling and simulation of photovoltaic cells. *Electronics (Basel)*, *49*(1), 42–47.

Podhurenko, V. S. (2007). *Netradytsiini dzherela enerhii: navch. posib* [Alternative energy sources: textbook]. NUC, Mykolaiv.

Popel, O. S. (1988). Issledovanie sistem teplosnabzheniya na osnove solnechnyh prudov [Research of systems of a heat supply on the basis of solar ponds]. Moscow: IVTAN. (in Russian)

Popel, O. S., & Fortov, V. E. (2015). Vozobnovlyaemaya ehnergetika v sovremennom mire [Renewable energy in the modern world]. Publishing house of Moscow Power Engineering Institute. (in Russian)

Popel', O. S., Frid, S. Y., Kolomiets, Y. G., Kiseleva, S. V., & Terekhova, E. N. (2010). Atlas resursov solnechnoj jenergii na territorii Rossii [Atlas of solar energy resources in Russia]. Moscow: OIVT RAN.

Popel, O. S., Kisileva, S. V., Morgunova, M. O., Gabderakhmanova, T. S., & Tarasenko, A. B. (2015). The use of renewable energy sources for energy supply to consumers in the Arctic zone of the Russian Federation. *Arctic: Ecology and Economics*, *1*(17), 64–69.

Popov, N. (2010). *Hlektrosnabzhenie. Rabochie rezhimy setej 0,38...10kV: uchebnoe posobie* [Power supply of the networks of 0,38...10kV: working modes]. Kostroma, Russia: KGSHA.

Prism Solar Technologies. (n.d.). *Prism Solar Technologies*. Retrieved from https://www.prismsolar.com/

Promes, E. J. O., Woudstra, T., Schoenmakers, L., Oldenbroek, V., Thattai, A. T., & Aravind, P. V. (2015). Thermodynamic evaluation and experimental validation of 253 MW Integrated Coal Gasification Combined Cycle power plant in Buggenum, Netherlands. *Applied Energy*, *155*, 181–194. doi:10.1016/j.apenergy.2015.05.006

Rafikova, Yu. (2015). *Geoinformational mapping of renewable energy resources (on the example of the South of Russia)* (PhD diss.). Lomonosov Moscow State University.

Rathi, M. R., & Mohan, N. (2005). *A novel robust low voltage and fault ride through for wind turbine application operating in weak grids.* Paper presented at the IEEE Industrial Electronics Society Annual Conference. 10.1109/IECON.2005.1569295

Rebecca, L. B. (2012). *Wind Power: The Industry Grows Up*. Tulsa, OK: PennWell.

Recommendations for the determination of climatic characteristics of wind energy resources. GGO. NGO "Vetroen". (1989). Leningrad: Gidrometeoizdat.

Rekioua, D., & Matagne, E. (2012). *Optimization of Photovoltaic Power Systems. In Green Energy and Technology, Modelization, Simulation and Control.* London: Springer-Verlag. doi:10.1007/978-1-4471-2403-0

Reliable Prognosis. (n.d.). Retrieved from http://rp5.ru

REN21 GSR. (2018). *Renewables 2018. Global status report. REN21Secretariat.* Paris: UNEP.

Renewable Energy Policy Network for the 21st Century. (2018). *Renewables 2018, Global status report.* Paris, France: REN21.

Republic of Uzbekistan, Large pumping stations. (2005). *Brief review of technical condition.* Tashkent: State Inspectorate for Control and Supervision over the Technical Condition and Safety of Large and Especially Important Water Facilities under the Cabinet of Ministers of the Republic of Uzbekistan.

Reznitsky, L. A. (1996). Teplovye akkumulyatory [Thermal accumulators]. Energoatomizdat. (in Russian)

Reznitsky, L. A. (1998). Teploakkumuliruyushchie veshchestva i processy [Heat accumulating substances and processes]. *Journal of Inorganic Chemistry, 43*(8), 1288–1298.

RMA "Південне". (1999). Vitroelektrychnyi ahrehat ВЕУ-500 [A wind power station ВЕУ-500]. Prospect. Dnipropetrovsk.

Rosgidromet. (2017). *Report on the peculiarities of climate on the territory of the Russian Federation for 2016.* Retrieved from http://www.meteorf.ru/press/news/13595/

Rozenfeld, V. E., & Staroselskii, N. A. (1975). *High-Frequency Noncontact Electrical Transport.* Moscow: Transport. (in Russian)

Ruiz-Arias, J., Terrados, J., Pérez-Higueras, P., Pozo-Vázquez, D., & Almonacid, G. (2012). Assessment of the renewable energies potential for intensive electricity production in the province of Jaén, southern Spain. *Renewable & Sustainable Energy Reviews, 16*(5), 2994–3001. doi:10.1016/j.rser.2012.02.006

Russkin, V.A., Semenov, S.M., & Dickson, R.K. (2016). Investigation of the search algorithms for the point of maximum power for the boost converter of the solar inverter voltage. *Izvestiya Tomsk Polytechnic University. Engineering of Geo-Resources, 4*, 78-87.

Ryhlov, A.B. (2011). Regularities in the variation of the average wind speed with altitude in the surface layer of the atmosphere at southeast European Russia for solving the problems of wind energy. *Scientific Notes of the Russian State Hydrometeorological University*, 89-99.

Saaty, T. (1989). *Group decision making and the AHP*. New York: Springer Verlag. doi:10.1007/978-3-642-50244-6_4

Sacchelli, S., De Meo, I., & Paletto, A. (2013). Bioenergy production and forest multifunctionality: A trade-off analysis using multiscale GIS model in a case study in Italy. *Applied Energy, 104*, 10–20. doi:10.1016/j.apenergy.2012.11.038

Salles, M. B., Cardoso, J. R., Grilo, A. P., Rahmann, C., & Hameyer, K. (2009). Control Strategies of Doubly Fed Induction Generators to Support Grid Voltage. *IEEE Transactions on Power Energy, 135*, 1551–1556.

Samarin, A. V., Rygalin, D. B., & SHklyaev, A. A. (2012). Sovremennye tekhnologii monitoringa vozdushnyh ehlektrosetej LEHP [Modern technologies for monitoring of power transmission lines]. *Estestvennye i tekhnicheskie nauki, 1.*

Sánchez-Lozano, J., Teruel-Solano, J., Soto-Elvira, P. L., & Socorro García-Cascales, M. (2013). Geographical Information Systems (GIS) and Multi-Criteria Decision Making (MCDM) methods for the evaluation of solar farms locations: Case study in south. *Renewable & Sustainable Energy Reviews, 24*, 544–556. doi:10.1016/j.rser.2013.03.019

Santarius, P., Krejci, P., Brunclik, Z., Prochazka, K., & Kysnar, F. (2015). Evaluation of power quality in regional distribution networks. In *Proceedings of the 23rd International Conference on Electricity Distribution: CIRED.* Lyon, France: IET.

Santos-Martin, D., Rodriguez-Amenedo, J., & Arnaltes, S. (2009). Providing ride-through capability to a doubly fed induction generator under unbalanced voltage dips. *IEEE Transactions on Power Electronics, 24*(7), 1747–1757. doi:10.1109/TPEL.2009.2016965

Sarbon, S., & Sc, M. (2013). *The unresolved patents of Nikola Tesla. The inventions that Tesla did not protect.* Nikola Tesla Museum.

Scaling - Up Renewable Energy Program. (2012). *Ethiopia Investment Plan (Draft Final).* Federal Democratic Republic of Ethiopia, Ministry of Water and Energy.

Scharmer, K., & Greif, J. (2000). *The European solar radiation atlas.* Les Presses Mines ParisTech.

Scott, G. W., Wilreker, V. F., & Shaltens, R. K. (1984). Wind turbine generator interaction with diesel generators on an isolated power system. *IEEE Transactions on Power Apparatus and Systems*, *PAS-103*(5), 933–937. doi:10.1109/TPAS.1984.318695

SEAR. (2017). *State of Electricity Access Report 2017.* International Bank for Reconstruction and Development / The World Bank. Retrieved from http://www.worldbank.org/en/topic/energy/publication/sear

Semenov, A. E., Selezneva, A. O., & Vinogradov, A. V. (2015). Sravnenie pokazatelej nadezhnosti vozdushnyh i kabel'nyh linij v gorodskoj i sel'skoj mestnosti. In *Proceedings of the VII All-Russian Scientific and Practical Conference: Osnovnye napravleniya razvitiya tekhniki i tekhnologii v APK.* Knyaginino, Russia: NGIEHU.

Semenov, A. A. (1968). *The Theory of Electromagnetic waves.* Moscow State University. (in Russian)

Sen, Z. (2008). *Solar Energy Fundamentals and Modeling Techniques, Atmosphere, Environment, Climate Change and Renewable Energy* (1st ed.). Springer.

Sergeev, V. D., Kuleshov, E. V., Proskurenko, S. S., & Chernyshova, A. S. (2000). Metodika rascheta magnitoelektricheskogo sinhronnogo vetrogeneratora [The methodology of calculating a magnetic and electrical synchronic wind generator]. *Trudyi dalnevostochnogo politehnicheskogo instituta im. V.V. Kuybyisheva,* (127), 181-185. Retrieved from https://elibrary.ru/item.asp?id=13042317

Sergeev, V.D., & Kuleshov, E.V. (2001). Optimizatsiya magnitoelektricheskih sinhronnyih vetrogeneratorov [The improvement of magnetic and electrical synchronic wind generators]. *Trudyi dalnevostochnogo gosudarstvennogo tehnicheskogo universiteta*, (129), 243-246.

Shang, L., & Hu, J. (2012). Sliding-mode-based direct power control of grid-connected wind-turbine driven doubly fed induction generators under unbalanced grid voltage conditions. *IEEE Transactions on Energy Conversion*, *27*(2), 362–373. doi:10.1109/TEC.2011.2180389

Shell. (2018). *Floating LNG.* Retrieved July 13, 2018, from https://www.shell.com/energy-and-innovation/natural-gas/floating-lng.html

Shishkin, N. D. (2005). Kombinirovanie i ehffektivnoe ispol'zovanie istochnikov teplovoj ehnergii v avtonomnyh teploehnergeticheskih kompleksah, vklyuchaya vozobnovlyaemye istochniki [Combination and effective use of thermal energy sources in autonomous heat and power complexes, including renewable sources]. Moscow: RSL. (in Russian)

Shokhzoda & Tyagunov. (2016). Assessment of solar energy potential on the territory of the Republic of Tajikistan. *Twenty-second international scientific and technical conference of students and postgraduates "Radio electronics, electrical technology and power engineering* (vol. 3). Moscow publishing house MPEI.

Shokhzoda & Tyagunov. (2017). Heat supply based on Solar Energy in Tajikistan. *Conference on Sustainable Energy Supply and Energy Storage Systems*, 181-186.

Shokhzoda B. T., & Kokin N. N. (2016). *Evaluation of hybrid systems electrical and heat supply based on solar energy.* Moscow: Publisher MPEI Vestnik.

Shokhzoda, B. T. (2016). *Solar mobile power station", International scientific and technical conference.* "Energy supply and energy saving in agriculture", Russia, Moscow.

Shokhzoda, B. T., & Kuznetsova, V. A. (2013). Development of power supply system for residential house in Tajikistan. *Nineteenth international scientific and technical conference of students and postgraduates "Radio electronics, electrical technology and power engineering* (vol. 4). Moscow publishing house MPEI (in Russian).

Shpil'rain, E. E., Malyshenko, S. P., & Kuleshov, G. G. (1984). *An Introduction to Hydrogen Power Engineering*. Moscow: Energoatomizdat.

Shyam, S. C., & Rajeev, K. A. (2011). Estimation of Hourly Solar Radiation on Horizontal and Inclined Surfaces in Western Himalayas. *Smart Grid and Renewable Energy*, *2*(01), 45–52. doi:10.4236gre.2011.21006

Siano, P. (2014). Demand response and smart grids – a survey. *Renewable & Sustainable Energy Reviews*, *30*, 461–478. doi:10.1016/j.rser.2013.10.022

Silva, S. B., de Oliveira, M. A. G., & Severino, M. M. (2010). Economic evaluation and optimization of a photovoltaic-fuel cell-batteries hybrid system for use in the Brazilian Amazon. *Energy Policy*, *38*(11), 6713–6723. doi:10.1016/j.enpol.2010.06.041

Simulink. (n.d.). *The MathWorks, Inc. Version 8.7 (R2016a).* retrieved from https://www.mathworks.com/products/simulink.html

Singh, Saw, & Kumar. (2012). Technical viability of holographic film on solar panels for optimal power generation. *International Journal of Advances in Engineering and Technology.*

Sipos Aktorik. (n.d.). Retrieved January 20, 2016, from http://www.sipos.de

Sister, V. (2001). *Solid household waste.* Publisher of the ACH.

Sobyanin, V. A. (2003). High-temperature solid oxide fuel cells and methane conversion. *Russian Chemical Journal*, *47*(60), 62–70.

Sokolov, E. Ya. (2001). *Heating and heating networks*. Moscow: MPEI.

Solimpeks. (n.d.). *Solimpeks Solar Energy corp*. Retrieved from http://www.solimpeks.com/

Solovyev, D.A. (2017). Small Energy in the Arctic: Adaptation Problems and Risks. *Jenergija: Jekonomika, Tehnika, Jekologija*, (11), 14–21.

Solovyev, D.A., Morgunova, M.O., & Gabderakhmanova, T.S. (2017). Adaptation of the energy infrastructure in the Arctic to climate change using renewable energy sources. *Jenergeticheskaja Politika*, (4), 72–80.

Sorensen, B. (2011). *Preobrazovanie, peredacha i akkumulirovanie ehnergii: uchebno-spravochnoe posobie* [Transformation, transmission and accumulation of energy: educational-reference manual]. Dolgoprudny: Intelligence. (in Russian)

Sourkounis, C., & Tourou, P. (2012). *Grid Code Requirements for Wind Power Integration in Europe*. Paper presented at the Power Options for the Eastern Mediterranean Region, Limassol, Cyprus.

Spiganovic, A. (2004). Obespecheniya kachestva ehlektroehnergii sistem ehlektrosnabzheniya [Ensure the power quality of power systems]. In *Proceedings of the konferencii «EHlektroehnergetika i ehnergosberegayushchie tekhnologii»*. Lipeck, Russia: LGTU.

Spiganovic, A., & Zakharov, K. (2007). *Vnutrizavodskoe ehlektrosnabzhenie i rezhimy: uchebnik* [In-Plant electricity and modes: tutorial]. Lipetsk, Russia: LGTU.

Sridhar & Kumar. (2012). Development of Solar MPPT System Using Boost Converter with Microcontroller. *International Journal of Advanced Research in Electrical. Electronics and Instrumentation Engineering*, *1*(4), 334–340.

Stepanov, V.M., & Shpiganovich, A.A. (2015). Opredelenie iskhodnyh dannyh dlya povysheniya ehffektivnosti funkcionirovaniya sistem ehlektrosnabzheniya [Determination of initial data for improving power supply system efficiency]. *Izvestiya Tul'skogo gosudarstvennogo universiteta. Tekhnicheskie nauki, 12*(2).

Stiller, P. H., Scott, G. W., & Shaltens, R. K. (1983). Measured effect of wind generation on the fuel consumption of an isolated diesel power system. *IEEE Winter Power Meeting*.

Strebkov & Tveryanovich. (2007). *Solar Radiation Concentrator*. VIESH.

Strebkov, D. S., & Nekrasov, A. I. (2018) Resonant Methods of Generation Transmission and Application of Electrical Energy. Moscow: FSAC VIM. (in Russian)

Strebkov, D. S., Avramenko, S. V., & Nekrasov, A. I. (1999). RF Patent 2136515. Method and Apparatus for Supplying Power to Electric Vehicles. *Inventor's Bulletin*, 25. (in Russian)

Strebkov, D. S., Avramenko, S. V., & Nekrasov, A. I. (2000). RF Patent 2158206. Apparatus for Supplying Power to a Mobile Electric Mechanism. *Inventor's Bulletin*, 30. (in Russian)

Strebkov, D. S., Korolev, V. A., Nekrasov, A. I., & Toporkov, V. N. (2006). RF Patent 2269854. Cable system for Power Supply of Mobile Agricultural Machine. *Inventor's Bulletin, 1*. (in Russian)

Strebkov, D. S., Nekrasov, A. I., Juferev, L. J., & Roshin, O.A. (2007). RF Patent 2297928. Method and Apparatus for Electric Power Supply of Electric Transport. *Inventor's Bulletin*, 12. (in Russian)

Strebkov, D.S., Nekrasov, A.I., Trubnikov, V.Z., & Nekrasov, A.A. (2018). Single-wire resonant electric power systems for renewable based electric grid. In *Handbook of Research on Renewable Energy and Electric Resources for Sustainable Rural Development*. Academic Press. DOI: . doi:10.4018.978-1-5225-3867-7

Strebkov, D. S., Avramenko, S. V., & Nekrasov, A. I. (2000). Single-Trolley System for Supplying Power to Mobile Electric Mechanisms, *Proceedings of the International Scientific-and-Technical Conference "Automation of Agricultural Production"*, 65–66.

Strebkov, D. S., Avramenko, S. V., & Nekrasov, A. I. (2000). System for Supplying Electric Power to Mobile Electric Mechanisms [in Russian]. *Drive Techniques*, *4*(26), 32–38.

Strebkov, D. S., Avramenko, S. V., Nekrasov, A. I., & Roshchin, O. A. (2002). Investigation of Electric Parameters of a Single-Conductor Power System. [in Russian]. *Drive Techniques*, *6*(40), 28–30.

Strebkov, D. S., Nekrasov, A. I., Avramenko, S. V., & Sakharov, D. N. (2002). Methods for Supplying Electric Power to Mobile Agricultural Vehicles, Electrification of the Agriculture. *Proceedings of the VIESH*, 88, 74–82. (in Russian)

Strebkov, D. S., & Tver'yanovich, E. V. (2007). *Concentrators of solar radiation*. Moscow: Academic Press.

Suhas, C. D., Nitin, R. L., & Pillai, R. (2015). Power quality monitoring with smart meters. In *Proceedings of the 23rd International Conference on Electricity Distribution: CIRED*. Lyon, France: IET.

Sun, T., Chen, Z., & Blaabjerg, F. (2005). Transient stability of DFIG wind turbines at an external short-circuit fault. *Wind Energy (Chichester, England)*, *8*(3), 345–360. doi:10.1002/we.164

Swanson. (n.d.). *Photovoltaic Concentrators*. Sun Power Corporation.

SWERA - Solar and Wind Energy Resource Assessment. (n.d.). Retrieved from https://openei.org/doe-opendata/dataset/swera

Szabó, S. (2011). Energy solutions in rural Africa: mapping electrification costs of distributed solar and diesel generation versus grid extension. Environ. Res. Lett., 6.

Takahashi, R., Tamura, J., Futami, M., Kimura, M., & Idle, K. (2006). A new control method for wind energy conversion system using doubly fed synchronous generators. *IEEE Transactions on Power Energy*, *126*(2), 225–235. doi:10.1541/ieejpes.126.225

Tesla, N. (1894). *Electric Railway System.* US Patent 614942.

Tesla, N. (1914). Science and Discovery are the greater forces which will lead to the consummation of the war. The Sun, Dec. 20. 1914. In *Nicola Tesla Selected works.* Beograd: Academic Press.

Tesla, N. (1915). The wonder world to be created by electricity manufactures record. In Nikola Tesla. Selected works. Beograd: Academic Press.

Tesla, N. (1978). *Colorado Springs Notes 1899 – 1900.* Beograd: Published by Nolit.

Tesla, N. (1999). *My Inventions. Electrical Experimenter, February, March, June and October. 1919. In Nikola Tesla. Selected works* (p. 56). Beograd: Articles.

Tesla, N. (1999). *The Problem of Increasing Human Energy. The Century Illustrated Monthly Magazine. June, 1900. In Nikola Tesla. Selected works* (p. 133). Beograd: Articles.

Tesla, N. (1999). *The Wonder World to be Created by Electricity. Manufactures. Record, Sept. 9. 1915. In Nicola Tesla. Selected works* (pp. 369, 377, 378). Beograd: Articles.

Tesla, N. (2008). *From Colorado to Long island: Research Notes – Colorado Springs 1899-1900 – New York. 1900-1901.* Nikola Tesla Museum.

Tesla, N. L. (1956). *Patents.* Beograd: N. Tesla Museum.

The Nobel Prize. (n.d.). *The Nobel Prize in Physics 1921.* Retrieved 2013-03-16 from https://www.nobelprize.org/prizes/physics/1921/summary/

Tian, H., Mancilla-david, F., Ellis, K., Jenkins, P., & Muljadi, E. (2012). *A Detailed Performance Model for Photovoltaic Systems Preprint. Solar Energy Journal.*

Tikhonov, A., Tugov, I. I., & Yuferev, L. Y. (2009). Technologies of renewable (solar) energy. Moscow: Publisher Research and Manufacturing Association "SODIS".

Tin, T., Sovacool, B. K., Blake, D., Magill, P., El Naggar, S., Lidstrom, S., ... Berte, J. (2010). Energy efficiency and renewable energy under extreme conditions: Case studies from Antarctica. *Renewable Energy*, *35*(8), 1715–1723. doi:10.1016/j.renene.2009.10.020

Transient processes of pumping stations. Calculations of hydraulic shock in pressure pipelines . (n.d.). Manual for VSN 33-2.2.12-87 "Reclamation systems and facilities: Pumping stations - Design standards". Retrieved January 29, 2016, from http://library.fsetan.ru/doc/perehodnyie-protsessyi-nasosnyih-stantsij-raschetyi-gidravlicheskogo-udara-v-napornyih-truboprovodah-posobie-k-vsn-33-2212-87-meliorativnyie-sistemyi-i-sooruzheniya-nasosnyie-stantsii-normyi-proektirovaniya

Trehub, M.I. (2004). *Lopat vitrodvyhuna* [The blade of a wind engine]. Patent of Ukraine, IPC F03D 1/06, F04D 29/30, N° 68628; publ. 16.08.2004. Bul. N° 8.

Trehub, M.I., Kozyrskyi, V.V., Hrebenikov, V.V., Pryimak, M.V. (2013). *Aksialnyi mahnitoelektrychnyi henerator* [An axial magnetic and electrical generator]. Patent of Ukraine, IPC H02K21/12, H02K21/24, N° 79084; publ. 10.04.2013. Bul. N° 7.

Troen, I., & Lundtang Petersen, E. (1989). *European Wind Atlas.* Roskilde: Risø National Laboratory.

Trojanowska, M. (2014). Bezpieczeństwo elektroenergetyczne terenów wiejskich [Electricity safety in rural areas]. *Agrotekhnika i ehnergoobespechenie, 1*(1), 468-475.

Twidell, D., & Weir, A. (1990). *Vozobnovlyaemye istochniki ehnergii* [Renewable energy sources]. Moscow: Energoatomizdat. (in Russian)

UNECA. (2017). *Africa Sustainable Development Report: Tracking Progress on Agenda 2063 and the Sustainable Development Goals. African Union, Economic Commission for Africa.* Addis Ababa, Ethiopia: African Development Bank and United Nations Development Programme.

UNEP. (2017). *Atlas of Africa Energy Resources.* Nairobi, Kenya: United Nations Environment Programme.

Unified scenario conditions of PJSC «RAO Energy Systems of the East» for 2014-2039 years, order of 10.10.2013 N°191.

Valve, V.-M., & Corp, M. (n.d.). Retrieved January 20, 2016, from http://www//valmatic.com

Vasilev, Yu. S., Elistratov, V. V., Mukhammadiev, M. M., & Pretro, G. A. (1995). Vozobnovlyaemye istochniki ehnergii i gidroakkumulirovanie [Renewable energy sources and pumped storage]. St. Petersburg: SPbSTU. (in Russian)

Vaskov, A. G., Deryugina, G. V., Tyagunov, M. G., & Chernov, D. A. (2012). Certificate of registration of the data basis N°2012620870. Specialized database (SDB) «Wind shear profile». National Research University "Moscow Power Engineering Institute".

Vin, J. (2013). *Study of solar energy efficiency for Autonomous power supply systems in the Republic of the Union of Myanmar* (Thesis). Moscow. MPEI.

Vinogradov, A. & Borodin, M. (2012). EHkspertnaya ocenka popravochnyh koehfficientov k stoimosti potreblennoj ehlektroehnergii v zavisimosti ot eyo kachestva [Expert evaluation of correction factors to the cost of electricity consumption depending on its quality]. *Vesti vysshih uchebnyh zavedenij CHernozem'ya, 3*, 14-19.

Vinogradov, A. V., Chernyshov V. A., Semenov A. E., & Moroz A. O. (2014). Sposob monitoringa tekhnicheskogo sostoyaniya ehlementov vozdushnoj linii ehlektroperedach [Method for monitoring the element technical condition of an overhead power line]. *Agrotekhnika i ehnergoobespechenie, 3*, 64-68.

Vinogradov, A., & Borodin, M. (2013a). EHkonomicheskaya ehffektivnost' vnedreniya sposobov i sredstv korrektirovki stoimosti potreblennoj ehlektroehnergii v zavisimosti ot ee kachestva [Economic efficiency of introduction of methods and means of adjustment of cost of the consumed electric power depending on its quality]. *Proceedings of the mezhdunarodnoj nauchno-prakticheskoj konferencii. "Osobennosti tekhnicheskogo i tekhnologicheskogo osnashcheniya sovremennogo sel'skohozyajstvennogo proizvodstva.*

Vinogradov, A., & Borodin, M. (2013b). *R.F. Patent No. 134338.* Patent Service of the Russian Federation.

Vinogradov, A., Borodin, M., & Bolshev, V. (2014). Sposob upravleniya kachestvom ehlektricheskoj ehnergii [Method for managing power quality]. *Tekhnika v sel'skom hozyajstve, 4*, 30-31.

Vinogradov, A., Vasiliev, A., Bolshev, V., Semenov, A., & Borodin, M. (2018). Time Factor for Determination of Power Supply System Efficiency of Rural Consumers. In V. Kharchenko & P. Vasant (Eds.), Handbook of Research on Renewable Energy and Electric Resources for Sustainable Rural Development (pp. 394–420). Academic Press. doi:10.4018/978-1-5225-3867-7.ch017

Vinogradova, A.V. (2014). Statisticheskaya kharakteristika sel'skikh elektricheskikh setey [Statistical characteristics of rural electrical networks]. *Agrotekhnika I energoobespecheniye, 1*(1), 419-423.

Vinogradova, A.V., Vinogradov, A.V., & Konstantinov, A.V. (2016). Primeneniye sektsionirovaniya liniy elektroperedachi 0,38 kV dlya povysheniya nadezhnosti elektrosnabzheniya sel'skikh potrebiteley [Application of sectionalizing of 0.38 kV transmission lines to improve power supply reliability for rural consumers]. *Teoreticheskiyi nauchno-prakticheskiy zhurnal "Innovatsii v APK: problemyiperspektivy", 2*(10), 12-27.

Vinogradov, A. V. (2007). *Sovershenstvovanie avtomaticheskogo rezervirovaniya na dvuhtransformatornyh podstanciyah* [Improvement of automatic backup on two-transformer substations]. Orel, Russia: FGBOU VPO Orel GAU.

Vinogradov, A. V., & Perkov, R. A. (2015). Analiz povrezhdaemosti ehlektrooborudovaniya ehlektricheskih setej i obosnovanie meropriyatij po povysheniyu nadezhnosti ehlektrosnabzheniya potrebitelej [Analysis of the damageability of electrical equipment in power networks and the rationale of measures to improve power supply reliability of consumers]. *Vestnik NGIEHI, 12*(55), 12–20.

Vinogradov, A. V., & Vinogradova, A. V. (2016). *Povyshenie nadezhnosti ehlektrosnabzheniya sel'skih potrebitelej posredstvom sekcionirovaniya i rezervirovaniya linij ehlektroperedachi 0,38 kV* [Improving the power supply reliability of rural consumers by the distribution sectionalizing and the backup of 0.38 kV power networks]. Orel, Russia: FGBOU VO Orel GAU.

Vinogradov, A. V., Vinogradova, A. V., & Bolshev, V. E. (2017). Ustroystva I sistema monitoringa nadezhnosti elektrosnabzheniya I otkloneniya napryazheniya v elektricheskikh setyakh 0,38 kV [Devices and monitoring system for power supply reliability and voltage deviation in 0.38 kV electrical networks]. *Vestnik of the Nizhniy Novgorod State Engineering and Economic Institute, 11*(78), 69–82.

Vinogradov, A. V., Vinogradova, A. V., & Kuchinov, A. A. (2017). Vremya osushchestvleniya tekhnologicheskogo prisoyedineniya k elektricheskim setyam kak factor effektivnosti system elektrosnabzheniya [The time of technological connection to electrical networks as a factor of the power supply system efficiency]. *Vestnik of the Nizhniy Novgorod State Engineering and Economic Institute, 6*(73), 54–60.

Vinogradov, A., Borodin, M., Volchenkov, Y. U., & Peshekhonova, Z. H. (2015). *Sovershenstvovanie deyatel'nosti po ehnergosberezheniyu i po osushchestvleniyu tekhnologicheskih prisoedinenij filiala OAO "MRSK Centra" - "Orelehnergo* [Improvement of the energy efficiency and the implementation of technological connection of the branch of JSC "IDGC of Center" - "Orelenergo"]. Orel, Russia: FGBOU VPO Orel GAU.

Vinogradov, A., Vasiliev, A., Semenov, A., Sinyakov, A., & Bolshev, V. E. (2017). Analiz vremeni pereryvov v energosnabzhenii sel'skikh potrebiteleyi metodov yego snizheniya za schet monitoringa tekhnicheskogo sostoyaniya liniy elektroperedach [Analysis of the power supply outage time for rural consumers and methods of its reducing due to monitoring power line technical condition]. *Bulletin of All-Russian Institute of Electrification of Agriculture, 2*(27), 3–11.

Vinogradova, A., & Skrobov, A. (2013). Ob uvelichenii poter' ehlektroehnergii v LEHP vsledstvie izmenenij v proektirovanii s uchetom uvelicheniya dopustimyh otklonenij napryazheniya soglasno GOST 54149-2010 [On the increase of electric power losses in transmission lines due to changes in design taking into account the increase in permissible voltage deviations in accordance with GOST 54149-2010]. *Vestnik NGIEHI, 6*(25), 16–19.

Vissarionov, Deryugina, Kuznetsova, & Malinin. (1998). *Calculation of solar energy resources, a textbook for courses "Theoretical foundations of energy renewable resources" "Design and operation of solar and wind farms"*. Moscow: Publishing house of MPEI.

Vissarionov, V.I. Burmistrov, A.A., Deryugina, G.V., Kuznetsova, V.A., Kunakin, D.N., Malinin, N.K., & Pugachev, R.V. (2008). *Solnechnaja jenergetika: Uchebnoe posobie dlja vuzov* [Solar energy: A textbook for high schools]. Moscow: Izdatel'skij dom MJeI.

Vissarionov, V. I., & Trusov, I. S. (1986). Matematicheskiye modeli perekhodnykh protsessov v nasosnoy stantsii pri samozapuske elektrodvigatelya posle kratkovremennogo pereryva pitaniya [Mathematical models of transient processes in a pumping station with a self-starting electric motor after a short-time power interruption]. *JournalEnergetika, 11*, 115–119.

Viswanadh, M.M.G., Sekhar, A.S.R. (2012). Minimization of power loss and improvement of voltage profile by optimal placement of wind generator in distribution network. *International Journal of Engineering Research and Applications, 2*(6), 987-993.

Volkova, N. G. (Ed.). (1989). Matematika i SAPR. Vychislitel'nye metody. Geometricheskie metody [Mathematics and CAD. Computational methods. Geometric methods]. Moscow: Mir.

Volovikov, V. V. (2002). *Use of mathematical modeling in professional work in radio electronics: proceedings / Information technologies in design, production and education*. KGTA.

Wegley, H.L., Ramsdell, J.V., & Orgill, M.M. (1980). *Siting Handbook for Small Wind Energy Conversion Systems*. Battelle: US DOE.

White, M., & Ahad, A. (2016). Solar PV Residential Grid-Tie System for Optimizing Installation and Performance. *International Journal of Energy Optimization and Engineering, 5*(1), 48–67. doi:10.4018/IJEOE.2016010104

Wilford. (1988). *Scientists, Using Mirrors, Match Energy Intensity of Sun's Surface*. Retrieved from https://www.nytimes.com/1988/03/11/us/scientists-using-mirrors-match-energy-intensity-of-sun-s-surface.html

Winston, R. (1970). Light Collection Within the Framework of Geometrical Optics. *Journal of the Optical Society of America, 60*(2), 245–247. doi:10.1364/JOSA.60.000245

Winter, C.-J. (1991). Solar Power Plants. Berlin: Heidelberg. doi:10.1007/978-3-642-61245-9

World Bank Data. (2016). *Access to electricity, rural*. Retrieved from https://data.worldbank.org/indicator/EG.ELC.ACCS.RU.ZS

Wu, Z., Sun, H., & Du, Y. (2014). A Large Amount of Idle Capacity Under Rapid Expansion: Policy Analysis on the Dilemma of Wind Power Utilization in China. *Renewable & Sustainable Energy Reviews, 32*, 271–277. doi:10.1016/j.rser.2014.01.022

Xiang, D., Ran, L., Tavner, P., & Yang, S. (2006). Control of a doubly fed induction generator in a wind turbine during grid fault ride-through. *IEEE Transactions on Energy Conversion, 21*(3), 652–662. doi:10.1109/TEC.2006.875783

Yahno, O.M., & Taurit, T.H., & Hrabar I.H. (2002). Vetroenergetika: konstruirovanie i raschet VEU: ucheb. posob [Wind energy: the design and calculation of WES: textbook]. *ZhSTU, Zhytomyr,* 255.

Yakovlev, B. V. (2008). Increase in efficiency of central heating and heat supply systems. Moscow: Novosti teplosnabzheniya.

Yan, X., Venkataramanan, G., & Wang, Y. (2009). *Grid fault tolerant operation of DFIG wind turbine generator using a passive resistance network.* Paper presented at the Energy Conversion Congress and Exposition (IEEEECCE), San Jose, CA.

Yang, J., Fletcher, E., & O'Reilly, J. (2010). A series dynamic resistor based converter protection schemes for doubly fed induction generator during various fault conditions. *IEEE Transactions on Energy Conversion*, *25*(2), 422–432. doi:10.1109/TEC.2009.2037970

Yang, K., Huang, G. W., & Tamai, N. (2001). A hybrid model for estimating global solar radiation. *Solar Energy*, *70*(1), 13–22. doi:10.1016/S0038-092X(00)00121-3

Yasin, K. (2016). A survey on smart metering and smart grid communication. *Renewable & Sustainable Energy Reviews*, *57*, 302–318. doi:10.1016/j.rser.2015.12.114

Yellowhair, J., Yellowhair, J., Ho, C. K., Ho, C. K., Ortega, J. D., Ortega, J. D., ... Andraka, C. E. (2015). Testing and optical modeling of novel concentrating solar receiver geometries to increase light trapping and effective solar absorptance. In *Proc. SPIE 9559. High and Low Concentrator Systems for Solar Energy Applications.*

Yuferev, L. Yu., Strebkov, D. S., & Roshin, O. A. (2010). *Experimental Models of Resonant Electric Power Transmission Systems.* Moscow: VIESH. (in Russian)

Zhao, Y., Sadhukhan, J., Lanzini, A., Brandon, N., & Shah, N. (2011). Optimal integration strategies for a syngas fuelled SOFC and gas turbine hybrid. *Journal of Power Sources*, *196*(22), 9516–9527. doi:10.1016/j.jpowsour.2011.07.044

Zhelezko, Y. (2002). O normativnyh dokumentah v oblasti kachestva ehlektroehnergii i uslovij potrebleniya reaktivnoj moshchnosti[About normative documents in the field of power quality and consumption conditions of jet capacity]. *EHlektricheskie stancii, 6*, 18–24.

Zhou, P., He, Y., & Sun, D. (2009). Improved direct power control of a DFIG-based wind turbine during network unbalance. *IEEE Transactions on Power Electronics*, *24*(11), 2465–2474. doi:10.1109/TPEL.2009.2032188

About the Contributors

Valeriy Kharchenko was born on the 18 December in 1938. The editor is a researcher and consultant in a wide range of energy, energy policy, RUE and RES. Doctor of Engineering, professor. Participation in a number International Scientific projects, reviewer of the International Journal Solar Energy, member of Editorial Boards in number Journal of different countries.

Pandian Vasant is a senior lecturer at Department of Fundamental and Applied Sciences, Faculty of Science and Information Technology, Universiti Teknologi PETRONAS in Malaysia. He holds PhD (UNEM, Costa Rica) in Computational Intelligence, MSc (UMS, Malaysia, Engineering Mathematics) and BSc (2nd Class Upper-Hons in Mathematics, UM, Malaysia) in Mathematics. He has co-authored research papers and articles in national journals, international journals, conference proceedings, conference paper presentation, and special issues lead guest editor, lead guest editor for book chapters, edited books, and keynote lecture. (218 publications indexed in SCOPUS). In the year 2009, Dr. Pandian Vasant was awarded top reviewer for the journal Applied Soft Computing (Elsevier), awarded outstanding reviewer in the year 2015 for ASOC (Elsevier) journal and Top reviewer for Sentinels of Science: Computer Science (Oct. 2015 - Sept. 2016). He has 26 years of working experience at the various universities from 1989-2017. Currently he is Editor-in-Chief of IJCO and IJEOE, Member of American Mathematical Society, MERLIN, and NAVY Research Groups.

* * *

Baba Babaev studied at the faculty of heat and power engineering of the Moscow State Engineering and Construction Institute, which graduated in 1982. After graduation worked at the Kola NPP. From 1990 to 1993, working in the Scientific Research Energy Institute G. M. Krzhizhanovskogo, Dagestan branch, increased scientific qualifications to a senior researcher. In 1996 defended the dissertation of the candidate of chemical sciences at the Kuban State Technological University. And in 2016 defended the dissertation of the doctor of technical sciences in "United Institute of High Temperatures". Specialist in the field of energy, development of energy systems based on renewable sources and energy storage, actively introduces new information technologies in various fields of energy in the educational process. More than 160 scientific works have been published, 17 patents and 8 author's certificates for fundamentally new developments in the field of energy, some of which are implemented in practice. At a high scientific and methodological level reads the special courses "Energy accumulators", "Physical principles of renewable energy sources", "Energy facilities of non-traditional renewable energy sources", holds open lectures, participates in the formation of the subject of course and diploma works, training of young specialists. Since 1996 has been working at the Federal State Budget Educational Educational Institution of Higher Education "Dagestan State University", Professor of the Department "Renewable Energy Sources".

Vadim Bolshev is a research officer with Federal Scientific Agroengineering Centre VIM and specializes in research in the field of power supply for rural consumers. Vadim studied at the Orel Technical College as an electrical technician from 2004 to 2008, at the Orel State Agrarian University as an electrical engineer from 2008 to 2012. In 2016 he graduated from the postgraduate course at the Orel State Agrarian University. Vadim has a great professional experience in the field of working with electrical equipment. So he worked in the water and sewer utility "Orelvodokanal" for 3.5 years as an electrician (2010 - 2013), in the scientific and clinical multidisciplinary center named after Z.I. Krugloi for 3 years as an engineer (2013-2016) and 1.5 years as chief engineer (2016-2018). To date, he works in the Federal Scientific Agro-Engineering Center of VIM as a research fellow. The field of scientific activity is to develop methods and tools aimed at improving power supply efficiency. This includes, first of all, the development of methods and devices for monitoring power quality and the technical state of power supply system elements as well as the automation of these systems.

Maxim Borodin is an Associate Professor of the Power Supply Department of the Orel State Agrarian University named after N.V. Parakhin and specializes in research into improving the performance of rural consumers. Maxim studied at the Orel State Agrarian University at the Faculty of Agrotechnics and Energy Supply, specializing in electrification and automation of agriculture, received a diploma with honors, and also has a second higher education in the specialty "Public and municipal finance." In 2014 he defended his thesis for the degree of Candidate of Technical Sciences with a degree in Electrotechnical Complexes and Systems. Maxim is an employee of the electrical laboratory the OrelSAU. The field of scientific activity is the development of methods and tools aimed at improving the efficiency of power supply systems. This includes, first and foremost, the development of methods and devices for monitoring the quality of electrical energy, the development of methods and means to ensure the quality of electrical energy. Author and co-author of more than 40 publications, 2 inventions, 2 monographs and 4 teaching aids.

Yangibay Charyev was born on March 16, 1951 in the village of Guinuk Denava district of the Lebap region of the Turkmen SSR. Parents - Father Dzhumaev Chary (20.02.1917) teacher of history of secondary school, mother Khodzhakuliyeva Mamagul (08.03.1928r.) Collective farmer. In 1974 he graduated with honors from the Turkmen State Pedagogical Institute, Physics Department with a degree in Physics and Technical Mechanics. Worked as an engineer as a junior researcher, senior research fellow, leading research associate, chief scientific officer of the Physical and Technical Institute of the Academy of Sciences of Turkmenistan (1978-1994). In 1984 he defended his thesis on the topic "Electric and photoelectric properties of GaP, InP MIS structure" and received his PhD in Physics and Mathematics. In 1993 he defended his thesis on the topic "Generation-Recombination Processes in Photo Transformations at High Light Excitations" and received a doctorate in physics and mathematics. Yangibay Charyev carries out research in the field of interaction with matter and transformation of electromagnetic radiation, development of methods and systems using renewable energy sources. He is the author of more than 100 scientific works, including articles, author's certificates; has graduate students. Since 1994-2005. worked in the Turkmen State Pedagogical Institute head. Department of Physics. Since 2005 he has been working at the Turkmen State Institute of Transport and Communications in the department of "Telecommunications".

Dmitry Chernov works in the field of renewable energy, viz., wind power, since mid-2013. Chernov began their work in the specialty at LLC "Activity", being on the training in the magistracy, which helped them a lot with the writing of the master's thesis. During this period, Dmitry gained valuable experience in the design of the wind farm and the mathematical modeling of the wind-diesel complexes. They took part in writing the feasibility study for investments in the construction of wind farms in the Volgograd and Murmansk regions, the Republic of Adygea, the Nenets Autonomous Okrug, the Chukotsky District, the Primorsky Territory and various sites in the Kamchatka Territory. At least 10 projects were implemented. The key customers were PJSC "Mobile Energy", LLC "LUKOIL", General Electric. Chernov has deepened their experience and now, along with estimates of wind resources for the purpose of a preliminary study of the construction site of a wind farm, they analyze the performance indicators and develop recommendations for optimizing further operation. Perfection of the methodology for determining the technical and economic potential of wind-diesel complexes on the basis of the analysis of operating data of the existing Far East District facilities is devoted to my Ph.D. thesis, the writing of which they have been doing since 2014.

Yuliia Daus was born in 1986. In 2008 they received the diploma of specialist in Energy Management in Zaporizhzhya National Technical University. In the period 2008 – 2014, Daus worked at the same university as an assistant of the Power supply of industrial enterprises department. From 2012 to 2015 they studied at post graduate school on Electrotechnical complexes and systems specialty. Now, they work as an editor of the scientific journal " Don Agricultural Science Bulletin ". In 2017 they got their master's degree in Electrical Engineering and power.

Galina Deryugina is a Senior Lecturer of the Department "Hydropower and renewable energy". Education: Moscow Power Engi-neering Institute (1988). Their research area is in development and substantiation of the parameters of wind-power stations in distributed power systems. And they have over 40 publications and 3 copyright certificates.

Alexei Dubinin is a doctor of technical sciences, professor at the «Power and heating equipment» college graduate. Awards: Medal of the order of merit for the fatherland, second degree. Education: Ivanovo power Institute (1965). Area of scientific interests: apparatus for the production of technological atmospheres from natural gas and solid fuels. He is the author of more than 250 articles, textbooks and manuals.

Vera Dyachenko was born in 1965. In 2008, they received the diploma of specialist in Power Engineering in Zaporizhzhya National Technical University. They work at the same university as associated professor of the Power supply of industrial enterprises department. In 2013, Dyachenko got degree of Candidate of Technical Sciences.

Haiyang He is a Ph.D. student in the Institute of Hydropower and Renewable Energy, Moscow Power Engineering Institute (National Research University), studied bachelor and master degrees in North China Electric Power University (Beijing, China), obtained three China national invention patents, published four papers. Research interests solar power generation, and the grid-connected technology of renewable energy.

Dmitry Ilyin received the M.Sc. degree in power engineering from the Moscow Power Engineering Institute, Moscow, Russian Federation, in 2016. Currently, he is working toward the Ph.D. degree at the Department of Renewable Energy Sources in Moscow Power Engineering Institute. His current job position in the industrial software company, in Microgrid analysis and modelling department as a head specialist. Involved in projects on the development of the wind power plants automated control system and distributed energy. His current research interests include the field of control theory, power electronics and wind turbine stability.

Nikita Karpov is a Teaching Assistant of the Department "Hydropower and renewable energy". Education: Moscow Power Engineering Institute (2018). Research area: development and substantiation of the parameters of wind-power stations in distributed power systems. Publications: over 15.

Aleksandr Khokhlov was born in 1950 in Samarkand city, Uzbekistan. September, 1967 – June, 1972: study at Tashkent Polytechnic Institute, Tashkent, Uzbekistan. Major: Hydropower sets, Engineer in hydro-electric engineering. September, 1972 - September, 1980: work at the Ministry of water resources, Tashkent, Uzbekistan. September, 1980 - February, 1990; work and post-graduate study at "SANIIRI" Research Institute, Tashkent, Uzbekistan. Major: Melioration. May, 1990: received the degree of PhD. Dissertation subject "Improvement of operating modes of pumping stations with long pipelines". February, 1990 – present: work in Research-and-production enterprise "Vodopodyomnik" ("Water lifting") as a director. He totally has 75 publications: one book chapter IGI-Global, four monographs, seventeen patents for invention, forty tree articles in scientific technical journals and others.

Vladimir Khokhlov was born in 1976 in Tashkent city, Uzbekistan. September 1993 – June, 1998: study at Tashkent Institute of Irrigation and Mechanization of Agriculture, Tashkent, Uzbekistan Major: Water economic building, Engineer in hydro-technics. June 1994 – present: work in Research-and-production enterprise "Vodopodyomnik" ("Water lifting") in series as a technician, engineer, researcher and deputy director. September,1998- September, 2001: post-graduate study at Tashkent State Technical University, Tashkent, Uzbekistan. Major: Hydro-technical and meliorative building. February 2002: received the degree of PhD. Dissertation subject "Improvement of Drainage Water Pumping from the Buildings of Hydropower Sets". January 2004 – December 2006: work and doctor of science degree study at Tashkent State Technical University. Major: Power systems and complexes. March 2009: Received the degree of Doctor of science in technics. Dissertation subject "Power Saving Operating Modes of Pumps and Pumping Stations with Long Pipelines". Since September 2011: Professor on Fluid mechanics and Hydraulics for the third academic level and for masters at Department of applied sciences of Turin Polytechnic University in Tashkent, Uzbekistan. Activities include lecturing, practice and laboratory training. He totally has 96 publications: one book chapter IGI-Global, two textbooks, six monographs, twelve patents for inventions, fifty-two articles in scientific technical journals, twenty two reports in scientific conferences and others.

Sophia Kiseleva is the senior scientific worker of Renewable Energy Sources Laboratory (Lomonosov Moscow State University, Faculty of Geography). Dr. Sc. (physical-mathematical). They are part of the faculty of Physics Lomonosov Moscow State University Research area: renewable energy sources: resource evaluation, ecology-geographical aspects of renewable energy; laboratory modeling of ocean and atmosphere's dynamical process Publications: over 80 articles, including patents.

Volodymyr Kozyrskyi is doctor of Engineering, Professor, Honored Scientist of Ukraine. He was graduated department of the electrification of Ukrainian Agricultural Academy (now The National University of Life and Environmental Sciences of Ukraine) in 1978. He was appointed director of the research institute of engineering and technology in September 2005. He was appointed director of the Education and Research Institute of Energetics, Automation and Energy Efficiency in 2009.

Tatyana Kudinova is an associate professor of the Foreign Languages Department at the Orel State Agrarian University named after N.V. Parakhin and specializes in teaching English to bachelors studying in the specialty "Power Supply", "Electrical Equipment". Tatyana has been studied at the Orel State University from 1994 to 1999 and received the qualification "Teacher of English and German". In 2006, Tatyana defended her thesis on specialty 10.02.19 "Language Theory". She has great professional experience in teaching English and working with scientific terminology in various fields of science and technology. Tatyana has been working at the Orel State Agrarian University since 1999.

Azamat Kurbonov was born in 1992 in Karshi city, Kashkadarya region, Uzbekistan. September 2010 – June 2016: study at Tashkent State University of Economics. June 2016 – present: work in "Syrdarya" Water Economics Association as engineer-economist, in Tashkent State Technical University, as PhD student.

Pavel Kuznetsov received a master's degree in 2010 in energy in the specialty "Alternative Energy Sources" at the Sevastopol National University of Nuclear Energy and Industry. In 2013 he graduated from the postgraduate course in the specialty "Conversion of Renewable Energy". In 2018 he defended his Ph.D. thesis on the specialty "Conversion of renewable energy" on the theme: "Increasing the energy efficiency of photovoltaic stations operating in conditions of uneven illumination."

Natalya Makhiyanova is an employee of the Power Supply Department of Orel State Agrarian University and specializes in research in the field of designing power supply systems. Natalia studied at the Orel State Agrarian University for a mechanical engineer from 1996 to 2001. In 2017, she underwent professional retraining in the direction of "Power Supply". He is a co-author of scientific publications on the optimization of the parameters of overhead transmission lines. The field of scientific activity is related to the improvement of calculation methods in the design of distribution electrical networks.

Maria Morgunova is currently a junior researcher at Joint Institute for High Temperatures, and a PhD student at KTH Royal Institute of Technology. Maria holds Degree of Licentiate of Engineering (KTH, May 2015) and has recently defended PhD in International Economics (Gubkin Russian State University of Oil and Gas, Moscow, 2018). Key interests: conventional and unconventional energy resources, energy markets and systems in transition, electricity markets, digitalization, energy geoeconomics and geopolitics, renewable energy, Arctic region, offshore technology, decentralized and remoted energy supply, system analysis, development blocks, industrial dynamics, multi-level perspective, socio-technical systems.

Jelena Muhamedijeva received the B.S. and engineer degrees in system engineering from Riga Technical University, Riga, Latvia, in 1993 and M.S. degree in computer systems from Institute of Information technologies High School, in 2008. From 2010 to 2013 she was a Project manager and researcher in IT research project in cooperation with Kazakh National Medical University. From 2012 she was a

Researcher in MON RK project & Almaty, Kazakhstan. Since 2014, she has been a Researcher in MON RK project" Development of methods and algorithms for heterogeneous data collection, analysis and interpretation in multilayer spatially distributed monitoring system models of renewable energy sources", Almaty, Kazakhstan. She is the co-author of two books and more than 40 articles. Her research interests include Information technologies, Multiagent systems, Big Data. She is co-author of two Certificates of state registration of rights to the copyright object.

Ravil Mukhamediyev received the engineer degree in radio-electronics from the Riga Civil Aviation Engineers Institute, Riga, Latvia, in 1983 and Ph.D. degree in information systems from RCAEI in 1987. From 1987 to 1995 he worked in the RCAII as Assistant, Lecturer, Assistant Professor. From 2004 to 2010, he was an Assistant Professor and Head of Department with Higher School of Information System Management, Riga, Latvia. Since 2010 he has been an Professor with different Higher Schools of Almaty, Kazakhstan. His current position is an Professor and Head of Department with Kazakh National Technical University. He is the author of three books, more than 200 articles, and Leading Researcher of four research projects. His research interests include machine learning, data processing, decision support systems and etc. He is an Editorial Boards member of the journals CMNT and CoS.

Renat Mustakayev was born on June 8, 1982 in Almaty, Kazakhstan. In 2003 got bachelor degree in Information systems from Kazakh National University. In 2005 got master degree in Information systems from Kazakh National University. There are 12 authored and coauthored publications. Have 15 years experience in enterprise information systems development.

Liudmila Nefedova is the Senior Researcher in the Laboratory of renewable energy sources at the Lomonosov Moscow State University, Department of Geography, Russia.

Alexey Nekrasov was born October 30, 1945, in 1973 graduated from the faculty of electrification of agriculture of the Kazakh GSHI on a speciality the electrical engineer of agricultural production. In 2007 he defended his doctoral thesis on the specialty 05.20.02 – Electrotechnology and electric equipment in agriculture. Currently conducts research on electrification of agriculture, for the effective use and operation of electrical equipment of agricultural enterprises. Directly involved in the development of a new resonance methods of transmission and use of electric energy, and also the implementation of other fundamental and most important applied scientific and technical programs. Nekrasov A. I. has over 275 scientific publications, including 19 books, monographs, guidelines and brochures, 50 patents and copyright certificates on inventions. Completed developments in agricultural production and recommended for use by scientists, design organizations and experts-energy, are used in the educational process.

Anton Nekrasov was born 04 November 1981 in Moscow.In 2004 graduated from the Moscow state Agroengineering University n. a. V. P. Goryachkin, for being awarded the qualification of engineer-electrician, specialty – electrification and automation of agriculture. He is the author of 65 published scientific works, including 3 guidelines and 6 patents, of which 35 papers in journals approved by VAK. Working as a researcher. In 2015, he defended his thesis.

Vladimir Panchenko studied at "Bauman Moscow State Technical University" at the faculty "Power Engineering " with qualification of engineer. From 2009 to 2012 studied at the graduate school of the ''All-Russian Scientific Research Institute of Agriculture Electrification'', specializing in "Renewable Energy Installations". Passed an internship program on the International Energy Program in the Solar Energy Center (SEC), the Certificate under the Indian Technical and Economic Cooperation (ITEC), the Program of the Ministry of External Affairs (MEA), the Government of India - India, Delhi, November-December 2011. Participated in the International Conference-Exhibition Intersolar Europe 2012, Germany, Munich, June 11 - 15, 2012 as part of the delegation of the Ministry of Energy of the Russian Federation Ph.D. thesis "Development and research of solar photovoltaic thermal module with a paraboloid-type concentrator" was protected on July 2, 2013. From 2013 to 2017 head of the laboratory ''Solar photovoltaic modules with an experimental - technological site''. The Council of Young Scientists of the VIESH was elected the Chairman of the Council of Young Scientists and was appointed a member of the Academic Council of the VIESH in 2014 - 2017. In 2014 was appointed as a contact person and a member of the International Solar Energy Society of the Russian section of the community. Since September 2015 associate professor at the ''Moscow State University of Railway Engineering''. Since 2017, a senior researcher at the ''Solar Energy Laboratory'' FSBSI "Federal Scientific Agroengineering Center VIM". Passed the advanced training in 2016 in ''Russian Engineering Academy of Management and Agribusiness '' for an additional professional program '' Automated Information Processing and Management Systems''. Passed the advanced training in 2017 in ''Moscow State University of Railway Engineering' the program "Using e-learning and distance educational technologies in the implementation of educational programs on distance learning". In 2016 awarded the grant ''Young teacher of Moscow State University of Railway Engineering''. In 2018 awarded the Scholarship of the President of the Russian Federation for young scientists and graduate students engaged in advanced research and development in priority areas for the modernization of the Russian economy. Has 3 medals of exhibitions, 3 diplomas of the winner of competitions, participated in more than 70 conferences and exhibitions of various levels, has more than 100 publications, including patents, monograph and educational publications. Area of interest includes three-dimensional modeling, solar energy and distance education.

Andrii Petrenko graduated with honors in 2005 Department of Energy and Automation, National Agricultural University (now National University of Life and Environmental Sciences of Ukraine) in specialty "Energy in Agricultural Production". In 2010 he defended his thesis for the degree of candidate of technical sciences, specialty 05.09.03 - Electrotechnical complexes and systems.

Sergey Shcheklein is an honoured power engineer of Russian Federation, professor, doctor of technical science, head of Department of the nuclear power plants and renewable energy of the Ural Federal University named after the first President of Russia Boris Yeltsin. Was born May 29, 1950 in Sverdlovsk. He graduated from the Urals Polytechnical Institute on a speciality «Nuclear power stations and installations». Worked in the Department of physical and technical problems of energy of the Ural scientific center of USSR Academy of Sciences. From 1975 onwards works in college graduate. In 1977 year has defended the dissertation on competition of a scientific degree of candidate of technical sciences on a speciality "Nuclear power plants". In 1987 he was elected to the post of head of the Department of Atomic Energy. In 1991 he defended the dissertation on competition of a scientific degree of the doctor of technical sciences. In 1992 he received the title of Professor. Sergey Shcheklein is an expert in the field of nuclear power plants, renewable energy sources and energy saving; he is the author of over 400 scientific papers, including 11 monographs and tutorial, 60 papers published in refereed scientific journals. He has 71 patents. His Hirsch index (h-index) is equal to 11. S.E. Shcheklein known for his research in the field of study of nonstationary stochastic processes and in two-phase steam-liquid flows; processes heat-fluid dynamics with boiling poorly overheated liquids in narrow channels; liquid phase inversion phenomena in curved channels when dispersed flow modes of gas (steam) fluid flows with respect to tasks of improving the safety and efficiency of NUCLEAR POWER PLANTS with nuclear reactors of different types. Is the founder of scientific direction-actively managing characteristics of fluid dynamics and heat transfer in a wave of gas (vapor) film trends. Is supervisor of works on creation of new types of equipment for thermo-convert solar energy systems, developing new designs of wind and hydropower installations; systems of alternative motor fuels from biomass using solar, hydraulic and wind power; study of processes and the creation of new designs installations for ultrasonic intensification of water electrolysis and hydrolysis of biomass for the production of hydrogen, ethanol and biogas; complex projects on the development and implementation of energy efficient buildings and integrate installations of renewable energy in housing. Led by the Department of nuclear power plants and renewable energy sources is recognised in the country Center for training highly qualified specialists for the development of new energy technologies, scientific and technical support for the operation of nuclear power plants to fast neutron reactors, perform complex repairs and removal of the NPP. Has a branch at Beloyarsk, training center for new technologies, repair and refurbishing of the plant. Since 1995, the Department carries out preparation of specialists in "renewable energy". For projects and inventions of employees and PhD students at the Department in 2014 organized and actively working « Euro-Asian centre of renewable energy and energy conservation». More than 60 graduates have defended doctoral thesis. Leader and a member of the organizing committees of several international and Russian conferences, is a member of editorial boards of scientific and technical journals «Izvestiya vuzov. Nuclear energy», «International scientific journal for alternative energy and ecology», proc. of the Odessa Polytechnic University (Ukraine); a member of a number of dissertations and academic councils; has been actively involved in organizing and conducting with 2001 onwards the annual all-student competition "Energy and resource saving", the all-Russian scientific and practical Conference and exhibition of technical creativity of students, postgraduates and young scientists «Energy and resource saving. Power supply. Non-conventional and renewable sources of energy». For 29 years the Chair has trained more than 1500 specialists for work in the field of nuclear energy, with 2000-2016 timeframe. in energy conservation and renewable energy sources, with the participation of the Chair has trained more than 20000 students College graduate institutes and faculties, and specialists of industrial enterprises of the Sverdlovsk region. Under his initiative and leadership, occurs implemented Master's Program «Power systems and complexes on the basis of renewable sources». S.E. Shcheklein has the State and industry awards: Honorary mark "For excellent successes in work" (1981), a gold and silver medal «For the safe operation of nuclear power plants» (2016), (2001), "Veteran nuclear power and industry of Russia" (2004), "50 years of Atomic Energy of Russia" (2004), «For merits in development of science and economy of Russia "(2002); awarded with diplomas of the Ministry of Atomic Energy of the Russia, Ministry of education and science of Russia, the Governor and the Government of the Sverdlovsk region. Member of international energy Academy (1996). The honoured power engineer of Russia (2001). For complex research in the field of renewable energy and energy efficiency in 2009 year was awarded the national environmental prize V.I. Vernadsky in nomination "Energy of the future". The authors of the study are perennial issues of energy efficiency, of particular relevance in countries and regions with severe climatic conditions. Features of the approach is to develop their own scientific, technical and organizational and political approaches to new energy technologies, synthesis and large-scale introduction of energy saving technologies available in the world and the involvement of the potential of renewable sources of energy in the practical sphere; research and development aimed at safe use of nuclear energy including brider nuclear reactors of new generation; involvement in research activities of young people (students, masters, postgraduates, young scientists), continuous transfer of knowledge to students, professionals in the fields of energy, construction, housing and communal services, budgetary and industrial enterprises by reading lectures, organize the work on improvement of professional skill and retraining. Project "Energy-efficient house with a complex of renewable energy sources in the year 2009, award-winning national environmental Vernadsky prize.

Tatiana Shestopalova teaches and does research work involved since 1982. Direction of scientific activity - electricity industry. Since 1995 - the autonomous power of consumers on the basis of renewable energy sources, has 95 scientific and methodical publications.

Bekhruzi Shokhzoda finished the bachelor's degree in National Research University "Moscow Power Engineering Institute" at Department of Hydro Power Engineering and Renewable Energy Sources in 2012. In 2014, he completed a master's degree at the same University and in 2015 continuing as a graduate student at the same University. Presently his area of interest is solar energy conversion technology. He is the author of two patents, winner of several competitions and had published more than 10 papers in different international journals based on his research work.

Alexander Solovyev is the head of the renewable energy sources Laborotory at Lomonosov Moscow State University, Department of Geography.

Dmitry Solovyev is a research associate of Sea Atmosphere Interaction and Climate Laboratory (SAIL) (https://sail.ocean.ru). He graduated from Physical Faculty, Lomonosov Moscow University, Researcher Sea-Air Interaction and Climate Laboratory (SAIL) Shirshov Institute of Oceanology, Russian Academy of Sciences and Department of distributed energy systems of Joint Institute for High Temperatures, Russian Academy of Sciences. The author of a number of articles focused on the ocean's role in climate variability and change, ocean's renewable energy, Geophysics & technological problem of the Energy sector.

Nikolay Sorokin is a senior lecturer of the Orel State Agrarian University named after N.V. Parakhin and specializes in research in the field of improving the reliability and informatization of 6-35 kV power networks. Nikolai studied at the Orel State Agrarian University for an electrical engineer from 1998 to 2003. In 2006 he graduated from the postgraduate course at Orel State Agrarian University. Since 2003 Nikolai has worked with the Orel State Agrarian University. During this time he held the following positions: Assistant of the Power Supply Department, Senior Teacher of the Power Supply Department, Head of the Scientific and Innovation Activity Sector of the OrelSAU Research Department. Currently, Nikolay works as a senior lecturer in the power supply department of the Orel State University of Agriculture and is responsible for the academic work at the Faculty of Agricultural Engineering and Power Supply.

Dmitry Strebkov is a Supervisor at VIESH, Academician of the Academy of Sciences of the Russian Academy of Agricultural Sciences, and is a Ph.D. Professor. Born March 11, 1937 in Vinnitsa. He graduated in 1959 the Faculty of electrification of the Moscow Institute of mechanization and electrification of agriculture. DS Strebkov world's first theorized and created a controlled light photovoltaic generator voltage of 30 kV for electrical processes. DS Strebkov - a prominent scientist in the field of electrification and automation of agriculture, renewable energy, new power transmission methods and new fuels, energy conservation. DS Strebkov - a prominent scientist of world renown in the field of renewable energy sources and new energy technologies. He has published more than 1,200 scientific papers, including 407 patents for the invention of the USSR and the Russian Federation and 60 foreign patents (USA, Germany, France, Great Britain, Japan and others.).

Janna Titova was born in 1976 in Tashkent city, Uzbekistan. September, 1993 – June, 1998: study at Tashkent Institute of Irrigation and Mechanization of Agriculture, Tashkent, Uzbekistan. Major: Mechanization of water economic and meliorative, Engineer-mechanic. June 2006 – present: work in Research-and-production enterprise "Vodopodyomnik" ("Water lifting") as research assistant, main researcher. January, 2014 – December, 2016: doctor of science degree study at Tashkent State Technical University, Tashkent, Uzbekistan. Major: Power systems and complexes. December, 2016: received the degree of Doctor of science in technics. Dissertation subject "Algorithms and Methods of Regulation of the Pumping Stations Operating Modes". Professor's assistant, docent on Heating and cooling of electrical machines and transformers for the second academic level at Tashkent State Technical University, Department of electrical machines. Activities include lecturing and practice training. She totally has 44 publications: one book chapter IGI-Global, two textbooks, five monographs, four patents for invention and software, sixteen articles in scientific technical journals, seventeen reports in scientific conferences and others.

Mykola Trehub was graduated department of the electrification of Ukrainian Agricultural Academy (now The National University of Life and Environmental Sciences of Ukraine) in 1979. In 1988 he defended his thesis for the degree of candidate of technical sciences, specialty 05.20.02 - Energy in Agricultural Production. Scientific interests are electrotechnologies, power and renewables in agriculture.

Mikhail Tyagunov is a doctor of Technical Sciences, Professor Tyagunov M. G. in 1970 he graduated from the Moscow power engineering Institute (MPEI), Faculty of Electricity. He was awarded the Medal "In memory of the 850 anniversary of Moscow", the Honorary worker of higher education of Russia. The scope of his interests and current activities is renewable energy. Author of 140 publications, including 3 author's certificates of the USSR for inventions.

Aleksey Vasiliev is doctor of technical sciences, professor. He is a graduate of the Faculty of Electrification of the Azov-Black Sea Institute of Agricultural Mechanization. he worked in this institute more than 25 years in posts from a junior research worker to a professor, a head of a department and a prorector. Since 2011 he has been a deputy Director of the All-Russian Institute of Agriculture Electrification, which entered the Federal Scientific Agroengineering Centre VIM in 2016. Direction of scientific work: "Automated control systems and electrotechnologies in Agricultural technological processes".

Scientific research is aimed at reducing energy intensity, increasing the productivity of processes and increasing the power supply reliability of the agro-industrial complex. He is the author of more than 230 scientific papers.

Alexander Vinogradov is a leading researcher at the Federal Scientific Agroengineering Center VIM and specializes in research into improving power supply system efficiency for rural consumers. Alexander studied at the Kostroma State Agricultural Academy from 1994 to 1999, majoring in Electrification and Automation of Agriculture. From 1999 to 2018 he worked as an assistant, senior lecturer, head of the department, associate professor of the Power Supply Department of the Orel State Agrarian University. Since 2018 he has been a leading researcher at the Federal Scientific Agroengineering Center VIM. Alexander has been the General Director of the LLC "Information and Energy Center AVPS-Innovation" since 2007 Alexander has experience in the preparation, organization and implementation of educational services, the development of teaching technologies, the development and creation of technical training facilitiesas well as experience in conducting research, developing theoretical and practical aspects of various kinds of interactions including social ones. He also has practical experience in conducting energy surveys, designing power supply systems, installing electrical equipment, managing the design organization and designing. He has 4 dan karate-do and 3 dan kobudo. Research interests: research is being carried out to develop new approaches to the creation of power supply systems, automation of electric networks, the development of energy accounting systems, the provision of quality of electricity supplied to consumers, energy saving, electrical modeling of human and society, educational technologies, use of renewable energy sources. He is an author and a co-author of more than 200 scientific and methodical works, including 33 patents, 9 monographs, 6 essays. He is also an author and a co-author of 4 collections of literary works. He is the author and the editor-in-chief of the newspaper "High Voltage Truth" and the editor-in-chief of the journal "Agrotechnics and energy supply".

Alina Vinogradova is a researcher at the Federal Scientific Agroengineering Center VIM and specializes in research in the field of power supply for rural consumers. Alina used to study at the Oryol State Agrarian University from 2003 to 2008. In 2015 she defended her thesis. Alina has great professional experience in the field of designing electrical networks. So she worked as a design engineer (electrician) in JSC Orelenergoremont in the period from 2007 to 2008; as an engineer for project work at JSC Orelenergoremont from 2008 to 2009; as an engineer of the design department of VL in JSC "Voronezhenergoproekt" from 2009 to 2010; as an assistant of the Department of "Power Supply" in OrelSAU from 2010 to 2011; as a senior lecturer of the department "Power supply" in OrelSAU from 2011 to 2016; as an associate professor of the Department of "Power Supply" in OrelSAU from 2016 to 2018. At the moment she works in the federal scientific agroengineering center VIM as a research assistant. The field of scientific activity is to develop methods and tools aimed at improving the efficiency of power supply. This includes, first of all, the development of methods and devices for monitoring the quality of electrical energy and the technical state of the elements of power supply systems as well as its automation.

Kirill Yakunin was born in 1992, Almaty, Kazakhstan and is currently he is a PhD student at KazNITU, Almaty, Kazakhstan.

Igor Yudaev was born in 1965 Volgograd Russian Federation. 2012 – defense of the thesis for competing the academic degree of Doctor of Technical Sciences on specialty 05.20.02 "Electrotechnology and electrical equipment in agriculture" on the topic "Electroimpulse energy-saving technology for combating weed vegetation" at Goryachkin Moscow State Agroengineering University. 2002 – defense of the thesis for competing the academic degree of Candidate of Technical Sciences at Goryachkin Moscow State Agroengineering University on the topic "Justification of the technological parameters of electroimpulsive destruction of weed vegetation". 2004-2014 – Dean of the Electrification of agriculture faculty (since 2012 of the Electric power faculty) of the Volgograd State Agricultural Academy (since 2012 the Volgograd State Agrarian University). 2010-2014 – Head of the Electrotechnology and electrical equipment in agriculture department of Volgograd State Agricultural Academy (since 2012 Volgograd State Agrarian University). 2014-present – Deputy Director for Science work of the Azov-Black Sea Engineering Institute of FSBEI HE "Don State Agrarian University". Since June 2005 – member of the Educational and Methodological Association (EMA) of agroengineering specialties of agrarian universities in Russia; since May 2011 – member of the educational and methodological council for electricity supply and the educational and methodological EMA commission for the relay protection of higher educational institutions of Russia in the field of energy and electrical engineering. Since August 2006 – Professor of the Russian Academy of Natural Sciences. Since June 2009 – Member of the Committee on Renewable Energy Issues of Russian Scientific Research Institute of Artificial Intelligence. since February 2011 – Corresponding Member of the Russian Engineering Academy in the Energy section. Since April 2011 – member of the Society of Biotechnologists of Russia named after Yu.A. Ovchinnikova. Since May 2014 - Academician (full member) of the Russian Engineering Academy in the Energy section. The main scientific areas are: the research of the application and use of high-voltage impulse actions, the effects of the arc discharge field, the electromagnetic field of high tension in the technological processes of production, processing and storage of crop production; study of environmental safety problems; studying the issues of the light culture of greenhouse plants; study of the renewable energy potential in rural areas and the prospects for introducing renewable energy sources into them. He has 157 publications, including 142 scientific works; 3 monographs; 10 teaching and methodical works; 5 patents.

Index

R

S

T

U

V

W

www.ingramcontent.com/pod-product-compliance
Lightning Source LLC
LaVergne TN
LVHW060258011026
849161LV00011B/2737